PRINCIPLES OF BIOREACTOR DESIGN

PRINCIPLES OF BIOREACTOR DESIGN

Binoy Ranjan Maiti
Formerly Professor of Chemical Engineering,
IIT, Kharagpur and Emeritus Professor of Biotechnology,
Heritage Institute of Technology, Anandapur, Kolkata

London • New Delhi

MV Learning
A Viva Books imprint

3, Henrietta Street
London WC2E 8LU
UK

4737/23, Ansari Road,
Daryaganj, New Delhi 110 002
India

ISBN: 978-93-87692-82-4

Printed and bound in India.

Contents

Preface

In recent years, there is an emerging trend of exploiting biochemical processes for useful products. These processes are characterized by low energy requirement, specificity of products and mild environmental conditions. Desired products are enzymes, antibiotics, solvents, proteins, antibodies, vaccines, hormones, insulin, interferons, etc. which are in great demand as commercial commodities.

All chemical or biochemical processes require a reaction vessel or a reactor along with other downstream processing equipments. The reactor is recognized as the heart of any reaction process and its efficiency determines the economic feasibility of the process. The performance of a reactor depends upon its rational design which requires the knowledge of appropriate kinetics, favourable hydrodynamic conditions, suitable mass transfer and heat transfer knowledge.

This book deals with fundamental principles of reactor design along with the basic kinetics for biological systems like enzymes, microbial cells, animal cells, plant cells, recombinant cells, etc.

The chapters are arranged around basic kinetics and reactor design for chemical systems and then extended to biological systems. The first three chapters are concerned with chemical kinetics and reactor design. The fundamental aspects of these topics form the basis of developing the design aspects of biological systems.

The chapters from 4 to 13 comprise the biological systems containing kinetics, mechanism, mass transfer, reactor dynamics, steady state conditions, etc. The topics include free-enzyme kinetics and reactor design (Chapter 4), kinetics and mass transfer for immobilized enzymes or cells (Chapter 5), rate processes of free microbial cells and their immobilized states, design and operation of reactors with special reference to CSTR (Chemostat) (Chapter 6), transport phenomena in bioreactors (Chapter 7), sterilization reactors (Chapter 8), design and analysis of bioreactors (Chapter 9), kinetics and reactors for animal cells (Chapter 10), membrane bioreactor design and analysis (Chapter 11), design and operation of classical bioreactors for waste water treatment (Chapter 12). The last (Chapter 13) contains brief review of unconventional bioreactors which are novel and innovative.

In the appendices topics included MCQ and short questions based on GATE, kinetic data for some biological systems, diffusivities in gels, dimensionless groups and some useful mathematical formulae.

The book is intended for undergraduate and postgraduate students for engineering course in biotechnology of Indian universities.

The special features of the book are theories have been explained by using large number of numerical problems. For practice by the students, numerical problems have been incorporated in each chapter. Some advanced topics are included such as transport phenomena in bioreactors, reactor dynamics, membrane reactors, unconventional reactors, etc.

I would like to acknowledge several colleagues and friends for their advice. I am especially thankful to Prof. B. B. Paira, Adviser, Heritage Institute of Technology, Kolkata for his continuous inspiration and suggestions. I am also thankful to my son, Dr. Aniruddha Maiti and daughter-in-law, Smt. Joyeeta Maiti, who scanned all figures and placed them at their appropriate places in the text and made online transactions.

Thanks are also due to Mr. Arunava Majumdar who took infinite pains to type the manuscript with all mathematical equations.

New Town (Kolkata) **B. R. Maiti**
2017

CHAPTER 1

Chemical Kinetics and Reactor Design

1.0 Introduction

This chapter contains the fundamental principles for homogenous and elementary reactions with different orders. The kinetic parameters in the form of specific rates or reaction rate constants have been evaluated from various reaction models. The mechanisms of some non-elementary reactions have been illustrated to develop a reaction rate.

Batch data have been fitted to proposed reaction rate model. The methods of kinetic data analysis are based on differential and integral techniques.

Equations for batch and plug flow reactors, stirred tank reactor, stirred tanks in series, plug flow with recycle, semi batch reactors have been developed for liquid and gas phase reactions.

1.1 Constant Volume Batch Reactor

Constant volume means constant density reaction system. For the above system the rate equation, ri, can be expressed as

$$r_i = \frac{1}{V}\frac{dN_i}{dt} = \frac{d\left(\frac{N_i}{V}\right)}{dt} = \frac{dC_i}{dt} \tag{1.1}$$

for ideal gases,

where $C_i = p_i / RT$ and the rate, ri is expressed as

$$r_i = \frac{1}{RT}\frac{dp}{dt} \tag{1.2}$$

where V = volume in m^3, Ni = k. moles of component i

C_i = concentration of the component i (k.moles/m^3),

pi = is the partial pressure of component i (bar)

R = Universal gas constant (bar-m^3/(kmole)(°C)

Defining fractional conversion, of component, A, X_A, the rate can be given as follows:

$$XA = \frac{N_{AO} - N_A}{N_{AO}} = 1 - \frac{\frac{N_A}{V}}{\frac{N_{AO}}{V}} = 1 - \frac{C_A}{C_{AO}} \tag{1.3}$$

Where N_{AO} is the initial moles of A in the batch reactor at time, t = 0 and N_A is the moles of A in the batch reactor after time t,

Now,
$$dC_A = -\, dC_A / C_{AO} \tag{1.3a}$$

where C_{AO} is initial concentration.

1.2 Analysis of Batch Reactor Data by Integral and Differential Method

1.2.1 Integral method

For unimolecular, irreversible first order reaction,

$$A \rightarrow \text{Products} \tag{1.4}$$

The rate can be given,

$$-r_A = -\frac{dC_A}{dt} = k\ C_A \tag{1.5}$$

Rearranging and integrating,

$$-\int_{C_{AO}}^{C_A} \frac{dC_A}{C_A} = k\int_0^t dt \tag{1.6}$$

$$-\frac{\ln C_A}{C_{AO}} = kt$$

or
$$C_A = C_{AO}e^{-kt} \tag{1.6a}$$

where k is the first order rate constant in sec^{-1}. The rate for first order reaction in terms of fractional conversion, X_A is expressed as

$$\frac{dX_A}{dt} = k(1 - X_A) \tag{1.7}$$

Integrating

$$\int_0^{X_A} \frac{dX_A}{(1-X_A)} = k\int_0^t dt \tag{1.8}$$

Or
$$-ln(1-X) = kt$$

For irreversible, bimolecular second order reaction

$$A + B \rightarrow \text{Products} \tag{1.9}$$

The corresponding rate equation is

$$-r_A = -\frac{dC_A}{dt} = -\frac{dC_B}{dt} = k_2 C_A C_B \tag{1.10}$$

in terms of fractional conversion, X_A at any time t, the amounts of A and B at t are $C_{AO} X_A$, $(C_{BO} - C_{AO} X_A)$ thus

$$-r_A = C_{AO}\frac{dX_A}{dt} = k_2(C_{AO} - C_{AO}X_A)(C_{BO}C_{AO}X_4) \tag{1.11}$$

Let $M = C_{BO} / C_{AO}$ (the initial molar ratio of reactants)
Integrating equation (1.11a),

$$\int_O^{X_A} \frac{dX_A}{(1-X_A)(M-Xk)} = C_{AO}k_2\int_O^t dt \tag{1.12}$$

After breaking into partial fractions, integrating and rearranging, we get,

$$\frac{\ln(M-X_A)}{M(1-X_A)} = CA_o(M-1)k_2 t \tag{1.13}$$

The equation may be plotted as follows in the form of a linear plot passing through (0, 0)

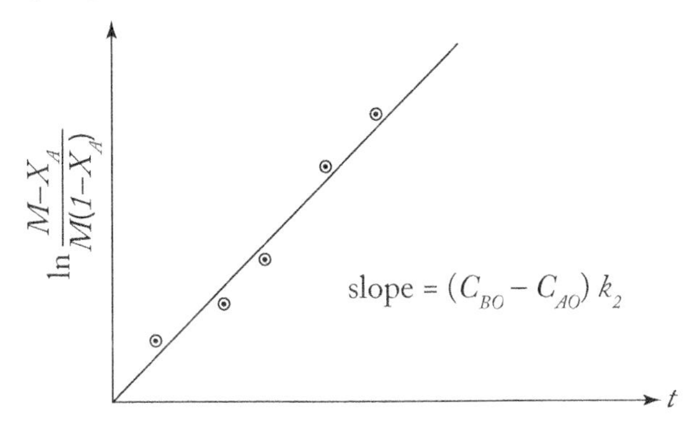

Fig. 1.1 The plot of 2nd order reaction

From the slope of the linear plot, $(C_{BO} - C_{AO})\, k_2$, the second order rate constant, k_2 can be evaluated, since the values of C_{AO} and C_{BO} are known.

For equimolal second order equation,

$$-r_A = k_2 C_A^2$$

Or
$$-\frac{dC_A}{dt} = kC_A^2 = kC_A^2(1 - X_A)^2$$

Or integrations

$$\frac{1}{C_A} - \frac{1}{C_{AO}} = \frac{1}{C_{AO}}\left(\frac{X_A}{1 - X_A}\right) = kt$$

Or $C_A = \dfrac{C_{AO}}{1 + C_{AO}kt}$

For Pseudo first order reaction

In the equation (1.11) if C_{BO} >>1 then

$$C_{AO}\left(\frac{C_{BO}}{C_{AO}} - 1\right) \approx C_{BO}$$

$$M - X_A \approx M$$

Then equation (11) reduces to

$$ln\frac{1}{(1 - X_A)} = kC_{BO}t = k't \tag{1.14}$$

The equation becomes Pseudo first order, $kC_{BO} = k'$ (Pseudo first order rate constant, time^{-1})

For the nth order irreversible reaction,

$A \rightarrow$ Products, $-r_a = kC_A^n$

The rate equation is,

$$dC_A/dt = k\, C_A^n \tag{1.15}$$

To find out the rate constant, k, by differential method, we follow as below:

1.2.2 Differential method

The batch data, C_A vs t are plotted and tangents are drawn to the curve, C_A vs t, in fig. 1.2 and different slopes are obtained as – (dC_A/dt)

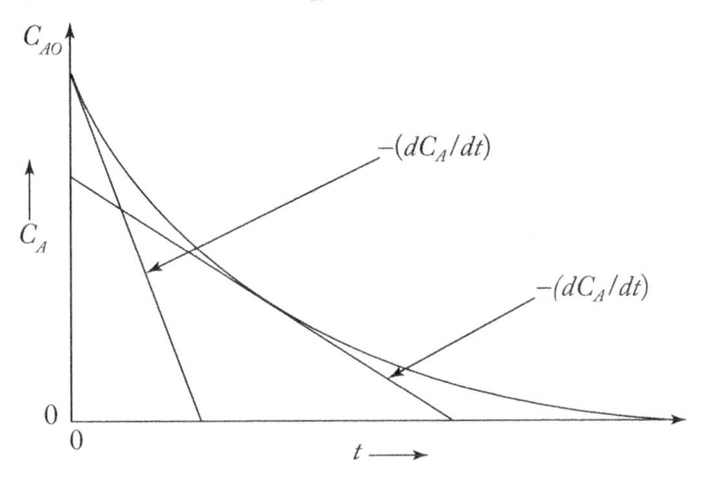

Fig. 1.2 Finding the derivative of C_A vs t

Now taking the logarithm of both sides of the equation (1.15), we get

$$\log(-dC_A/dt) = \log k + n \log C_A \tag{1.16}$$

From the linear plot of log ($-dC_A/dt$) VS log C_A, the slope will give the value of n, the order of reaction and the intercept will give log k, from which k can be evaluated. Equation (1.15) can also be integrated as

$$\int_{C_{AO}}^{C_A} C_A^{-n} dC_A = -k\int_0^t dt$$

or $$\frac{1}{-n+1}\left(C_A^{-n+1} - C_{AO}^{-n+1}\right) = kt$$

Rearranging

$$C_A(t) = C_{AO}[1+(n-1)kC_{AO}^{n-1}]^{\frac{1}{1-n}} \tag{1.17}$$

When $n \neq 1$

Putting $n = 2$

$$C_A(t) = C_{AO} / (1 + kC_{AO}\, t) \tag{1.18}$$

The above equation is the integrated form of second order reaction.

1.2.3 Zero order and negative order reaction

Putting n = 0 in the equation (1.15), we get,

$dC_A/dt = -k$

and after integration, the equation (1.19) is obtained as

$$C_A = C_{AO} - kt \quad (1.19)$$

$$C_A = 0 \text{ at } t = C_{AO}/k$$

The concentration becomes negative for larger times

1.2.3.1 Complex reactions

For a simple catalytic reaction of the type,

$A \rightarrow$ Products

The rate may be expressed as

$$r_A = k\,KC_A / (1 + K\,C_A) \quad (1.20)$$

where k, K are temperature dependent constants. k is the reaction rate constant and K is an adsorption constant.

Case I: When $KC_A >> 1$, the rate becomes $r_A \approx k$ and the rate becomes zero order.

Case II: When $KC_A << 1$, the rate becomes, $r_A = kKC_A$, giving rise to first order kinetics

Case III: For a bimolecular catalytic reaction

2A → products

The rate may be given as

$$r_a = k\,K\,C_A / (1 + K\,C_A)^2 \quad (1.21)$$

the rate indicates negative order, when $KC_A >> 1$ and the rate becomes

$$r_A = 1/kC_A \quad (1.22)$$

but the equation approaches first order if $KC_A <<1$ when

$$r_A = k\,K\,C_A \quad (1.22a)$$

1.2.4 Half – life time

From half-life data, the order of reaction can be determined. Consider the reaction,

$A^a + B^b \rightarrow$ Products

We can express rate as

$$-r_A = kCA^a\,CB^b \quad (1.23)$$

If the reactants are present in their stoichiometric ratios, that ratio will be maintained throughout the reaction and the overall rate can be given in term of nth order equation,

$$-dC_A/dt = k\,C_A^n \tag{1.24}$$

Integrating, we get,

$$C_A^{1-n} - C_{AO}^{1-n} = (n-1)kt/k \tag{1.25}$$

where n ≠ 1

Defining the half-life time as the time required for the concentration of reactant to drop to one half of the original value. Thus we obtain,

$$t_{\frac{1}{2}} = \frac{(0.5)^{1-n}-1}{k(n-1)} C_{AO}^{1-n} \tag{1.26}$$

Taking logarithm of both sides, we have

$$\log t_{\frac{1}{2}} = \log\left[\frac{(0.5)^{1-n}-1}{k(n-1)}\right] + (1-n)\log C_{AO} \tag{1.27}$$

A plot of log $t_{½}$ vs log C_{AO} gives a straight line with a slope of (1 – n) from which we can find out the order of reaction n.

Performing a series of runs each at different initial concentrations, C_{AO}, the half-life time data may be plotted as in fig 1.3.

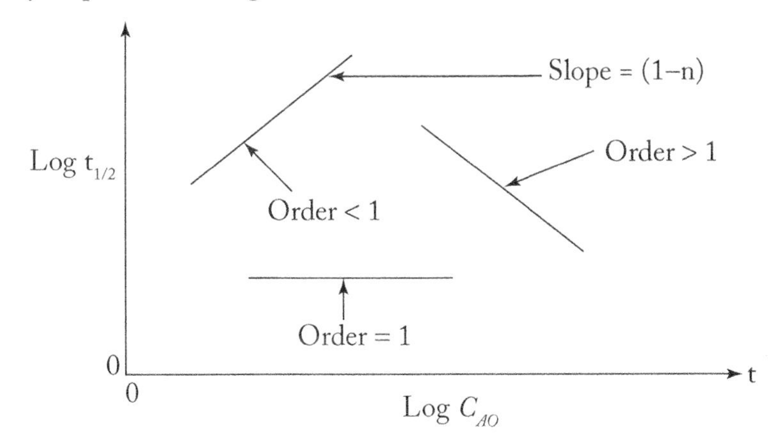

Fig. 1.3 Overall order of reaction from a series of half-life experiments

1.3 Autocatalytic Reactions

In this case one of the products of the reaction acts as a catalyst. The reaction can be shown as,

$$A + R \rightarrow R + R \tag{1.28}$$

where $-rA = -dC_A/dt = kC_A\,C_R$ (1.29)

Since the total number of moles of A & R remain constant as A is consumed,
$C_O = C_A + C_R = C_{AO} + C_{RO}$ = constant

$$-dC_A / dt = k\, C_A (C_O - C_A) \tag{1.30}$$

Rearranging and making partial fractions

$$-\frac{dC_A}{C_{AO}(C_O - C_A)} = -\frac{1}{C_O}\left(\frac{dC_A}{C_A} + \frac{dC_A}{C_O - C_A}\right) = kdt$$

which on integration between the limit (C_{AO}, C_A)

$$\frac{\ln(C_{AO}(C_O - C_A))}{C_A(C_O - C_{AO})} = \frac{\ln\left(C_R / C_{RO}\right)}{C_A / CA_O} = C_O\, kt = (C_{AO} + C_{BO})kt \tag{1.31}$$

Putting $M = C_{RO} / C_{AO}$ and using fractional conversion of A, X_A, the above equation may be written as

$$\frac{\ln(M + X_A)}{M(1 - X_A)} = C_{AO}(M + 1)kt \tag{1.32}$$

The linear plot of ln $(M + X_A) / M(1 - X_A)$ vs t will pass through (0, 0) with a slope of $(C_{AO} + C_{RO})$ k, from which the rate constant, k can be evaluated.

The plot of X_A vs t will give a S-shaped curve which is very characteristic of auto catalytic reaction.

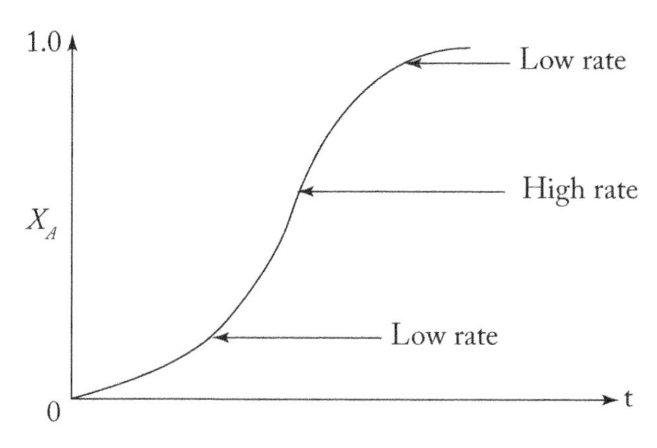

Fig. 1.4 x_A vs t for auto catalytic reaction

The classical example of auto catalytic reaction is the microbial cell reaction.

First order reversible Reactions

Consider the following example,

$$A \underset{k_2}{\overset{k_1}{\rightleftarrows}} R, \tag{1.33}$$

K_C = Equilibrium constant

$\Delta G_R^O = RT \ln Kc$, ΔGR_0 is standard free energy change of the reaction.

Assume M concentration ratio,

$M = C_{RO}/C_{AO}$, $C_R = C_{RO} + C_{AO} X_A$

$$\frac{dC_R}{dt} = \frac{dC_A}{dt} = k_1 C_A - k_2 C_R = k_1 (C_{AO} - C_{AO} X_A) - k_2 (MC_{AO} + C_{AO} X_A) \tag{1.34}$$

Now at equilibrium, $dC_A/dt = O$, then

$$Kc = k_1/k_2 = C_{Re} / C_{Ae} = (M + X_{Ae}) / (1 - X_{Ae}) \tag{1.35}$$

Writing the rate in terms of conversion, X_A equation (1.34) becomes

$$C_{AO} \frac{dX_A}{dt} = k_1 \left(C_A - \frac{C_R}{K_C} \right) \tag{1.36}$$

Or $$\frac{dX_A}{dt} = k_1 \left[1 - X_A - \frac{(M + X_A)}{\left(\frac{M + X_{A_e}}{1 - X_{A_e}} \right)} \right]$$

Where X_{Ae} is the equilibrium conversion

Simplifying,

$$\frac{dX_A}{dt} = \frac{k_1 (M+1)(X_{Ae} - X_A)}{(M + X)} \tag{1.37}$$

Integrating between the limits (O, X_A) and (O, t) we get

$$-\ln \left(1 - \frac{X_A}{X_{A_e}} \right) = -\frac{\ln (C_A - C_{Ae})}{C_{Ae} - CA_e} = \frac{(M+1)}{M + X_{Ae}} k_1 t \tag{1.38}$$

If $C_R = 0$, $M = 0$, the equation (1.38) reduces to

$$-\ln\left(1-\frac{X_A}{X_{Ae}}\right)=\frac{k_1}{X_{Ae}}t \tag{1.39}$$

A plot of $-\ln(1 - X_A/X_{Ae})$ vs t, from equation (1.38) will give a straight line with a slope of $(M + 1)^{k}{}_1 / (M + X_{Ae})$ from which k_1 can be evaluated with the values of M and X_{Ae} the equilibrium conversion.

1.4 Series Reactions with First Order Kinetics

Let us consider the following consecutive unimolecular reactions

$$A \xrightarrow{k_1} B \xrightarrow{k_2} C \tag{1.40}$$

The following rate equations may be written for the three components (A, B, C)

$$r_A = \frac{dC_A}{dt} = -k_1C_A \tag{1.41}$$

$$r_B = \frac{dC_B}{dt} = k_1C_A - k_2C_B \tag{1.42}$$

$$r_C = \frac{dC_C}{dt} = k_2C_C \tag{1.43}$$

at $t = 0$, $C_A = C_{AO}$, $C_{BO} = 0$, $C_{CO} = 0$.

From the first equation, integrating we have,

$-\ln C_A/C_{AO} = k_1 t$ or $C_A = C_{AO}\, e^{-k1t}$

From the material balance of B

$$\frac{dC_B}{dt} + k_2C_B = k_1C_{AO}e^{-kt} \tag{1.44}$$

The above is a first order linear differential equation, using an integrating factor, I.F,.

$$\int_e k_2 dt = e^{k_2 t}$$

Multiplying both sides by $e^{k_2 t}$, we have,

$$\frac{dC_B}{dt}e^{k_2t} + k_2C_Be^{k_2t} = k_1CA_{Oe}^{+(k_2-k_1)t}$$

Or $$\frac{d(C_{Be}^{k_2t})}{dt} = k_1 C_{AO} e(k_2 - k_1)t \tag{1.45}$$

Integrating the above

$$C_{Be}^{k_2t} = \frac{k_1 C_{AO}}{k_2 - k_1} e^{(k_2-k_1)} + C \tag{1.46}$$

where C is the constant of integration.

The boundary condition, at $t = 0$, $C_B = 0$.

$$O = \frac{k_1 C_{AO}}{(k_2 - k_1)} + C$$

Or $$C = -\frac{k_1 C_{AO}}{(k_2 - k_1)} \tag{1.47}$$

Substituting the value of C in equation (1.46), we have

$$C_{B_e}^{k_2t} = \frac{k_1 C_{AO}}{k_2 - k_1} e(kt_2 - k_1) - \frac{k_1 C_{AO}}{k_2 - k_1}$$

Or $$C_B = \frac{k_1 C_{AO} e^{-k_1t}}{k_2 - k_1} - \frac{k_1 C_{AO} e^{-k_2t}}{k_2 - k_1} \tag{1.48}$$

Or $$C_B = k_1 C_{AO} \left[\frac{e^{-k_1t}}{k_2 - k_1} + \frac{e^{-k_2t}}{k_1 - k_2} \right]$$

Now, $$C_{AO} = C_A + C_B + C_C \tag{1.48a}$$

$$C_C = CA_O - C_A - C_B$$

$$= CA_O - C_{AO_e}^{-k_2t} - k_1 C_{AO} \left[\frac{e^{-k_1t}}{k_2 - k_1} + \frac{e^{-k_2t}}{k_1 - k_2} \right] \tag{1.49}$$

Or $$C_C = C_{AO} \left[1 + \frac{k_2}{(k_1 - k_2)} e^{-k_1t} + \frac{k_1}{(k_2 - k_1)} e^{-k_2t} \right] \tag{1.50}$$

Case 1: If $k_2 >> k_1$
$C_C = C_{AO}(1 - e^{-k_1 t})$
Case II: if $k_1 >> k_2$
$C_C = C_{AO}(1 - e^{-k_2 t})$
Case III: CB_{max} can be obtained by differentiating the equation (1.48) and putting $dC_B/dt = 0$.
The time for maximum concentration of B is given as

$$t_{max} = \frac{\ln \frac{k_2}{k_1}}{k_2 - k_1} = \frac{1}{k_{logmea}} \tag{1.51}$$

and the maximum concentration of B, CB_{max} is obtained by combining equation (1.48) & (1.51), we get

$$\frac{CB_{max}}{C_{AO}} = \left(\frac{k_1}{k_2}\right)^{k_2/(k_2 - k_1)} \tag{1.52}$$

Case IV when $k_1 = k_2$
We know, $C_A = C_{AO} e^{-k_1 t}$

$$\frac{dC_B}{dt} + k_1 C_A - k_1 C_R$$

Or

$$\frac{dC_B}{dt} + k_1 C_B = k_1 C_A = k_1 C_{AOe}{}^{-k_2 tm} \tag{1.53}$$

Integrating,

$$C_B e^{-k_1 t} = k_1 C_{AO} t + c \tag{1.54}$$

at $t = 0$, $C_B = 0$
So, $C = 0$

$$C_B/C_{AO} = k_1 t\, e^{-k_1 t} \tag{1.55}$$

To find out CB_{max}, $dC_B/dt = 0$. Differentiating equation (1.55) we have

$$\frac{d\left(\frac{C_B}{C_{AO}}\right)}{dt} = (k_1 t)(-k_1) e^{-k_1 t} + k_1 e^{-k_1 t} = 0$$

$$tk_1^2 e^{-k_2 t} = k_1 e^{-k_1 t}$$

Or
$$t_{max} = \frac{k_1}{k_1^2} = \frac{1}{k_1} \quad (1.56)$$

Now
$$\frac{CB_{max}}{C_{AO}} = k_1 \cdot \frac{1}{k_1} e^{-k_1 \times \frac{1}{k_1}} = e^{-1} \quad (1.57)$$

1.5 Variable Volume Batch Reactor

The change in the number of moles in the stoichiometry of the equations leads to variable volume system in the gaseous phase.

If V_o = initial volume of the reactor and

V = the volume at time t after certain conversion of the reactant A, X_A = fractional conversion of A

Then $V = V_O(1 + \in_A X_A)$

$\in_A$ = expansion factor

Where $XA = (V - V_0)/ V_0 \in_A$ (1.58)

So,
$$dX_A = \frac{dV}{V_O \in_A} \quad (1.59)$$

And
$$\in_A = \frac{V_{XA=1} - V_{XAO}}{V_{X_O}} \quad (1.59a)$$

Consider an isothermal gas phase reaction

$A \rightarrow 3B$

$$€A = (3-1) / 1 = 2 \quad (1.60)$$

with 50% inerts, using 1 mole of A and 1 mole of inerts

$$\in_A = \frac{4-2}{2} = 1.0$$

Now $N_A = N_{AO}(1 - X_A)$

$$C_A = \frac{N_A}{V} = \frac{N_{AO}(1-X_A)}{V_O(1+\in_A X_A)} = C_{AO}\frac{1-X_A}{1+\in_A X_A}$$

Or

$$\frac{C_A}{CA_O} = \frac{1-X_A}{1+\in_A X_A} \tag{1.61}$$

The rate is expressed in general as given below,

$$-r_A = \frac{1}{V}\frac{dN_A}{dt} \tag{1.62}$$

Expressing in terms of X_A & C_A

$$-r_A = \frac{CA_O}{(1+\in_A X_A)}\frac{dX_A}{dt} \tag{1.63}$$

In terms of volume

$$-r_A = \frac{CA_O}{V\in_A}\frac{dV}{dt} = \frac{C_{AO}}{\in_A}\frac{d\ lnV}{dt} \tag{1.64}$$

1.5.1 Zero order reaction

$$-r_A = \frac{C_{AO}}{\in_A}\frac{d\ln V}{dt} = k \tag{1.65}$$

Integrating we have

$$CA_0 / €A\ln V/V_0 = kt \tag{1.66}$$

A linear plot of ln V/Vo vs t yield a slope of $\frac{\in_A k}{C_{AO}}$ from which k can be calculated.

1.5.2 First order reaction

$$-r_A = \frac{C_{AO}}{\in_A}\frac{d\ln V}{dt} = kC_A = kC_{AO}\left(\frac{1-X_A}{1+\in_A X_A}\right) \tag{1.67}$$

replacing X_A in terms of V, Vo and $\in_A$ we have,

$$X_A = \frac{V-V_O}{\in V_O} \tag{1.68}$$

Integrating equation (1.67) using equation(1.68) we get,

$$-ln\left(1-\frac{\Delta V}{\in_A V_O}\right)=kt \tag{1.69}$$

The plot of $-ln\left(1-\frac{\Delta V}{\in_A V_O}\right) vs\ t$ will give rise to a linear plot and the rate constant k may be evaluated from the slope.

For the equation (1.60), $\in_A = 2$ for pure reactant. The rate equation can be expressed in integrated form,

$$-\ln\left(1-\frac{\Delta V}{2V_O}\right)=kt \tag{1.69a}$$

We know that

$$\frac{dN_A}{dt}=-VkC_A=V_O(1+2X_A)kC_{AO}\frac{(1-X_A)}{(1+2X_A)} \tag{1.70}$$

From the equation (1.60)

Since $dN_A = -N_{AO}\,dX_A$, $C_{AO}=\frac{N_{AO}}{V_O}$

Equation (1.70) reduces to

$$\frac{dX_A}{dt}=k(1-X_A)\frac{(1+2X_A)}{1+2X_A}=k(1-X)$$

Integrating between the limits (0,t) and (0, X_A)

$$t=\int_{X=0}^{X}\frac{dX}{k(1-X)}=-\frac{1}{k}\ln(1-X) \tag{1.71}$$

So,

$$X(t)=1-e^{-kt} \tag{1.72}$$

Substituting back into the equation (1.61), we get

$$C_A(t)=C_{AO}\frac{1-X}{1+2X}=C_{AO}\frac{e^{-kt}}{1+2(1-e^{-kt})}$$

Or,

$$C_A(t)=C_{AO}\frac{e^{-kt}}{(3-2e^{-kt})} \tag{1.73}$$

From the above equation, $C_A(t)$ can be calculated at any time provided k is known.

Problem 1.1

When a concentrated solution of urea is stored in a metalic container, it slowly decomposes to biuret by the following elementary reaction,

$$2NH_2CONH_2 \rightarrow NH_2\text{–}CO\text{–}NH\text{–}CONH_2 + NH_3$$

An urea solution of concentration, C_{AO} = 10 kmol/m^{-3} was stored at 100°C and after 8 hours it was found that 1 mole% has turned into biurent. Find the rate equation for this reaction.

Solution: Assume the reaction to be of constant density and second order with respect to urea stoichiometry,

$$-\frac{dC_A}{dt} = k_2 C_A^2$$

$$\text{Or } C_{AO}\frac{dX_A}{dt} = k_2 C_{A_O}^2 (1 - X_A)^2$$

$$\text{Or } \int_O^{XA} \frac{dX_A}{(1 - X_A)^2} = k_2 C_{AO} \int_O^t dt$$

$$\text{Or } \frac{1}{1 - X_A} - 1 = k_2 C_{AO} t$$

$$\text{Or } k_2 = \frac{X_A}{1 - X_A}\frac{1}{C_{AO}t} = \frac{0.01}{0.99 \times 10 \times 8} = 1.26 \times 10^{-4} \frac{m^3}{(kmole)(hr)}$$

So the rate equation is

$$-\frac{dC_A}{dt} = 1.26 \times 10^{-4} CA \frac{kmole}{(m^3)(hr)}$$

Ans

Problem 1.2

The second order irreversible saponification reaction is given as,

$$NaOH(aq) + CH_3CO_2C_2H_5(aq) \rightarrow CH_3CO_2Na(aq) + C_2H_5OH(aq)$$

The reaction is carried out in a well stirred batch reactor which is charged with an aqueous solution containing NaOH and ethyl acetate with equal concentration ($C_{AO} = C_{BO}$ = 0.1 kmol/m³).

After 15 minutes the conversion of ethyl acetate is 18%

(a) Find out the rate constant of the reaction

(b) For an initial charge containing NaOH and ethyl acetate in equal concentration, $C_{AO} = C_{BO}$ = 0.2 kmol/m³.

  (i) What time would be required to achieve 30% conversion in a stirred batch reactor

  (ii) What volume is necessary to produce 50kg of sodium acetate?

Solution:

(a) Given: $C_{AO} = C_{BO} = 0.1$ kmol/m^3. Batch reactor design equation,

$$-\frac{dC_A}{dt} = kC_AC_B = kC_A^2$$

Or, $$C_{AO}\frac{dX_A}{dt} = kC_{AO}^2(1-XA)^2$$

Or, $$\int_O^{XA}\frac{dX_A}{(1-X_A)^2} = kC_{AO}\int_O^t dt$$

Or, $$\frac{1}{1-X_A} - 1 = kC_{AO}t$$

Or, $$\frac{X_A}{1-X_A} = kC_{AO}t \qquad \text{(A)}$$

Given at t = 15 min, $x_A = x_B = 0.18$ substituting in the above equation (A)

0.18/ (1–0.18) (0.1) 15 = k

or k = 0.1463 m^3/(kmole) (min)

So the rate equation is

$-dC_A/dt = 0.1463\ C_AC_B$ k.moles/(m^3)(min). Ans

(b) (i) In the second case $C_{AO} = 0.2$ kmole, $C_{BO} = 0.2$ kmole at t = 0,

$x_A = x_B = 0.3$

$X_A/(1-X_A) = kC_{AO}t$

[0.3/(1–0.3)] /(0.2)(0.1463) = t

or t = 14.65 minutes Ans.

(ii) Reactor volume for Na-acetate:

Mol. Weight of CH_3CO_2Na = 82

50 kg = 50/82 = 0.61 kmole of Na–acetate

at 30% conversion

NB0, ethyl ester needed = 0.61/0.30 = 2.033 k.moles

$C_{BO} = N_{BO}/V$ or $V = N_{BO}/C_{BO}$ = 2.033 kmole/ 0.2 kmole/m^3

or V = 10.65 m^3. Ans

Problem 1.3

Thermal decomposition of nitrous oxide is carried in gas phase at 1030K in a batch reactor and the half–life data are given below:

p^o_A, mm H_g	52.5	139	290	300
$t_{½}$, sec	800	470	256	212

Find out the order of reaction, n and the rate constant k.

Solution:

For nth order reaction

$$-r_A = -\frac{dC_A}{dt} = k_C C_A^n$$

$$t_{\frac{1}{2}}(half\ life\ time) = \frac{(0.5)^{1-n}}{k_c(n-1)} \cdot C_{A_O}^{1-n}$$

For constant volume, $C_{AO} = \frac{p_{AO}}{RT}$

So $t_{1/2}$ in terms of p_{AO},

$$t_{\frac{1}{2}} = \frac{(0.5)^{1-n}}{k_C(n-1)} \cdot \frac{p_{AO}^{n-1}}{(RT)^{n-1}}$$

Taking logarithm of both sides we have

$$\log t_{\frac{1}{2}} = \log\left[\frac{(0.5)^{n-1}}{k_c(n-1)(RT)^{n-1}}\right] + (1-n)\log p_{AO}$$

$\log t_{½}$	2.9	2.67	2.40	2.32
$\log p'_{AO}$	1.72	2.14	2.46	2.47

From the linear plot of $\log t_{½}$ vs $\log p_{AO,}$

Slope = 1 – n = – 0.85 or n = 1.85 ≈ 2 Ans

I, (Intercept) = 4.75

$I = \log (0.5)^{1-2} / k_c (2 - 1) (RT)^{1-2}] = 4.75$

$\log [\ 2/k_c/RT] = 4.75$

$2/ (k_c/RT) = 56.23$

$2 / 56.23 = k_c/RT = k_p = 3.55\ mm^{-1}sec^{-1}$ Ans

The rate equation is,

$-dP_A/dt = 3.55p_A^2$ in(mm Hg)/sec. Ans

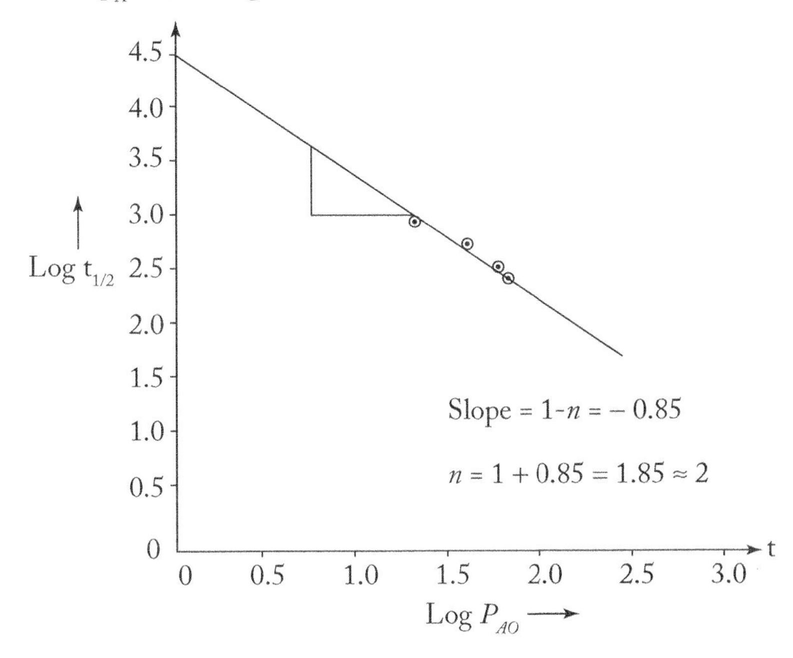

Fig. 1.5 Half–life time plot vs. p_{AO} (in log–log)

Problem 1.4

The rate of decomposition of acetaldehyde, expressed as mm Hg per min, corresponding to the various percent decomposition of CH_3CHO.

% decomposition	0	5	10	15	20	25	30	35	40	45	50
rate, dP/dt	8.53	7.49	6.74	5.90	5.14	4.69	4.31	3.75	3.11	2.67	2.29

(a) Determine the order of reaction and (b) the rate constant if initial pressure is 200 mm Hg

Solution:

The decomposition of acetaldehyde can be given as,

$$CH_3CHO \rightarrow CH_4 + CO$$

$1-x_A \qquad x_A \qquad x_A$

Total moles $= 1 - x_A + x_A + x_A = 1 + x_A$

$C_A = C_{AO}(1 - x_A)/1 + x_A$

or $p_A = p_{AO} (1 - x_A) / 1 + x_A$

$- dp_A/dt = k_p p_A^n = k_p p_{AO}^n [(1 - x_A) / 1 + x_A]^n$, taking log of both sides,

$\log (-dp_A/dt) = \log (k_p P_{AO}^n) + n \log (1 - x_A) / (1 + x_A)$

Assuming first order, n = 1

$-dp_A / dt = r_A = (k_p P_{AO}) (1 - x_A) / (1 + x_A)$

x_A	0	0.05	0.10	0.15	0.20	0.25	0.30	0.35	0.40
$r_A = dp_A/dt$	8.53	7.49	6.74	5.90	5.14	4.69	4.31	3.75	3.11
$(1-x_A)/(1+x_A)$	1.0	0.90	0.82	0.74	0.67	0.60	0.538	0.48	0.428

$\log (r_A) = \log (k_P Q_{AO}) + n \log 1 - x_A / 1 + x_A$

$\log r_A$	0.93	0.874	0.828	0.710	0.67	0.634	0.57	0.49	0.426
$\log (1 - x_A) / (1 + x_A)$	0	−0.045	−0.082	−0.174	−0.22	−0.269	−0.318	−0.368	−0.42

From the linear plot of (–dp/dt) vs $(1 - x_A) /(1 + x_A)$

Slope = $(k_p P_{AO}) = 8.18$ mm/min

$k_p = 8.18 / P_{AO} = 8.18 / 200 = 0.041 \text{ min}^{-1}$

The linear plot of log (–dp/dt) vs $\log (1 - x_A)/ (1 + x_A)$ gives the slope = 1.1

and the intercept $(\log k_p p_{AO}^n) = 0.93$

or $k_p p_{AO} = 8.51$ mm Hg / min

$kp = 8.51 / 200 = 0.042 \text{ min}^{-1}$

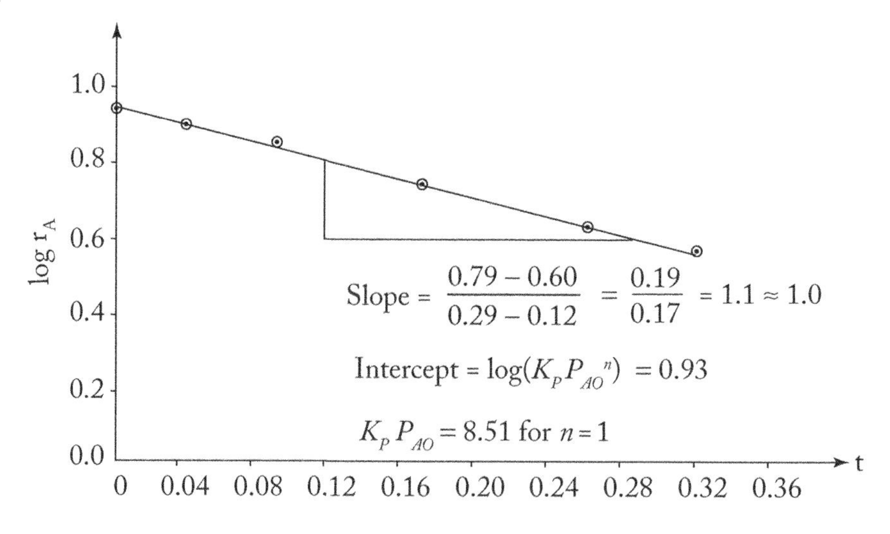

$\log [(1 - x_A)/ (1 + x_A)]$

Plot of log r_A vs log $[(1 - x_A)/(1 + x_A)]$

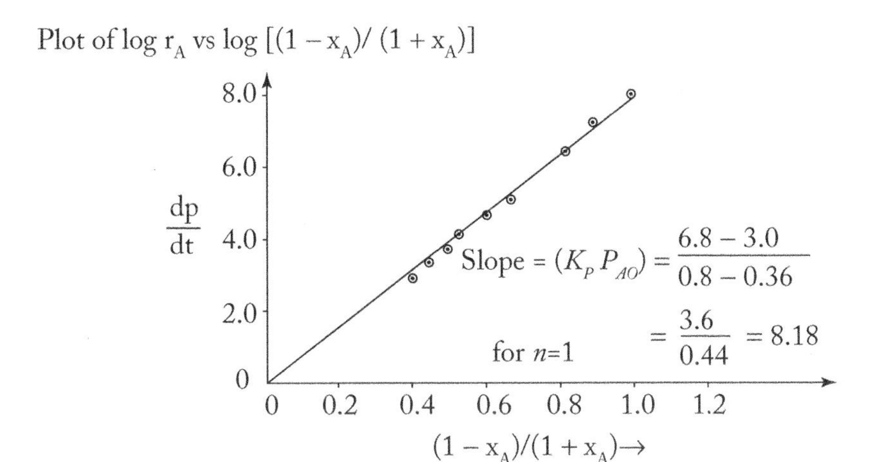

Plot of dp/dt vs $(1 - x_A) / (1 + x_A)$

Problem 1.5

One solution contains substance, A of concentration, C_{AO} = 1.6 kmole/m^3 and another substance B of concentration, C_{BO} = 1.0 kmole/m^3. Equal amounts of these solutions are mixed quickly and measurements of C_A vs time, t are made:

The reaction is reversible

$$A \underset{k_2}{\overset{k_1}{\rightleftarrows}} B$$

Find the rate equation of the reaction:
Data given are:

t, hr	0	0.5	1.0	1.5	2.0	3.0	4.0	5.0	10.0
C_A, kmole/m^3	0.800	0.670	0.600	0.563	0.543	0.527	0.522	0.520	0.520

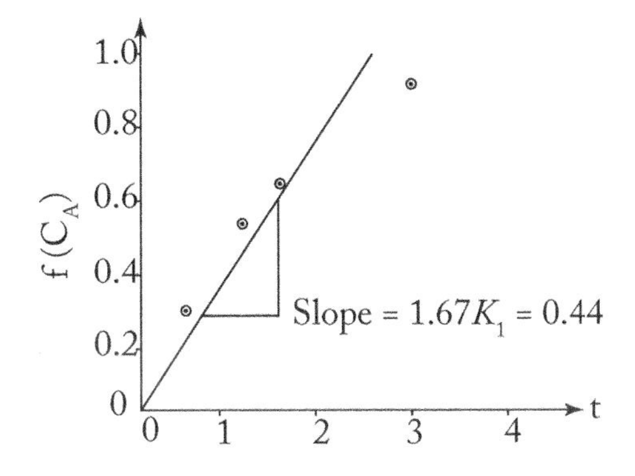

Plot of f (C_A) vs t

Solution:

$$A \underset{k_2}{\overset{k_1}{\rightleftarrows}} B$$

$C_{AO} = 0.8$, $C_{BO} = 0.5$ k.mole/m^3

$C_B - = C_{BO} + C_{AO} - C_A$

$C_{Be}/C_{Ae} = k_1/k_2 = K = (C_{BO} + C_{AO} - C_{Ae}) / C_{Ae}$

$= 0.5 + 0.8 - 0.52 / 0.52$

$= 1.5$

$-dC_A/dt = k_1 [C_A - C_B/K] = k_1/1.5 [1.5 C_A - C_{AO} + C_A]$

integrating and rearranging

$f(C_A) = \ln[(1.5 C_{AO}) / (2.5 C_A - C_{AO})] = 2.5 / 1.5 k_1 t = 1.67 k_1 t$

To plot the above equation in a linear one, the following table is made

t	0	0.5	1.0	1.5	3.0
C_A	0.8	0.67	0.60	0.563	0.527
$f(C_A)$	0	0.315	0.539	0.68	0.84

Plot is shown on the previous page

Slope $= 1.67 k_1 = 0.63 - 0.30 / 1.5 - 0.75 = 0.33 / 0.75 = 0.44$

$k_1 = 0.44/1.67 = 0.263$ hr^{-1}

$k_2 = k_1/K = 0.263 / 1.5 = 0.175$ hr^{-1}

Problem 1.6

Thermal decomposition of nitrous oxide on gold at 900°C and the initial pressure of 230 mmHg is carried out in a constant volume batch reactor and yields the following data:

t,min	30	53	100
% decomposition	32	50	73

Find out a rate expression for the system.

Solution: The nitrous oxide decomposes as follows

$N_2O(A) \rightarrow N_2 + ½ O_2$

Basis: 1 mole of N_2O

$1 - x_A$ x_A $0.5x_A$

$N_t = 1 + 0.5\,x_A$

If $A \equiv N_2O$

Then, $C_A = C_{AO}(1 - x_A)/(1 + 0.5\,x_A)$

Let the rate equation be

$-dC_A/dt = kC_A^{\,n}$

Assuming n = 1, the rate equation becomes

$-dC_A/dt = kC_A$

or $C_{AO}/(1+0.5x_A)\,dx_A/dt = kC_{AO}(1 - x_A)/(1 + 0.5x_A)$

or $dx_A/(1 - x_A) = kdt$

Integrating, we have $-\ln(1-x_A) = kt$

x_1	0.32	0.50	0.73
$-\ln(1-x_A)$	0.385	0.693	1.31
t, min	30	53	100
k, min^{-1}	0.0128	0.013	0.013

So the mean k is 0.0129 min^{-1}.

So the rate equation is

$-dC_A/dt = 0.0129\,C_A$ Ans

In terms of partial pressure of nitrous oxide,

$-dP_A/dt = 0.0129\,P_A$ Ans.

1.6 Effect of Temperature on Specific Rate: Arrhenius Equation

The dependency of the rate constant, k on temperature for an elementary reaction is given by Arrhenius equation

$$k = Ae^{-E/RT} \tag{1.74}$$

where A is the frequency factor and E is the activation energy.

The Arrhenius equation was originally derived from thermodynamic consideration. For an elementary reaction where rates are rapid to achieve a dynamic equilibrium, the Vanit Hoff equation gives,

$$dlnK/dT = \Delta H^{o}/RT^{2} \quad (1.75)$$

For a reaction of type,

$$A + B \underset{k_2}{\overset{k_1}{\rightleftarrows}} C$$

with k_1 and k_2 are the rate constant of the forward and reverse reaction and the equilibrium constant, K is related to k_1 and k_2 as

$$K = k_1/k_2$$

Then from the equation (1.71)

$$dlnk_1/dT - dlnk_2/dT = \Delta H/RT^2 \quad (1.76)$$

The right hand side of equation (1.72) can be divided into to enthalpy changes, ΔH_1 and ΔH_2, such that,

$$\Delta H = \Delta H_1 - \Delta H_2 \quad (1.77)$$

Then equation (1.72) may be split into two equations as follows:

$$dlnk_1 / dT = \Delta H_1/RT^2$$

$$(1.78)$$

$$dlnk_2/dT = \Delta H_2/RT^2$$

Integrating either equation and introducing an integration constant A, we have an equation of Arrhenius type,

$$k = A\ e^{-\Delta H/RT} \quad (1.79)$$

At two different temperatures, having the same concentration, the ratio of two rates may be given by Arrhenius equation as

$$\ln r_2/r_1 = \ln k_2/k_1 = E/R\,(1/T_1 - 1/T_2) \quad (1.80)$$

where $T_2 > T_1$ & E, (the activation energy.)

COLLISION AND TRANSITION STATE THEORIES give k in the simplest form:

$$k = A'\ T^{m}\ e^{-E/RT},\ 0 \leq m \leq 1 \quad (1.81)$$

Since the exponential terms are very temperature sensitive, the pre–exponential term, $A'T^m$, the variation of the later with temperature is effectively masked and we get ultimately $k = Ae^{-E/RT}$ which is the classical Arrhenius equation.

1.6.1 Significance of activation energy

The following factors are to be noted in connection with activation energy

(a) From Arrhenius Law, the plot of lnk vs 1/T yields a straight line with a negative slope of −E/R, from which the activation energy, E, can be calculated.

(b) Reactions with high activation energy are temperature sensitive and the rate has been found to be reaction rate controlling

(c) Reactions with low activation energy are relatively temperature insensitive and the rate becomes mass transfer controlling.

Problem 1.7

The gas phase chlorination of methane is carried out in a constant pressure isothermal vessel at a pressure of 1.013 bar and temperature of 25°C

CH_4 (g) + $2Cl_2$ (g) → CH_2Cl_2 + 2 HCl
(A) (B)

The feed is stoichiometric. The rate constant at 25°C,
k = 0.2 mol / (kmole)^2 (s)/ m^6 /(kmole)^2(s)

(a) What is the time required for 60% conversion of CH_4

(b) If the frequency factor of the rate constant is 2.0 x 10^{12} m^6 / (kmole)^2 (s), evaluate the activation energy

(c) Calculate the value of the rate constant at 50°C

(d) If the temperature of 50°C is used to carry out the reaction, what is the time necessary for 60% conversion of A?

Now $C_{AO} = p_A/RT = 1.013 / (0.0831)(298)$

$= 0.04$ kmol/m^3

k = 0.2 m^6 / (kmole)^2 (s) at 25°C

$C_{BO} / C_{AO} = 2.0$

$CH_4 + 2Cl_2 \rightarrow CH_2Cl_2 + 2HCl$

Basis: 1 mole of CH_4 & 2 moles of Cl_2

Material balance:

Moles of CH_4 $= 1 - x_A$

Moles of $2Cl_2$ $= 2 - 2x_A$

Moles of CH_2Cl_2 $= x_A$

Mole of HCl $= 2x_A$

Nt (total moles) $= 3$

$C_A = C_{AO}(1 - X_A)/3$, $C_B = C_{AO}(2 - 2x_A)/3$

The rate is given as

$$C_{AO}\frac{dX_A}{dt} = kC_AC_B{}^2 = k\frac{C_{AO}}{3}(1-X_A)4^{C_{AO}{}^2}\frac{(1-X_A)^2}{9}$$

Or $\dfrac{dXA}{dt} = \dfrac{4k^{C_{AO}{}^2}}{27}(1-X_A)^3$

$$\int_O^{X_A}\frac{dX_A}{(1-X_A)^3} = \frac{4k}{27}C_{AO}{}^2\int_O^t dt$$

$$\frac{1}{2}\left[\frac{1}{(1-X_A)^2}-1\right] = \frac{4k}{27}C_{AO}{}^2 t$$

$$\frac{1}{(1-X_A)^2}-1 = \frac{8k}{27}C_{AO}{}^2 t$$

At $X_A = 0.6$, $k = 0.2\dfrac{m^6}{(Kmole)^2 S}, C_{AO} = 0.04\dfrac{kmole}{m^3}$

$1/(1-0.6)^2 - 1 = 8/27\ (0.2)\ (0.4)^2 t$

$5.25 = 9.48 \times 10^{-5} t$, solving for t

We have the reaction time, t

t = 55,379 sec = 15.38 hrs. Ans

(b) Calculation of activation energy from Arrhenius equation

ln k = ln A – E/RT

Given, A = 2.0×10^{12}

k = 0.2 R = 8314 KJ/ (kmole)(k)

T= 298

$\ln 0.2 = \ln (2.0 \times 10^{12}) - E / (8314)(298)$

$-1.6094 = 0.963 + 12 \times 2.3025 - E \times 4.03 \times 10^{-7}$

$E = 30.20 / (4.03) \times 10^7 = 7.49 \times 10^7$ KJ/Kmole

(a) We have the rate constant

$k = 2.0 \times 10^{12} \exp(-7.47 \times 10^7/RT)\ m^6 / (kmole)^2(s)$

The value of k at 50°C: T = (50 + 273) = 323

$k = 2 \times 10^{12} \exp[-7.47 \times 107/(8314)\, 323]$

$= (2 \times 10^{12})\, e^{-27.81}$

$= (2 \times 10^{12})\,(0.836 \times 10^{-12})$

$= 1.672\ m^6/(kmole)^2(s)$

(b) So for 60% conversion, time required is:

$t = 5.25 \times 27 / (8k\, C_{AO}^{\ 2}) = 5.25 \times 27 / (1.672)\,(0.04)^2 = 6623$ sec $= 1.84$ hrs Ans

1.7 Chemical or Biochemical Reactors

Reactors can be operated in batch or flow modes. Their classification may be based on (i) mode of operation, (ii) degree of mixing, (iii) multiphase systems.

A simple classification may be given as

1. Batch
2. Semi batch
3. Flow

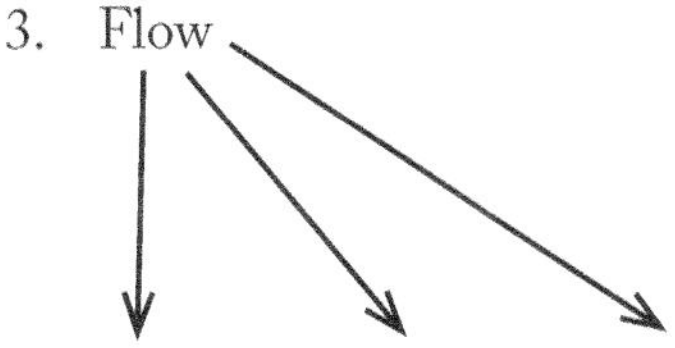

PFTR (Plug flow tubular reactor) | Partially mixed | CSTR (Continuous stirred tank reactor)

The flow reactors may function in either steady state or transient mode.

1.7.1 Multiphase reactors

These reactors may be of the following types:

1. Gas solid e.g. Packed bed catalytic reactor like SO_2 converter
2. Liquid – Solid: e.g. Packed bed with immobilized enzymes or cells, a CSTR with fermentation (Chemostat or turbidosat)
3. Gas–liquid: e.g. Bubble column reactor, airlift fermenter
4. Gas–liquid–solid: e.g. fludized bed; trickle bed
5. Special types containing multiple phases
   (a) Membrane reactor
   (b) Micro–carriers
   (c) Micro–encapsulation

This class of reactors is designed for specific enzymes or cell reactions.

We shall discuss the general characteristic features of some basic reactors where simple chemical reactions take place. In latter chapters, we will discuss these reactors where biochemical reactions with enzymes or cells are carried out.

Batch Reactor

Before we discuss the basic aspects of PFR, it can be shown that they are equivalent to batch reactors, where time of batch reactor, t, is replaced by Z/u where Z is the axial length and u is the linear axial velocity. For example, a batch reactor may be given as,

$-dC_A/dt = r_A$

Now replacing t = Z/u or dt = dZ/u, where linear velocity, u, in the PFR is constant and z is the axial length,

we have

$$-u\, dC_A/dZ = r_A \tag{1.82}$$

Assuming first order equation for the rate and integrating we have,

$C_A = C_{AO}\, e^{-kL/u} \qquad = C_{AO} e^{-k\tau}$

where τ (the residence time) = L/u = V/F

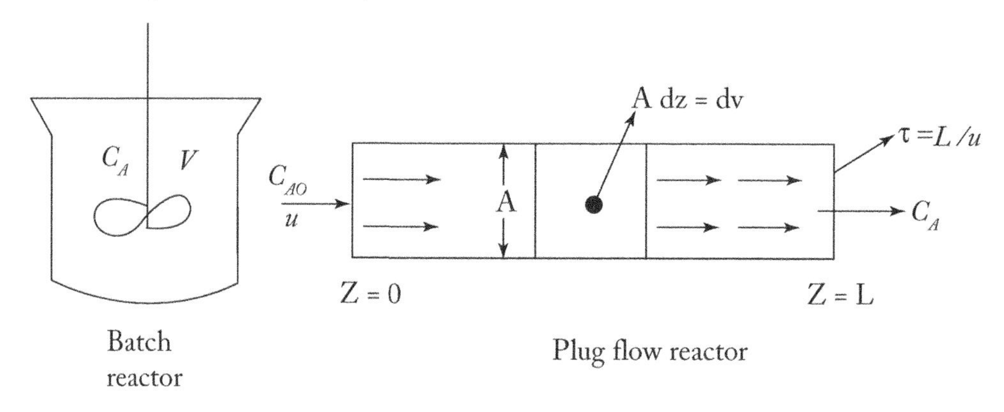

Fig. 1.6 Correspondence between time in a batch reactor and position in a Plug Flow Reactor

1.7.2 Mass balance in a flow reactor (plug flow reactor)

A general mass balance of a flow reactor may be given as

Input of reactant A into the volume element (dV) – Output of reactant A out of the volume element

± rate of accumulation or consumption with the volume element = rate of accumulation of reactant in the volume element (1.83)

When fluid moves through a long pipe with turbulent flow (Re > 2100), the system behaves as a plug flow without variation of the axial velocity, u, along the axis and cross section of the tube.

A differential mass balance over an elemental volume of AdZ where Z is the axial distance and A is the cross–section of the tube, can be given as

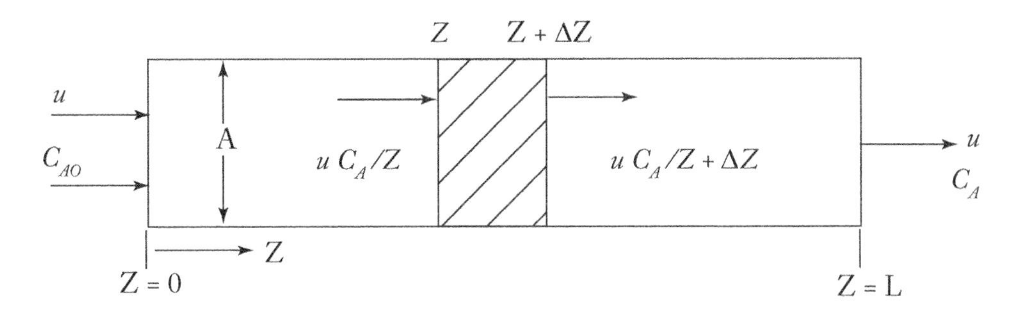

Fig. 1.7 Model of a plug flow reactor

$$\mathrm{A}u C_A \text{ at } Z - \mathrm{A}u C_A \text{ at } Z+\Delta Z - \mathrm{A}\Delta z r\mathrm{A} = \mathrm{A}\Delta z\, dC_A/dt \tag{1.84}$$

where r_A is the rate of decomposition of the reactant A and C_A is the concentration of the reactant A. Rearranging and dividing by AΔZ, we have

Or,
$$-u\frac{\delta C_A}{\delta Z} - r_A = \frac{\delta C_A}{\delta t} \tag{1.85}$$

The equation (1.82) represents ideal plug flow reactor model for unsteady state condition.

For steady state, $dC_A / dt = 0$

The equation (1.82) reduces to

$$-u\, dC_A/dZ = r_A \tag{1.85a}$$

Writing the above equation in terms of flow rate, F and elemental volume dV, we get

$$-F\, dC_A/dV = r_A \tag{1.86}$$

where F = uA, dV = AdZ

where A is the cross–section of the reactor. Now integrating the above equation

$$\frac{V}{F} = -\int_{C_{AO}}^{C_A} \frac{dC_A}{r_A} \tag{1.87}$$

In terms of fractional conversion, X_A

$$\frac{V}{F_{AO}} = \int_{O}^{X_A} \frac{dX_A}{r_A} \tag{1.88}$$

where $F_{AO} = FC_{AO}$ (molal flow rate) and $C_A = C_{AO}(1 - X_A)$

For a first order reaction:

$$r_A = kC_A = kC_{AO}(1-x_A) \quad (1.89)$$

$$\frac{V}{F_{AO}} = \int_O^{X_A} \frac{dX_A}{kC_{AO}(1-X_A)} \quad (1.90)$$

where $V/F = \tau$ (residence time)

Or,

$$k\tau = \ln\frac{1}{(1-X_A)} \quad (1.91)$$

or,

$$X_A = 1 - e^{-k\tau} \quad (1.92)$$

For a second order reaction

$$\frac{\tau}{C_{AO}} = \int_O^{X_A} \frac{dX_A}{kC_{AO}^2(1-X_A)^2} \quad (1.93)$$

Or,

$$kC_{AO}\tau = \frac{1}{1-X_A} - 1 = \frac{X_A}{1-X_A}$$

Or,

$$X_A = \frac{kC_{AO}\tau}{1+kC_{AO}\tau} \quad (1.94)$$

For a third order equation

$$\frac{\tau}{C_{AO}} = \int_O^{X_A} \frac{dX_A}{kC_{AO}^2(1-X_A)^3}$$

$$kC_{AO}^2\tau = \frac{1}{2}\left[\frac{1}{(1-X_A)^2} - 1\right] \quad (1.95)$$

For a gas phase first order reaction with a change in the number of moles, $\in$ the plug flow reactor design equation is given as

$$\frac{\tau}{C_{AO}} = \int_O^{X_A} \frac{dX_A}{r_A} \quad (1.96)$$

The rate, r_A can be expressed for a first order reaction in terms of fractional conversion, x_A and $\in_A$, $-r_A = kC_A = kC_{AO}(1-x_A)/(1+\in_A X_A)$ and $\in_A = (N - N_{AO})/N_{AO}$

Equation (1.93) becomes,

$$\tau = \int_O^{X_A} \frac{(1+\in_A X_A)dX_A}{k(1-X_A)} \tag{1.97}$$

Or

$$k\tau = \int_O^{X_A} \frac{(1+\in_A)dX_A}{(1-X_A)} - \in_A \int_O^{X_A} dX_A$$

Or,

$$k\tau = -(1+\in)ln(1-X_A) - \in_A X_A \tag{1.98}$$

For a second order irreversible reaction

A + B → Products, $-r_A = kC_A C_B$ with equimolar feed

$-r_A = kC_A^2$ & $C_A = C_{AO}(1-x_A)/(1+\in_A X_A)$

Now $$\tau/C_{AO} = \int_O^{X_A} \frac{(1+\in_A X_A)^2 dX_A}{kC_{AO}^2(1-X_A)}$$

Or,

$$kC_{AO}\tau = \int_O^{X_A} \frac{(1+\in_A X_A)^2 dX_A}{(1-X_A)^2} \tag{1.99}$$

For integration, we have to use partial fractions for the expression under integral

$$\frac{(1+\in_A X_A)^2}{(1-X_A)^2} = \frac{1+2\in_A X_A + \in_A^2 X_A^2}{(1-X_A)^2}$$

Or

$$\frac{1+2\in_A + \in_A^2}{(1-X_A)^2} - \frac{2\in_A(1-X_A)}{(1-X_A)^2} - \frac{\in_A^2(1-X_A^2)^2}{(1-X_A)^2}$$

Or,

$$\frac{(1+\in_A)^2}{(1-X_A)^2} - \frac{(2\in_A + 2\in_A^2)}{(1-X_A)} + \in_A^2 \tag{1.100}$$

Now integrating

$$\int_O^{X_A} \frac{(1+\in_A)^2 dX_A}{(1-X_A)^2} - \int_O^{X_A} 2\in_A(1+\in_A)\frac{dX_A}{1-X_A} + \int_O \in_A^2 dX_A \tag{1.101}$$

Or,

$$(1+\in_A)^2 \frac{X_A}{(1-X_A)} + 2\in_A(1+\in_A)ln(1-X_A) + \in_A^2 X_A = kC_{AO}\tau \tag{1.102}$$

The above equation is the integrated form of a PFR for a second order and variable volume reaction.

Problem 1.8

Acetaldehyde is decomposed at 500°C and 1.013 bar pressure in a plug flow reactor of volume, V = 5m³.

The reaction is $CH_3CHO \rightarrow CH_4 + CO$

The flow rate, was varied and the corresponding conversion was obtained as follows:

Flowrate kg/hr, F_{AO}	130	50	20	10
Conversion, x_A	0.05	0.15	0.25	0.40

Find out a satisfactory rate equation for the reaction.

Solution:

For material balance, based on 1 kgmole of CH_3CHO,

Moles of CH_3CHO $= 1 - x_A$

Moles of CH_4 $= x_A$

Mole of CO $= x_A$

N_t $= 1 + x_A$

Now $C_A = N_A/V = N_{AO}(1-x_A)/V_o(1+x_A) = C_{AO}\, 1 - x_A/(1+x_A)$

plug flow reactor design equation,

$$\frac{V}{F_{AO}} = \int_O^{X_A} \frac{dX_A}{-r_A}$$

$$\text{Now } -r_A = kC_A^{\ 2} = kC_{AO}^{\ 2}\frac{(1-X_A)^2}{(1+X_A)^2}$$

$$\text{So } kC_{AO}^{\ 2}\frac{V}{F} = \int_O^{X_A} \frac{(1+X_A)^2 dx}{(1-X_A)^2}$$

Integration may be carried by using particle fractions of RHS

$$\frac{(1+2x_A+x_A^{\ 2})}{(1-x_A)^2}$$

$$= \frac{1}{(1-x_A)^2} - \frac{2(1-x_A)}{(1-x_A)^2} - \frac{(1-x_A^{\ 2})^2}{(1-X_A)^2} + \frac{1}{(1-X_A)^2}$$

$$= \frac{4}{(1-x_A)^2} - \frac{2}{(1-X_A)} - \frac{(1+X_A)}{(1-X_A)^2}$$

$$= \frac{4}{(1-X_A)^2} - \frac{4}{(1-X_A)} + 1$$

$$I = \int_O^{X_A} \frac{4\,dx}{(1-x_A)^2} - 4\int_O^{X_A} \frac{dX_A}{1-X_A} + \int_O^{X_A} dX_A$$

$$= \frac{4}{(1-X_A)} - 4 + 4ln(1-X_A) + X_A$$

Rearranging

$4/(1 - x_A) + 4 \ln (1 - X_A) + X_A - 4 = f(X_A)$

$f(x_A) = kC_{AO}^2\, V/F_{AO}$

F_{AO} kg mole/hr	2.95	1.136	0.454	0.227
V/F_{AO}	1.695	4.40	11.013	22.02
x_A	0.05	0.15	0.25	0.40
$f(x_A)$	0.053	0.2458	0.4325	1.023

Now $C_{AO} = P_t/RT = 1.013 / (0.0831)(773) = 0.0157$ kmole/m³.

Now $f(X_A)$ is plotted against V/F_{AO} in fig. 1.8 and from the linear plot, the slope will give kC_{AO}^2 from which k can be evaluated.

Slope $= kC_{AO}^2 = 0.046$

or $k = 0.046 / (0.0157)^2 = 186.6$ m⁶ / (k.mole)²(hr)

So the rate equation is

$-r_A = 186.6\, C_A^2$ (kmoles)/(m³)/(hr) at 500°C

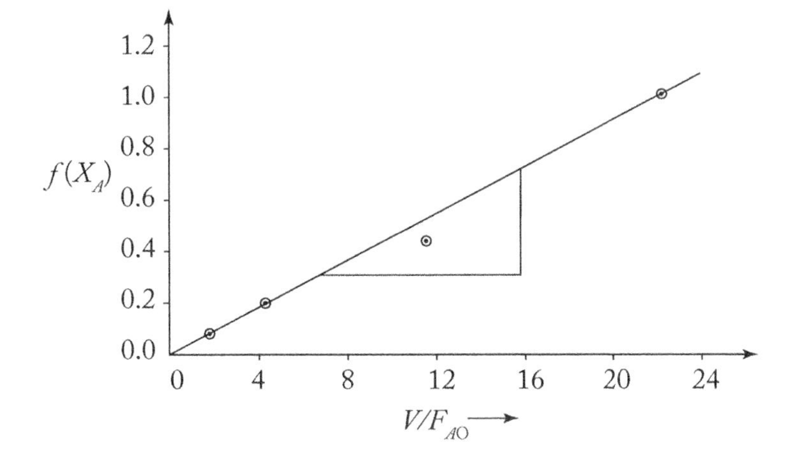

Fig. 1.8 Plot f(x) vs V/F_{AO}

Problem 1.9

A plug flow reactor of volume, V = 2m^3 processes an aqueous feed with a flow rate of 0.1m^3/min containing a reactant A (C_{AO} = 0.1 kmole/m^3). A reversible reaction takes place as follows:

$$A \underset{k_2}{\overset{k_1}{\rightleftarrows}} B \quad -r_A = k_1 C_A - k_2 C_B$$

where $k_1 = 0.04\ min^{-1}$ and $k_2 = 0.01\ min^{-1}$

Find (i) the equilibrium conversion, x_{Ae} (ii) actual conversion in the reactor

Solution:

(i) Equilibrium conversion, x_{AC}
$K = k_1/k_2 = 0.04 / 0.01 = 4 = C_{Be} / C_{Ae} = X_{Ae} / (1-X_{Ae})$
Solving $X_{AO} = 0.8$ Ans

(ii) Plug flow reactor design equation

$$\frac{V}{F_{AO}} = \frac{\tau}{C_{AO}} = \int_0^{X_A} \frac{dX_A}{-r_A}$$

Now $-r_A = k_1(C_A - C_B/K) = k_1 C_{AO}\ [(1 - x_A) - x_A / K]$

$= k_1 C_{AO}/K\ [\ K - (1 + K)\ x_A]$

Substituting the expression of r_A in the design equation of PFR

$$\frac{\tau}{C_{AO}} = \frac{K}{k_1 C_{AO}} \int_0^{X_A} \frac{dX_A}{K-(1+K)X_A}$$

Integrating

$$(k_1\tau)/K = 1/((1+K))[\ln[K(1+K)X_\downarrow A]_\downarrow (X_\downarrow A)^\uparrow O$$

Or, $$\frac{k_1\tau(1+K)}{K} = \ln\frac{K}{K(1+K)X_A}$$

Now putting the values

$k_1 = 0.04\ min^{-1}$, t = 2m^3/ 0.1m^3/min = 20 min

So LHS = (0.04) (20) (1+4) / 4 = 1.0

$1 = \ln 4 /(4 - 5\ x_A)$

Solving, $x_A = 0.5$

So the conversion in the reactor is 0.5 Ans

Problem 1.10

(a) An enzymatic reaction is carried out in a CSTR with reactant, A and the rate equation is given as follows:

Enzyme

A -----------> P

$-r_A = 0.1\ C_A / 1 + 0.5\ C_A$ kmol/(m^3)(min)

Calculate the volume of CSTR for 95% conversion of A with a feed stream of 2.5 m^3/min with initial reactant concentration. $C_{AO} = 2$ kmole/m^3 and the enzyme concentration.

(b) If the exit concentration of A, $C_A = 0.01$ mole/m^3 in a plug flow reactor, what will be the volume of the reactor if the flow rate and initial reactant concentrations of A & E are the same.

Solution:

(a) Given q = 2.5 m^3/min. $C_{AO} = 2$ Kmol/m^3, at 95% conversion of A,

$C_A = 0.05 \times 2 = 0.1$ kmol/m^3.

Design equation of CSTR:

$V_B/q = (C_{AO} - C_A) / -r_A = (C_{AO} - C_A)\ (1 + 0.5C_A)/0.1\ C_A$

$$V_B = q\left(\frac{(C_{AO} - C_A)(1 + 0.5C_A)}{0.1C_A}\right)$$

$$V_B = 2.5\left[\frac{(2.0 - 0.1)(1 + 0.5 \times 0.1)}{(0.1)(0.1)}\right] = 498.75\ m^3$$

(b) For a plug flow reactor

$$\frac{V}{q} = \int \frac{-dC_A}{-r_A} = -\int_{C_{AO}}^{C_A} \frac{(1 + 0.5C_A)dC_A}{0.1C_A}$$

$$= -10\int_{C_{AO}}^{C_A} \frac{dC_A}{C_A} - \frac{0.5}{0.1}\int_{C_{AO}}^{C_A} dC_A$$

$$= 10\ln\left(\frac{C_{AO}}{C_A}\right) + 5(C_{AO} - C_A)$$

$$\text{Or, } V = 2.5\left[\frac{10\ln 2.00}{0.10} + 5(2 - 0.1)\right] = 98.64\ m^3$$

1.8 Continuous Stirred Tank Reactor (CSTR)

A continuous stirred tank reactor is a completed by mixed system. The reactant is assumed to be constant in the reactor as the rate of reaction is also constant throughout the reactor. A material balance for the reactor of volume, V using a reactant, A can be shown as:

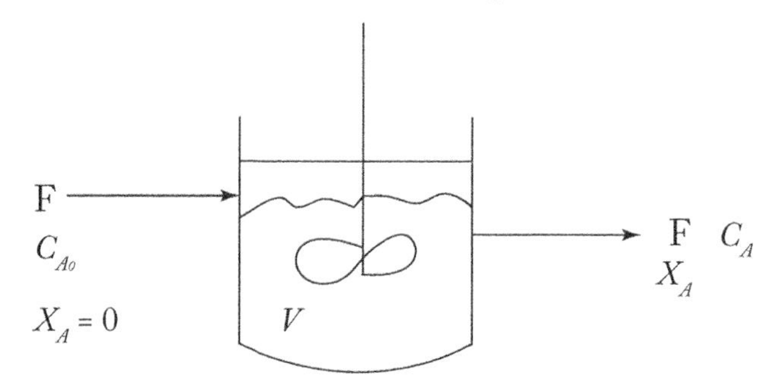

Fig. 1.9 Material balance in a CSTR

$$F\,(C_{AO} - C_A) - (r_A)V = V dC_A / dt \tag{1.103}$$

For a steady state dCA/dt = 0

$$V/F = (C_{AO} - C_A) / r_A = C_{AO} X_A / r_A \tag{1.104}$$

Or $\tau = \dfrac{C_{AO} X_A}{r_A}$ ....(95) Where, $\dfrac{V}{F} = \dfrac{\text{Reactor volume}}{\text{flow rate}} = \tau$

or $V/F_{AO} = X_A/r_A$ where F_{AO} = molal flow rate
where τ is called residence time.

For a first order reaction,

$r = kC_A$

$$\tau = \frac{V}{F} = \frac{C_{AO} X_A}{-r_A} = \frac{C_{AO} X_A}{k\, C_{AO}(1 - X_A)}$$

Or,

$$k\tau = \frac{X_A}{1 - X_A} \tag{1.105}$$

Or,

$$X_A = \frac{k\tau}{1 + k\tau} \tag{1.106}$$

For a second order reaction

$$\frac{\tau}{C_{AO}} = \frac{X_A}{kC_{AO}^{\;2}(1 - X_A)^2}$$

Or,

$$kC_{AO}\tau = \frac{X_A}{(1-X_A)^2} \tag{1.107}$$

For a first order variable volume,

$$V = V_O(1+\epsilon_A X_A), \frac{C_A}{C_{AO}} = \frac{1-X_A}{1+\epsilon_A X_A}$$

$$k\tau = \frac{X_A(1+\epsilon_A X_A)}{1-X_A} \tag{1.108}$$

For a second order reaction with constant volume $-r_A = kC_A^2$ $\epsilon_A = 0$

$$k\tau = \frac{C_{AO} - C_A}{C_A^2} \tag{1.109}$$

Or,

$$k\tau\ C_A^2 + C_A - C_{AO} = 0$$

Or,

$$C_A = \frac{-1+\sqrt{(1+4kC_{AO})}}{2k\tau} \tag{1.110}$$

Problem 1.11

Ethyl acetate is saponified in a CSTR of volume, V = 5m³. NaOH solution is fed to the reactor at the rate of 0.2m³/min with initial NaOH concentration, C_{AO} = 0.1 Kmol/m³ and the second feed of ethylacetate (B), C_{BO} = 0.22 kmole/m³ is fed at the same rate of 0.2 m³/min. The reaction is 2nd order and is carried out at 20°C and the rate constant, is k = 92 m³/(K.mole)(min). Determine the exit concentration of ethyl acetate.

Solution:

Assume A = NaOH, B = ethyl acetate

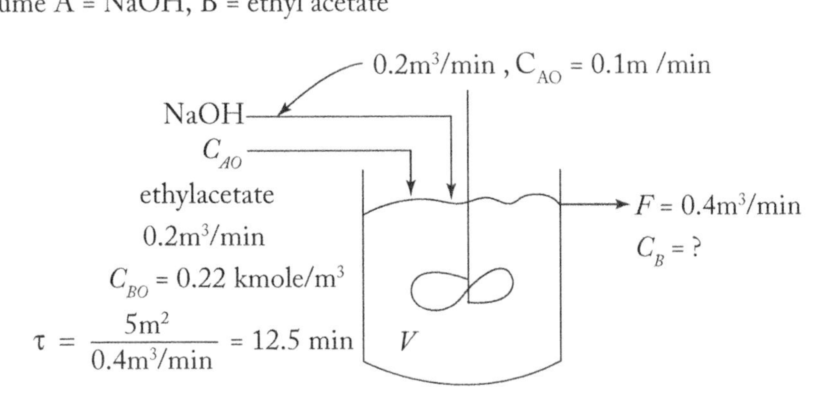

Since the two equal feeds are mixed together, the effective initial concentration of NaOH and ester are C_{AO} = 0.05 k.mol/m³, C_{BO} = 0.11 kmol/m³.

The design equation of CSTR is
$V/F_{AO} = X_A/r_A = X_A/kC_AC_B$
Now $C_A = C_{AO}(2 - X_A)$
$C_B = C_{BO} - C_{AO}X_A = C_{AO}(M - X_A)$
where $M = C_{BO}/C_{AO}$

$$\frac{\tau}{C_{AO}} = \frac{X_A}{kC_{AO}^2(1-X_A)(M-X_A)}$$

$$\text{Or } k\tau\ C_{AO} = \frac{X_A}{(1-X_A)(M-X_A)} \quad \text{.......(A)}$$

$$M = \frac{C_{BO}}{C_{AO}} = \frac{0.11}{0.05} = 2.2$$

$k = 92\ m^3/(k.mole)(min)$

$$k\tau\ C_{AO} = (92)(12.5)(0.05) = 57.5$$

Now the equation, A can be written as
$57.5 = x_A/(1 - X_A)(2 - X_A)$
Solving by trial and error, $X_A \approx 0.985$
$C_B = C_{AO}(M - X_A) = 0.05(2.2 - 0.985) = 0.06075$ k.mols/m^3. Ans
where C_B is the exit concentration of ethyl acetate

1.9 Fed Batch (Semibatch) Reactor

In the fed batch reactor, feed is added to a reactor having some initial volume,V0 and the product is withdrawn continuously.

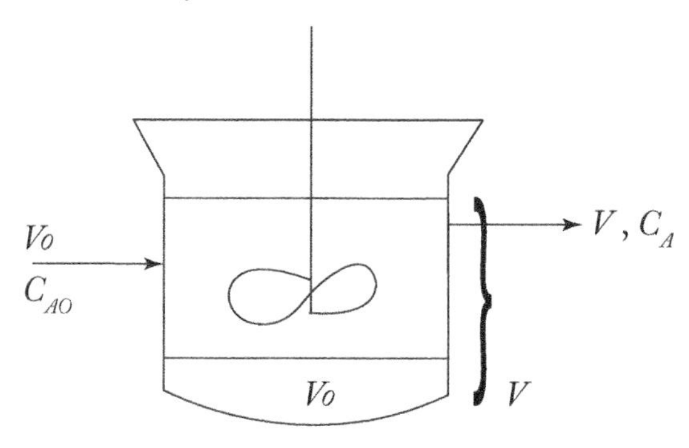

Fig. 1.10 Fed batch reactor

For a constant density and constant volume reactor, the mass balance is given as

$$\mathrm{V}dC_A/\mathrm{dt} = \mathrm{v_o}C_{AO} - \mathrm{v}C_A - \mathrm{Vr_A} \tag{1.111}$$

Where v_o and v are the inlet and outlet flow rate.

Now this is exactly transient CSTR. But in the semibatch reactor, the reactant A may be fed continuously when nothing is removed and the reactor volume increases with flow. Further, once the reactor is charged, the reactant, A (unreacted) may be removed continuously, so the volume decreases with time.

In the first case, equation (1.101) reduces to

$$\mathrm{V}dC_A/\mathrm{dt} = \mathrm{v_o}C_{AO} - \mathrm{r_A V} \tag{1.112}$$

Now the feed rate, vo = dV/dt

Or

$$\mathrm{V} = \mathrm{V_o} + \mathrm{v_o t} \tag{1.113}$$

Where Vo is the initial volume at t = 0.

Introducing V into equation (1.112) we get

$$(\mathrm{V_o} + \mathrm{v_o t})dC_A/\mathrm{dt} = \mathrm{v_o}C_{AO} - \mathrm{r_A}(\mathrm{V_o} + \mathrm{v_o t}) \tag{1.114}$$

The above is a simple first order differential equation and the variables are not separable. Rearranging the equation (1.104) and introducing first order kinetics ($r_A = kC_A$), we have,

$$dC_A/\mathrm{dt} = \mathrm{v_o}C_A/\mathrm{V_o} + \mathrm{v_o t} - \mathrm{k}C_A \tag{1.115}$$

The above equation can be integrated after rearranging the equation (1.115) and with the aid of an integrating factor,

$$dC_A/\mathrm{dt} + \mathrm{k}C_A = -\mathrm{v_o}C_{AO}/\mathrm{V_o} + \mathrm{v_o t} \tag{1.116}$$

Using an integrating factor, e^{kt}

$$d(C_A e^{kt}) = -\,e^{kt}\,\mathrm{v_o}C_{AO}/\mathrm{V_o} + \mathrm{v_o t}\,\mathrm{dt} \tag{1.117}$$

Integrating equation (1.117), we have

$$C_{AO}\,e^{kt} = -\frac{v_O C_{AO}}{V_O + v_o t}\cdot\frac{1}{k}e^{kt} - e^{kt} - e^{kt}\left[\frac{V_O C_{AO}}{v_O} ln(V_O + v\cdot t)\right] + P \tag{1.118}$$

Where P is the integration constant.

Putting t = 0, $C_A = C_{AO}$ in equation (1.118) we have,

$$C_{AO} = -\frac{v_o C_{AO}}{k(V_O)} - C_{AO} ln V_O + P$$

Or, $$P = C_{AO} + \frac{v_o C_{AO}}{kV_O} + C_{A_0} lnV_O \tag{1.119}$$

Substituting the value of P in equation (1.108), we get,

$$C_A = C_{AO} + \frac{v_o C_{AO}}{k}\left[\frac{1}{V_O} - \frac{1}{(V_O + v_o t)}\right] - C_{AO} ln\left(\frac{V_O + v_o t}{V_O}\right) \tag{1.120}$$

Where C_A is the concentration in the fed–batch reactor at any time t.

Problem 1.12

In a semibatch reactor acetic anhydride (A) is to be hydrolyzed by initially charging the stirred tank with 1m^3 of an aqueous solution containing 0.1 kmol/m^3 anhydride.

The vessel is heated to 40°C and at that time a feed solution containing anhydride C_{AO} = 0.1 kmol/m^3 is fed at the rate of 0.2 m^3/min. Product is not withdrawn and the volume of the reactor (liquid volume) goes on increasing. The reaction rate is r_A =0.38 C_A (kmol)/(m^3)(min)

Calculate the variation of C_A & liquid volume vs time in the reactor.

Solution:

Given: $V_o = 1.0$ m^3,

$v_o = 0.2$ m^3/min, $C_{AO} = 0.1$ kmole/m^3

$k = 0.38$ min^{-1}, $V = V_o + v_o t$

The design equation can be given in differential form:

$(V_o + v_o t)\, dC_A/dt = V_o C_{AO} - r_A(V_o + v_o t)$

or $dC_A/dt = V_o C_{AO} / V_o + v_o t - kC_A$

or $dC_A/dt + kC_A = v_o C_{AO}/V_o + v_o t$

Integrating the above equation using an integrating factor, e^{kt},

$$\int d(C_A e^{kt}) = -\int_O^t e^{kt} \frac{V_O C_{AO}}{(V_O + v_o t)} dt + \rho$$

$$\text{Or, } C_A e^{kt} = -\frac{v_o C_{AO}}{k(V_O + V_O t)} e^{pt} - e^{kt}[C_{AO} ln(V_O + v_o t)] + P \quad \text{.............................. (A)}$$

To evaluate P, uses the initial condition, at t=0, $C_A = C_{AO}$, then
$P = C_{AO} + v_o C_{AO}/kV_o + C_{AO}\ \text{ln} V_o$
Substituting the value of P into the equation (A)

We have,

$$C_A = C_{AO} + \frac{v_O C_{AO}}{k}\left[\frac{1}{V_O} - \frac{1}{(V_O + v_o t)}\right] - C_{AO} ln\left(\frac{V_O + v_o t}{V_O}\right)$$

Now substituting the value of
$C_{AO} = 0.1$ kmole/m^3, $V_o = 1.0$m^3,
$V_o = 0.2$ m^3/min, k = 0.38 min^{-1}
We have,

$$C_A = 0.1 + \frac{(0.2)(0.1)}{0.38}\left[1 - \frac{1}{1+0.2t}\right] - 0.1\ln(1+0.2t)$$

or $C_A = 0.1 + 0.52\ (0.2t/1 + 0.2t) - 0.1\ \ln\ (1 + 0.2t)$
The variation of C_A & V is given in the following table

t, min	C_A mol/m^3	X_A	V, m^3 (vo + 0.2t)
0	0.10	0	1.0
1.0	0.09	0.1	1.2
2.0	0.076	0.23	1.4
3.0	0.072	0.27	1.6
5.0	0.043	0.43	2.0
10	0.0248	0.75	3.0

1.10 Plug Flow Reactor with Recycle

Plug flow recycle reactor can be schematically represented by the following flow diagram:

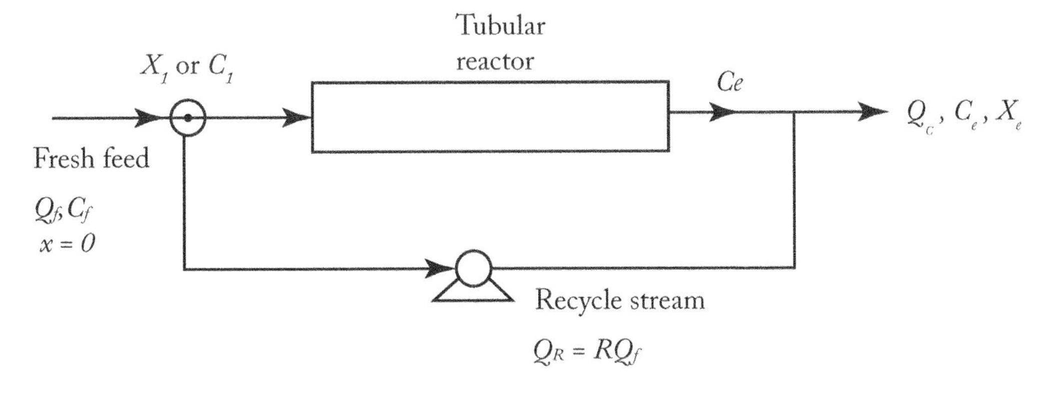

Fig. 1.11 Plug flow with recycle

A fresh feed of the reactant with flow rate of Q_f and concentration of, C_f is introduced to the reactor and the exit flow rate is Q_e with concentration of C_e. From the exit stream, a recycle stream of Q_R with concentration of C_e is fed to the inlet of the reactor. The limiting reactant is A and F_{AO} represents the feed rate of A corresponding to A in the fresh feed plus the reactant A in the recycle stream. Let us assume that the volumetric flow rate is unaffected by the reaction and $Q_e = Q_f$. Then $F_{AO} = Q_R C_{Af} + Q_f C_{Af} = C_{Af} Q_f (R+1)$. (1.111), if there were no conversion in the recycle.

$R = Q_R / Q_f$, Q_f = fresh feed flow rate & Q_R = recycle flow rate.

So the plug flow design equation will be

$$\frac{V}{Q_f} = -C_{Af}(R+1)\int_{X_{A1}}^{X_{Ae}} \frac{dX_A}{r_A} \tag{1.121}$$

where X_{A1} represents the conversion entering the reactor and X_e is the conversion of the effluent. X_1 or C_{A1} is a function of the recycle ratio, R,

Now $X_A = (C_A f - C_A)/C_A f$

$dX_A = - dC_A/C_A f$

with the substitution of the above equation, the equation (1.121) becomes

$$\frac{V}{Q_f} = (R+1)\int_{C_{A_1}}^{C_{Ae}} \frac{dC_A}{r_A} \tag{1.122}$$

The mixed feed concentration, C_{A1} is obtained by a mass balance

$(Q_R + Q_f)\, C_{A1} = Q_f C_{Af} + Q_R C_{Ae} = Q_f\,(C_{Af} + RC_{AC})$

Or $$C_{A1} = Q_f\,(C_f + RC_{Ae})/RQ_f + Q_f = C_{Af} + RC_{Ae}/(R + 1) \tag{1.123}$$

With the limits of equation (1.113), we have

$$\frac{V}{Q_f} = (R+1)\int_{\left(\frac{C_{Af}+R\,C_{Ae}}{R+1}\right)}^{C_A f} \frac{dC_A}{kC_A}, \tag{1.124}$$

Where $r_A = k\,C_A$

Integrating we have,

$$\frac{k\tau}{(R+1)} = \ln\left[\frac{C_{AO} + R\,C_{Af}}{(R+1)C_{Af}}\right] \tag{1.125}$$

For second order reaction:

$2A \rightarrow$ Products, $r_A = kC_A^2$, the design equation is,

$$\frac{k\,C_{AO}\tau}{R+1} = \frac{C_{AO}(C_{AO} - C_{Af})}{C_{Af}(C_{AO} + R\,C_{Af})} \tag{1.126}$$

Case I: If R = 0, we have ideal plug flow reactor from equation (1.116)

$$k\tau = \ln\left(\frac{C_{Af}}{C_{A0}}\right)$$

Case II: if R → ∞

The equation (1.113) reduces to CSTR

$$\frac{V}{Q_f} = (R+1)\frac{C_{Ae} - C_{A1}}{r_A} \tag{1.127}$$

$$\frac{V}{Q_f} = \frac{(R+1)}{r_A}\left[\frac{C_{Ae}(R+1) - (C_{Af} + R\,C_{Ae})}{R+1}\right] \tag{1.128}$$

Or

$$V / Q_f = (C_{Ae} - C_{A}f) / r_A \tag{1.129}$$

The above equation is identical with the design equation of ideal stirred tank reactor.

Problem 1.13

For an irreversible first order liquid phase reaction is carried in a plug flow reactor with recycle ratio, and with initial concentration, C_{AO} = 5 kmole/m^3, 80% conversion is achieved in a plug flow reactor. If a recycle stream of R is used and the throughput of the whole recycle system is kept constant, what are the exit concentrations of the reactant A, for R = 0.33 and 0.67?

Solution:

For a simple plug flow reactor with first order kinetics

$$k\tau = \ln\left(\frac{1}{1-X_A}\right) = \frac{\ln 1}{1-0.8} = 1.61$$

Now the equation for a PFR with recycle R

$$\frac{k\tau}{R+1} = ln\left[\frac{C_{AO} + RC_{Ae}}{(R+1)C_{Ae}}\right]$$

where C_{Ae} is the exit concentration

with R = 0.33, C_{AO} = 5 kmole/m^3, kτ = 1.61

we have

$1.61/1.33 = \ln [\, 5 + 0.33\, C_{Ae} / 1.33\, C_{Ae}]$

Solving C_{Ae} = 1.21 kmole/m^3 (x_A = 0.758)

2nd case, R = 0.67

$1.61/1.67 = \ln [5 + 0.67\, C_{Ae} / 1.67\, C_{Ae}]$

Solving $C_{Ae} = 1.348$ kmole/m^3 ($x_A = 0.73$)

1.11 Kinetics of Non-Elementary Reactions

Under this category we have three classes of reactions:

1) Free radical or chain reactions
2) Non–chain reactions
3) Gas–solid catalytic reaction based on adsorption, desorption and surface reaction

Chain Reactions

The reactions proceed through three steps such as

1) Initiation:
   Reactant → (intermediate)*
2) Propagation:
   (Intermediate)* + Reactant → (Intermediate)* + Product
3) Termination:
   (Intermediate)* → Products

1.11.1 Mechanisms of the decomposition of N_2O_5

The overall the reaction is:

$$2N_2O_5 \rightarrow 4NO_2 + O_2 \tag{1.130}$$

The reaction is not likely to be elementary and is assumed to occur through several steps of chain reactions.

The steps are:

1) N_2O_5 ....>............. $NO_2 + NO_3$ (1.131)

2) $NO_2 + NO_3 \xrightarrow{k_2} NO + O_2 + NO_2$ (slow) (1.132)

3) $NO + NO_3 \xrightarrow{k_3} 2NO_2$ (1.133)

It is assumed that the net rate of formation of intermediates (NO_2, NO, NO_3) are zero. This is known as Pseudo steady state approximation.

If the second step is much slower, the controlling rate for N2O5 decomposition is expressed as,

$$-\tfrac{1}{2}\, d(N_2O_5)/dt = d(O_2)/dt = k_2\, (NO_2)\, (NO_3) \tag{1.134}$$

Now the net rates of formation of the intermediates (NO_3, NO, NO_2) are assumed to be zero.

$$d(NO_3)/dt = 0 = k_1(N_2O_5) - k_2(NO_2)\,(NO_3) - k_1'(N_{O2})(NO_3) - k_3(NO)(NO_3) \tag{1.135}$$

Solving for (NO_3)

$$NO_3 = k_1(N_2O_5)/k_2(NO_2) + k_3(NO) + k_1'(NO_2) \quad (1.136)$$

$d(NO)/dt = 0 = k_2(NO_2)\,(NO_3) - k_3\,(NO)(NO_3)$
Solving for (NO)

$$(NO) = k_2/k_3\,(NO_2) \quad (1.137)$$

Using equations (1.126) and (1.128) we have,

$$(NO_3) = k_1(N_2O_5)/\,(2k_2 + k_1')\,NO_2 \quad (1.138)$$

Substituting the equation (1.129) into equation (1.125), the decomposition rate of N2O5 can be expressed as

$$-½\,d(N_2O_5)/dt = k_2(NO_2)\,k_1(N_2O_5)/(2k_2 + k_1')\,(NO_2)$$

Or
$$-\frac{1}{2} r_{N_2 v_S} = \frac{k_1 k_2}{(2k_2 + k_i)}(N_2 O_S) \quad (1.139)$$

The first order kinetics given by equation (1.130) for the above reaction are consistent with the experimental data.

1.11.2 Non chain reactions

In this type of reactions, an intermediate compound is formed from the reactant and then the intermediate compound decomposes to products.

The steps are:

Reactants → (Intermediate)*

(Intermediate)* → Products

The net rate of (intermediate) formation is assumed to be zero. The classical example of this type of reaction is Michaelis–Menten equation which is based on the above mechanism which is illustrated below.

Mechanism of Enzymatic reaction
An enzymatic reaction may be presented as

$$E + S \underset{k_{-1}}{\overset{k_1}{\rightleftarrows}} ES \xrightarrow{k_2} E + P \quad (1.140)$$

where E is the enzyme, S is the substrate, ES is the intermediate compound, P is the production. The second step is slow, the rate of production of P can be given as

$$v = dP/dt = -\,ds/dt = k_2[ES] \quad (1.141)$$

Assuming the Pseudo steady state assumption i.e. the net rate of production of the intermediate, [ES] is zero.

$$d(ES)/dt = 0 = k_1[E][S] - k_{-1}[ES] - k_2[ES] \quad (1.142)$$

again

$$E = [E_0] - [ES] \quad (1.143)$$

where E is the free enzyme, E_0 is the initial concentration of enzyme.
Combining the equations (1.133) & (1.134), we get

$$[ES] = k_1[E_0] - [ES][S] / k_{-1} + k_2 \quad (1.144)$$

Solving for [ES], we have

$$[ES] = \frac{[E_O][S]}{\frac{k_{-1}+k_2}{k_1} + S} \quad (1.145)$$

Substituting the equation (1.136) into equation (1.132), we obtain

$$v = \frac{dP}{dt} = -\frac{dS}{dt} = \frac{k_2[E_O][S]}{\frac{k_{-1}+k_2}{k_1} + S} \quad (1.146)$$

Or

$$v = \frac{v_m[S]}{K_M + [S]} \quad (1.147)$$

Where, $K_M = (k_{-1} + k_2)/k_1$
And $v_m = k_2[E0]$
The equation (1.147) is known as Michaelis–Menten equation. The kinetic parameter, K_M is called Michaselis–Menten constant and the other parameter, v_m is the maximum enzymatic rate and v is the enzymatic reaction rate.

1.12 CSTRs in Series

We consider several CSTRs in series as shown below:

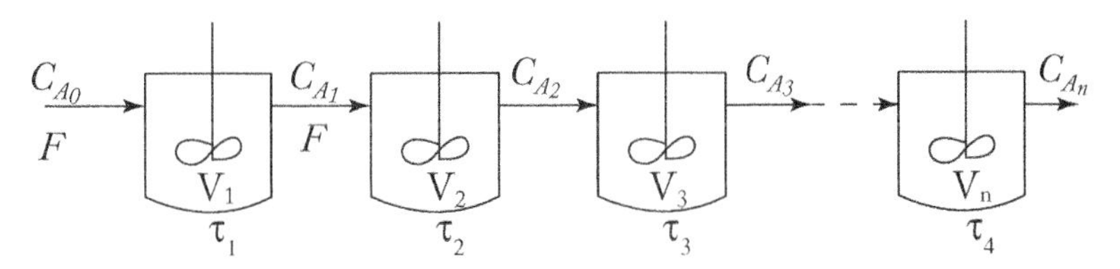

Fig. 1.12 CSTRs in series

The feed is introduced to the first reactor with concentration of a reactant A, C_{AO}. The effluent concentration of the other reactors are C_{A2}, C_{A3} .... C_{An}.

With first order kinetics at a constant temperature the mass balance equations can be shown as,

$$C_{AO} - C_{A_1} = \tau_1 r(C_{A_1})$$

$$C_{A_1} - C_{A_2} = \tau_2 r(C_{A_2}) \tag{1.148}$$

$$C_{A_2} - C_{A_3} = \tau_3 r(C_{A_3})$$

For first order irreversible reactions, the above equations can be solved sequentially yield,

$$C_{A_1} = \frac{C_{AO}}{1 + k\tau_1} \tag{1.149}$$

$$C_{A_2} = \frac{C_{A_1}}{1 + k\tau_2} = \frac{C_{AO}}{(1 + k\tau_1)(1 + k\tau_2)} \tag{1.150}$$

Similarly

$$C_{A_3} = \frac{C_{AO}}{(1 + k\tau_1)(1 + k\tau_2)(1 + k\tau_3)} \tag{1.151}$$

For nth reactor

$$C_{An} = \frac{C_A n - 1}{1 + k\tau_n} = \frac{C_{AO}}{\pi_k (1 + k\tau_k)} \tag{1.152}$$

If each reactor has the same residence time τ, the total residence time τ

$\tau = \sum \tau_n = n\tau$ and the concentration from the nth reactor is given as

$$C_{An} = \frac{C_{AO}}{(1+k\tau_n)^n} \tag{1.153}$$

It can be shown that a series of CSTRs has a performance closer to that of PFTR. Since the CSTR is cheaper and easier to maintain, the use of some smaller CSTRs in series will approach the performance of the PFTR having the same volume. However, the cost of several smaller reactors is usually greater than one large reactor. So the optimal design will depend upon the total reactor volume and cost.

1.12.1 Graphical method for tanks in series system

To find out the performance of a tanks–in–series system, a graphical method may be used. The advantage of this system is that the concentration of each reactor can be determined from the graph. In this technique, a graph of the rate of reaction, r_A is plotted against reactant concentration, C_A.

Rearranging the first stage of the tanks in series system, we have

$$\frac{V_1}{Q} = \tau_1 = \frac{C_{AO} - C_{A_1}}{r_{A_1}} \tag{1.154}$$

where V_1 = volume of the first tank, Q = flow rate. Further rearranging the above equation in a linear function between r_A and C_A.

$$r_{A_1} = \left(\frac{Q}{V_1}\right)C_{AO} - \left(\frac{Q}{V_1}\right)C_{A_1} \tag{1.155}$$

For the well stirred reactor, r_{A1} and C_{A1} will represent a point on the rate curve. To locate this point, C_{AO} is fixed on the abscissa point and then a straight line is to be drawn with a slope of $1/\tau_1$ through C_{AO}.

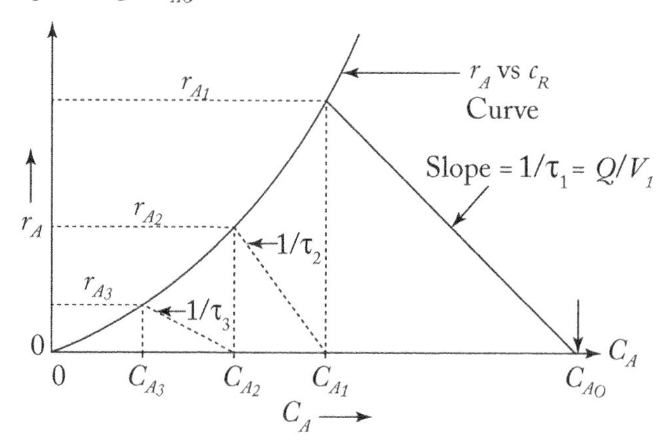

Fig. 1.13 Graphical method for CSTRs in series

The straight line cuts the rate–curve at a point which corresponds to C_{A1}. For a three stage system, the second stage is obtained by locating C_{A1} on the abscission and then drawing a straight line from C_{A1} with a slope of $1/\tau_2$ which intersects the rate curve at a point which corresponds to the reactant concentration. C_{A2}. Similarly the outlet concentration from the third stage C_{A3} is located on the abscissa. The stages may be continued unless desired outlet concentration is reached.

Problem 1.14

Acetic anhydride is hydrolysed in three stirred tank reactor operated in series. The feed flows to the first reactor ($V=1m^3$) at a rate of 0.4 m^3/min. the second and the third reactors have volumes of 2 and 1.5 m^3 respectively. The temperature is 25°C at which the first order irreversible rate constant is 0.160 min^{-1}. Use a graphical method to calculate the fraction hydrolysed in the first, second and third reactors.

Solution:

Assume $C_{AO} = 1.0$ kmole/m^3.

Assume several C_A values from 0 to 1.0 kmole/m^3

$r_A = kC_A = 0.16\ C_A$, $\tau_1 = 1/0.4 = 2.5$ min $\tau_2 = 2/.4 = 5$ min, $\tau_3 = 1.5/.4 = 3.75$ min

C_A	0	0.2	0.4	0.6	0.8	1.0	1.2
r_A	0	0.032	0.064	0.096	0.128	0.16	0.192

$1/\tau_1 = 0.4/1 = 0.4$

$1/\tau_2 = 0.4/2 = 0.2$

$1/\tau_3 = 0.4/1.5 = 0.27$

From the graph

$C_{A1} = 0.70\ (X_A = 0.3)$

$C_{A2} = 0.38\ (X_A = 0.62)$

$C_{A3} = 0.23\ (X_A = 0.72)$

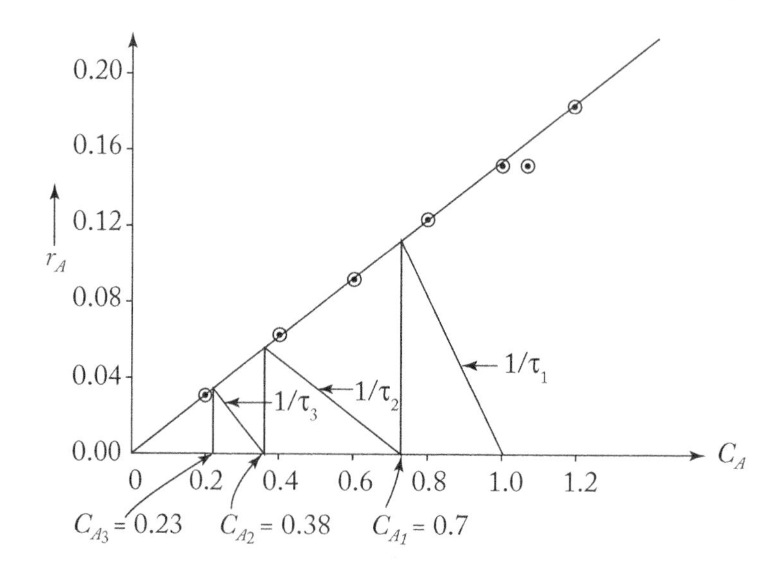

Fig. 1.14 Graphical method for CSTRs in series

A straight line is drawn with a slope of 0.4 from C_{AO} in the abscissa, which cuts the rate curve at $C_{A1} = 0.7$. From C_{A1}, another line is drawn with a slope of 0.2 which cuts the rate curve at $C_{A2} = 0.38$ and the procedure is repeated to get $C_{A3} = 0.23$

1.13 Combination of PFTR and CSTR in Series

We consider a combination of reactors involving both CSTRs and PFTRs.
For a PFTR followed by a CSTR as shown below, the exit concentration of the second reactor can be given for a first order kinetics.

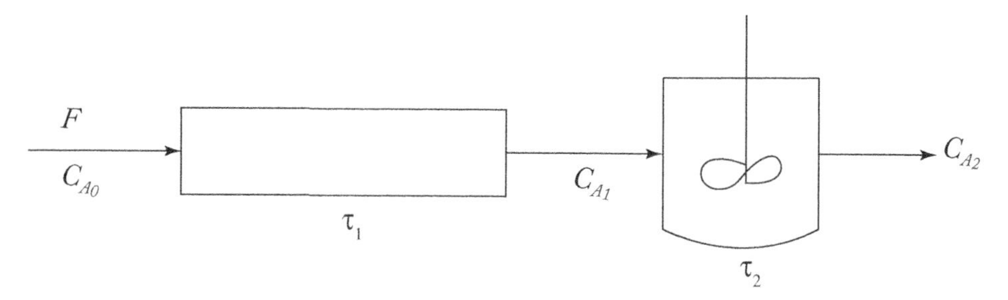

Fig. 1.15 Combination of a PFTR and a CSTR

$$C_{A_2} = \frac{C_{A_1}}{1 + k\tau_2} = \frac{C_{AO}e^{-k\tau_1}}{1 + k\tau_2} \tag{1.156}$$

If the reactor sequence is reversed, that is CSTR is followed by PFTR, we obtain

$$C_{A_2} = C_{A_1} e^{-k\tau_2} = \frac{C_{AO} e^{-k\tau}}{(1 + k\tau_1)} \tag{1.157}$$

For $\tau_1 = \tau_2$ (equal volume reactors) and with first order kinetics, the performance expressions are same. But in general it is a common strategy to use a CSTR first when the conversion is low, and then switch to PFTR as the conversion becomes high to minimize the total reactor volume.

1.14 Summary

The chapter deals with the simple kinetics of homogeneous systems with constant and variable volumes. The kinetic parameters have been evaluated by differential and integral methods using linearized system.

The basic reactor models have been discussed with their appropriate models for constant and variable volume systems. Other reactor models have also been discussed with model equations based on the basic reactor systems. No heterogeneous kinetics or non–ideal reactors have been discussed. These two topics have been treated in subsequent chapters.

Exercise (Problems)

1.1 A mechanism for the decomposition of ozone (O_3) in the pressure of chlorine (cl_2) is given in terms of clo_2 and clo_3, functioning as the chain carrier. Show that the mechanism is described by the rate,

$$-\frac{d(O_2)}{dt} = 2k_2 \sqrt{\frac{k_1}{k_4}} (cl_2)^{0.5} (O_2)^{1.5}$$

The steps are:

$$cl_2 + O_3 \xrightarrow{k_1} clO + clO_2$$

$$clO_2 + O_3 \xrightarrow{k_2} clO_3 + O_2$$

$$clO_3 + O_3 \xrightarrow{k_3} clO_2 + 2O_2$$

$$clO_3 + clO_3 \xrightarrow{k_4} cl_2 + 2O_2$$

1.2 Thermal decomposition of diethyl ether is postulated to proceed by the chain mechanism:

$$C_2H_5OC_2H_5 \xrightarrow{k_1} CH_3^{\cdot} + CH_2OC_2H_5^{\cdot}$$

$$CH_3^{\cdot} + C_2H_5OC_2H_5^{\cdot} \xrightarrow{k_2} C_2H_6 + CH_2OC_2H_5^{\cdot}$$

$$C_2H_5OC_2H_5^{\cdot} \xrightarrow{k_3} CH_3^{\cdot} + CH_2CHO^{\cdot}$$

$$CH_3^{\cdot} + C_2H_5OC_2H_5 \xrightarrow{k_4} \text{end of CH}$$

Show that the rate equation is,

$$-\frac{d(C_2H_5OC_2H_5)}{dt} = k_2\sqrt{\frac{k_1k_2}{k_2k_4}}(C_2H_5OC_2H_5)$$

Problem 1.3

The reactants A and B react to form a product P. The rate data for the formation of P are given at 100°C. Equal quantities of A & B are initially present. The data are: $C_{AO} = C_{BO} =$ 0.5 kmoles/m^3

C_A, kmoles/m^3	0.131	0.125	0.121	0.117	0.111	0.101
t,min	8	10	20	30	40	60

Find out the order of the reaction and the rate constant.

Problem 1.4

The total pressure (π_0) versus time, t are given for dimethyl ether decomposition:

$$CH_3OCH_3 \rightarrow CH_4 + CO + H_2$$

π_0 bar	0.410	0.540	0.640	0.740	1.025	1.225
t,min	0	3.5	12.0	20.0	52.0	α

Find out the rate equation for the reaction.

1.5 Nitrous oxide decomposes according to second order rate equation,

$$2N_2O \rightarrow 2N_2 + O_2$$

The specific reaction rate is 0.977 m^3/(kmole)(s) at 895°C. Calculate the fraction decomposed at 10 seconds and 10 minutes. The reaction is irreversible. The initial pressure is 1.013 bar.

1.6 The production of propionic acid is carried out at by acidification of water solution of the sodium salt according to reaction,

$$HCl + C_2H_5COONa \rightleftarrows C_2H_5COOH + NaCl$$

(A) (B) (C) (D)

The reaction rate may be represented by second order reversible equation. The batch data were obtained by estimating the acid by standard NaOH solution. Equal molal feed rates of A & B are used. The laboratory data given are:

t,min	0	10	20	30	50	∞
C_A,kmoles/m^3	2.7	1.65	1.21	0.973	0.746	0.540

Find out the rate constants of the reaction.

1.7 A second order irreversible liquid phase reaction, A + B → Product, is carried out in a plug flow reactor. Two streams, of A, B are feed to the reactor, each flow is 0.5m^3/min and $C_{AO} = C_{BO} = 1$ kmole/m^3. For 80% conversion of A, what is the volume of the reactor? Given $k_2 = 0.005$ m^6/(kmole)^2(min).

1.8 A CSTR is used to carry out an irreversible second order reaction,
$A + B \rightarrow R$
The flow rate is 0.5m^3/min, $C_{AO} = C_{BO} = 0.2$ kmole/m^3, V = 2m^3, Calculate the concentration of A at the exit of the reactor, Given $k_2 = 20$ m^6/(kmol)^2(min)

1.9 A CSTR is used to determine the kinetics of a reaction A → R. Various flow rates of aqueous solution of A ($C_{AO} = 1.0$ kmole/m^3) are introduced to a one m^3 reactor and for each run, exit concentration of A is measured. Find a rate equation from the following data:

F, m^3/min	0.2	0.5	0.8
C_A, Kmole/m^3	0.16	0.33	0.44

1.10 A tubular flow reactor of volume (0.6m^3) was used for the decomposition of acetone (A) at 520°C and 1.013 bar pressure. The reaction is
$CH_3COCH_3 \rightarrow CH_2{=}CO + CH_4$
The following data were obtained.

Flow rate of acetone, m^3/min	2.43	0.91	0.394	0.235
Conversion of acetone, X_A	0.050	0.130	0.240	0.350

Find out a rate equation to fit the following data.

1.11 The following irreversible first order reactions in series occur at constant density

$$A \xrightarrow{k_1} B \xrightarrow{k_2} C$$

$k_1 = 0.15$ min^{-1}, $k_2 = 0.05$ min^{-1}
The reaction system is to be analysed in a continuous flow reactor with a volumetric feed rate of 5 m^3/min and feed composition.
$C_{AO} = C_{AO}$ (1kmole/m^3), $C_{BO} = 0$, $C_C = 0$.

What are the exit concentration of B, the desired product, if we use
(a) a single stirred reactor of V = 10m^3
(b) a PFR of volume = 10 m^3.

References

Aris, Rutherford, Elementary Chemical reactions, Prentice Hall, 1969

Denbigh, Kenneth, G., Chemical Reactor Theory, Cambridge University Press, 1965

Fogler, H.Scott, Elements of reaction Engineering, 2nd ed. Prentice Hall, 1992

Frost, A.A. and R. G. Pearson, Kinetics and Mechanisms, John Wiley & Sons, Inc. N.Y., 1961

Hill, Charles G. Jr., An introduction to Chemical Engineering Kinetics and Reactor Design, Wiley, 1977

Laidler, Keith. J., Chemical Kinetics, 3rd ed, Harper & Row, 1987

Lanny D. Schmidt, The engineering of chemical reactions, Oxford University Press, 2010

Levenspiel, Octave, Chemical Reaction Engineering, 3rd ed, John Wiley & Sons, 1999

Pilling, Michael, J. and Paul W. Seakins, Reation Kinetics, Oxford University Press, 1995

Smith, J.M., Chemical Engineering Kinetics, 3rd ed. Mc. Graw Hill, 1981

Walas, Stanley M., Reaction Kinetics for chemical processes, Mc Graw Hill Book Co., 1959

CHAPTER 2

Gas-Solid Catalytic Reactions

2.0 Introduction

Heterogeneous catalytic processes employ solid catalysts like V_2O_5 in SO_3 formation from SO_2 and O_2; alumina (Al_2O_3) is **used** in cracking reactions. Fluid phase catalytic reactions are non-elementary, have complex mechanisms, and are assumed to occur according to at least five steps:

1. Diffusion of reacting molecules from the bulk of the gas to the slid catalytic surface
2. Adsorption of reactants on the surface of the catalysts
3. Reaction on the surface
4. Desorption of products from the surface
5. Diffusion of the products into the bulk of the fluid.

Since the consideration of all of these steps will lead to complex overall rates of reaction, which is very difficult to use, it is presumed that one or two of the above steps are the controlling rates.

With respect to reaction kinetics with heterogeneous catalysts, the following facts have been found to be important:

1. Adsorption is a necessary preliminary step to reaction catalysed by solid surfaces.
2. Chemisorption is invariably the type of adsorption involved in this case with the molecules of the solid surface
3. Catalytic surfaces are heterogeneous, and chemisorption takes place on the active sites of the surface

2.1 Adsorption Rates and Equilibrium

Many equations for adsorption equilibrium are available on the basis of experiments and theory. One such simplest equation has been given by Langmuir (1916) based on some idealized conditions. These are:

(a) There is no interaction between adsorbed molecules or sites
(b) The surfaces are smooth and of uniform adsorptive capacity
(c) Only a unimolecular layer of adsorbate is formed.

When applied to chemical reactions, solid catalysts are highly porous. They are chemically prepared as fine solids having high internal surface area. The fine catalyst particles may be pelleted and sintered at high temperature to make them porous and active.

2.2 Chemisorption and Physical Adsorption

The oldest theory of catalysis as proposed by Faraday (1825) assumes that adsorption of reactants must occur first and the reaction proceeds in the adsorbed state.

Adsorption is due to an attraction between the molecules of the solid surface, called the adsorbent and those of the fluid, called the adsorbate. In some cases, the attraction is mild and of the same nature as that between like molecules and this phenomena is called physical adsorption. In other cases the force of attraction is nearly equal to the forces involved in the formation of chemical bonds and the process is called chemisorption.

This methodology is known as Langmuir – Hinselwood mechanism (Hinshelwood, 1926). Systematic applications of these mechanisms have been presented by Hougen and Watson (1947).

2.3 Adsorption Isotherms

Several adsorption isotherms have been applied to catalytic gas-solid reactions.

2.3.1 BET equation

The Langmuir's adsorption isotherm was extended to multi– layer adsorption by Brunauer, Emmet and Teller and their equation is designated as BET equation (Brunauer et.al. 1938) which is very useful for the measurement of surface-area of particles.

$$\frac{p}{C(p_o - p)} = \frac{1}{v_m C} + \frac{(C-1)p}{C v_m p_o} \tag{2.1}$$

where p_0 is the saturated vapour pressure of gas (N_2). C is a constant for a particular temperature and gas-solid systems, v_m is the volume of the gas adsorbed on the surface of the solid to form a complete monolayer.

2.3.2 Freundlich adsorption isotherm

The above adsorption isotherm is given as

$$\theta = k\, p^n \tag{2.2}$$

where θ is the fraction of surface covered by a reactant gas, k, n are constants, n is usually a fraction. The isotherm is used in developing catalytic rates for some reaction systems.

Using the above isotherm, the catalytic rate of phosgene ($COCl_2$) gas formation from CO and Cl_2 in presence of active carbon has been given as:

$$r_{co} = kP_{co}\left(P_{cl_2}\right)^{1/2} \tag{2.3}$$

2.3.3 Langmuir adsorption isotherm

Langmuir adsorption isotherm has been extensively used to develop rate equations for catalytic gas-solid reactions. Most surface reactions involve adsorption of Langmuir type.

Its mechanism is based on the assumption that the process of chemisorption involves a reaction between the adsorbate, G and active site on the surface, σ. The process can be illustrated as

$$G + \sigma \underset{k_{-1}}{\overset{k_1}{\rightleftharpoons}} G\sigma \tag{2.4}$$

For diatomic molecule

$$G_2 + 2\sigma \rightleftharpoons 2G\sigma \tag{2.5}$$

Adsorption processes are in general reversible and attain equilibrium.

(i) The Langmuir isotherm can be developed as follows:
The rate of adsorption, $r_1 = k_1 p\ (1-\theta)$
The rate of desorption, $r_{-1} = k_{-1}\theta$
where θ is the fraction covered by the adsorbate, p is the partial pressure.
At equilibrium, the rates of adsorption and desorption (r_1 & r_{-1}) are equal. Therefore,

$$k_1 p\ (1-\theta) = k_{-1}\ \theta \tag{2.6}$$

or

$$\theta = K\,p\,/\,(1 + K\,p) \tag{2.7}$$

where $K = k_1 / k_{-1}$ is the equilibrium constant.
In terms of concentration, Langmuir adsorption isotherm may be expressed as,

$$\theta = v/v_m = K_c\,C_A\,/(1 + K_c\,C_A) \tag{2.8}$$

where v is the volume of the adosrbate adsorbed per unit mass of adsorbent (solid) v_m is the volume of the adsorbate per unit mass of adsorbent to make a complete mono layer on the porous surface and K_c is the equilibrium constant and C_A is the concentration of the gaseous reactant, A in the bulk gas.

(ii) When dissociation occurs with adsorption, $A_2 + 2\sigma \rightarrow 2A\sigma$, the rates of adsorption (r_1) and desorption (r_{-1}) are

$$r_1 = k_1\, p_A\, (1 - \theta_A) \tag{2.9}$$

$$r_{-1} = k_{-1}\, \theta_A^{\,2} \tag{2.10}$$

where A_2 is a diatomic molecule.
At equilibrium, the rates are equal and θ_A is,

$$\theta_A = \frac{k_A^{1/2}\, p_A^{1/2}}{1 + K_A^{1/2}\, p_A^{1/2}} \tag{2.11}$$

Where K_A is adsorption constant and θ_A is the fraction of the surface covered by the adsorbate, A.

(iii) When two monatomic gases are adsorbed on the same surface and θ_A and θ_B are the fractions of the surfaces covered by the molecules of A & B, the fraction of the uncovered surface is $(1-\theta_A-\theta_B)$, then

$$r_1A = k_{1A}\, p_A\, (1-\theta_A-\theta_B) \tag{2.12}$$

$$r_{-1}A = k_{-1A}\, \theta_A \tag{2.13}$$

$$r_1B = k_{1B}\, p_B\, (1-\theta_A-\theta_B) \tag{2.14}$$

$$r_{-1}B = k_{-1B}\, \theta_B \tag{2.15}$$

At equilibrium, the adsorption and the desorption rates are equal, solving for θ_A and θ_B, we get

$$\theta_A = K_A\, p_A / (1 + K_A p_A + K_B p_B) \tag{2.16}$$

$$\theta_B = K_B\, p_B / (1 + K_A p_A + K_B p_B) \tag{2.17}$$

where K_A, K_B are equilibrium adsorption constants for A & B; p_A, p_B are the partial pressures of A and B.

(iv) For two gases (A & B) with the gas B being diatomic, the θ's can be given as

$$\theta_A = \frac{K_A p_A}{1 + K_A p_A + \sqrt{K_B p_B}} \tag{2.18}$$

$$\theta_B = \frac{K_B p_B}{1 + K_A p_A + \sqrt{K_B p_B}} \tag{2.19}$$

(v) For a reversible reaction with an inert. (I) as shown below,

$$A + B \rightleftharpoons R + S + I$$

and all components are adsorbed and I is an inert component, the θ's can be given as

$$\theta_A = \frac{K_A p_A}{1 + K_A p_A + K_B p_B + K_R p_R + K_S p_S + K_I p_I} \quad (2.20)$$

$$\theta_B = \frac{K_A p_A}{1 + K_A p_A + K_B p_B + K_R p_R + K_S p_S + K_I p_I} \quad (2.21)$$

Briefly for the component i

$$\theta_i = \frac{K_i p_i}{\sum 1 + K_i p_i} \quad (2.22)$$

(vi) When the surface is sparsely covered, $1 - \theta \approx 1.0$, so,
we have $\theta = P$

(vii) When the surface is largely covered, then

$\theta \rightarrow 1$ and $1 - \theta = 1 / KP$

2.4 Development of Rate Equations Based on Langmuir Adsorption Isotherm or Hougen — Watson Models

The rates of reactions catalysed by solid surfaces can be defined as moles reacted per unit time per unit weight of the catalyst,

$$r_A = -\frac{1}{W}\frac{dN_A}{dt} = \frac{N_{AO} dX_A}{W\, dt} \quad (2.23)$$

in terms of conversion, x_A and molal flow rate F_A for a flow reactor.
The rate is given as,

$$r_A = dX_A / d(W/F_A) \quad (2.24)$$

where W is the weight of catalyst and X_A is the fractional conversion of A. The following cases may be considered for developing the rate of catalytic processes.

2.4.1 Surface-reaction-rate controlling

In this case the adsorption equilibrium is maintained at all time and the overall rate of reaction is governed by the rate of combination on the surface.

The rate of a reaction occurring on the surface is proportional to the amounts of the reactants in terms of the fractions of the surfaces covered by them. For example:

Case I:

$A + B \rightarrow R + S$

$$r_A = -\frac{1}{W_c}\frac{dN_A}{dt} = k_1\theta_1\theta_2 = \frac{k_1 K_A P_A K_B p_B}{\left(1 + K_A P_A + K_B p_B\right)^2} \tag{2.25}$$

Where p_A, p_B are the partial pressures of the reactants A and B and only A & B are adsorbed, N_A is the number of molesof A and Wc is weight of the catalyst. Lumping the constants,

The equation (2.25) becomes

$$r_A = \frac{k \cdot p_A p_B}{\left(1 + K_A p_A + K_B P_B\right)^2} \tag{2.26}$$

where $k = k_1 K_A K_B$

Case II.

For a reversible reaction

$$A + B \underset{k_{-1}}{\overset{k_1}{\rightleftharpoons}} R + S$$

The rate can be given as,

$$r_A = k_1\theta_A\theta_B - k_{-1}\theta_R\theta_S = k_1\left(\theta_A\theta_B - \frac{\theta_R\theta_S}{K_r}\right) \tag{2.27}$$

where K_r is the equilibrium constant for the reaction. Substituting for θ's in terms of particle pressures and adsorption equilibrium constants, we have,

$$r_A = \frac{k_1\left(K_A K_B p_A p_B - K_R K_S p_R p_S / K_r\right)}{\left(1 + K_A p_* + K_B p_B + K_k P_S + K_S P_S\right)^2} \tag{2.28}$$

For further simplification, substituting,
$K = K_A K_B K_r / K_R K_S$ and $k = k_1 K_A K_B$, the rate equation will be expressed as,

$$r_A = \frac{k\left(p_A p_B - p_R p_S / K_r\right)}{\left(1 + K_A p_A + K_B p_B + K_R P_R + K_S P_S\right)^2} \tag{2.29}$$

where K_r is the reaction equilibrium constant.

The generalized form of all these equations is of the type as given below:

$$r_A = \text{(kinetic terms) (driving force) / (adsorption terms)} \tag{2.30}$$

In the equation (2.29), k is the kinetic term, ($p_Ap_B - p_Rp_S/K$) is the driving force and the denominator contains the adsorption terms.

Case III.

For the reaction, $A \rightleftharpoons R + S$, a dual site mechanisms is to be introduced. Here the molecule, A, which is adsorbed on one active site will form an intermediate compound on the vacant adjacent site, σ.

The intermediate then dissociates into Rσ + Sσ, and each desorbed into R and S, leaving each adsorbed active site of its own. Thus the mechanism can be presented as follows

$A + \sigma \rightarrow A\sigma$
$A\sigma + \sigma \rightarrow A\sigma_2$
$A\sigma_2 \rightarrow R\sigma + S\sigma$
$R\sigma \rightarrow R + \sigma$
$S\sigma \rightarrow S + \sigma$

The net rate of reaction can be given as

$$r_A = k_1\theta_a\theta_u - k_{-1}\theta_R\theta_S \tag{2.31}$$

$$= \frac{k\left(p_A - p_R p_S / K\right)}{\left(1 + K_A p_A + K_R p_R + K_S P_s\right)^2}$$

where θ_v is the vacant site and is given as

$$\theta_v = \frac{l}{\left(1 + K_A p_A K_B p_B + K_R P_R + K_S p_S\right)} \tag{2.32}$$

Case IV:

$A_2 + B \rightleftharpoons R + S$

The rate of forward reaction can be expressed as

$r_1 = k_1\, \theta_a^{\,2}\, \theta_b$

and the reverse rate, $r_{-1} = k_{-1}\, \theta_R\theta_S\theta_v$.

For dissociation, θ_A is given as

$$\theta_A = \frac{\sqrt{K_A p_A}}{1 + \sqrt{K_A p_A} + K_B p_B + K_R p_R + K_S p_S} \tag{2.33}$$

The overall rate is then given as,

$$r_A = k_1\theta_A^2\theta_B - k_{-1}\theta_R\theta_S\theta_v \quad (2.34)$$

$$= \frac{k\left(p_A p_B - p_R p_S\right)}{\left(1++\sqrt{k_R p_A} + K_B p_B + K_R p_S + K_S p_S\right)^3}$$

Case V:

If the reaction between two reactants which are adsorbed on two different kinds of active sites on the same surface, the mechanism is called dissimilar sites. Each reactant follows the adsorption equilibrium.

For a reaction, A + B → R + S where the products, R & S are not adsorbed, the rate can be given as

$$r_A = \frac{k_1 K_A p_A K_B p_B}{\left(1+K_A p_A\right)\left(1+K_B p_B\right)} = \frac{k p_A p_B}{\left(1+K_A p_A\right)\left(1+K_B p_B\right)} \quad (2.35)$$

where $k = k_1\, k_A$

2.4.2 Adsorption rate controlling

The rate of reaction on the surface may be so rapid in some cases, that adsorption equilibrium is not reached and the rate of adsorption of one of the participants for the reactants or products becomes controlling as the reaction of that component is slower than the others.

For the reaction, A + B → R + S,

It is assumed that the rate of surface reaction is rapid and the reaction equilibrium is quickly attained. Consequently adsorption of A proceeds slowly and adsorptive equilibrium is not established. But an amount of that component is present corresponding to its chemical equilibrium between the participants reacting on the surface. If p_A^* is the partial pressure corresponding to the chemical equilibrium, the fraction of the surface covered by A is given as before,

$$\theta_A = \frac{K_A p_A^*}{\left(1+K_A p_A^* + K_B p_B + K_R p_R + K_S p_S\right)} \quad (2.36)$$

At the chemical equilibrium, the rate, r, can be written as,

$$r = k_1\, \theta_A\, \theta_b - k_{-1}\, \theta_R\, \theta_S = 0 \quad (2.37)$$

then

$$\theta_R\, \theta_S / \theta_A\, \theta_B = K \quad (2.38)$$

where K is the chemical equilibrium constant for the surface reaction.

Expressing θ's in terms of partial pressures and rearranging we have,

$$p_A^* = p_R\, p_S / p_B\, K \tag{2.39}$$

where $K = KK_AK_B / K_RK_S$
Replacing p_A^* from equation (2.36) by equation (2.39) we have,

$$\theta_A = \frac{K_A p_R p_S / K p_3}{1 + \frac{K_A p_R p_S}{K p_B} + K_B p_B + K_R p_R + K_S P_S} \tag{2.40}$$

Now the rate of reaction is the net rate of adsorption and is given as,

$$r = k_1 p_A \theta_v - k_{-1}\theta_A = k_1\left(p_A\theta_v - \frac{\theta_A}{K_A}\right) \tag{2.41}$$

where θ_v (the fraction of the vacant sites) $= 1 - \theta_A - \theta_B - \theta_k - \theta_s$,
K_A (adsorption equilibrium constant) $= k_1 / k_{-1}$
Now substituting θ_A & θ_v, the rate of reaction of A is presented as,

$$r_A = \frac{k(p_A - p_R p_S / K p_B)}{(1 + K_A p_R p_S / K p_B + K_b p_B + K_R P_S + K_S P_S)} \tag{2.42}$$

Case II.
Consider the reaction $A_2 \rightleftharpoons R$ where the adsorbtion, A_2 dissociates on adsorption.
Then, the fraction of the surface covered by A, θ_A can be given as,

$$\theta_A = \frac{\sqrt{K_A p_R / K}}{1 + \sqrt{K_A p_R / K} + K_R p_R} \tag{2.43}$$

where $p_A^* = p_R / K$
and the rate is expressed as,

$$r_A = k_1 p_A \theta_v^2 - k_{-1}\theta_A^2 = \frac{k(p_A - p_A / K')}{\left(1 + \sqrt{k_A p_R / k} + K_R p_R\right)^2} \tag{2.44}$$

Where $K' = \dfrac{k_1 K}{k_{-1} K_A}$

Case III.

For an irreversible reaction of the type and chemical reaction controlling
$A + B \rightarrow R + S$
Here the component A is not adsorbed, but B, R, S are adsorbed. The rate of reaction of A, r_A is proportional to the rate of impact of gas molecules of A on adsorbed moles of B and it can be given as,

$$r_A = k_1 p_A \theta_B = \frac{k p_A p_B}{\left(1 + k_B p_B + k_R p_R + k_S p_S\right)} \tag{2.45}$$

Where $$\theta_B = \frac{k_B p_B}{\left(1 + k_B p_B + k_R p_R + k_S p_S\right)} \tag{2.45a}$$

and $$k = k_1 K_B \tag{2.45b}$$

It is evident from the above cases, various possibilities of reaction rates for different controlling mechanisms and the status of adsorption, are possible and those have been tabulated by Hougen – Watson (1947).

2.5 Rate Equations Based on Freundlich Adsorption Isotherm

We know that Freundlich adsorption isotherm can be given in terms of fraction of surface adsorbed, θ, as

$\theta_A = k\, p_A^{\,n}$

where k and n are constants and n is normally a fraction. A typical example of this type of model may be presented for the reaction between CO and Cl_2 in presence of catalyst of active carbon, producing phosgene gas, ($COCl_2$), which is a nice chlorinating agent. Its rate has been established on the above mechanism in the following form:

$$r_{co} = k p_{co} \left(pcl_2\right)^{1/2} \tag{2.46}$$

which is compatable to the established Hougen-Watson model:

$$r_{co} = \frac{k\, k_{co} k_{cl_2} \left(p_{co}\right)\left(p_{cl_2}\right)}{\left(1 + k_{cl_2} p_{cl_2} + K_{cocl_2} p_{cocl_2}\right)} \tag{2.47}$$

Where only Cl_2 and $COCl_2$ are adsorbed on catalytic surface

2.6 Determination of Kinetic Parameters for Catalytic Reactions

Determination of rate parameters of a model based on the mechanism of a surface catalysed or adsorption controlling system depends upon the linearization of the equation with number of constants which may contain lumped rate constants. The experimental data should be used to evaluate the constants which should all be either zero or positive.

Negative numbers which may arise are not acceptable and assumed mechanism leading to such negative number should be discarded. If proposed two parallel models give rise to positive parameters, the models are discriminated using the statistical technique. Some classical methods may be mentioned below:

2.6.1 Least square method / Linear regression

Consider the rate equation (2.44) which can be linearized as in equation (2.48)

$$r = \frac{kp_A p_B}{1 + k_B p_B + k_R p_R + k_S p_S}$$

Rearranging

$$\left(\frac{p_A p_B}{r}\right) = \frac{1}{k} + \frac{K_B}{k} p_B + \frac{K_R p_R}{k} + \frac{K_S}{k} p_S \tag{2.48}$$

$$y = a + b\, p_B + c\, p_R + d\, p_S \tag{2.49}$$

where a, b, c, d are constatns related to the constants of the original equation. For different values of p_A, p_B, p_R & p_S, the rates should be determined from the flow reactor data, where $r_A = dx_A / d(W/F_A)$, W is the weight of the catalyst and F_A is the molal flow rate of A.

The constants a, b, c, d can be obtained by linear regression.

2.6.2 Initial rates

The number of possible mechanisms may be reduced when the initial rates are expressed as a function of concentration or total pressure of the system. Such rates can be obtained by intrapolation of data over a wide range or directly from the data of differential reactor with small conversion.

Consider a first order reversible reaction with surface reaction controlling. Two mechanisms are proposed such as

$A \rightleftharpoons R$

And $A_2 \rightleftharpoons R$

The first case is simple adsorption and the second one is adsorption with dissociation. The corresponding initial rates (r_0) without considering the product are:

$$r_0 = \frac{kp_A}{1 + K_A p_A} \tag{2.50}$$

$$r_0 = \frac{kp_A}{\left(1 + \sqrt{K_A p_A}\right)^2} \tag{2.51}$$

In the absence of reaction products, p_A is proportional to the total pressure, π and the above equations may be modified in terms of total pressure, π, and r_0. The initial rates may be given as

$$r_0 = \frac{k'\pi}{1 + K'\pi} \tag{2.52a}$$

$$r_0 = \frac{k'\pi}{\left(1+\sqrt{K'\pi}\right)^2} \tag{2.52b}$$

Now a study of the effect of total pressure on the initial rate of the reaction will distinguish the two mechanisms. Several experiments at different initial total pressures are required. Typical curves for several for first and second order reactions have been illustrated by Hougen & Watson (1947).

The two equations, (2.52) and (2.53) may be linearized as follows:

$$\frac{\pi}{r_0} = \frac{1}{k} + \frac{K'}{k}\pi \tag{2.53}$$

$$\left(\frac{\pi}{r_0}\right)^2 = \frac{1}{k'} + \frac{\sqrt{K'}}{k'}\sqrt{\pi} \tag{2.54}$$

From the slopes and intercepts, the kinetic parameters may be evaluated. The more reliable parameters may be determined by statistical analysis.

Some simple catalytic reactor design problems are discussed below.

Problem 2.1

For a catalytic reaction, A→ R, the kinetic data were obtained in an experimental packed bed reactor, using various amounts of catalyst-and a fixed feed rate,

F_{Ao} = 10 kmoles/hr,

W, Kg Cat	1	2	3	4	5	6	7
X_A	0.12	0.20	0.27	0.33	0.37	0.41	0.44

(a) Find the reaction rate at 40% conversion

(b) For a large packed bed reactor with feed rate F_{Ao} = 400 kmoles/hr, how much catalyst is needed for 40% conversion?

(c) How much catalyst is required if the above reactor used a very large recycle stream?

Solution:

Evaluation of the rate by graphical method from the plot of X_A vs. W/F_{Ao}

X_A	0.12	0.20	0.27	0.33	0.37	0.41	0.44
W / F_{Ao}	0.10	0.20	0.30	0.40	0.50	0.60	0.70

By drawing tangents to the curve of X_A vs W/F_{Ao} at different X_A value, we get r_A from the slopes and their values are tabulated.

X_A	0	0.10	0.20	0.30	0.40
r_A kmoles / (ms) (hr)	1.25	0.47	0.33	0.166	0.05
1/r	0.80	2.12	3.0	6.02	20
	y_0	y_1	y_2	y_3	y_4

(a) The reaction rate at 40% from the plot of x_A vs W/F_{Ao} at $x_A = 0.4$
$r_A = 0.05$

(b) Weight of the catalyst required for the flow rate, F = 400 k moles/hr, can be obtained from the following equation

$$\frac{W}{F_A} = \int_O^{xA} \frac{dx_A}{r_A}$$

The integral on the right hand side can be evaluated by the numerical method (Simson's 1/3 rule)

$$I = \frac{0.1}{3}\left[0.8 + 4(2.12 + 6.02) + 2(3.0) + 20\right] = 1.36$$

$$W = F_{A_0} \times I = 400 \times 1.36 = 544Kg$$

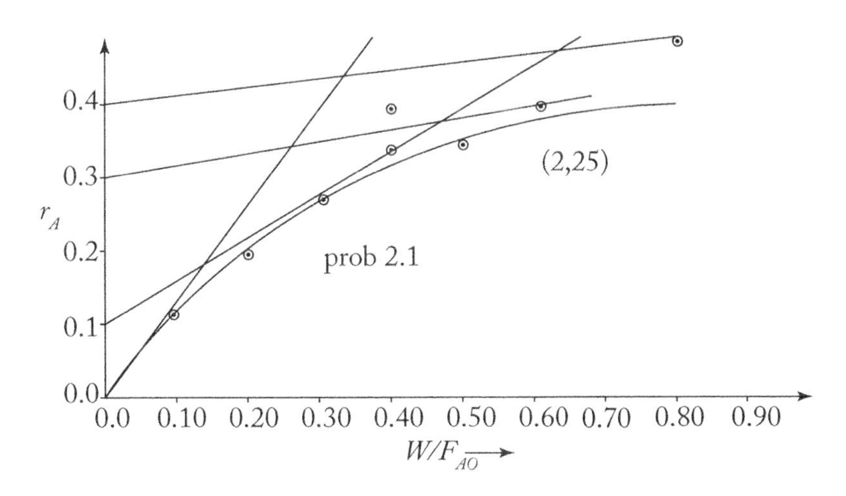

Fig. 2.1 Catalytic rates (r_A) from the plot of x_A vs W/F_{Ao}

(c) A packed bed reactor with a large recycle is equivalent to a backmix reactor and the design equation is
$W/F_{Ao} = x_A / r_A$
$F_{Ao} = 400$ k. moles/hr
$x_A = 0.4$ & $r_A = 0.05$
W = 400 x 0.4 / 0.05 = 3200 kg of catalyst

Problem 2.2

The cracking of cumene takes place in presence of alumina as follows:

$C_6H_5CH\,(CH_3)_2 \rightleftharpoons C_6H_6 + C_3H_6$

$A \rightleftharpoons R + S$

Examination of the experimental data showed that on the basis of single site mechanism the adsorbed cumene decomposes to adsorbed benzene and unadsorbed propylene. The rate of reaction can be given on the basis of surface reaction controllimg

$A\sigma \rightleftharpoons R\sigma + S$

$$r_A = \frac{k\left(P_A - p_R p_S / K\right)}{1 + K_A p_A + K_R p_R}$$

The initial rates (r_o) were studied by the measurement of total pressure, π

$$r_o = \frac{kp_A}{1 + K_A p_A} \approx \frac{a\pi}{1 + b\pi}$$

The cracking data were given below at 510°C

r_o, kmoles /(hr) (kg cal)	4.3	6.5	7.1	7.5	8.1
π, bar	1.0	2.65	4.32	7.0	14.36

Find out the constants of a and b from the data

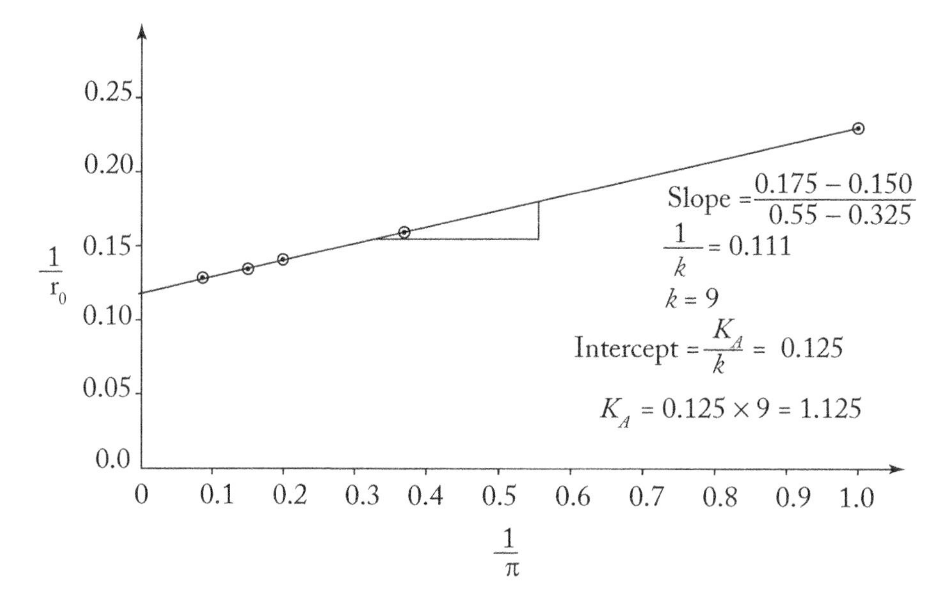

Fig. 2.2 Plot of $1/r_o$ vs $1/\pi$

$$r_0 = \frac{kp_A}{1 + K_A p_A}$$

$$A \rightleftharpoons R + S$$

$$N_{AO} - x_A \quad x_A \quad x_A$$

$$N = N_{AO} - x_A + x_A + x_A = N_{AO} + X_A$$

$$C_A = \frac{p_A}{RT} = \frac{N_A}{V} = \frac{N_A - X_A}{V} = \frac{NA_0}{V} - \frac{(N - NA_0)}{V}$$

$$p_A = 2\pi_0 - \pi$$

$$r_O = \frac{k(2\pi_O - \pi_O)}{1 + K_A(2\pi_O - \pi_O)} = \frac{k\pi_O}{1 + K_A\pi_O}$$

$$\frac{1}{r_O} = \frac{1}{k}\frac{1}{\pi_O} + \frac{K_A}{k}$$

$\frac{1}{r_O}$ is plotted again $\frac{l}{\pi_O}$, from the linear plot,

Slope $= \frac{1}{k} = 0.111$ or k = 9

Intercept $= \frac{K_A}{k} = 0.125$

Or $K_A = 0.125 \times 9 = 1.125$

So the rate equation is

$$r_o = \frac{9\pi_0}{1 + 1.125\pi_0}$$

Problem 2.3

The reaction, $CO_2 + H_2 \rightleftharpoons H_2O + CO$

$A + B \rightleftharpoons R + S$

was carried out at 538°C at atmospheric pressure in presence of Fe-Cu catalyst. The controlling step is assumed to be the reaction of adsorbed CO_2 with a molecule of H_2 in the gas phase to produce a CO molecule in the adsorbed state and H_2O molecule in the gas phase. The rate equation has been given as

$$r_A = \frac{k\left(p_A p_B - p_R p_S / K\right)}{1 + K_A p_A + K_S p_S} \frac{k.mols}{(hr)(kg\,cat)}$$

where the equilibrium constant K = 0.267 at 538°C.
The data obtained at 538°C and atmospheric pressure with the feed containing 20% CO_2.

x_A	0.604	0.586	0.482	0.399	0.099	0.037
W/F_A kg cat/ kgmole/hr moles of CO_2 / hr	121	70.0	30.2	19.1	5.5	2.5

Evaluate the following
(a) The partial pressures of all participants
(b) The rate $r = dx_A / d(W/F)$ as a function of x_A
(c) The constants of the rate equation, k, K_A, K_s
Show that the values evaluated are k = 0.595, K_A = 4..46, K_s = 41.65

Solution:

(a) $A + B \rightleftharpoons R + S$
0.2 0.2
x_A x_A
$0.2\,(1 - x_A)$, $0.8 - 0.2\,x_A$
Feed 0.2 mole of CO_2 and 0.8 mole of H_2
$N_t = 0.2\,(1 - x_A) + 0.8 - 0.2x_A + 0.2\,x_A + 0.2\,x_A = 1.0$
$p_A = 0.2\,(1 - x_A) / 1 \times 1$ atm
$p_B = (0.8 - 0.2x_A) / 1 \times 1$ atm
$p_R = 0.2\,x_A / 1 \times 1$ atm
$p_S = 0.2\,x_A / 1 \times 1$ atm

x_A	0.604	0.586	0.482	0.099	0.037
W/F	121	70.0	30.2	5.5	2.5

x_A vs W/F is plotted in a graph and tangents are drawn at $x_A = 0$, $x_A = 0.1, 0.2, 0.3, 0.4, 0.5$ and the evaluated rates are

x_A	0	0.1	0.2	0.3	0.4	0.5
r_A	0.026	0.020	0.010	0.005	0.0025	0.0007
$r_A \times 10^3$	26	20	10	5	2.5	0.7

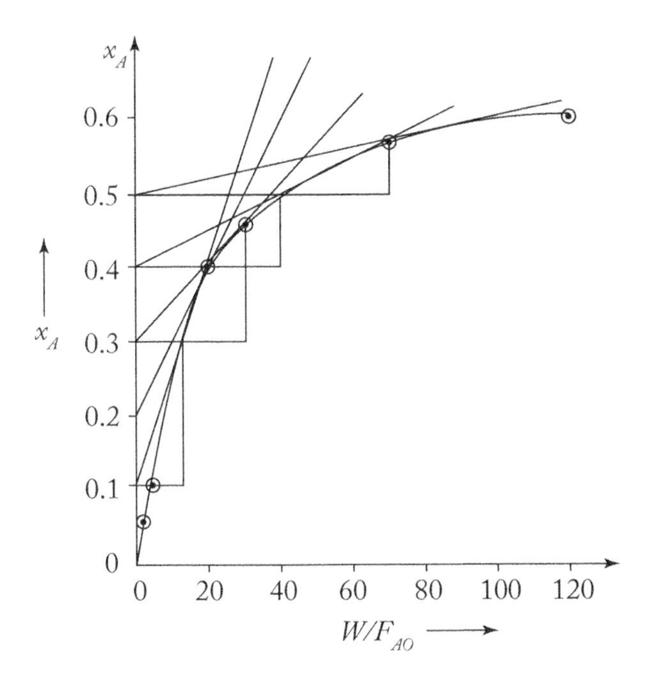

Fig. 2.3 Tangent to the curve of x_A vs W/F_A

x_A	0.0	0.2	0.4	
$p_A = 0.2\,(1-x_A)$	0.2	0.16	0.12	$\sum P_A = 0.48$
$p_B = 0.8 - 0.2x_A$	0.8	0.76	0.72	$\sum P_B = 2.28$
$p_R = 0.2\,x_A$	0	0.04	0.08	$\sum P_R = 0.12$
$p_S = 0.2\,x_A$	0	0.04	0.08	$\sum P_S = 0.12$
r	0.026	0.020	0.0025	
$y = p_A p_B - p_R p_S / 0.267$	6.15	5.78	25.0	$\sum y = 36.9$

Linearizing

$$\left(\frac{p_A p_B - p_R p_s / K}{r}\right) = \frac{1}{k} + \frac{K_A}{k} p_A + \frac{K_S}{k} P_S$$

Or $y = A + B\,p_A + C\,P_S$

Where $A = 1/_k$, $B = \frac{K_A}{k}$, $C = \frac{K_S}{k}$

$$A + B\,\Sigma\,p_A + C\,\Sigma\,p_S - y = 0 \qquad (\alpha)$$

$$nA + B\sum P_A^2 + C\sum P_S p_A - \sum P_A y = 0(\beta)$$

$$A\sum P_S + B\sum P_A P_S + C\sum P_S^2 - \sum P_S y = 0(\gamma)$$

Putting the values of the summation components in the equation (α), we get,
3A +.48B +0.12C – 36.9 = 0

Similarly the equation (β) & (Y) can be evaluated in terms of A, B & C. From these 3 equations, A, B & C have been evaluated

Problem 2.4

The oxidation of NO by oxygen was carried out in presence of catalyst, active carbon at 30°C and the rate is given as

$$r = \frac{(p_{NO})^2 (p_{O_2})}{a + b(p_{NO})^2 + Cp_{NO_2}}, \frac{kg\ moles\ of\ NO}{(Kgcat)(hr)}$$

where the values of the constants are
a=0.0001619, b = 4.842, c = 0.001352
P = the partial pressures in bar

Find the volume of the reactor for converting 50 tons / day of NO to NO_2 when air-NO is mixture containing 1.5 mole % of NO is used.

The other data are: bulk density of the catalyst is 480 kg/m³ and the total pressure is 3 bar.

Solution:

$NO + ½\, O_2 \rightarrow NO_2$

Basis: 0.015 mole of NO and 0.985 mole of O_2

Material balance:

NO = 0.015 $(1-x_A)$

$O_2 = 0.985 - 0.5\,(0.015x_A) = 0.985 - 0.0075x_A$

$NO_2 = 0.015x_A$

Adding, $N_t = (1-0.0075x_A)$

Let us find out the rate at three values of $x_A = 0, 0.45, 0.90$

at $x_A = 0$

$$p_{NO} = \frac{0.015(1-x_A)\pi}{1-0.0075x_A} = \frac{0.015\times 3}{1} = 0.4$$

$$(p_{NO})^2 = (.045)^2 = 0.002025$$

$$b(P_{NO})^2 = (4.842)(.002025) = 0.00980$$

$$P_{O_2} = \frac{(0.985-0.0075x_A)^3}{1-0.0075x_A} = 0.985\times3 = 2.955$$

$$P_{NO_2} = \frac{0.015x_A}{1-0.0075x_A} = \frac{0.015(0)}{1-0} = 0$$

at $x_A = 0$

$$r_A = \frac{(P_{NO})^2(P_{O_2})}{a+b(P_{NO})^2+C(p_{NO_2})}$$

$$r_A = \frac{(.002025)(2.955)}{0.0001619+0.00930} = 0.6064$$

$$\frac{1}{r_A} = 1.6489$$

At $x_A = 0.45$

$$P_{NO} = \frac{0.015(1-0.45)\times 3}{1-0.0075(.43)} = 0.02483$$

$$(p_{NO})^2 = 0.0006167, b(P_{NO})^2 = 0.002986$$

$$P_{O_2} = \frac{[0.985-0.0075(0.45)]^3}{1-0.0075(45)} = 0.3$$

$$P_{NO_2} = \frac{0.15(.45)^3}{1-0.0075(.45)} = 0.02031$$

$$cP_{NO_2} = (0.001352)(0.02031) = 0.00002746$$

$$r = \frac{0.008958}{0.0001619 + 0.002986 + 0.00002746} = 2.2810$$

$$\frac{1}{r} = 0.3544$$

At $x_A = 0.90$

$$P_{NO_2} = \frac{0.015(1-0.9)3}{1-0.0075(0.9)} = 4.50 \times 10^{-3}$$

$$(P_{NO_2})^2 = 2.0277 \times 10^{-5}$$

$$b(P_{NO_2})^2 = (4.842)(2.0277) \times 10^{-5} = 9.818 \times 10^{-5}$$

$$P_{O_2} = \frac{0.985 - 0.0075(0.90) \times 3}{1-0.0075(0.90)} = 2.9817$$

$$P_{NO_2} = \frac{0.015(0.9) \times 3}{1-0.0075(0.9)} = 0.040527$$

$$cP_{NO_2} = (0.001352)(0.040527) = 5.4793 \times 10^{-5}$$

$$r_A = \frac{(P_{NO})^2 (P_{O_2})}{a + b(P_{NO})^2 + C(p_{NO_2})}$$

$$= \frac{(2.0277 \times 10^{-5})(2.9817)}{0.0001617 + 9.818 \times 10^{-5} + 5.4793 \times 10^{-5}} = 0.192$$

$$\frac{1}{r_A} = 5.208$$

$$\frac{W}{F_{AO}} = \int_0^{0.90} \frac{dx_A}{r_A}$$

$$I = \frac{h}{3}\left[1.6489 + 2(0.3544) + 5.208\right] = 0.9121$$

$$W = F_{AO} \times 0.9121 = (138.9)(0.9121) = 126.69 \text{ Kg catalyst}$$

Volume of the catalyst

$$= \frac{w}{\rho_B} = \frac{126.69(kg)}{480\left(\frac{kg}{m^3}\right)} = 0.264m^3$$

Problem 2.5

In a differential reactor the rate of formation of CH_4 from CO and H_2 in presence of a nickel catalyst has been established in the following form:

$$CO + 3H_2 \rightarrow CH_4 + H_2O$$

$$r_{CH_4} = \frac{0.01983\, p_{H_2}^{1/2}\, p_{CO}}{1 + 1.5P_{H_2}} \frac{kmole}{(Kgcat)(s)} \text{ at } 260°C$$

For the production of 20 tons of CH_4 per day, how much catalyst is necessary for 80% conversion of CO if we use:

(a) a fixed reactor
(b) a fluidized bed reactor

The feed contains 75% H_2 and 25% CO at 260°C and a pressure of 10 bar.

Solution:

$CO + 3H_2 \rightarrow CH_4 + H_2O$

Basis: 0.25 mole of CO and 0.75 mole H_2

Moles of CO = 0.25 $(1-x_A)$

Moles of H_2 = 0.75 – 3 $(0.25x_A)$

Moles of CH_4 = 0.25 x_A

Moles of H_2O = 0.25 x_A

....

$N_t = 1 - 0.5x_A$ where Nt is the total number of moles of CO, H_2, CH_4, H_2O.

We evaluate the rates at different conversion and use the numerical method to obtain the value of W/F.

at $x_A = 0.0$

$$p_{co} = \frac{0.25(1-0)\times 10}{1-0.5\times 0} = 2.5 bar$$

$$p_{H_2} = \frac{(0.75-0.75\times 0)\times 10}{(1-0.5\times 0)} = 7.5 bar$$

$$r_{CH_4} = \frac{0.0183(p_{H_2})^{1/2} P_{CO}}{1+1.5(P_{H_2})}$$

$$= \frac{0.0183(7.5)^{1/2}(2.5)}{1+1.5(7.5)} = 0.01022$$

$$\frac{1}{r_{CH_4}} = 97.77$$

At $x_A = 0.4$

$$p_{co} = \frac{0.25(1-0.4)\times 10}{1-0.5\times .4} = 1.875 bar$$

$$p_{H_2} = \frac{[0.75-0.75(0.40)]\times 10}{1-0.5\times .4} = 5.625 bar$$

$$r_{CH_4} = \frac{0.0183(5.625)^{1/2}(1.875)}{1+1.5(5.625)} = 0.008623$$

$$\frac{1}{r_{CH_4}} = 115.96$$

At $x_A = 0.8$

$$p_{CO} = \frac{0.25(1-.8)\times 10}{1-0.5(0.8)} = 0.833 bar$$

$$p_{H_2} = \frac{(0.75 - 0.75 \times 0.8)10}{1 - 0.5(0.8)} = 2.5 bar$$

$$r_{CH_4} = \frac{0.0183(2.5)^{1/2}(0.833)}{1 + 1.5(2.5)}$$

$$\frac{1}{r_{CH_4}} = 197$$

$$\frac{W}{F_{AO}} = \int_0^{x_A = 0.8} \frac{dx_A}{r_A}$$

$$F_{AO} = \frac{20 \times 1000}{16 \times .8 \times 24 \times 3600} = 0.018 k.moles\ co / second$$

$$I = \frac{0.4}{3}\left[97.77 + 2(115.9) + 197\right] = 70.22$$

$$W = F_{AO} \times I = 0.018 \times 70.22 = 1.264 kg$$

(b) $W = F_{AO}\dfrac{x_A}{r_A} = .0\ 018 \times 0.8 \times 197 = 2.8368 Kg$

Problem 2.6

Catalytic hydration of ethylene in the vapour phase at 136 bar using catalyst pellets of phosphoric acid on diatomaceous earth at 270°C.

The rate equation for the reaction, $C_2H_4 + H_2O \rightleftharpoons C_2H_5OH$

$$r = \frac{kK_A K_B (P_A P_B - P_R / K)}{(1 + K_a p_A + K_B p_B)^2} \frac{kmoles}{(Kgcat)(hr)}$$

The values of the constants at 270°C are

k = 0.00665
$K_A = K_B = 0.00880$
K (equilibrium constant) = 206.7 bar^{-1}.

Calculate the amount of catalyst required to convert 20% of C_2H_4 under the isothermal conditions at 270°C when the total feed rate is 100 K.moles / hr containing equimolar proportion of ethylene and water.

Solution:

$A + B \rightleftharpoons R,$

Basis 0.5 mole A

+ 0.5 mole B

Material balance for A,B, R

$A = 0.5(1 - X_A)$, $B = 0.5(1 - X_A)$ $R = 0.5\,X_A$,

$N_t = 0.5\,(1 - x_A) + 0.5\,(1 - x_A) + 0.5\,x_A$

$= 1 - 0.5\,x_A$

where N_t is the total number of moles

$$p_A = \frac{0.5(1-x_A)136}{1-0.5x_A},\ p_B = \frac{0.5(1-x_A)136}{1-0.5x_A}$$

$$p_R = \frac{0.5x_A(136)}{1-0.5x_A}$$

At $x_A = 0$, $p_A = 68$ bar $p_B = 68$ bar $p_R = 0$.

$$r = \frac{(0.00665)(0.0088)^2(68^2-0)}{[1+(0.0088\times 68)\times 2]^2} = 4.93\times 10^{-4}$$

$$\frac{1}{r} = 2.028\times 10^3$$

At $x_A = 0.1$, $p_A = p_B = \dfrac{0.5(1-0.1)136}{1-0.5\times .1} = 64.42$

$$p_R/K = \frac{0.25(0.1)136}{(1-0.5\times 0.1)206.7} = 0.0173$$

$$r = \frac{(0.00665)(7.744\times 10^{-5})(64.42^2-0.017)}{[1+(.0088)(64.42)\times 2]^2} = 4.693\times 10^{-4}$$

$$\frac{1}{r} = 2.18\times 10^3$$

At $x_A = 0.2$

$$p_A = p_B = \frac{0.5(1-0.2)136}{1-0.5\times0.2} = 60.44bar$$

$$p_R/K = \frac{0.5(0.2)\times136}{(1-0.5\times0.2)206.7} = 0.073bar$$

$$r = \frac{(0.00665)(.0088)(60.44^2-0.073)}{(1+.0088\times60.44\times2)^2} = 4.41\times10^{-4}$$

$$\frac{1}{r} = 2.26\times10^3$$

$$\frac{W}{F_{AO}} = \int_0^{0.2} \frac{dx_A}{r_A}$$

Using simpsons 1/3 rule

$$I = \frac{0.1}{3}[2.028 + 2\times2.18 + 2.26]\times10^3 = 2.88\times10^2$$

$$W = (100\times.5)(2.88\times10^2)Kg = 14.4\times10^3 Kg = 14.4 \text{ tons of catalyst}$$

2.7 Determination of Some Engineering Properties of Solid Catalysts

The major engineering properties of catalysts are:

1. External surface area, S_o
2. Internal pore surface area, S_g
3. Void volume ($\in_P$) solid densities (ρ_S and ρ_P)
4. Pore size and pore size distribution
5. Physical structure of commercial catalysts

2.7.1 External surface area

The external surface area, S_o per unit volume of a spherical particle of diameter d_P in cm is

$$S_o = \frac{\pi d_P^2}{\frac{\pi d_P^3}{6}} = \frac{6}{d_P} cm$$

If the particle densities ρ_P, the surface area per gm of particles, S_g^o can be given as

$$Sg^0 = \frac{6}{\rho_P d_P}, \frac{cm^2}{gm} \tag{2.55}$$

2.7.2 Internal (porous) surface area by BET method

The standard method for estimating porous surface area of a catalyst is their BET method based on the amount nitrogen gas adsorbed at equilibrium at the normal boiling point (–195.8°C).

BET (Brunauer, Emmett and Teller) Equation is expressed as,

$$\frac{p}{v(P_O - p)} = \frac{1}{v_m C} + \frac{(C-1)p}{Cv_m P_O} \tag{2.56}$$

p_o is the saturation or vapour pressure of the component adsorbed and C is a constant for a particular temperature and gas-solid system, v is the volume of the gas adsorbed and v_m is the volume of the gas adsorbed per gm of catalyst to form a complete monolayer on the surface of the catalyst,

A plot of $p / v(P_o - p)$ vs p / p_o would give a straight line with intercept,

$I = 1/v_{mc}$ at $P/p_0 = 0.0$

and slope, $S = (C - 1) / v_m c$

Solving these equation

$v_m = 1/(I + S)$...where I and S are the intercept and slope respectively (2.57)

If α is the surface area covered by one adsorbed molecule, the total surface area is given as,

$$S_g = \left[\frac{v_m N_O}{V}\right] \propto \tag{2.58}$$

where N_o is the avogadro's number, V is 22,400 cm^3/gmole.

Assuming that adsorbed molecules are spherical, α can be expressed as,

$$\propto = 1.09\left[\frac{M}{N_O \rho}\right] \tag{2.59}$$

where M is the Molecular wt and ρ is the density of the adsorbed molecules. For N_2 gas at –195.8°C, $\rho = 0.808$ g/cm^3 and the area per molecule is 16.2×10^{-16} cm^2. Putting these values in the equation (2.59), the surface area per gm is

$$S_g = 4.35 \times 10^4 \ cm^2/g \text{ solid adsorbent} \tag{2.60}$$

2.7.3 Void volume and solid density

The void volume is accurately measured by Helium –Mercury displacement method.

If m_p is the mass of catalyst placed in the chamber, volume of helium displaced by the sample is V_{He} and that displaced by mercury is V_{Hg}, the pore volume,

$$Vg = (VHg - VHe) / mp \tag{2.61}$$

The helium volume is a measure of the true density of the solid catalysts, ρs

$$\rho s = mp / VHe \tag{2.62}$$

Now the porosity of the particle, $\in_p$ is calculated by the equation

$\in_p$ = void volume of particle / total volume of particle

$$= \frac{m_p v_g}{m_p vg + m_p(1/\rho_S)} = \frac{V_g \rho_S}{1 + v_g \rho_S} \tag{2.63}$$

Again,
$$\in_p = \frac{void\ volume}{total\ volume} = \frac{V_g}{1/\rho_p} = \rho_p V_g \tag{2.64}$$

The sintered and pelleted catalyst contains two types of void regions. The void space within the particles are commonly termed micropores and the void space between the particles are called macropores. Such materials are said to contain bidisperse spores systems.

Two types of porosity may be defined as

$$\in_M = macro\ porous\ void\ fraction$$

$$= \frac{macropore\ volume}{total\ volume} = \frac{(V_g)M}{1/\rho_P} \tag{2.65}$$

Microporosity,
$$\in_\mu = \frac{(V_g)\mu}{1/\rho_p} = (V_g)_\mu \rho_P \tag{2.66}$$

Solid fraction, $\in_S$ is then obtained by the relation

$$\in_5 = 1 - \in_M - \in\mu \tag{2.67}$$

Void fraction of particles, $\in_P$ is

$$\in_P = \frac{(V_g)_P}{1/\rho_P} = (V_g)_p \rho_P \tag{2.68}$$

2.7.4 Pore size and pore size distribution

Very often the void spaces inside a particle are not uniform in size and length and normally interconnected. Void spaces have complex and random geometry.

According to a simplified model of the pore structure, the void spaces are simulated as cylindrical pores. The size of the void space is interpreted as a radius, a of a cylindrical pore. The distribution of pore volume is determined in terms of the pore radius (a). There are two method for measuring the distribution of pore volume. The Mercury porosimeter determines the pore size distribution from 100 A to 10,000 A.

The nitrogen adsorption method determines the pores not larger than 200 A.

Average pore size, $\bar{a}$ may be determined from the total surface area and pore volume in a hypothetical particle. They can be expressed as

$m_p S_g = (2\pi a L)^n$

$m_p V_g = (\pi a^2 L)^n$

where mp and n are the mass and the number of particles

$$\text{So, } \bar{a} = 2V_g / S_g \tag{2.69}$$

The volume average pore radius, $\bar{a}$ is

$$\bar{a} = \int_0^{vg} a\,dv / V_g \tag{2.70}$$

using pore volume data in the form of pore volume distribution (dV/dlog a) vs. a in $\dot{A}$.

Problem 2.7

Low temperature (–195.8°C) nitrogen adsorption data were obtained for Fe-Al_2O_3 ammonia catalyst.

The BET data for 50.4 gm sample are:

Press, mm Hg	8	30	50	102	139	148	233	258	330	442	480	507	350
Volume adsorbed, cm^3 at 0°C and 1 atm	103	116	130	148	159	163	188	198	221	270	294	316	365

BET equation given as,

$$\frac{p}{v(p_0 - P)} = \frac{1}{v_m C} + \frac{(C-1)P}{C v_m P_O}$$

Solution:

$P \times 10^4 / v (760 - p)$	3.54	10.47	14.83	25.9	51.48	63.41
P/760	0.039	0.134	0.1947	0.3394	0.5815	0.667

Intercept from the plot of $P \times 10^4 / v (760 - p)$ vs p / P_o

$I = 1.5 \times 10^{-4}$, $S = 66.66 \times 10^{-4}$

$v_m = 10^4 / (1.5 + 66.66) = 146.75 \text{ cm}^3 / 50.4 = 2.91 \text{ cm}^3 / \text{gm}$ of cat.

$S_g = 4.35 \times 10^4 (2.91) \text{ cm}^2/\text{gm}$

$= 12.66 \times 10^4 \text{ cm}^2/\text{gm} = 12.66 \text{ m}^2/\text{gm}$ of catalyst

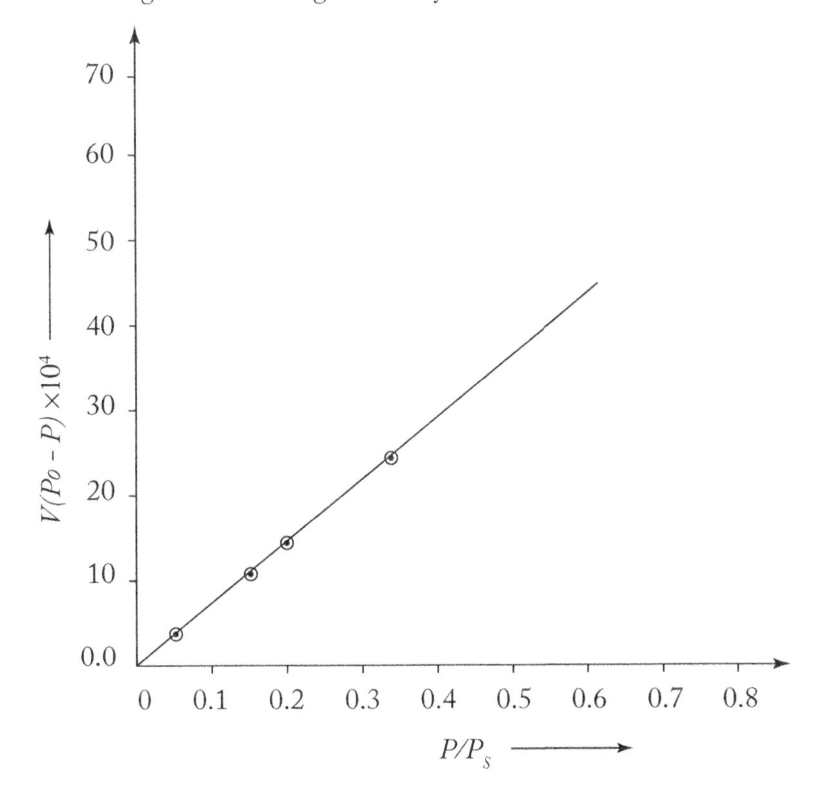

Fig. 2.4 Plot of BET equation

Problem 2.8

The true density of the solid material in activated alumina particles is 3.67 gm/cm^3. The density of the solid determined by mercury displacement is 1.547 gm/cm^3. The surface area by adsorption measurement is 175 m^2/gm.

The bulk density of a bed of the alumina particles is 250 cm^3 graduate is 0.81 gm/cm^3.

Calculate

(a) The pore volume per gram, V_g (b) porosity of the particles, $\in_P$, (c) the mean pore radius ($\bar{a}$)
(d) what fraction of the total volume of bed is void space between the particles ($\in_b$)
(e) what fraction is void space within the particles.

Solution:

(a) $$V_g = \frac{1}{\rho_P} - \frac{1}{\rho_S} = \frac{1}{1.547} - \frac{1}{3.675} = 0.6465 - 0.2721 = 0.3744\frac{cm^3}{gm}$$

(b) $$\in_P = \frac{pore\ volume}{total\ volume} = \frac{V_g}{1/\rho_P} = \rho_P V_g = (1.547)(0.3744) = 0.579$$

(c) $$\bar{a} = \frac{2V_g}{S_g} = \frac{2(0.3744)}{175\times10^4} = 42.47\times10^{-8}\ cm = 42.47\dot{A}$$

(d) Volume of the bed of the particles

$$V_b = \frac{1}{\rho_b} = \frac{1}{0.81} = 1.234\frac{cm^3}{gm}$$

$$\text{Volume of the particles} = \frac{1}{\rho} = \frac{1}{1.547} = 0.6464\frac{cm^3}{gm}$$

$$\text{Void volume of the bed} = 1.234 - .6464 = 0.5876\frac{cm^3}{gm}$$

$$\in_b \text{ fractional void space between the particle} = \frac{0.5876}{1.234} = 0.4761$$

(e) Fractional void space within the particle, $\in_P = 0.81(\rho_P)\times0.374(V_g) = 0.3032$

2.8 Effect of External Transport in Solid Catalyst

When catalytic reactions are carried out in a packed bed reactor with pelleted catalyst particles, there may arise concentration and temperature difference between the bulk of the reactant fluid and the catalyst surface. Let us consider the external diffusion and hear transfer effects in catalysts.

2.8.1 External diffusion and heat transfer

Consider an irreversible reaction on a solid catalyst of first order. At the steady state, the rate, r_p, expressed as moles per unit mass of the pellet per unit time, may be expressed in terms of diffusion rate from the bulk of the gas to the catalyst surface, and this rate is equal to the reaction rate at the catalyst surface:

$$r_p = k_m a_m (C_b - C_s) = k_s C_s \tag{2.71}$$

where C_b and C_s are the concentrations in the bulk and at the catalyst surface, k_m is the mass transfer coefficient, a_m is the external surface area of the pellet per unit weight of thecatalyst (cm²/gm), ks (cm/sec) is the surface reaction rate constant.

Now solving for C_s, we have

$$C_s = \frac{k_m a_m}{k_S + k_m} C_b \tag{2.72}$$

Or

$$r_P = k_O C_b = \frac{1}{1/k_S + 1/K_m a_m} C_b \tag{2.73}$$

Where

$$\frac{1}{k_O} = \frac{1}{k_s} + \frac{1}{k_m am} \tag{2.74}$$

where r_p is called the global rate where reaction and diffusion are combined, k_o is the overall reaction rate constant.

Now the mass transfer coefficient, $k_m a_m$ can be evaluated from the correlation in terms of jd factor which is a function of particle Reynolds number dp G/μ.

$$J_D = \frac{k_m \rho}{G}\left(\frac{am}{at}\right)\left(\frac{\mu}{\rho D}\right)^{0.67} \tag{2.75}$$

And

$$J_H \text{ or } J_D = \frac{0.458}{\epsilon_B}\left(\frac{dpG}{\mu}\right)^{-0.407} \tag{2.76}$$

The a_m/a_t is the ratio of effective mass transfer area and total external area of the particles.

where G = mass velocity based on cross section area of empty reactor).

dp = diameter of the catalyst particle

ϵ_B = void fraction of the bed

μ = viscosity of fluid

ρ = density of the fluid

D = molecular diffusivity

Heat transfer between a fluid and particle surface in a packed bed due to molecular and convective process can be given in terms of jH factor,

$$j_H = \frac{h}{c_p G}\left(\frac{am}{at}\right)\left(\frac{CP\mu}{kf}\right)^{0.67} \quad (2.77)$$

j_H is also a function of particle Reynolds number as given by equation (2.76)

2.8.2 Heat transfer effect in the reaction of nonporous catalysts

The exothermic reactions may create the difference between bulk temp, T_b and catalyst surface temperature, T_s.

The unknown surface temperature can be predicted from the following equations by balancing the heat generated due to reactions to the heat removal at the steady state.

$$k_m am(C_b - C_s)(-\Delta H) = ha_m(T_S - T_b) \quad (2.78)$$

Eliminating km and h from equations (2.67) and (2.68) we have,

$$T_S - T_b = (C_b - C_S)\Delta H\left(\frac{C_p\mu / kf}{\mu / \rho D}\right)^{0.67}\left(\frac{J_D}{J_H}\right) \quad (2.79)$$

Further simplification may be made when Lewis number, the ratio of Prandtle & Schmidt number and the ratio of J_D / J_H are approximately one.

$$(T_S - T_b) = \frac{-\Delta H}{C_P\rho}(C_b - C_S)$$

where C_p and ρ are the specific heat and density of the fluid, ΔH is the heat of reaction.

By calculating $(C_b - C_s)$ and $(T_s - T_b)$ from the previous equations, we can predict whether there are significant effects on the rate of the reaction.

Knowing the global rate, or measured rate, r_p and other properties of the fluid, we can estimate the concentration and temperature gradients in the reaction system. The working equations may be given from the previous equations:

$$C_b - C_s = r_P \frac{(\mu / \rho D)^{0.67}}{at(G/\rho)J_D} \quad (2.80)$$

and

$$T_s - T_b = \frac{r_P}{a_t}\frac{(-\Delta H)(C_P\mu / k_f)}{J_H C_P G} \quad (2.81)$$

If the concentration, $(C_b - C_s) / C_b$ is less than 0.1, diffusion effect is negligible.

Since the exponential effect of temperature on the rate is large, small values of $(T_s - T_b)$ may have significant effect on the rate.

Problem 2.9

The reaction: $H_2 + ½ O_2 \rightarrow H_2)$ takes place on a catalyst of platinum on aluminia in the form of a spherical pellet of diameter of 1.8 cm (weight of each pellet is 2 gm). The reaction rate has been given as

$$rp = 0.327 (p_{o2})^{0.804 Exp (} - 22{,}237/RT) \text{ in kmoles } O_2 / (\text{kg cat}) (s)$$

where p_{O_2} is in bar and T is oK.

(a) Calculate the global reaction rate at a location in a packed bed where the bulk condition are

$$\left(p_{O_2}\right)_b = 0.06\,bar,\ \left(p_{H_2}\right)_b =.0\ 94\,bar$$

$P_t = 1.0$ bar, $T_b = 100^0C$

(b) Are the external mass and heat transfer negligible? The following data are given $G = 0.338$ kg/(m^2) (s), DAB = 1.15×10^{-4} m^2/s

The properties of gas mixture are those of hydrogen at T = 373 K

$$C_{PH_2} = 18.8KJ / kgK, \mu_{H_2} = 10^{-5} P_a S \ -\Delta H = 2.42 \times \frac{10^5 kJ}{kmol}, kf = 0.214 J / (m)(S) / (\ )$$

Solution:

(a) Assuming Ts = 373°C

$$r_p = 0.327(0.06)^{0.804} e \frac{-22{,}237}{(8.314)(373)}$$

$$= 2.617 \times 10^{-5} \frac{kmole\, O_2}{(kgCat)(S)}$$

(b) To estimate the concentration and temperature difference between the bulk and catalyst surface, we have to use the two following equations.

$$C_b - C_S = \frac{r_P}{at} \frac{(Sc)^{2/3}}{\left(G/\rho\right)^{J_D}}$$

$$T_S - T_b = \frac{r_P}{a_t} \frac{(-OH)(P_r)^{2/3}}{J_H C_P G}$$

$$\rho_{H_2}(373°K) = \frac{2}{22.4} \times \frac{273}{373} = 0.0645 \frac{kg}{m^3}$$

$$R_{e_p} = \frac{dPG}{\mu} = \frac{(1.8 \times 10^{-2})(0.338)}{10^{-5}} = 608$$

$$J_D = J_H = \frac{0.458}{0.5}(R_{e_p}) - 0.407$$

$$= \frac{0.458}{0.4}(608)^{-0.407}_{(0.0736)} = 0.083$$

$$P_r = C_P \mu / K_f$$

$$= \frac{(18.8)(10^{-5})}{0.214 \times 10^{-3}} = 0.878$$

$$S_C = \frac{\mu}{\rho DAB} = \frac{10^{-5}}{(0.0645)(1.15 \times 10^{-4})} = 1.348$$

$$mp / at = 4\pi r^2 / mp = 4\pi (0.9)^2 / 2 = 5.08 \frac{m^2}{gm}$$

$$= 0.508 m^2 / kg$$

$$P_r = \frac{C_P \mu}{k_f} = \frac{(18.8)(10^{-5})}{0.214 \times 10^{-3}} = 0.878$$

$$C_b - C_S = \frac{r_p}{at} \frac{(S_C)^{2/3}}{(G/P)^{J_D}} = \frac{(2.617 \times 10^{-5})(1.348)^{2/3}}{(0.508)\left(\frac{0.338}{0.0648}\right)(.083)}$$

$$= \frac{\left(2.617\times10^{-5}\right)1.2214}{(0.508)(5.216)(.083)} = 1.458\times10^{-4}$$

$$C_b = \frac{0.06}{(.08311)(373)} = 1.935\times10^{-3}\,\frac{k.moles}{m^3}$$

$$\frac{(C_b - C_s)}{C_b} = 0.075$$

Again $(p_b - p_s) = RT\,(C_b - C_s)$

$$= (0.08311)(373)\left(1.458\times10^{-4}\right)bar$$

$$= 4.52\times10^{-4} = 0.000452\,bar$$

So, the external mass transfer is negligible.

Now, $T_s - T_b = \frac{r_p}{at}\frac{(-\Delta H)(P_r)^{2/3}}{J_H\,C_P\,G}$

$$= \frac{\left(2.617\times10^{-5}\right)\left(2.4\times10^{5}\right)(0.878)^{2/3}}{(0.508)\times(0.083)(18.8)(0.338)} = 21.5$$

So, Ts = 373 + 21.5 = 394.5

$$r_P = 0.327(0.06)^{0.804}\,e\frac{-22,237}{(8.314)(394.5)}$$

$$= 0.327\times(.1041)\left(1.136\times10^{-3}\right) = 3.867\times10^{-5}$$

$$Y = \frac{r_P}{r_{P_0}} = \frac{3.867\times10^{-5}}{2.617\times10^{-5}} = 1.47$$

Where Y is the ratio of rp to rp0

So the heat transfer resistances are not negligible.

2.9 Effect of Internal Transport in Porous Solid Catalysts

When a reaction takes place in a porous catalyst pellet there may be concentration and temperature gradients inside the porous pellet. It is necessary to evaluate the average rate per pellet based on the concentration and temperature difference between the surface and the inner pore of the solid. To predict the concentration gradients & temperature gradients, it is necessary to predict two important transport properties – the effective diffusivity, De and the effective thermal conductivity, ke.

2.9.1 Determination of effective diffusivity, De

The diffusivity inside pores may be of two types depending on the pore structure of the pelleted catalyst.

If the pores are larger than 200 Á, **the molecular diffusivity,** D_{AB} become predominant and for the pores above 200 Á, **the K nudsen diffusivity,** Dk is important. For bidisparse catalysts system, both molecular diffusivity and Knudsen diffusivity are to be taken into consideration.

The molecular or bulk diffusivity, D_{AB} is given by Chapman and Enskog formula at moderate temperature and pressure,

$$D_{AB} = 1.8583 \times 10^{-3} \frac{T^{3/2}\left(\frac{1}{MA} + \frac{1}{MB}\right)^{1/2}}{P_t \sigma_{AB}^{2} \Omega_{AB}} \tag{2.82}$$

where D_{AB} = bulk diffusivity cm²/sec, T = temperature, K, M_A and M_B = molecular weights of gases A and B. P_t = total pressure in atm, σ_{AB}, Ω_{AB}, are constants in the Lennard – Jones potential energy function for molecular pair, AB, σ_{AB} in Á.

Ω_{AB} = Collision integral and is a function of $k_B T/\epsilon_{AB}$ for real gas and kB = Boltzman constant.

For evaluating the Knudsen diffusion coefficient which involves collision between the molecules and the wall of the pores, the following equation has been used for a cylindrical pore of radius, $\bar{a}$

$$D_{KA} = 9.70 \times 10^{3} \bar{a} \left(\frac{T}{M_A}\right)^{1/2} \tag{2.83}$$

where D_{KA} = Knudsen diffusivity in cm²/sec, $\bar{a}$ is the average pore radius in cm.

The effective diffusivity, De depends upon the porosity of the pellet, ϵ_p and the tortuosity factor, δ, an adjustable parameter, defined as the ratio of tortuous path of the molecule to the shortest path of the molecule. According to the parallel pore model of Wheeler used to represent the monodisperse pore size distribution, the average pore radius, $\bar{a}$ is given as

$$\bar{a} = \frac{2V_g}{5g}$$

where V_g and S_g are the pore volume and pore surface area per gm of catalyst respectively. According to Wheeler, it is assumed that complex pore is an assembly of cylindrical pores of radius, $\bar{a}$

The tortuosity factor, δ is defined as,

$$\delta = x_L / r \tag{2.84}$$

where xL is the diffusion path and r is the radial coordinate in the resultant direction of diffusion. Now the effective diffusivity,

De is defined as,

$$D_e = \frac{\epsilon_P D_K}{\delta} \tag{2.85}$$

Fo poly disperse pore structure containing micropores and macropores, the Random pore model of Wakao & Smith gives the effective diffusivity in the following form

$$D_e = D_M \epsilon_M^2 + \frac{\epsilon_\mu^2 (1 + 3\epsilon_M)}{1 - \epsilon_M} D_\mu \tag{2.86}$$

$$\frac{1}{DM} = \frac{1}{D_{AB}} + \frac{1}{(D_k)_M} \tag{2.87}$$

$$\frac{1}{D\mu} = \frac{1}{D_{AB}} + \frac{1}{(D_k)_M} \tag{2.88}$$

where ϵ_M = macroporosity,

$\epsilon\mu$ = microporosity, D_M & D_μ are the combined diffusivities for macro and micropores respectively

2.9.2 Effective thermal conductivity, ke

The small values of ke is caused by the various void spaces that hinder heat transfer in porous catalyst. Experimental value of ke have been evaluated for different void fraction of macropores. A simple empirical equation for ke has been given as,

$$k_e = k_5 \left(\frac{k_f}{k_s}\right)^{1-\epsilon_M} \tag{2.89}$$

where k_s, k_f are the thermal conductivity of the solid and the fluid, respectively, $\in_M$ is the macro porosity.

2.10 Diffusion and Reaction in Porous Catalyst Pellets

The effect of diffusion in the rate of reaction in a catalyst pellet is quantitatively determined by a term known as effectiveness factor at isothermal condition, η. It is defined as

η = actual rate for the entire pellet / rate at the condition of the catalyst surface

$$= r_p/r_s \quad (2.90)$$

where rp is the observed rate per unit mass of catalyst and rs is the reaction rate at the surface condition.

$$r_p = \eta\, r_s = \eta\, f(Cs) \quad (2.91)$$

2.10.1 Effectiveness factor

The effectiveness factor, η can be determined in terms of effective diffusivity, De and specific surface reaction rate, k, size of the pellet (r_0).

For quantitative evaluation of effectiveness factor, η for a spherical particle, first order kinetics at constant temperature, we make the following mass balance:

For a spherical pellet of radius, r_0 and at the steady state, the rate of diffusion of the reactant, A to the volume element, $4\pi r^2 \Delta r$ is equal to the rate of reaction, we can write,

$$\left(-4\pi r^2 D_e \frac{dC_A}{dr}\right)_{at_r^-} \left(-4\pi r^2 D_e \frac{de}{dr}\right)_{\substack{at \\ r+Ar}} = 4\pi r^2 \Delta r\, \rho_P k_1 C_A \quad (2.92)$$

Now taking the limit $\Delta r \rightarrow 0$ and simplifying, we get the following differential equation,

$$\frac{d^2C_A}{dr^2} + \frac{2}{r}\frac{d.C_A}{dr} - \frac{k_1 \rho_P C_A}{D_e} = 0 \quad (2.93)$$

with the boundary conditions, at r = 0, dc/dr = 0 (2.94)

and at $r = r_o$, $C_A = C_{AS}$

The above equation (2.94) with the given boundary conditions has been solved as

$$\frac{C_A}{C_{A_s}} = \frac{r_0}{r} \frac{Sin\,h(3\phi_S / ro)}{Sin\,h\,3\phi_S} \quad (2.95)$$

where C_{AS} is the concentration of A on the external surface of the catalyst pellet.

where ϕ_s is a dimensionless characteristic parameter, known as Thiele modulus, defined for a spherical particle.

$$\phi_S = \frac{r_o}{3}\left(\frac{k_1 \rho_P}{De}\right)^{1/2} \tag{2.96}$$

Now the diffusion rate of reaction into the pellet at r = r_o is dC_{AS}/dr given as

$$r_P = \frac{1}{mp} 4\pi r_o^2 De\left(\frac{dC_A}{dr}\right)_{r=0} = \frac{3\ De}{r_S\ \rho_P}\left(\frac{dC_{A_S}}{dr}\right) \quad r = ro \tag{2.97}$$

Now mp (mass of the pellet) is

$$mp = \frac{4}{3}\pi r_o^3 \rho_P, r_S = 4\pi r_o^3 k_1 C_{A_S}$$

So $\eta_s = \frac{r_P}{r_S} = \frac{3De}{N_O \rho_P k_1 C_{A_S}}\left(\frac{dC_A}{dr}\right)$ at r = r_o (2.98)

Differentiating equation (2.95) with respect to r

$$\frac{1}{C_{A_S}}\left[\frac{dC_A}{dr}\right] = \frac{r_o}{Sin\,h\,3\phi_3}\left[\frac{1}{r}\cdot\frac{3\phi_s}{r_o} Sin\,h\left(3\phi_s \frac{r}{r_o}\right) - \frac{1}{r^2} Sin\,h\left(3\phi_S \frac{r}{ro}\right)\right] \tag{2.99}$$

Evaluating $\left[\frac{dC_A}{dr}\right] at\ r = r_o$

$$\left[\frac{dC_A}{dr}\right]_{r_o} = \frac{C_{A_S}}{r_o}\left[3\phi_S Cot\,h(3\phi_S)\right] \tag{2.100}$$

Substituting the value of $(d_{CA}/d_r)r = r_o$ in equation (2.99), we get an expression for effectiveness factor for a spherical pellet, η_s

$$\eta_s = \frac{1}{(3\phi_S)^2}\left[3\phi_s \cdot \frac{1}{\tan h\,3\phi_s} - 1\right] = \frac{1}{\phi_s}\left[\frac{1}{\tan h\,3\phi_s} - \frac{1}{3\phi_S}\right] \tag{2.101}$$

Effectiveness factor, η has been plotted for various values of ϕ (Thiele parameter) in fig 2.1. It can be shown that for small values of ϕ_s, $\eta \rightarrow 1.0$

For $\phi_s > 5$, a good approximation for η is

$$\eta_s = 1/\phi_s \tag{2.102}$$

2.10.2 Effectiveness factor for cylindrical pellet (ηc)

The reactant concentration inside the pellet with first order kinetics can be given using a shell balance

$$\frac{d^2C_A}{dr^2} + \frac{1}{r}\frac{dC_A}{dr} - \frac{k_1\rho_P C_A}{D_e} = 0 \tag{2.103}$$

Using the same type of boundary conditions (2.86a), the effectiveness factor for a cylindrical pellet can be deduced as

$$\eta_C = \frac{1}{\phi_C}\frac{I_1(\phi_C)}{I_0(\phi_C)} \tag{2.104}$$

Where I_1 = modified Bessel function of first kind, I_o = modified Bessel function of zero order and ϕ_c = Thiele modulus for a cylindrical pellet.

It is defined as

$$\phi_c = \frac{r_o}{2}\left(\frac{k_1\rho}{D_e}\right)^{1/2} \tag{2.105}$$

where r_o is the radius of the cylinder.

2.10.3 Effectiveness factor for a flat plat (η_L)

The reactant concentration inside a flat plate, can be expressed using a shell balance

$$\frac{d^2C_A}{dx^2} - \frac{k_1\rho_P C_A}{D_e} = 0 \tag{2.106}$$

where x is the thickness of the plate.

The appropriate boundary conditions are

at $x = 0$, $d_{CA}/dx = 0$, at $x = L/2$, $C_A = C_{AS}$ (2.107)

where L is the total thickness of the plate and the effectiveness factor, η_L may be deduced as before

$$\eta_L = 1/\phi_L \tan h\phi_L \tag{2.108}$$

and the Thiele modulus, ϕ_L has been defined as

$$\phi_L = \frac{L}{2}\left(\frac{k_1\rho}{D_e}\right)^{1/2} \tag{2.109}$$

Generalized Thiele modulus for the nth order reaction for a spherical particle may be given as

$$\phi_n^2 = \frac{r_0^2}{9} \cdot \frac{k_n \rho_P}{D_e} C_S^{n-1} \tag{2.110}$$

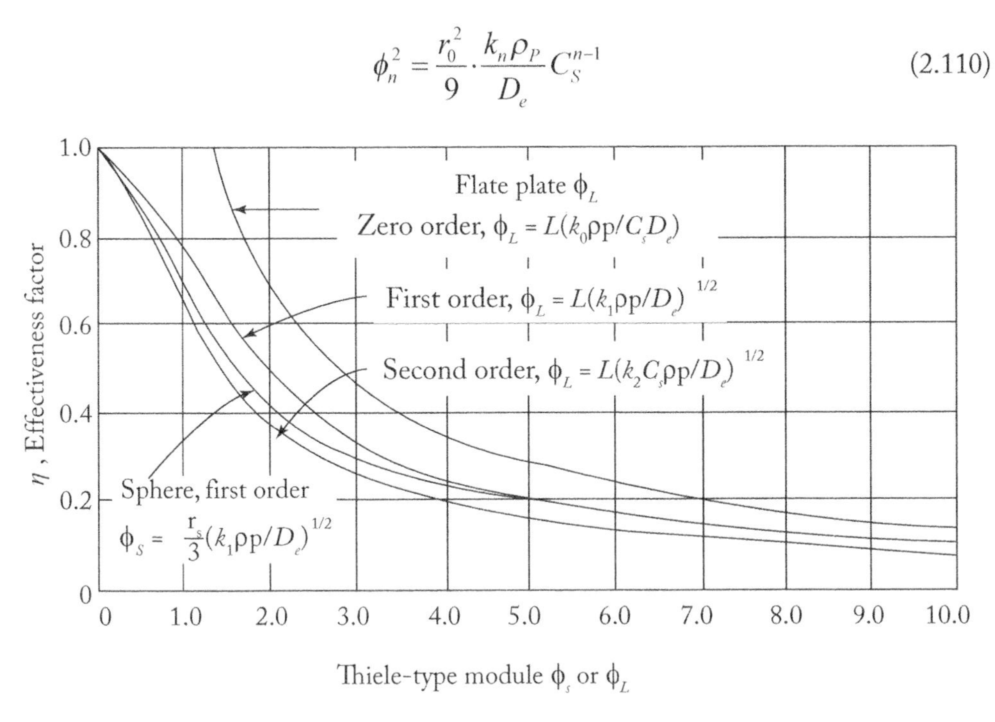

Fig. 2.5 Effectiveness factor for various pellet shapes and kinetic equations (Smith, 1981, p.480)

2.10.4 Prediction of intrapellet diffusion

Weisz has provided a criterion whether the intrapellet diffusion is significant or not

If $\phi_s \leq 1/3$ then $\eta \rightarrow 1.0$

Further, the criterion for negligible intrapellet diffusion can also be given as

$$r_0 \left(\sqrt{\frac{k_1 \rho_P}{D_e}} \right)^{1/2} \leq 1$$

Or

$$\frac{r_0^2 k_1 \rho_P}{D_e} \leq 1 \tag{2.111}$$

Eliminating k_1 from the measured rate at $\eta = 1.0$, $r_p = k_1 C_{AS}$ then the criterion of equation (2.112) can be given as

$$\frac{r_0^2 r_p \rho_P}{\epsilon_P D_e} < 1 \tag{2.112}$$

2.10.5 Effect of intrapellet diffusion on the kinetics

The effect of intrapellet diffusion is significant on the order of reaction, n and the activation energy, E. It can be shown by mathematical analysis that the reaction order and the activation energy are reduced as below.:

$$n' = \frac{1+n}{2}$$

$$\text{Eapp} = \text{Etrue}/2 \tag{2.113}$$

In terms of effectiveness factor, η, and surface concentration, CAs, the global rate, r_A' can be given as

$$r_A' = \eta\ (r_A's) = \eta\ (k_n\ C_{AS}) \tag{2.114}$$

For large values of Thiele modulus

$$\eta = \left(\frac{2}{n+1}\right)^{1/2} \frac{3}{\phi} = \left(\frac{2}{n+1}\right)^{1/2} \frac{3}{R}\left(\frac{D_e CA_s^n}{k_n \rho_P}\right)^{\frac{1-n}{2}} \tag{2.115}$$

and

$$r_A' = \frac{3}{R}\left(\frac{2}{n+1}\right)^{1/2}\left(\frac{D_e}{k_n \rho_P C_s^{n-1}}\right)^{1/2} \cdot k_n C_A^n \tag{2.116}$$

$$= \frac{3}{R}\left(\frac{2}{n+1}\cdot\frac{D_e}{\rho_P}\right)^{1/2} k_n^{1/2} CA_s^{(1+n)/2} \tag{2.117}$$

If the true reaction rate is,

$$r_{AS} = kn'\ C_{AS} \tag{2.118}$$

Equating equations (2.118) and (2.119)

$$\frac{3}{4}\left(\frac{2}{n+1}\cdot\frac{D_e}{\rho_P}\right)^{1/2} k_n^{1/2} CA_s^{(1+n)/2} = k_n' CA_s^{n'} \tag{2.119}$$

Comparing

η' = (1+η)/2

Problem 2.10

The catalytic cracking of gas oil was carried out at 500°C in a fixed bed reactor with catalyst particles of diameter of 0.32 cm.
The properties of the catalyst are: mean pore radius of the particles,
$\bar{r} = 30\,\dot{A}$, $V_g = 0.35$ cm³/g. Catalyst, Mol. Wt of gas oil = 120. The rate data obtained at 500°C and 1 atmospheric pressure, give the first order rate constant, $(k)_{exp} = 0.25$ cm³/(s) (g) catalyst. The tortuously factor, δ = 3.0 Knudsen diffusion is controlling. Calculate the effectiveness factor, η of the particles.

Solution:

T = 500°C = 773°K, Mol. Wt.of gas oil =120,

$\bar{r} = 30\,\dot{A} = 30 \times 10^{-8}$ cm.

$$D_{KA} = 9.7 \times 10^3 \bar{r} \sqrt{T / MA}$$

$$= 9.7 \times 10^3 \left(30 \times 10^{-8}\right) \sqrt{\frac{773}{120}}$$

$$= 7.385 \times 10^{-3}\, cm^2 / Sec$$

$V_g = 0.35$ cm²/g. cat

$\rho_p = 1/\, 1\text{–}0.35 = 1.53$ gm/cm³

$\epsilon_p = \rho_p V_g = (1.53)\,(0.35) = 0.53$

$$De = \frac{DK_A \epsilon_p}{\tau} = \frac{\left(7.385 \times 10^{-3}\right)(0.53)}{3} = 1.30 \times 10^{-3}\, cm^2 / sec$$

$$\phi_S = \frac{r}{3}\sqrt{\frac{k_1 \rho_P}{D_e}} = \frac{0.16}{3}\sqrt{\frac{(0.25)(1.53)}{1.30 \times 10^{-3}}}$$

$$= 0.91, \quad 3\phi_s = 2.73$$

$$\eta = \frac{1}{\phi}\left(\frac{1}{\tan 3\phi_s} - \frac{1}{3\phi_s}\right) = \frac{1}{0.91}\left[\frac{1}{\tan h 2.73} - \frac{1}{2.73}\right]$$

$$\tan h\, 2.73 = \frac{e^{2.73} + e^{-2.73}}{e^{2.73} - e^{-2.73}} = \frac{15.3329 + 0.06521}{15.3329 - .06521} = \frac{15.3981}{15.2677} = 1.0085$$

η = 0.70 Ans.

Problem 2.11

The dehydrogenation of butane was carried out in presence of chromia-alumine catalyst pellets (d_p = 0.32 cm) at 530°C and 1 atmospheric pressure.
The first order rate constant of the reaction, k_1 = 0.94 $cm^3/(s)(g\ catalyst)$
V_g = 0.94 cm^3/gm, Pore radius, $\bar{r} = 110\,\mathring{A}$, the tortuosity factor, δ = 3.0. Find out the effectiveness factor, η of the catalyst.

Solution:

Data given:

Mol. Wt of butane (C_4H_{10}) = 58, dp = 0.32 cm, T = 530°C = 530 + 273 = 803 k, $\bar{r} = 110\,\mathring{A}$

k = 0.94 $cm^3/(s)\ (gm)$ catalyst, V_g = 0.35 cm^3/gm

$\rho_p = 1/0.655 = 1.53\ gm/cm^3$, δ = 3.0

$$D_{KA} = 9.7\times10^3\left(110\times10^{-8}\right)\sqrt{\frac{803}{58}} = 3.97\times10^{-2}\,cm^2/Sec$$

$$\epsilon_P = \rho_P V_g = (1.53)(0.35) = 0.535$$

$$D_e = \left(3.97\times10^{-2}\right)(0.535)/3 = 7.0\times10^{-3}\,cm^2/sec$$

$$\phi_s = \frac{0.16}{3}\sqrt{\frac{(0.94)(1.53)}{7.0\times10^{-3}}} = 0.776$$

$$3\phi_s = 2.33$$

$$\eta = \frac{1}{0.776}\left[\frac{1}{\tan 2.33} - \frac{1}{2.33}\right] = 0.76$$

2.11 Mass and Heat Transfer with Chemical Reaction within the Porous Catalyst Pellet

Since the temperature rises inside the catalyst due to heat of reaction (ΔH), there is temperature gradient inside the catalyst pellet.

2.11.1 Non-isothermal effectiveness factor

Let us consider an exothermic first order reaction with heat of reaction (–ΔH), the temperature distribution in a spherical catalyst of radius, r_o can be expressed, using shell balance,

$$-\left(-4\pi r^2 k_e \frac{dT}{dr}\right)_{at\ r=r} - \left(-4\pi r^2 k_e \frac{dT}{dr}\right)_{at\ r=Ar} = \left(4\pi r^2 \Delta r\right)\rho_P k_1 l\left(\Delta H\right) \tag{2.120}$$

Taking the limit $\Delta r \rightarrow 0$ and simplifying

$$\frac{d^2T}{dr^2} + \frac{2}{r}\frac{dT}{dr} = \frac{k_1 \rho_P C}{k_e} A\left(\Delta H\right) \tag{2.121}$$

with boundary conditions are at r = 0, dT/dr = 0 (2.122)

at $r = r_o$, T = Ts

The temperature distribution equation (2.114) and the concentration distribution equation (2.94) along with their boundary conditions would give the temperature and concentration profile. But numerical solution is necessary for the system as the rate constant is non-linear i.e. k = $A\overline{e}^{E/RT}$.

The two non-linear equations of concentration & temperature can be coupled to obtain significant relations between them.

From the two equations, $k_1\rho_P C_A$ may be eliminated to give

$$D_e\left(\frac{\delta^2 C_A}{dr^2} + \frac{2}{r}\frac{\delta CA}{\delta r}\right) = \frac{k_e}{\Delta H}\left(\frac{\delta^2 T}{\delta r^2} + \frac{2dT}{rdr}\right)$$

or

$$D_e \frac{d}{dr}\left(r^2 \frac{dC_A}{dr}\right) = \frac{k_c}{\left(\Delta H\right)}\frac{d}{dr}\left(r^2 \frac{dT}{dr}\right) \tag{2.123}$$

Integrating twice using the previous boundary conditions as given below:
at r=0 (at the centre), dc/dr = dT/dr = 0
at $r = r_o$ (at the pellet source)
$C_A = C_{AS}$, $T = T_s$
After integration, we obtain

$$T_C - T_S = \frac{\left(\Delta H\right)D_e}{k_e}\left(C_{AO} - C_{AS}\right) \tag{2.124}$$

The equation (2.124) is valid for all kinetics.
Putting $C_{AS} = 0$ at the centre of the pellet,
we **get the maximum temperature rise**

$$\left(T_C - T_S\right)_{max} = -\frac{\left(\Delta H\right)D_e}{k_e}C_{AO} \tag{2.125}$$

where T_C & T_S are the temperatures at the centre, and external surface of the pellet.

Heat of reaction parameter, β from equation (2.118), β is defined as,

$$\beta = \frac{(-\Delta H) D_e C_{AS}}{k_e T_S} \tag{2.126}$$

Weisz and Hicks (1962...) numerically determined the concentration profile inside the pellet and plotted non-isothermal, η′ vs $3\phi_s$ where ϕ_s (the Thiele parameter) is evaluated at the surface temperature (T_s) in fig. 2.6, η′ is non-isothermal effectiveness factor.

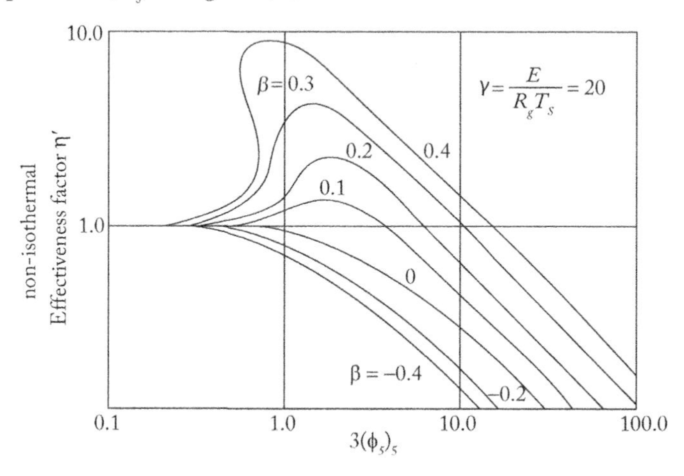

Fig. 2.6 Non-isothermal effectiveness factor for first order reaction in spherical pellets at different values of β & ϕ. (Smith, 1981, p.502)

They also used two dimensionless parameters such as Arrhinius number, γ and the heat of reaction parameter, β, defined as

$\gamma = E / RT_S$ and β has been given by the equation (2.119)

Thiele parameter, $(\phi_s)s$ is evaluated at the surface temperature, defined as,

$$(\phi_s)s = \frac{r_0}{3}\left[\frac{(k_1)_s \rho_P}{D_e}\right]^{1/2} \tag{2.127}$$

The plot of non-isothermal effectiveness factor, η′ versus $3(\phi_s)s$ is presented in fig. 2.6. The highest value of η′ for a given β curve would correspond to a steep temperature profile in the pellet and indicate the region of hot spot. Under these conditions physical processes dominate throughout the whole pellet, if temperature is not controlled, catalyst deactivation may take place.

Carberry's equation:

Carberry has shown that if $(\phi_s)_s > 2.5$, the non-isothermal effectiveness factor, η′ can be expressed in terms of $(\phi_s)_s$, β and γ,

$$\eta' = 1/(\phi s)s\, e^{\beta\gamma} \tag{2.128}$$

and
$$\beta\gamma = \frac{-\Delta H\, D_e C_s}{k_e T_s} \cdot \frac{E}{RT_s} \tag{2.129}$$

Weisz and Hicks extended their criterior for the combined heat and mass transfer. For a first order reaction, the criterion has been given as

$$\Phi = \frac{r_0^2 r_P \rho_P}{C_S D_e} e^{\beta\gamma/(1+\beta)} < 1 \tag{2.130}$$

where β=0, the criterion reduces to the isothermal conditions.

If Φ ≤ 1, the non-isothermal effectiveness factor will be near unity and the intrapellet mass and heat transfer gradients may be neglected.

Problem 2.12

Determine the effectiveness factor for a non-isothermal spherical catalyst pellet in which a first order isomerization is occurring.

Data given:

Interfacial area, a = 100m²/m³, ΔH_R = -800,000 J/mole

D_e = 8.0 x 10^{-8} m²/s, C_{AS} = 10 mole/m³

T_s = 400K, E = 120,000 J/mole

K_e (thermal conductivity of the pellet) = 0.004 J/m.s.k

r_p = 0.005m, specific reaction rate, k_1 = 10^{-1} m/s at 400K.

Density of catalyst pellet = 1.1 x 10^3 gm/m³

Solution:

$$\phi_S = \frac{r_O}{3}\sqrt{\frac{k_1 S_a}{D_e}} = \frac{0.005}{6}\sqrt{\frac{10^{-1}\left(\frac{m}{S}\right)\times 100\frac{m^2}{cm^2}}{8.0\times 10^{-8}}} = 11.18$$

When $\phi_s > 2.5$, carberry criterion gives, $\eta = \frac{1}{(\phi_s)_S} e^{\beta\gamma/5}$

$$\beta = \frac{(-\Delta H) D_E C_{AS}}{k_e T_S} = \frac{800{,}000(J/m\mu)\times 8.0\times 10^{-8}\frac{m^2}{s}\times\frac{10\,moles}{m^3}}{0.004\,J/(m)(s)(K)\times 400(K)} = 0.4$$

$$\gamma = \frac{E}{RT_s} = \frac{120,000 J/mol}{8.314\left(\frac{J}{mole\, K}\right)\times 400^K} = 36.0$$

$$\frac{\beta\gamma}{5} = \frac{(0.4)\times 36.0}{5} = 2.88$$

$$\eta' = \frac{1}{(\phi_S)_s} e^{\beta\gamma/5}$$, where η' is the non-isothermal effectiveness factor

$$\eta' = \frac{1}{11.18} e^{2.88} = 1.59$$

2.12 Summary and Comments

The basic mechanisms of catalytic gas solid reactions have been briefly presented. The rate models have been developed on the basis of Langmuir – Hinshelwood mechanisms for surface reaction controlling and adsorption controlling systems. The effects of transport properties like diffusion and heat transfer on the rate have been presented in terms of isothermal and non-isothermal effectiveness factors. The effects of diffusion on the kinetic parameters like order of reaction and activation energy has been demonstrated by mathematical analysis.

Packed bed reactor design has been illustrated with rate equation based on Hougen Watson model under isothermal condition. The complicated non– isothermal reactor design is not presented in the text.

Problems (Exercise)

2.1 A second order decomposition reaction is carried out

$A \rightarrow R + S$

in a tubular packed bed reactor packed with catalyst pellets of 0.4 cm diameter. The reaction is internal diffusion controlled. Pure reactant A enters the reactor at a superficial velocity of 3m/s at a temperature of 250°C and a pressure of 5 bar. The surface reaction controlling specific reaction rate is 5×10^{-9} $m^6/(g)(s)$ (mole).

(a) Calculate the length of the bed necessary to achieve 80% conversion

(b) If the reactor diameter is 5.08 cm, what is the amount of catalyst required?

Data given:

Effective diffusivity, $D_e = 2.66 \times 10^{-8}$ m^2/s

Bed porosity, $\epsilon_B = 0.4$

Pellet density = 2×10^6 g/m^3

Bed diameter = 1.5 m

2.2 The catalytic hydrogenation of carbon dioxide to produce methane

$CO_2 + 4H_2 \rightleftharpoons CH_4 + 2H_2O$

was carried out in a fixed bed catalytic reactor at a total pressure of 30 bar and 300°C with a feed rate of 100 kg moles of CO_2/hr and the stoichiometric ratio of hydrogen. The rate equation is expressed as

$$r_{EH_4} = \frac{k\,p_{CO_2}\left(p_{H_2}\right)^4}{\left(1 + K_1 p_{H_2} + K_2 P_{CO_2}\right)^5} \; \frac{kmoles\ of\ CH_4}{(hr)(Kgcat)}$$

where p_{CO_2} and p_{H_2} are partial pressures in bar.

At Pt = 30 bar and T = 300°C, the values of the constants are:

$k_1 = 7.0$ kmoles of CH_4 / (kg cat) (hr) bar^{-1}

$k_2 = 1.73$ bar^{-1}, $k_2 = 0.30$ bar^{-1}

Assume that diffusional, thermal and axial dispersion effects are negligible.

Calculate the weight of catalyst required for 20% conversion of CO_2.

2.3 The catalytic dehydration of butanol–1 was established as surface reaction controlling system. The initial rate equation is expressed as

$$r_0 = \frac{kK_A f}{\left(1 + K_A f\right)^2} \; \frac{Kg\ moles}{(hr)(kg\ catalyst)}$$

The following data are given:

r_o	0.27	0.54	0.70	0.76	0.52
p bar	15	465	915	3845	7315
f/p	1.00	0.88	0.74	0.43	0.46

where f is the fugacity of butanol–1. Evaluate the constants k and K_A.

2.4 Ethylchloride was formed from ethylene and hydrochloride in presence of Zircomium oxide on silica gel catalyst at 177°C,

$C_2H_4 + HCl \rightleftharpoons C_2H_5Cl$

If the surface reaction between adsorbed ethylene and adsorbed HCl controls the kinetics,

Develop a rate equation for the above system. Given the equilibrium constants for the overall reaction at 177°C is 35. Total pressure is 10 bar.

2.5 (a) Calculate the gas-to-particle mass and heat transfer coefficients for a gas whose properties are those of air at 373°K and 1.013 bar flowing through a fixed bed ($\in_b = 0.4$) of catalysts ¼ inch (21 cm) particles. The superficial mass velocity of the gas is 1.42 kg/m2s.

(b) Catalyst particles are used for a reaction on the outer surface at a rate of 3.0×10^{-7} gmoles of product / (gm cat) (s). The heat of reaction is –20,000 cal/ (gmole) (product). Estimate the concentration and temperature differences between the bulk and surface, $C_b - C_s$ and $T_b - T_s$ for a fixed bed reactor.

The properties of air:

Viscosity of air.

$\rho_{air} = 0.94\ kg/m^3$

$\rho_{air} = 10^{-5}\ \rho_{as}$, $D_{air} = 1.15 \times 10^{-7}\ m^2/s$

$C_{Pair} = 30$ J/gmole °C, kf = 0.138 J/m.s.k

Given:

$$\text{(a)}\quad J_D = J_H = 1.77\left[\frac{dPG}{\mu(1-\in_B)}\right]^{-0.44}$$

$$\text{When}\quad J_D = \frac{R_m\rho}{G}(\mu/\rho D)^{2/3}$$

$$J_H = \frac{h}{C_pG}\left(\frac{am}{at}\right)\left(\frac{C_P\mu}{k_f}\right)^{2/3}$$

$$\text{(b)}\quad C_b - C_s = r_P\frac{(\mu/\rho D)^{2/3}}{at(G/P)^{J_D}}$$

$$T_s - T_b = \frac{r_p(-\Delta H)}{at}\frac{(C_P\mu/k_f)^{2/3}}{J_H C_P G}$$

2.6 The pyrolysis of n-octane C_8H_{18} at 450°C yielded an apparent first order, irreversible rate constant, k = 0.25 cm³/(gm)(s). The reaction rate is given as
$r = k\,C_s$ in mol / (gcat) (s)
The data are obtained at 1atm pressure with a mono disperse silica catalyst whose average pore size is 30Á. Other properties of the ½ inch spherical catalyst are:
$S_g = 230\ m^2/g$ catalyst, V_g (pore volume) = 0.35 cm³ /(g. catalyst), tortuisity factor, δ = 2.0.
Using the parallel pore model, determine the effectiveness factor for catalytic reaction system.

2.7 At constant total pressure the first order reversible reaction

$O{-}H_2 \leftarrow p{-}H_2$

At –100°C and 1 atm pressure is carried out using NiO on Vycor catalyst. For particles with an average diameter of 50 micron, rate measurement gave $r_p(y_o - y_p) = 5.29 \times 10^{-5}$ gmole/ (g cat)(s).

Pellets of 1/8 inch long and ½ inch in diameter, contained in a cylindrical reactor and, the rate is $r_p = 2.18 \times 10^{-5}$ gmoles/ (gcat)(s). The density of the pellet is 1.46 g/cm³.

(a) From the experimental rate data, evaluate the effectiveness factor for the pellet.

(b) Using the random pore model to estimate D_e, predict the effectiveness factor for comparison with the answer to part (a). Only micropores ($\bar{a} = 45\bar{A}$) exist in Vycor and the porosity of the pellet was $\in_p = 0.304$.

$$\text{Hint (b)} = \text{De} = D_K \in_P^2, DK = 9.7\times10^3\bar{a}\sqrt{\frac{T}{MA}}$$

$$\phi_C = \frac{r_0}{2}\sqrt{\frac{k_1\rho_p}{D_e}}$$

$$\eta = \frac{1}{\phi_C}\frac{I_1(\phi_C)}{I_0(\phi_c)}$$

2.8 A reaction A→ R takes place on a porous catalyst pellet (spherical pellet), dp = 6mm and the effective diffusivity of the catalyst, De = 10^{-6} m²/s.

How much is the rate decreased by the pore diffusion resistance when the reactant concentration, $C_{AO} = 0.2$ kmole/m³ and the diffusion free kinetics are given as

$$r_A = 0.2C_A^2 \frac{kmole}{m^3\,S}$$

Hint k = 0.2 m³/(kmole)(s)

$$\phi = \frac{r_0}{3}\sqrt{\frac{kC_{A_0}}{D_e}}$$

$$\eta = \frac{1}{\phi}\left[\frac{1}{\tan h3\phi} - \frac{1}{3\phi}\right]$$

2.9 A small experimental packed bed reactor (W=1Kg) using very large recycle of product stream gives the following kinetic data:

$A \rightarrow R$

$C_{AO} = 0.01\ kmol/m^3$

X_A	0.9	0.8	0.7	0.6	0.35
$q \times 10^3\ m^3/hr$	5	20	65	133	540

Find the amount of catalyst needed for 75% conversion for a flow rate of 1 kmole of A/hr with $C_{AO} = 0.08\ kmol/m^3$ in the feed stream.

(a) In a packed bed reactor with no recycle of exit stream

(b) In a packed bed reactor with very high recycle

[: assume 2nd order]

References

Brunauer, S. Ph.H. Emmet, E. Teller, J. Am. Chem. Soc. 60, 309, 1938

But, John B, Reaction Kinetics and Reactor Design, Prentice Hall, 1980

Carberry, James J. Chemical and Catalytic Reaction Engineering, Mc. Graw Hill, 1976

Doraiswamy, L.K. and M.M. Sharma, Heterogenous Reactions, Vol – I, Gas Solid and Solid-Solid reactions, Wiley, 1984.

Doraiswamy, L.K. and M.M. Sharma, Heterogenous Reactions, Vol – II, Gas Fluid – Fluid Solid reactions, Wiley, 1984.

Farrauto, R.J. and C.H. Bartholomew, Fundamentals of Catalytic Processes, Chapman and Hall, 1997

Fogler, H.Scott, Elements of Reaction Engineering, 2nd ed Prentice Hall, 1992.

Froment, Gilbert F. and Kenneth B.Bischoff, Chemical Reactor Analysis and Design, 2nd ed Wiley, 1990.

Gates, Bruce, C. Kitzer, R. James and G.C. Schuit, Chemistry of Catalytic processes, Mc Graw Hill, 1979.

Hill, Charles G.Jr., An Introduction to Chemical Engineering Kinetics and Reactor Design, Wiley, 1977.

Hougen O and K.M. Watson, Chemical Process Principles, Volume III, Kinetics and Catalysis, Wiley 1947

Hinselwood, C.N. and T.E. green, J. Chemical Soc: 129, 730, 1926

Langmuir, I,. J. Am. Chem. Soc. 38, 221, 1916

Lanny D. Schmidt, The Engineering of Chemical Reactions, Oxford University Press, 2010

Lee, Hong H., "Heterogeneous reactor Design, Butterworth, 1985.

Levenspiel, Octave, Chemical Reaction Engineering, John Wiley & Sons, 3rd ed, 1999.

Mc Ketta, John J. Encyclopaedia of Chemical Processing and Design, Marlel Dekker, 1987

Missen, Ronald R, Charles A Mims Bradley A, Savile, Introduction to Chemical Reaction Engineering and Kinetics, Wiley, 1999.

Peterson, E.E. Chemical Reaction Engineering, Prentice Hall, 1965

Pilling, Michael J. and Paul W. Seakin, Reaction kinetics Oxford University Press, 1995

Ramachandran, P.A. and R. V. Chaudhari, Three phase catalytic reactors, Gordon and Breach, 1983

Satterfield, Charles N., Heterogeneous Catalysis in practice, Mc Graw Hill, 1980.

Smith, J. M. Chemical Enginering Kinetics, 3rd edition, 1981, Mc Graw Hill

Thomas, J.M. and W.J. Thomas, Introduction to the principles of Hetergenous Catalysis, Academic Press, 1967

Walas, Stanley M, Reaction kinetics for chemical processes Mc Graw Hill Book Cp, 1959

Westerterp, K.R., W.P.M. Van Swaaij, AACM Beenackers, Chemical Reactor Design and Operation,1993.

Weisz, P.B. and.S. Hicks, Chem. Eng. Sc 17, 265, 1962

CHAPTER 3

Residence Time Distribution, RTD

3.0 Introduction

We have defined in Chapter 2 two ideal reactors-Plug flow and backmix reactor. The first one is based on the assumption that there is no axial mixing or dispersion and in the latter, complete mixing behaviour has been presumed.

The deviation from the above two ideal flow reactors generally occurs due to channeling, recycling of the fluid or by the creation of stagnant zones in the vessel. This type of flow usually reduces the performance of different process equipments like heat exchangers, packed columns and reactors.

In order to measure the extent of mixing in a flowing vessel, we use the concept of residence time distribution or age distribution function.

The fluid elements entering a vessel have ages ranging from 0 to infinity, ∞, the distribution of these ages or residence times for the stream of fluid leaving the reactor is called residence time distribution (RTD), or exit age distribution function, E. This RTD is expressed in such a way that the area under curve is unity.

$$\int_0^\infty E dt = 1 \tag{3.1}$$

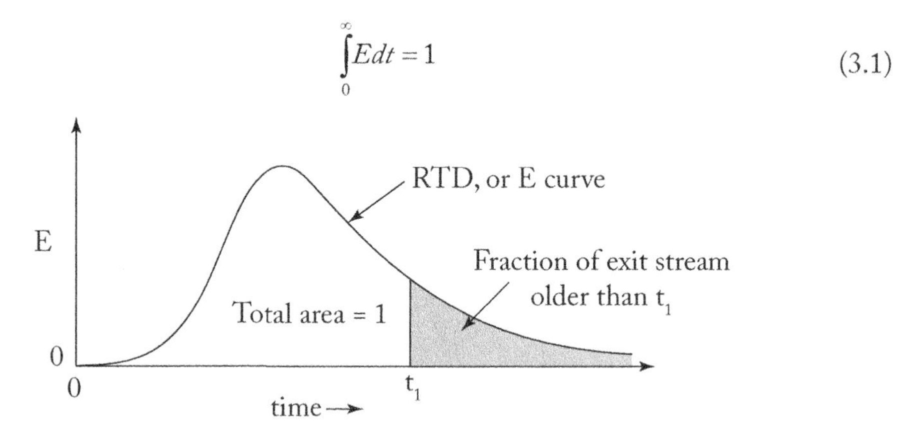

Fig. 3.1 Residence time distribution

The fraction of the fluid younger than age t_1 is,

$$\int_0^t E dt \tag{3.1a}$$

and the fraction of the fluid older than t_1 is

$$\int_{t_1}^{\infty} E\, dt = 1 - \int_0^{t_1} E dt \tag{3.2}$$

3.1 Experimental Method for Determining E

Two general methods based on Pulse input and Step input are employed with a non-reactive tracer like a dye or sodium salt solution which can be detected from the exit stream. In a pulse experiment small amount of tracer (say dye) is injected.

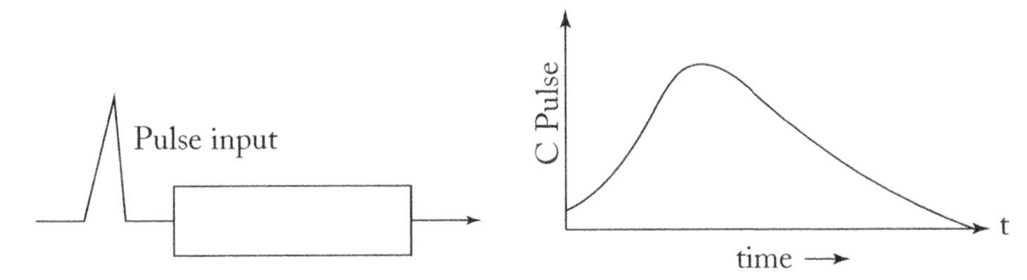

Fig. 3.2 Pulse input and Response curve

In the inlet flow of the vessel and the tracer concentration is recorded from the outlet stream with time; C vs t, is known as C-Curve.

The area under the C-Curve is Q,

$$Q = \int_0^{\infty} C\, dt = \sum C_i \Delta t \tag{3.3}$$

where Q = total amount of dye.

Mean of the C-Curve, $\bar{t}$ (mean residence time) is given as

$$\bar{t} = \frac{\int_0^{\infty} t\, C\, dt}{\int_0^{\infty} C\, dt} = \frac{\sum t_i c_i \Delta t}{\sum c_i \Delta t} \tag{3.4}$$

and the E curve is given as

$$E = \frac{C_{pulse}}{Q} = \frac{C_{pulse}}{\sum c_i \Delta t} \tag{3.5}$$

Dimensionless RTD function, E_θ is

$$E_\theta = \bar{t}E$$

In the step experiment, while the fluid is moving with a flow rate v m^3/s through the vessel, the flow at time, t = 0 is switched from ordinary fluid to another fluid with tracer concentration, C_0 and the outlet tracer concentration, Cstep, is measured and plotted as F curve in the form of (C/C_0) step vs time, t.

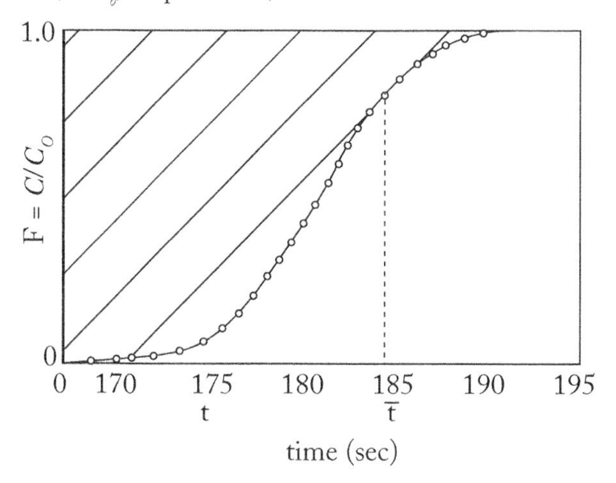

Fig. 3.3 F-Curve

Shaded area in the figure 3.3 = $C_0\bar{t}$

$$\text{or } \bar{t} = \frac{\int_0^{C_0} t\, dC_{step}}{\int_0^{C_0} dC_{step}} = \frac{1}{C_o}\int_0^{C_0} t\, dC_{step} \tag{3.6}$$

- **Relation between the F and E curves:**

  At any time t>0 only coloured fluid (red fluid) in the exit stream is younger than age t. Thus it may be shown

  (tracer of red fluid in the exit stream) = (tracer of exit stream younger than age t)

  writing in terms of F & E, we get

$$F = \int_0^t E\, dt \tag{3.7}$$

and differentiating equation (3.7), we have,

$$dF/dt = E \tag{3.7a}$$

The equation (3.7a) gives the relation between E and F.

Problem 3.1

The following RTD data were obtained for a reactor vessel by using a pulse tracer introduced at the inlet flow. Plot E curve and F curve.

RTD data

t,min	0	5	10	15	20	25	30	35
C, tracer concentration g/cm^3	0	3.0	5.0	5.0	4.0	2.0	1.0	0

Solution: $\Sigma C\Delta t = (3.0 + 5.0 + 5.0 + 4.0 + 2.0 + 1.0)\ 5 = 10.0$ g. min cm^{-3}

$E = C / C\Delta t$	0	0.03	0.05	0.05	0.04	0.02	0.01	0
$F = \Sigma E\Delta t$	0	0.15	0.40	0.65	0.85	0.95	1.0	0

E vs t is plotted in the figure 3.4 and in the same graph F vs t is plotted with F in the right side ordinate as C/C_o.

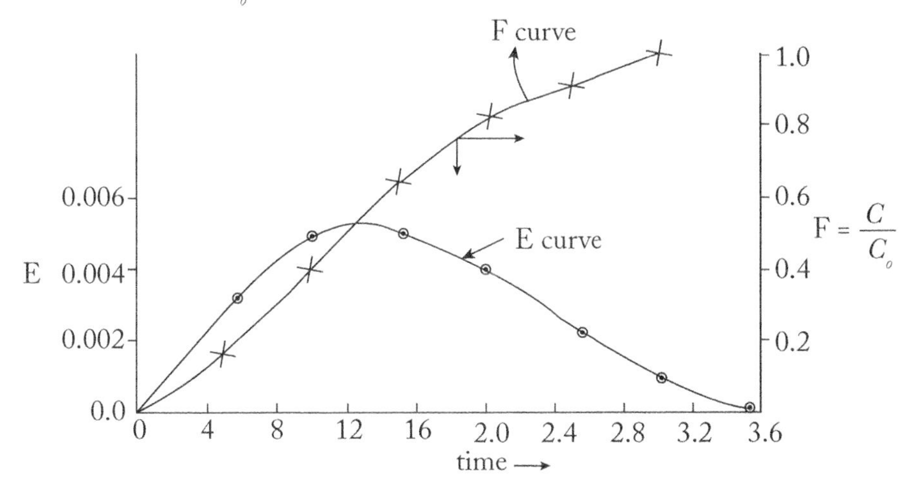

Fig. 3.4 E vs T and F vs t

3.2 Conversion from RTD in Flow Reactors (Segregated Model)

Assuming the fluid to be a microfluid, design equations for plug flow and backmix reactor have been developed.

For macrofluids, it is presumed that clumps of fluid stay in the reactor for different lengths of time, characterized by E-curve. If reaction takes place, each clump is imagined to behave as a batch reactor. Thus the fluid elements have different compositions.

The mean concentration in the exit stream will be function of kinetics and RTD of all the fluid elements.

The outlet concentration $\bar{C}_A$, can be given as

$$\frac{\bar{C}_A}{C_{A0}} = \int_0^\infty \left(\frac{C_A}{C_{A_0}}\right)_{batch} E\,dt \tag{3.8}$$

where C_{AO} is the inlet concentration

For discrete values,

$$\frac{\bar{C}_A}{C_{A0}} = \sum \left(\frac{C_A}{C_{A_0}}\right)_{batch} E\,dt \tag{3.9}$$

For first order,

$$\left(\frac{C_A}{C_{A_0}}\right)_{batch} = e^{-kt} \tag{3.10}$$

For second order

$$\left(\frac{C_A}{C_{A_0}}\right)_{batch} = \frac{1}{1 + k\ CA_0 t} \tag{3.11}$$

3.2.1 Dirac delta function, δ(t-t$_o$)

E function for plug flow can be shown as

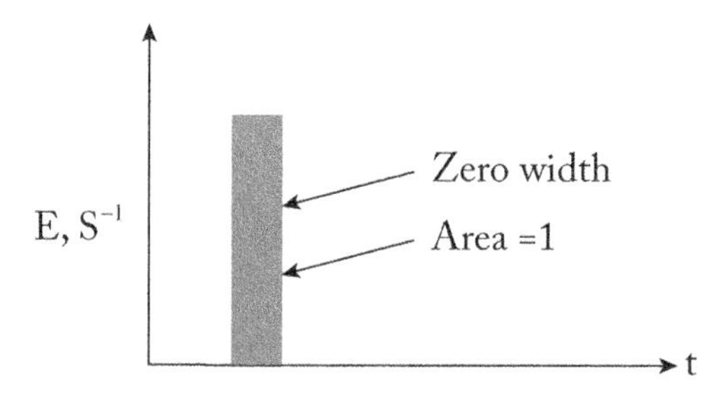

Fig. 3.5 The E-function for plug flow

E curve may be called the Dirac Delta function

Area under the curve is given as

$$\int_0^\infty \delta(t - t_0)\,dt = 1 \tag{3.12}$$

Integration of a function, f(t) with Dirac Delta function as

$$\int_0^\infty \delta(t - t_0) f(t)\,dt = f(t_0) \tag{3.13}$$

It is easier to integrate with a del function than any other.

Problem 3.2

Using the RTD data given in the problem 3.1 and assuming that a first order reaction is taking place in the same vessel, $A \rightarrow R$, $k = 0.05 min^{-1}$.

(a) Calculate X_A from the segregated model using the RTD data
(b) Compare the values with those predicted from ideal plug flow & ideal backmix reactors.

Solution:

RTD data

t, min	0	5	10	15	20	25	30	35
C, g/cm^3	0	3.0	5.0	5.0	4.0	2.0	1.0	0

Working equation for segregated model

$$\frac{\overline{C}_A}{C_{A0}} = 1 - \overline{X_A} = \sum e^{-kt} E(t) \Delta_t$$

Solution:

From the problem 3.1

t, min	0	5	10	15	20	25	30	35
E, min^{-1}	0	0.03	0.05	0.05	0.04	0.02	0.01	0

Given $k = 0.05 min^{-1}$, $\Delta t = 5$ mins, we prepare the following table

t	**e^{-kt}**	**E(t)**	**e^{-kt} E (t) Δt**
0	0	0	0
5	0.778	0.03	0.1167
10	0.6065	0.05	0.1517
15	0.472	0.05	0.1180
20	0.368	0.04	0.0796
25	0.2865	0.02	0.02865
30	0.2231	0.01	0.01115
35	0.1730	0	0

$C_A/C_{A0} = 1 - X_A = \Sigma e\text{-}ktE\Delta t$

$= (0.0 + 0.1167 + 0.1517 + 0.1180 + 0.0796 + 0.02865 + 0.01115 + 0.0) = 0.5068$

Or $X_A = 1 - 0.5068 = 0.4932$

(b)
$$\bar{t} = \frac{\sum t_i C_i}{\sum C_i}$$

or
$$\bar{t} = \frac{(0+15+50+75+80+50+30+0)}{(0+3+5+5+4+2+1+0)} = \frac{300}{20} = 15\,mins$$

For an ideal plug flow with first order kinetics

$$\frac{C_A}{C_{A_0}} = 1 - X_A = e^{-kt}$$

Assuming, $\bar{t} = \tau$

$$1 - X_A = e^{-.05\times 15} = 0.472$$

$$\text{Or } X_A = 0.528$$

For an ideal backmix reactor,

$$\frac{C_A}{C_{A0}} = \frac{1}{1+k\tau}$$

$$1 - X_A = \frac{1}{1+.05\times 15} = 0.5714$$

$$X_A = 0.428$$

So the average conversion from RTD is 0.494 which is in between the values of ideal plug flow & ideal backmix reactor.

3.3 Non-ideal Reactor Models

Different types of models are available to represent whether the flow is close to plug or mixed or somewhere in between the two.

These models are:

1. Dispersion model
2. Tanks-in-series model
3. Two parameter model (Combination of ideal reactors).

3.3.1 Dispersion model

In the RTD experiments the pulse tracer spreads through the vessel and this is characterized by dispersion which is different from molecular diffusion.

Mixing process involves redistribution of material either by slippage or eddies. Its nature is equivalent to molecular diffusion as given by Fick's Law

$$\frac{\delta C}{\delta t} = \mathfrak{D}\frac{\delta^2 C}{\delta x^2} \tag{3.14}$$

where $\mathfrak{D}$ is the diffusion coefficient and x is the axial distance. when this equation is used for mixing or dispersion D is called dispersion coefficient or longitudinal dispersion coefficient.

Mathematically, for tracer concentration C in a vessel for variable t and Z (axial distance) can be expressed as

$$\frac{\delta C}{\delta t} = D\frac{\delta^2 C}{\delta Z^2} \tag{3.15}$$

In dimensionless form, where
$z = (ut + x)/L$ $\theta = t/\bar{t} = tu/L$

$$\frac{\delta C}{\delta \theta} = \left(\frac{D}{uL}\right)\frac{\delta^2 C}{\delta Z'^2} - \frac{\delta C}{\delta Z} \tag{3.16}$$

where D/uL, the dimensionless group, is called the vessel dispersion number or the reciprocal of it, uL/D is called the Peclet number, Pe.

This parameter represents the extent of axial dispersion.

When

$$\left.\begin{array}{l}\frac{D}{uL} \rightarrow 0 \\ P_e \rightarrow \infty\end{array}\right\} \text{ negligible dispersion leading to plug flow}$$

$$\left.\begin{array}{l}\frac{D}{uL} \rightarrow \infty \\ P_e \rightarrow 0\end{array}\right\} \text{ large dispersion, hence mixed flow}$$

Typical RTD curves will indicate the extent of mixing in terms of D/uL, as shown in fig. 3.6.

The D or D/uL is determined from the distribution curve quantitatively from the mean and variances.

$$\bar{t} \text{ (mean passage of time) or first moment } = \frac{\int_0^t tC\,dt}{\int_0^t C\,dt} = \frac{\sum t_i C_i \Delta t}{\sum C_i \Delta t} \tag{3.17}$$

$$\sigma^2 \text{ (variance or second moment) } = \int_0^\infty \frac{(t-\bar{t})^2\,Cdt}{\int_0^\infty C\,dt} = \frac{\int_0^\infty t^2 C\,dt}{\int_0^\infty C\,dt} - \bar{t}^2 \tag{3.18}$$

$$\text{Or } \sigma_\theta^2 = \frac{\sum (t-\bar{t})^2\, C_i \Delta dt}{\sum C_i \Delta dt} = \frac{\sum t_i^2 C_i \Delta dt}{\sum C_i \Delta dt} - \bar{t}^2 \tag{3.19}$$

The variance represents the square of the spread of distribution. The larger the variance, greater is mixing and vice-versa and is shown in fig. 3.6

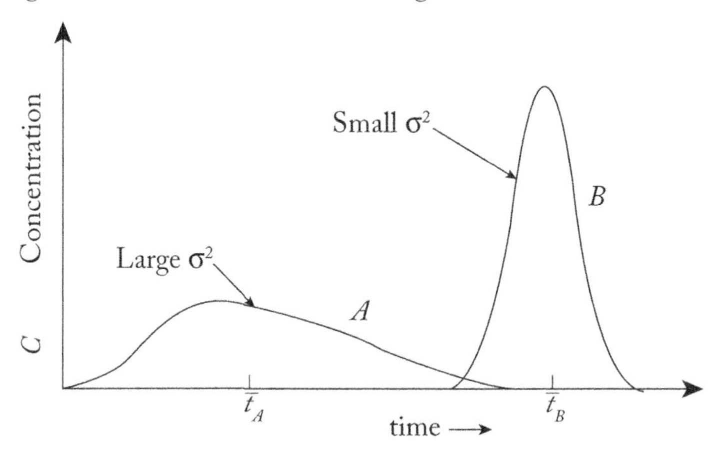

Fig. 3.6 C- Curve with different variances

For small values of D/uL, the dispersion model equation (3.16) in dimensionless parameters is solved with following boundary conditions,

at Z = 0, for θ ≥ 0,

$D\dfrac{dc}{dz}$ – uc = uco

at Z = L for θ ≥ 0, dc/dz = 0.

The concentration is given as a function of D/uL and θ in the following form and θ = L/u

$$C = \frac{1}{2\sqrt{\pi(D/uL)}} exp\left[-\frac{(1-\theta)^2}{4(D/uL)}\right] \tag{3.20}$$

Now E_θ (dimensionless RTD) has been given as,

$$E_\theta = \bar{t}E = \frac{1}{\sqrt{4\pi(D/uL)}} exp\left[-\frac{(1-\theta)^2}{4(D/uL)}\right] \tag{3.21}$$

The above equation represents a family of Gaussian curves called normal curves as shown below for various values of D/uL in fig. 3.7

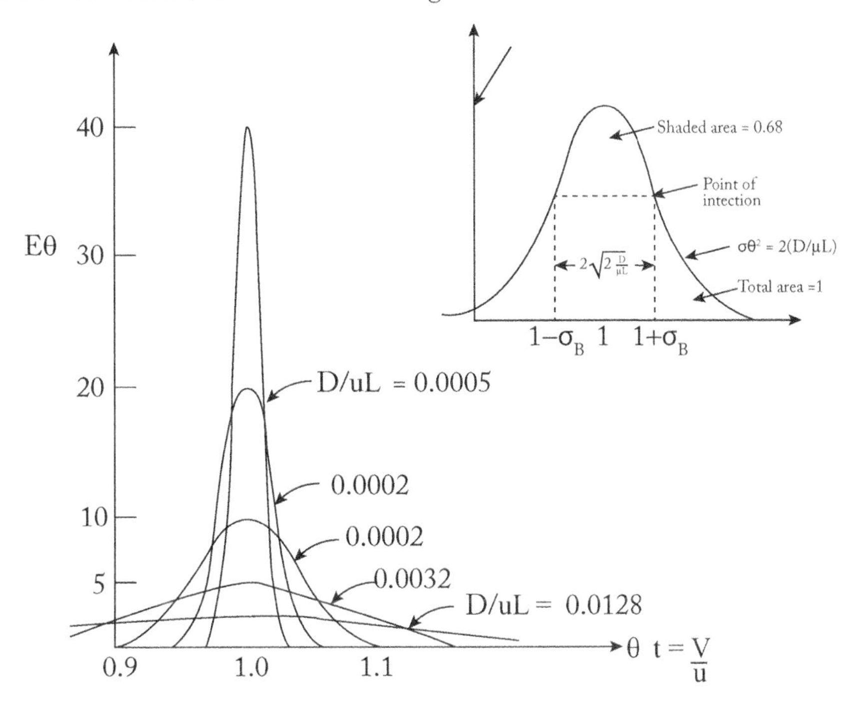

Fig. 3.7 Variation of RTD (E) with t at different values of D/uL (Levenspiel, 1999, p.297)

The dimensionless RTD, Eθ is plotted against dimensionless time, θ for various values of vessel dispersion number, D/uL. As the dispersion number decreases, the distribution curve becomes, skewed and skewed, indicating towards the ideal plug flow as shown in fig. 3.6.

The dispersion number, D/uL can be calculated from the C-curve by calculating its mean and variance or by measuring its maximum height or its width at the point of inflexion or by finding that width which includes 68% of the area.

3.3.2 Large deviation from plug flow when D/uL > 0.01

For two common boundary conditions such as closed vessel and open vessel, D/uL values have been calculated from dimensionless variance, $\sigma\theta^2$ ($\sigma\theta^2 = \sigma^2 / t^{-2}$) for **closed vessel**, the inlet and outlet conditions are plug flow in nature.

D/uL(dispersion number) can be calculated from the following equation,

$$\sigma_\theta^2 = \frac{\sigma^2}{\bar{t}^2} = 2\left(\frac{D}{uL}\right) - 2\left(\frac{D}{uL}\right)^2\left(1 - e^{\frac{-uL}{D}}\right) \quad (3.22)$$

Open vessel:

The open vessel conditions is that flow behaviour is non-ideal at the inlet and outlet and inside the vessel. A section of long pipe may show open vessel behaviour. However, D/uL(dispersion number) can be calculated for open vessel condition from the following equation,

$$\sigma_\theta^2 = \frac{\sigma_\theta^2}{\bar{t}^2} = 2D/uL + 8\left(\frac{D}{uL}\right)^2 \quad (3.23)$$

Limitations:

If the flow deviates greatly from the plug flow i.e. for large values of D/uL, the dispersion model may not be suitable, particularly, when D/uL >1. Moreover, if C-curve is not symmetric and has long tail or double peaks, dispersion model should not be used.

Problem 3.3

Using the RTD data given in problem 3.1, calculate the dispersion number or Peclet number.

t,min	0	5	10	15	20	25	30	35
c, gm/cm3	0	3.0	5.0	5.0	4.0	2.0	1.0	0

Solution:

$\bar{t}$ (mean residence time) $= \frac{\sum t_i C_i \Delta t}{\sum C_i \Delta t} = \frac{\sum t_i C_i}{\sum C_i \Delta t}$

$$= \frac{(0+15+50+75+80+50+30+6)}{(0+3+5.0+5.0+4.0+2.0+1.0)} = \frac{300}{20} = 15 mins$$

$$\text{or } \sigma^2 = \frac{(5\times 25)}{}$$

$$\text{or } \sigma^2 = \frac{(5\times 25)+(5\times 100)+(5\times 225)+(4\times 400)+(2\times 625)+(1\times 900)}{20} - (15)^2 \min^2$$

$$\sigma^2 = \frac{(125+500+1125+1600+1250+900)}{20} - 225$$

$$\sigma^2 = \frac{5500}{20} - 225 = 50(minutes)^2.$$

$$\sigma_\theta^2 = \frac{\sigma^2}{\bar{t}^2} = \frac{50}{225} = 0.22$$

For a closed vessel

$$\sigma_\theta^2 = 2\left(D/_{uL}\right) - 2\left(\frac{D}{uL}\right)^2\left(1 - e^{-\frac{uL}{D}}\right)$$

For a first trial assume D/uL = 0.11

$$RHS = 0.22 - .0242(0.9998) = 0.195 \neq 0.22$$

Assume D/uL = 0.12

RHS = 0.24 – 0.0288 (0.9997) = 0.21

The above value of D/ u L nearly satisfies the equation

Now D/u L is known as Dispersion Number

Peclet number = 1 / Dispersion No. =1/0.12 = 8.33

Since the D/uL > 0.01, there is sufficient backmixing.

3.3.3 Correlation for axial dispersion

The vessel dispersion number can be expressed as the product of two numbers,

$$D/uL = (D/udt)\ (d/L) \tag{3.24}$$

where D/udt is called intensity of dispersion and d/L is a geometric parameter of the vessel.

Now

$$D/udt = f(\text{Schmidt No.})\ (\text{Reynolds number}) \tag{3.25}$$

Reynolds number may be based on dt or dp. Thus $Sc = \mu / \rho D_{AB}$ $Re = dtu\rho/\mu$

When D_{AB} is molecular diffusivity. Experimental evidence indicates that dispersion model can represent flow in packed beds or in pipes. Levenspiel has given correlation-plot of D/udt as a function of dt $u\rho/\mu$, with different values of Schmidt numbers for Laminar and turbulent flow and for gases and liquids as given in figure. 3.8.

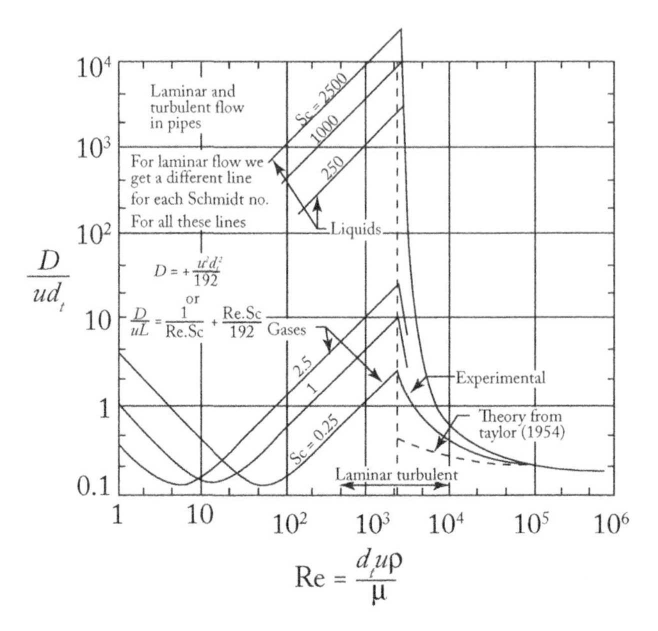

Fig. 3.8 Correlation for the dispersion of fluids flowing in pipes with Re and Sc (Levenspiel, 1999, p.310)

Intensity of dispersion, D/udt, for streamline flow in pipes has been plotted in fig. 3.9 versus Bodenstein number (Re.Sc) according to Taylor model. The pipe should be long to attain radial uniformity of a pulse of tracer. For liquids the molecular diffusion prominently affects the rate of dispersion in laminar flow.

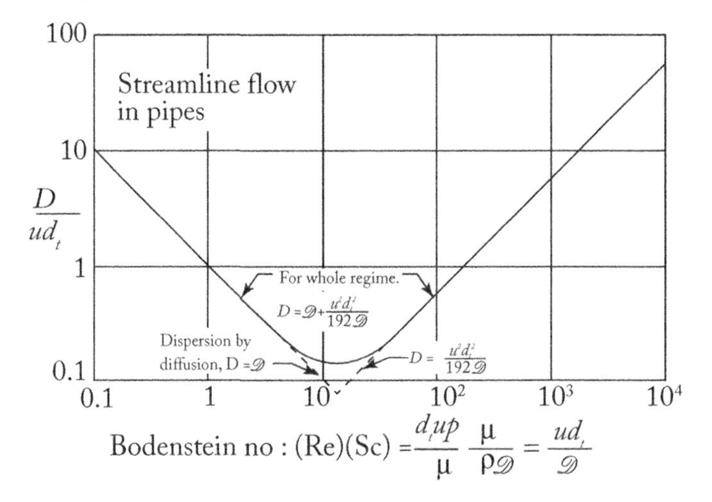

Fig. 3.9 Plot of D/udt vs. Bodenstein number (Levenspiel, 1999, p.311)

Correlations are also available for flow in a bed of porous solids or coiled tubes, for pulsating flow, for non-Newtonian fluids. Fig. 3.10 shows the plot of

D€/udp(intensity of dispersion). Vs particle Reynolds number, dp uρ/μ

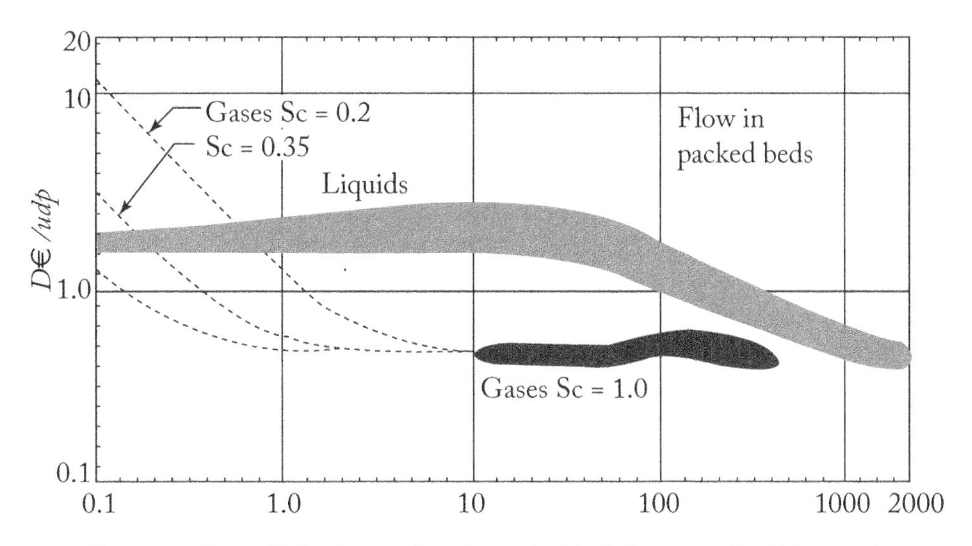

Fig. 3.10 Plot of D€/udp…vs Rep for packed bed (Levenspiel, 1999, p.310)

3.3.4 Chemical reaction and dispersion in a tubular flow reactor

Let us assume that there is reaction occurring in a tubular flow reactor along with dispersion.

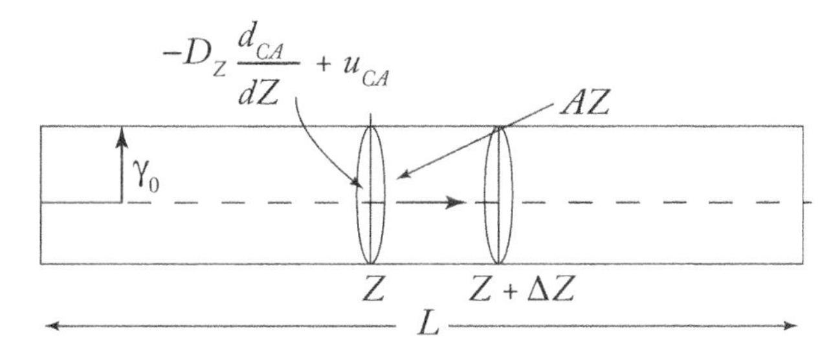

Fig. 3.11 Mass balance in a flow reactor and dispersion & reaction

A steady state mass balance is written for an element volume, $\pi r_o^2 \Delta Z$ with upflow at a uniform velocity

$$\pi r_0^2 \left(-D_Z \frac{d_{CA}}{d_Z} + u_{CA} \right)_{atZ} - \pi r_0^2 \left(-D_Z \frac{d_{CA}}{d_Z} \right)_{x+\Delta Z} - r_c \left(\pi r_0^2 \Delta Z \right) = 0 \qquad (3.26)$$

where rc is the rate of reaction. Dividing by Δz and taking the limit $\Delta Z \to 0$, we will get

$$D_Z \frac{d^2{}_{CA}}{dZ^2} - u \frac{d_{CA}}{dZ} - r_c = 0 \qquad (3.27)$$

For nth order reaction, r = kC_A^n and using dimension less form $Z' = Z / L$ $\tau = L / u = V / q$ the above expression (3.20) becomes

$$\frac{D_Z}{uL}\frac{d^2C_A}{dZ^2} - \frac{dC_A}{dZ'} - k\tau C_A^n = 0 \tag{3.28}$$

The equation (3.30) can be solved for first order reaction with the given boundary conditions,

$$\frac{D_Z}{uL}\frac{d^2C_A}{dZ'^2} - \frac{dC_A}{dZ'} - k\tau C_A = 0 \tag{3.29}$$

At $Z' = 0$, $$-D_z\frac{dC_A}{dZ'} + uC_A = uC_{AO} \tag{3.30}$$

At $Z' = 1.0$, $\frac{dC}{dZ} = 0$

The solution of equation (3.29) is given as

$$\frac{C_A}{C_{AO}} = 1 - X_A = \frac{4q\,exp\left(\frac{1}{2}\frac{uL}{D_Z}\right)}{(1+q)^2\,exp\left(\frac{1}{2}\cdot\frac{uL}{D_Z}\right)} \tag{3.31}$$

Where q =(1+ 4k τ (Dz/u L) ½ (3.32)

in terms of Damköhler number (Da) and Peclet number (Pe), the equation (3.33) becomes,

$$\frac{C_A}{C_{AO}} = \frac{4q\,exp(P_e/2)}{(1+q)^2\,exp\left(P_e\cdot\frac{q}{2}\right) - (1-q)^2\,exp\left(-P_e\frac{q}{2}\right)} \tag{3.33}$$

$$q = \sqrt{1 + 4Da/\rho e} \tag{3.34}$$

Where $D_a = k\tau$ and $P_e = uL / D_Z$

Da physically represents the ratio known as Damköhler number

$$D_a = \frac{(rate\ of\ consumption\ of\ A\ by\ reaction)}{(rate\ of\ transport)} = \frac{kC_{AO}^n}{uC_{AO}} = k\tau C_{AO}^{n-1} \tag{3.35}$$

For small deviation from plug flow, D/uL becomes small, E curve becomes Gaussian and expanding exponential and dropping the higher order terms, the equation (3.35) reduces to,

$$\frac{C_A}{C_{AO}} = exp\left[-k\tau + (k\tau)^2 \frac{D_Z}{uL}\right] \tag{3.36}$$

Comparing the performance of an ideal plug flow reaction,

$\frac{V}{V_P} = 1 + (k\tau) D_Z / uL$ or same CA out in the exit of the reactor and for same

V or τ (3.37)

$$\frac{C_A}{C_{AP}} = 1 + (kT)^2 D_Z / uL \tag{3.38}$$

where V_P = volume of an ideal plug flow reaction
C_{AP} = the outlet concentration from an ideal plug flow reactor.
$k\tau$ = Damköhler number

3.4 Tanks in Series Model

Tanks in series model represent a single parameter (N). The RTD is analysed to determine the number of ideal tanks in series (N) that will give approximately the same RTD as the non-ideal reactor. The RTD will be analysed from a tracer pulse injected into the first three equally sized stirred reactors.

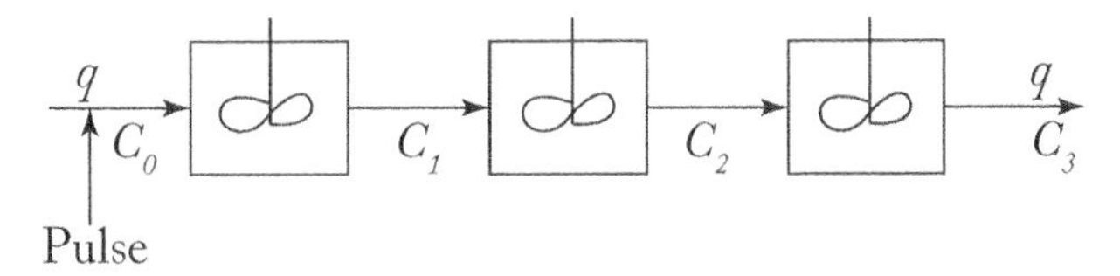

Fig. 3.12 Tanks in series

According to the definition of RTD the fraction of material leaving the system of three reactors that has been in the system between t and t + Δt.

$$E(t)\Delta t = \frac{qC_3(t)\Delta t}{No} = \frac{C_3(t)\Delta t}{\int_0^\infty C_3(t)dt} \tag{3.39}$$

Or

$$E(t) = \frac{C_3(t)}{\int_0^\infty C_3(t)dt} \tag{3.40}$$

Material balance of the tracer in the first tank,

$$V_1 \frac{dC_1}{dt} = -qC_1 \tag{3.41}$$

Integrating the above equation (3.43) to get C_1

$$C_1 = Co\,e^{-qt/V_1} = Co\,e^{-t/t_1} \tag{3.42}$$

where $t_1 = V_1 / q$

The volumetric flow rate, q is constant and all the reactor volumes are equal,

$$V_1 = V_2 = V_3.$$

Material balance of the tracer in the 2nd tank is

$$V_i \frac{dC_2}{dt} = qC_1 - qC_2 \tag{3.43}$$

or

$$\frac{dC_2}{dt} + \frac{C_2}{\bar{t}_1} = \frac{C_O}{\bar{t}_1} e^{-t/\tau i} \tag{3.44}$$

Integrating by using the integrating factor, $e^{t/\tau}$ along with the initial condition, $C_2 = 0$ at t = 0 gives,

$$C_2 = \frac{C_O t}{\tau_i} e^{-t/\tau} \tag{3.45}$$

Proceeding in the same way we have for third tank,

$$C_3 = \frac{C_O t^2}{2\tau_i^2} e^{-t/\tau} \tag{3.46}$$

Substituting equation (3.48) in equation (3.42), we have

$$E(t) = \frac{C_3(t)}{\int_0^{\infty} C_3(t)\,dt}$$

$$\text{Or } E(t) = \frac{t^2}{2\tau_i^3} e^{-t/\tau} \tag{3.47}$$

Extending this to a series of N CSTRs, the RTD, E(t) for N is,

$$E(t)=\frac{t^{N-1}}{(N-1)!\tau^{N}}e^{-t/\tau} \tag{3.48}$$

Now the total reactor volume is NV_i, $\tau_i=\tau/N$, $\theta=t/\tau$ then

$$E(\theta)=\frac{N(N\theta)^{N-1}}{(N-1)!}e^{-N\theta} \tag{3.49}$$

$$\text{Now, } 6_{\theta}^{2}=\frac{\sigma^{2}}{\bar{t}^{2}}=\int_{0}^{\infty}(\theta-1)^{2}E(\theta)do+\int_{0}^{\infty}E(\theta)d\theta=\int_{0}^{\infty}\theta^{2}E(\theta)d\theta-1 \tag{3.50}$$

Putting the values of E(θ) from equation (3.49)

$$\sigma_{\theta}^{2}=\int_{0}^{\infty}\theta^{2}\frac{N(N\theta)^{N-1}}{(N-1)!}e^{-N\theta}-1 \tag{3.51}$$

$$\sigma_{\theta}^{2}=\frac{N^{N}}{(N-1)!}\int_{0}^{\infty}\theta^{N+1}e^{-N\theta}d\theta-1=\frac{N^{N}}{(N-1)!}\left[\frac{(N+1)!}{N^{N+2}}\right]-1=\frac{1}{N} \tag{3.52}$$

So the number of tanks in series, N is given as,

$$N=\frac{1}{\sigma_{\theta}^{2}}=\frac{\bar{t}^{2}}{\sigma^{2}} \tag{3.53}$$

Equation (3.51) may be graphically presented as

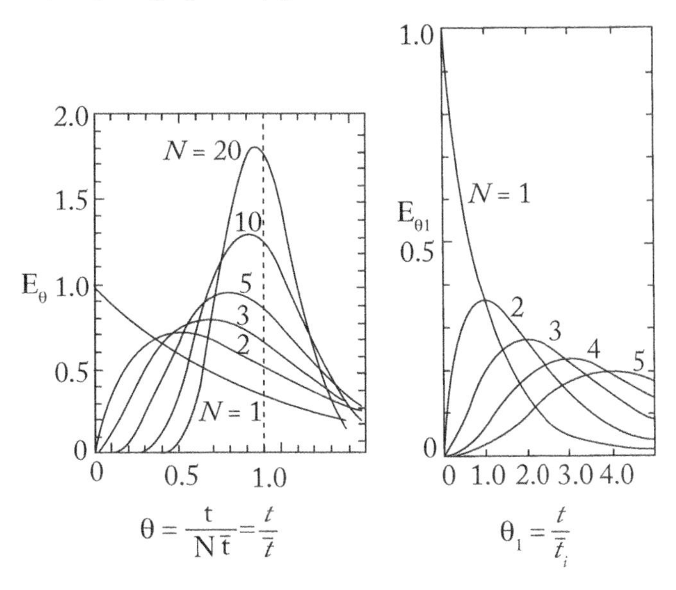

Fig. 3.13 RTD curves for tanks in series model with varying N (Levenspiel, 1999, p.323)

As N increases, the curves become skewed and skewed approaching towards plug flow (N >∞, and there is no dispersion, the system becomes plug flow)

when N = 1; The system becomes an ideal back-mix reactor.

For ideal CSTR based on first order reaction, the performance equation is given as,

$$\frac{C_A}{C_{AO}} = \frac{1}{1+k\tau} \tag{3.54}$$

For N tanks in series,

$$\frac{C_{A_{1\,Ntanks}}}{C_{AO}} = \frac{1}{\left(1+\frac{k\tau}{N}\right)^N} \tag{3.55}$$

where $\tau = \frac{V}{q}$ = average residence time

For small deviations from plug flow (for large N), comparison may be made with plug flow reactor.

For same volume:

$$\frac{C_A Ntanks}{C_{A_P}} = 1 + \frac{(k\tau)^2}{2N} \tag{3.56}$$

For same C_A outlet

$$\frac{V_N tanks}{V_P} = \left(1+\frac{k\tau}{2N}\right) \tag{3.57}$$

Where C_{AP} is the outlet concentration in the ideal plug flow reactor, V_P is the volume of ideal plug flow reactor, V_N tanks is the volume of the reactor, modelled by N number of tanks.

Problem 3.4

Using the RTD data given in Problem 3.1 and assuming a first reaction taking place in the same vessel,

k = 0.05 min^{-1}, $A \xrightarrow{k} R$

(a) Calculate the conversion of A, X_A from dispersion model
(b) Calculate X_A for the reaction assuming that there is small deviation from plug flow
(c) Also calculate the conversion on the basis of tanks-in-series model.

Given dispersion number (D/uL) = 0.12, τ = 15 minutes, τ is the residence time.

Solution:

(a) Conversion from dispersion in model

$$\frac{C_A}{C_{AO}} = 1 - X_A = \frac{4q\,exp\left(\frac{1}{2}P_e\right)}{(1+q)^2\,exp\left(\frac{q}{2}P_e\right) - (1-q)^2\,exp\left(-\frac{q}{2}P_e\right)}$$

Where $P_e = 8.33\left(\frac{1}{Dispersion\ number} = \frac{1}{0.12}\right) = 8.33$

D_a (Damköhler No.) = $k\tau$

$= k\tau = 0.05 \times 15 = 0.75$

$$q = \left(1 + \frac{4Da}{Pe}\right)^{1/2} = \left(1 + \frac{4(.75)}{8.33}\right)^{1/2} = 1.16$$

Now,

$$1 - X_A = \frac{4(1.16)exp\left(\frac{8.33}{2}\right)}{(1+1.16)^2\,exp\left(\frac{1.16}{2}\times 8.33\right) - (1-1.16)^2\,exp\left(\frac{-1.16}{2}\times 8.33\right)} = \frac{298.78}{584.3 - 0.0002} = 0.511$$

Or $X_A = 0.488$

(b) For small deviation from plug flow

$$\frac{C_A}{C_{AP}} = 1 - X_A = exp\left[-D_a + \frac{D_a^{\,2}}{P_e}\right] = exp\left[-0.75 + \frac{0.75^2}{8.33}\right] = 0.505$$

Or $X_A = 0.495$

(c) $N = 1/\sigma\theta^2 = \frac{1}{0.22} = 4.54$

$$\frac{C_A, Ntanks}{C_{AO}} = 1 - X_A = \frac{1}{\left(1 + \frac{Da}{N}\right)^N} = \frac{1}{\left(1 + \frac{0.75}{4.54}\right)^{4.54}} = 0.4995$$

Or $X_A = 0.505$

Problem 3.5

A homogenous gas phase reaction is carried out in a tube packed with inert particles to increase turbulence. The dispersion model fits flow in packed beds as well as flow in empty tubes but the dimensionless parameter, Du/dp should be used. At the flow rate used, the particle Reynolds number is 50. At this Re for gases D/udp = 1.0. The particle diameter is 1.0 cm. To obtain nearly plug flow, it is assumed that when RTD corresponds to that for ten stirred tanks in series.

(a) What length of tube is required?
(b) If we were satisfied with more axial dispersion corresponding to an RTD of that for five stirred tanks in series, how much shorter would be the tube length?
(c) What are the residence times in both cases?

Solution:

(a) $\sigma_\sigma^2 = 1/N = 1/10 = 0.1$

corresponding D/uL is calculated,

$$0.1 = 2\left(\frac{D}{uL}\right) - 2(D/uL)^2\left(1 - e^{-\frac{uL}{D}}\right)$$

by trial and error

D/uL = 0.053

$$\frac{D}{uL} = \frac{D}{udp} \cdot \frac{dp}{L} \qquad \text{(A)}$$

Now D/udp = 1.0, dp = 1 cm. Putting these values in equation (A)

$0.055 = 1 \times 1.0/L$

or L = 1.0/0.055 = 18.18cm

(b) $\sigma_\theta^2 = \frac{1}{5} = 0.2$

$$\text{So, } \sigma_\theta^2 = 2(D/uL) - 2\left(\frac{D}{uL}\right)^2\left(1 - e^{-\frac{uL}{D}}\right) = 0.2$$

By trial and error, $D/uL = 0.12$

$$D/_{uL} = \frac{D}{udp} \cdot \frac{dp}{L}$$

$$0.12 = 1.0\frac{1}{L}$$

$$L = \frac{1}{.12} = 8.33\,cm$$

(c) Rep = dpuρ/μ, Rep = 50, dp = 1cm, μ = 0.01 gm-cm/sec

So, u = (50*0.01)/1.0 = 0.5cm/sec

Now L_1 = 18.18cm and L_2 = 8.33cm

$\tau_1 = 18.18\,cm / 0.5\,cm / \sec = 36.36\,sec$

$\tau_2 = 8.33\ cm / 0.5\,cm = 16.66\,sec$

Problem 3.6

After introducing a small quantity of tracer in the feed stream of a continuous reactor, the following data are obtained as the concentration of tracer in the effluent at various times

t, min	0.1	0.2	1.0	2.0	5.0	10	30
Tracer, C, mg/L	0.20	0.17	0.15	0.125	0.07	0.02	0.001

(a) What type of ideal reactor does the actual vessel closely approach?
(b) If an isothermal first order reaction (k_1 = 0.175 min^{-1}) occurs in the vessel, what conversion may be expected?
(c) Calculate X_A for ideal CSTR, and ideal plug flow
(d) Calculate X_A from A dispersion model

Solution:

(a) Since all the data are not collected at equal time internal, by graphical interpolation, two sets of data are selected

min, t	0	1	2	3	4	5	10	15	20	25	30
C, mg/L	0	0.15	0.125	0.097	0.075	0.070	0.020	0.015	0.0065	0.0075	0.001

t, min	**c. mg/L**	**tiCi'**	**E(t)**	**$(t-t_m)^2$**	**$(t-t_m)^2$ E**
0	0	0	0	0	0
1	0.15	0.15	0.20	252.8	50.6
2	0.125	0.25	0.17	222.0	37.74
3	0.097	0.291	0.13	193	25.09

t, min	c. mg/L	tiCi'	E(t)	$(t-t_m)^2$	$(t-t_m)^2$ E
4	0.075	0.30	0.10	166	16.61
5	0.070	0.35	0.09	141.6	12.744
		Σ1.34			

$\Sigma Ci = 0.517$

$\Sigma Ci\Delta t = 0.517,\ \overline{t}_1 = 2.54$

10	0.20	0.20	0.027	47.0	1.269
15	0.0115	0.1725	0.0158	3.61	0.057
20	0.0065	0.130	0.0090	9.61	0.0865
25	0.0025	0.625	0.0034	65.61	0.223
30	0.0010	0.030	0.0013	171.6	0.222
		Σ0.595			

$\Sigma Ci = 0.415$

$\Sigma Ci\Delta t = 0.2075,\ \overline{t}_2 = 4.33$ min

$\underline{Q}t = 0.7245$

$\overline{t} = 2.54 + 14.33 = 16.87 = 16.9\text{min}$

$$\sigma^2 = \frac{1}{3} 1(0) + 4(50.6) + 2(37.74) + 4(25.09) + 2(16.61) + 12.744]\ (141.28) +$$

$$5/3\ [1(1.269) + 4(.057) + 2\ (.0865) + 4(.223) + 2(0.222) + 0]$$

$$= 146.28 \approx 146$$

$$\sigma_\theta^2 = \sigma^2 / \overline{t}^2 = 146 / (16.9)^2 = 0.51$$

$$N = \frac{1}{\sigma_\theta^2} = \frac{1}{0.51} = 1.96 \approx 2$$

So the actual vessel closely resembles to ideal CSTR.

(b) Assuming 2 tanks in series conversion for a first order reaction ($k_1 = 0.175\ \text{min}^{-1}$) can be calculated as

$$\frac{C_A}{C_{A_0}} = 1 - X_A = \frac{1}{\left[1 + \frac{(0.175)(16.9)}{2}\right]^2} = 0.16$$

Or $X_A = 0.84$

(c) Ideal CSTR:

$$\frac{C_A}{C_{A_0}} = 1 - X_A = \frac{1}{1 + (0.175)(16.9)} = 0.25$$

$x_A = 0.75$

ideal plug flow: $1 - x_A = e^{-(.159)\times 16.91} = .060(0.060)$

$x_A = 0.93$

(d) From Dispersion model: for $\sigma_\theta^2 = 0.511, D/uL = 0.4$

Pe = 2.5, Da = (0.175 × 16.9) = 2.95, q = $(1+ 4Da/Pe)^{1/2}$ = 2.39

$$X_A = 1 - \frac{4(2.39)exp(2.5/2)}{(1+2.39)^2 exp(1.2\times 2.5) - (1-2.39)^2 exp(-1.2\times 2.5)} = (1-0.144) = 0.85$$

So the dispersion and tanks-in series model are equivalent.

Problem 3.7

A tubular reactor of L = 3m and cross-section of 25 cm^2 is used to process a first order irreversible reaction

$A \rightarrow R$

with 98% conversion with a flow rate of 0.03 m^3/s.

Now a pulse tracer test is made and the following data were obtained: mean residence time, $\bar{t}$ = 10 secs, Variance, σ^2 = 65 sec^2. What conversion can be expected from the reactor using (a) tanks-in series model & (b) Dispersion model.

Solution:

(a) Reactor volume, V = $\pi r^2 L = 25 \times 10^{-4}\ m^2 \times 3m = 7.5 \times 10^{-3}\ m^3$

$\tau = V/q = 7.5 \times 10^{-3}\ m^3 / 0.03\ m^3/s = 0.25$ seconds

Assuming plug flow & first order

$\ln 1/(1 - X_A) = kt$

or $k = 1/t \ln 1/(1 - X_A) = 1/0.25 \ln 1/ 1 - 0.98 = 15.64\ sec^{-1}$

From tracer data:

$\sigma^2 = 65\ sec^2$, $\bar{t}$ = 10 sec

$\sigma\theta^2 = 65 / 100 = 0.65$

For tanks-in-series model, the parameter, N, is obtained

$N = 1/\sigma\theta^2 = 1 / 0.65 = 1.53$

(b) Conversion from tanks-in-series model.

The performance Equation is

$$\frac{C_A}{C_{A_0}} = 1 - \overline{x}_A = \frac{1}{\left(1+\frac{k\overline{t}}{N}\right)^N} = \frac{1}{\left[1+\frac{(15.6)(0.25)}{1.53}\right]^{1.53}} = 0.144$$

Or $\overline{x}_A = 1 - 0.144 = 0.856$

(c) For the dispersion model we calculate first D/uL

$$\sigma_\theta^2 = 0.65 = 2\left(\frac{D}{uL}\right) - 2\left(\frac{D}{uL}\right)^2\left(1 - e^{-\frac{uL}{D}}\right)$$

Solving by trial and error

$$\frac{D}{uL} = 0.7,\ P_e = 1.42$$

$$Da = k\tau = (15.6)(0.25) = 3.9$$

$$q = \left(1 + \frac{4Da}{Pe}\right)^{½} = \left[1 + \frac{(4)(0.7)}{1.42}\right]^{½}$$

$$\frac{C_A}{C_{A_0}} = (1 - X_A) = \frac{4(3.31)\,exp\left(\frac{1.42}{2}\right)}{(1+3.31)^2\,exp\left(\frac{3.31}{2}\times.42\right) - (1-3.31)^2\,exp\left(-\frac{3.31}{2}\times 1.42\right)}$$

$= 0.1387$

Or $X_A = 0.86$

3.5 Two Parameters Models

A real reactor may be modelled by one of two different combinations of ideal reactors. There are two popular models viz. (i) Real CSTR modelled using bypassing and dead space, (ii) Real CSTR modelled with an exchange volume. Let us discuss only the first one.

3.5.1 Real CSTR with by-passing and dead space

A real CSTR is modelled as a combination of an ideal CSTR of volume, V_s, a dead zone of volume, V_d, and a bypass with a volumetric flow rate, v_b. Tracer experiments may be used to evaluate the parameters of the model, V_s and V_d. If the total volume and volumetric flow rate are known and once V_s and V_d are found, v_b and V_d can be calculated. The real reactor system and model system are shown below in figs. 3.13 and 3.14.

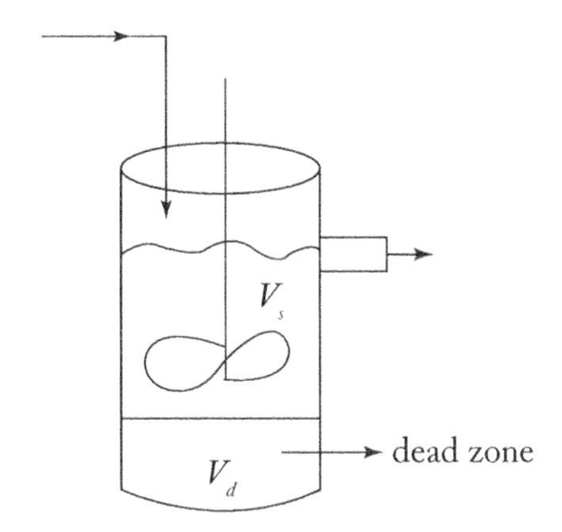

Fig. 3.14 CSTR with bypass and dead space

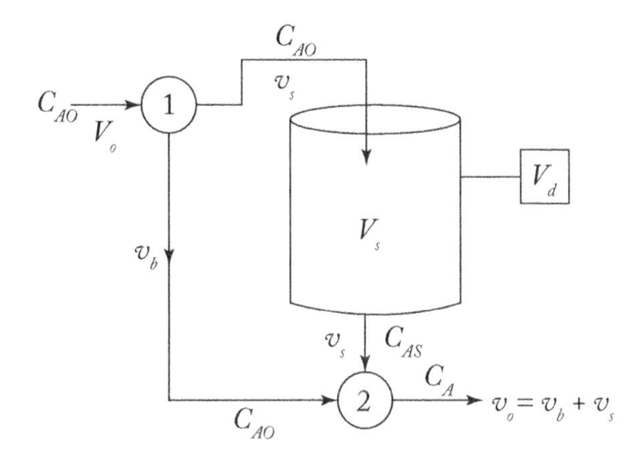

Fig. 3.15 Material balance for the real reactor

The bypass stream and effluent stream from the reactor volume and mixed at point 2. Material balance on the reactant species A for the first order reaction A → R, $r_A = k_{CA}$, around the point 2 is,

$$C_{AO}v_b + C_{AS}\,v_s = C_A\,(v_b + v_s) = C_A v_o \tag{3.58}$$

C_A (the concentration of A leaving the reactor) can be obtained from the above equation,

$$C_A = [v_b\,C_{AO} + v_s\,C_{AS}]\,/\,v_o \tag{3.59}$$

Let $\alpha = V_s / V$ and $V = V_s + V_d$

$$\beta = v_b / v_o,\; C_A = \beta\,C_{AO} + (1\text{-}\beta)\,C_{AS} \tag{3.60}$$

For a first order reaction, a mole balance on Vs gives

$$v_S C_{AO} - v_s\,C_{AS} - k\,C_{AS}\,V_s = 0 \tag{3.61}$$

Rearranging the above equation for CAS and using α & β, we get

$$C_{AS} = C_{Ao} (1 - \beta) v_o / [(1 - \beta) v_o + \alpha kV] \quad (3.62)$$

Substituting equation (3.62) into equation (3.61), effluent concentration, C_A can be obtained from the equation (3.63)

$$\frac{C_A}{C_{A_0}} = 1 - X_A = \beta + \frac{(1-\beta)^2}{(1-\beta) + \alpha k\tau} \quad (3.63)$$

If α and β are known, conversion can be predicted from equation (3.64). However, α and β can be determined from RTD experiments.

Problem 3.8

An ideal CSTR gives 75% conversion for a first order reaction A → R with residence time, τ = 6 min. But the real reactor is suspected to have bypassing and dead space. The reactor is modelled as a combination of ideal CSTR of volume, V_s and the dead volume, V_d and a bypass flow rate, vb. If V and v_o are the actual reactor volume and actual flow rate, respectively, the ratios, $\beta = v_b/v_o$ and $\alpha = V_s / V_o$ are obtained from RTD experiments. Let $\alpha = 0.8$, $\beta = 0.2$

Find out the conversion in the actual reactor.

Solution:

For an ideal CSTR,

$$k\tau = \frac{C_{A_0} - C_A}{C_A} = \frac{1}{0.25} - 1 = 3$$

$K = 3/6 = 0.5 \text{ min}^{-1}$

Where $\tau = 6\,minutes$

The performance equation for the system is given as,

$$(1 - X_A) = \beta + \frac{(1-\beta)^2}{(1-\beta) + \alpha k\tau}$$

Given: $\beta = 0.2$, $\alpha = 0.8$

$$1 - X_A = 0.2 + \frac{(1-0.2)^2}{(1-0.2) + (0.8)(0.5)(6)} = 0.4$$

Or $X_A = 0.6$

3.6 RTD in a Tubular Reactor with Laminar Flow (Krammers et al 1963)

A tubular reactor with laminar flow is assumed to have segregated flow and the velocity profile is parabolic. The velocity for laminar flow is given as

$$u(r) = \frac{2Q}{\pi r_0^2}\left[1 - \left(\frac{r}{r_O}\right)^2\right] \tag{3.64}$$

r is the radial position and r_o is the tube radius, Q is the volumetric flow rate. Since u is a function of r, the residence time, t, also varies with r,

$$t = \frac{L}{u} = \frac{\pi r_0^2}{2Q}\frac{L}{\left[1 - \left(\frac{r}{r_0}\right)^2\right]} \tag{3.65}$$

$$\text{or } t = \frac{V / Q}{2\left[1 - \left(\frac{r}{r_0}\right)^2\right]} = \frac{\bar{t}}{2\left[1 - \left(\frac{r}{r_0}\right)^2\right]} \tag{3.66}$$

where $V = \pi r_0 L, \bar{t} = V / Q$

We will show how the volume fraction of the effluent varies with radius r. the above equation is used to replace r with t.

The fraction of the effluent with a radius between r and r +dr is

$$dF(r) = E(t) = \frac{u(2\pi r dr)}{Q} \tag{3.67}$$

The fraction E(t) will have residence times between t and t +dt.

Substituting equation (3.65) for u and simplifying gives

$$dF(t) = \frac{4}{r_0^2}\left[1 - \left(\frac{r}{r_O}\right)^2\right] rdr \tag{3.68}$$

To replace r with t in this expression, we first differentiate equation (3.67) and then solve for r dr, we get,

$$rdr = \frac{\bar{t}}{4}\frac{r_0^2}{t^2}dt \tag{3.69}$$

Substituting $[1 - (r/r_o)^2]$ from equation (3.67) and r dr from equation (3.70) into equation (3.69), we obtain

$$dF(t) = \frac{1}{2}\left(\frac{V}{Q}\right)^2 \frac{dt}{t^3} = \frac{1}{2}\bar{t}^2 \frac{dt}{t^3} \tag{3.70}$$

Now,

$$E = \frac{dF}{dt} = \frac{1}{2}\frac{\bar{t}^2}{t^3} \tag{3.71}$$

Then

$$E_\theta = \frac{Et}{\bar{t}} = \frac{1}{2}\frac{1}{\theta^3} \tag{3.71a}$$

for $\theta \geq \frac{1}{2}$ and $\theta = t/\bar{t}$

The minimum residence time is not zero, but corresponds to the maximum velocity at the centre of the tube.

From equation (3.67), θmin is given as

$$\theta_{min} = \frac{1}{2}\frac{V}{Q} = \frac{1}{2}\bar{t} \tag{3.72}$$

Integration of equation (3.70) from t_{min} to t gives

$$F = \frac{1}{2}\bar{t}^2 \int_{\bar{t}/2}^{t} \frac{dt}{t^3} = 1 - \frac{1}{4}\left(\frac{t}{\bar{t}}\right)^{-2} \tag{3.73}$$

$$\text{or } F = 1 - 1/4\theta^2 \tag{3.73a}$$

E_θ may be plotted on the basis of the equation (3.71a) and F may be graphically presented from equation (3.74a).

Several modified E curves such as E*, E** have been defined.

E* is defined as the response cuve with correction for planar boundary and is expressed as

$$E^* = \frac{\bar{t}}{2t^2} \text{ for } \bar{t} => \frac{1}{2} \tag{3.74}$$

E** is the response curve with two corrections – one for entrance and the other for exit and mathematically presented as

$$E^{**} = 1/2\theta \text{ for } \theta \geq ½ \tag{3.75}$$

For the F curve, we have already deduced

$$F = 1 - \frac{1}{4\theta^2} \text{ and } \theta \geq \frac{1}{2}$$

and F* is given as

$$F^* = 1 - 1/2\theta \text{ for } \theta \geq ½ \ldots\ldots \tag{3.76}$$

There are simple relations between, E, E*, E** as given below

$$E^{**} = \frac{t}{\bar{t}} E^* = \frac{t^2}{\bar{t}^2} E \tag{3.77}$$

E curves for Non-Newtonian and non-circular channels have been given in literature.

Since Non-Newtonian fluids are often very viscous, they should be treated by the convective model. The E, E*, E** curves for various situations like power low fluids-Bingham plastics, falling films have been developed.

3.6.1 Conversion in laminar flow reaction

Assuming segregated flow in a laminar flow reactor, each fluid element follows its own streamline with no intermixing with the neighbouring elements, the conversion equation can be given as

$$\frac{C_A}{C_{A_0}} = \int_0^\infty \left[\frac{C_A}{C_{A_0}}\right]_{element\ of\ flid} E\,dt \tag{3.78}$$

Now (C_A/C_{AO}) element of fluid, can be given for different orders of reactions.

For n = 0, $$\frac{C_A}{C_{A_0}} = 1 - \frac{kt}{CA_0} \text{for } t \leq \frac{CA_0}{k} \tag{3.79}$$

For n = 1, $$\frac{C_A}{C_{AO}} = e^{-kt} \tag{3.80}$$

For n = 2 $$\frac{C_A}{C_{AO}} = \frac{1}{1 + kC_{AO}t} \tag{3.81}$$

Now for laminar flow

$$\text{E} = \frac{\bar{t}^2}{2t^3} \tag{3.82}$$

For n = 0, equation (3.79) can be integrated using equation (3.82) and we get,

$$\frac{C_A}{C_{A_0}} = \left(1 - \frac{kt}{2C_{AO}}\right)^2 \tag{3.83}$$

For n = 1 for laminar flow in a pipe

$$\frac{C_A}{C_{AO}} = \int_{\bar{t}/2}^{\infty} \left(e^{-kt}\right)\left(\frac{\bar{t}^2}{2t^3}\right) dt \tag{3.84}$$

$$\frac{C_A}{C_{AO}} = \frac{\bar{t}^2}{2} \int_{\bar{t}/2}^{\infty} \frac{e^{-kt}}{t^3} dt = y^2 ei(y) + (1-y)e^{-y} \tag{3.85}$$

where $y = \frac{k\bar{t}}{2}$, ei(y) is the exponential integral, defined as

$$ei(x) = \int_{x}^{\infty} \frac{e^{-u}}{u} du = -0.577 - lux + x - \frac{x^2}{2.2!} + \frac{x^3}{3.3!} \tag{3.86}$$

For n = 2 for a laminar flow reaction, the equation (3.79) may be integrated using equation (3.82) & 3.83 and we get,

$$\frac{C_A}{C_{AO}} = 1 - kC_{AO}\bar{t}\left[1 - \frac{kC_{AO}\bar{t}}{2} \ln\left(1 + \frac{2}{kC_{AO}\bar{t}}\right)\right] \tag{3.87}$$

The comparison of the performance of the laminar flow reactor with that of plug flow reactor for nth order reaction shows that at high conversion, X_A, convective flow does not drastically lower reactor performance.

The result also does not correspond to dispersion or tanks-in-series model.

Problem 3.9

A tubular reactor of 5 m^3 with laminar flow is used to carry out a first order reaction, k = 0.1min^{-1} with a flow rate of 0.5m^3/min.

Calculate the conversion, x_A in the reactor.

Assume first order liquid phase reaction.

Solution:

The reactor performance with residence time distribution is given as

$$\frac{C_A}{C_{AO}} = \int_{\bar{t}/2}^{\infty} e^{-kt} E(t) \, dt \tag{A}$$

For laminar flow

$$E(t) = ½ \frac{\bar{t}}{t^3}$$

Where $\bar{t} = V/Q$

Putting the value of E(t) in the equation (A), we get

$$\frac{C_A}{C_{AO}} = \frac{\bar{t}^2}{2}\int_{\bar{t}/2}^{\infty} \frac{e^{-kt}}{t^3} dt$$

or $C_A/C_{AO} = y^2$ ei (y) + (1 - y) e^{-y}

where y = $k\bar{t}/2$, ei(y) is the exponential function.

Now putting the values of $\bar{t}$, k and y.

k = 0.1 min^{-1}, $\bar{t}$ = $5m^3/0.5m^3/min$ = 10 min

y = (0.1) (10)/2 = 0.5

so $(1 - X_A) = C_A/C_{AO} = (0.5)^2$ ei (5) + (1 – 0.5) × $e^{-0.5}$

= 0.25 (0.5598) + 0.33 = 0.44

where ei (5) = 0.5598 from the exponential integral table

So $X_A = 0.56$

For comparison, if we assume plug flow

$$\frac{C_A}{C_{AO}} = e^{-kt}$$

Or $1 - x_A = e^{-0.1(10)} = 0.367$

Or $X_A = 0.633$

3.7 Summary

Simple methods have been presented to determine RTD in a flowing vessel by tracer study with pulse input or step input. From RTD data, model parameters of dispersion model or tanks-in-series model can be accurately determined and may be used to calculate actual conversion in the reactor. The residual error will be due to uncertainty in the extent of micro maxing.

If tank-in-series model or dispersion model cannot fit the actual RTD, a combination of plug flow and ideal stirred tank reactor can be chosen to fit the RTD data.

In laminar flow molecular diffusion in the axial direction causes little deviation if the reactor is reasonably long with respect to its diameter. Radial diffusion tends to make the reactor behave more like an ideal tubular flow unit.

For turbulent flow, axial dispersion including eddy diffusion can be significant. Its importance may not be significant if the Reynolds number is greater than 10^4 and the length is at least 50 times the diameter of the reactor.

In turbulent flow, as in laminar flow, radial diffusion tends to improve the accuracy of the plug flow model. Although the influence of non-ideal flow on conversion may be a small percentage but for high conversion level, the effect on subsequent separation units may be significant.

Problems (exercise)

3.1 For a reactor vessel RTD data are obtained from a step function for a given flow rate:

t,sec	0	15	25	35	45	55	65	75	∞
C, g/cm^3	0	.05	1.0	2.0	4.0	5.5	6.5	7.0	8.0

(a) Plot E vs t and F vs t
(b) Find the Peclet number and comment on the extent of mixing

3.2 To evaluate the RTD of a reactor vessel, a tracer is suddenly introduced into the inlet of the flowing stream and the outlet concentration of the tracer is measured as follows:

t,min	0.1	0.2	1.0	2.0	5.0	10	30
C, mg/L	0.20	0.17	0.15	0.125	0.07	0.02	0.001

(a) What type of ideal reactor does the actual reactor closely approach?
(b) If an isothermal first order reaction (k=0.2min^{-1}) occurs in the reactor, what conversion is achieved in the reactor?

3.3 For a first order reaction in a reaction vessel, the rate constant, k = 0.1 min-1. A pulse input tracer data are given below:

t, min	0	10	20	30	40	50	60	70
C, mg/L	35	38	40	40	39	37	36	35

(a) Find E vs t
(b) Calculate the Peclet or Dispersion number
(c) Calculate X_A from the dispersion model
(d) Calculate X_A from the tanks-in-series model
(e) Calculate X_A from segregated model
(f) Calculated X_A from ideal plug flow & CSTR

3.4 The following tracer data due to pulse input of tracer is given below:

t,min	0	5	10	15	20	25	30
C, mg/L	0	90	67	47	32	20	10

If a first order reaction A $\rightarrow$ R takes with the rate constant, $k = 0.15\ min^{-1}$,

(a) Calculate the conversion expected in the reactor, using dispersion model

(b) Also use tanks-in-series model to estimate conversion of A.

3.5 A liquid phase reaction is carried out currently in a series of four equal volume stirred tank reactors. It is planned to replace these reactors with a single tubular reactors to obtain the same conversion as the stirred tank reactor for the same residence time (V/q). Calculate the value of D/uL for the tubular reactor which indicates dispersion.

Problem 3.6

The RTD data for a reaction vessel are given below. A second order reaction takes in the vessel. The reaction rate constant, $k_2 = 100\ cm^3$ / (gmole) 9s) and the initial concentration of both reactants is the same, $C_{AO} = C_{BO} = 1$ gmole/L

The reaction is A + B $\rightarrow$ R

(a) Calculate the conversion of A using the RTD data

(b) Compare the conversions from ideal plug flow and the ideal CSTR, using residence time from the mean of distribution

RTD data:

t,sec	0	5	10	15	20	25	30
C, gm/cm^3	0	3.0	5.0	5.0	3.0	2.0	0.0

Problem 3.7

RTD data for a step function are obtained as follows:

$\theta = t/\bar{t}$	0	0.5	0.7	0.9	1.1	1.5	2.0	2.5
F (θ) or C/Co	0	0.10	0.25	0.45	0.6	0.85	0.9	0.99

Calculate E (θ)

Problem 3.8

One metre length of pipe packed with 2 mm particles is used to carry out a first order reaction, $k = 0.2\ min^{-1}$ and the fluid velocity is 0.2 m/min. Assume constant bed voidge and the intensity of dispersion, D/udp = 2.

1. Calculate the conversion of reactant A from dispersion model assuming small deviation from plug flow.

   A $\rightarrow$ R, $n = 0.2\ C_A$
2. Calculate conversion assuming plug flow

Problem 3.9

A second order reaction A + B → R with k = 0.1 m^3 / (k mole) (s) takes place in a laminar flow tubular reactor with residence time, $\bar{t}$ 10 seconds. The feed concentration of both reactants is same, $C_{AO} = C_{BO}$ = 1 kmole/m^3. The flow is segregated having no molecular diffusion.

(a) Calculate the conversion in the reactor and
(b) Compare the value assuming lug flow.

Hint:

$$\frac{C_A}{C_{AO}} = 1 - kC_{AO}\bar{t}\left[1 - \frac{kC_{AO}\bar{t}}{2}\ln\left(1 + \frac{2}{kC_{AO}\bar{t}}\right)\right]$$

Problem 3.10

From a pulse input to a vessel, the following RTD data are obtained

t,min	1	3	5	7	9	11	13
C, gm/L	0	0	10	10	10	10	0

If we want to represent the flow by tank-in-series model, calculate the number of tanks.

Problem 3.11

Tracer data for pulse input in a vessel are obtained as follows:

t,min	2	4	6	8	10	12	14	16	18	20	30	40	45
C, mg/L	57	90	86	77	67	40	45	48	30	32	15	7	0

(a) Calculate conversion assuming dispersion model
(b) Calculate conversion by tanks in series model
(c) Find the conversion by the direct use of tracer data

Assume first order reaction k = 0.05 min^{-1}

Problem 3.12

A tubular reactor of di = 5cm and length (L) = 5m is used to carry out a first order reaction $A \xrightarrow{k} R$ with k = 0.1 sec^{-1} with a feed rate of 10^{-3} m^3/sec. The physical properties of the system are ρ = 1000 ks/m^3, μ = 1.1 × 10^{-5} Pa, C_{AO} = 1 kmole/m^3.

Calculate concentration of A at the exit of the reactor

References

Ananthakrishnan, V., W.N. Gill and A.J. Barduhn, AIChEJ. 11, 1063, 1965
Aris, R., Chem. Eng. Sci., 9, 266, 1959
Bischoff, K.B. Chem. Eng. Sci 16, 131, 1961
Bischoff, K.B. and O. Levenspiel, Chem. Eng. Sci. 17, 245, 1962
Choi, C.Y. and D.D. Perlmuter, Chem. Eng. Sci. 31, 250, 1976
Cleland, F.A. and R.. Wilhelm, AIChE J. 2, 489, 1956
Dankwerts, P.V., Chem. Eng. Sci, 8, 93, 1958
Decker, W.D. and E.A. Mahlmon, Chem. Eng. Sci. 31, 1231, 1976
Denbigh, K.G., Chemical Reactor Theory, P. 50–51, Cambridge Uni. Press, 1965
Harell, J.E. and J.J. Perona, Ind. Eng. Chem. Process Design Develop. 7, 466 (1968)
Johnson, M.M. Ind. Eng. Chem. Fundamental, 9m 681, 1970
Kramers, H., K.R. Westerterp, Elements of Chemical Reactor design and operation, P. 87, Academic Press,. N.Y., 1963
Levenspiel, O. and WK. Smith, Chem. Eng. Sci. 6, 227, 1957
Levenspiel, O. Ind.. Eng. Chem. 56, 343, 1958
Levenspiel, O. Chemical Reaction Engineering, 3rd ed. Chaps 11-15, Wiley & Sons, 1999
MacMullin, R.B. and M. Weber, Jr. Trans. AIChE J. 31, 4096, 1935
Smith, J.M. Chemical Engineering Kinetics, 3rd ed. Chap 6, Mc Graw Hill Book Co., 1983
Taylor, G.I., proc. Ro. Soc. (London) 219A, 186, 1953
Vonken, R.K., D.B. Holmes and H.W. denHartog, Chem. Eng. Sci, 19, 209, 1964
Wehner, J.F. and R.H. Wilhelm, Chem. Eng. Sci. 6, 89, 1959
Westerterp, K.R., W.P.M. Van swaij, and A.A. R. Beenakers, Chemical Reactor Design and Operation, Joh Wiley & Sons, New York, 1984

CHAPTER 4

Enzymatic Reactions and Bioreactor Design

4.0 Enzymes, their Classification and Applications

Enzymes are protein molecules and act as biological catalysts. They are produced by living cells of animals, plants and microorganisms. Enzymes are specific, versatile and effective catalysts, generating higher reaction rates compared to chemically catalysed reactions under ambient conditions.

A common measure of activity of enzyme is the turnover number, which is the net number of substrate molecules reacted per catalyst site per unit time.

Enzymes are named by adding suffix-age to the end of the substrate viz urease, alcohol dehydrogenase, glucose isomerase etc. Some enzymes require a non-protein group for their activity, such as Mg, Zn, Mn, Fe or a coenzyme like NAD, FAD, CoA or some vitamins. An enzyme containing a non-protein group is known as a holoenzyme. The protein part of the enzyme is called apoenzyme. Thus holoenzyme is a combination of apoenzyme and co-factor.

Commercial enzymes have been classified into three major groups such as

1. Industrial enzymes: amylase, protease, glucose isomerase, lipases, penicillium acylases etc
2. Analytical enzymes: glucose oxidase, galactose oxidase, alcohol dehydrogenase, hexokinase, cholesterol oxidase etc.
3. Medical enzymes: asperginase, protease, lipases, streptokinase

High-fructose corn syrup (HFCS) is produced from corn starch using several enzymes as shown below:

$$\text{Corn starch} \xrightarrow{\alpha\text{-amylase}} \text{Thinned starch} \xrightarrow{\text{Gluco-amylase}} \text{Glucose} \xrightarrow{\text{Glucose-isomerase}} \text{HFCS}$$

HFCS contains fructose, sweeter than glucose and is used in soft-drinks.

4.1 Enzyme Kinetics

The mechanism of enzymatic reaction is mainly based on enzyme-substrate interaction. The formation of ES complex is effected by weak van der waal's forces or hydrogen binding. The substrate binds to a specific site on the enzyme, called the active site. The simplest model describing the interaction is the lock and key model in which the enzymes behave as the lock and the substrate (a much smaller molecule) represents the key.

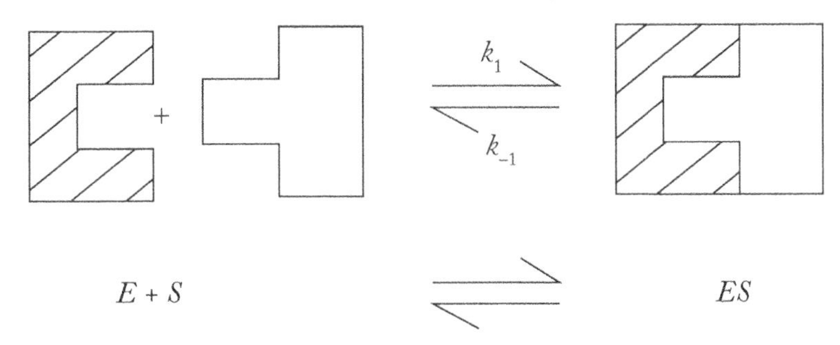

Enzyme catalysis is affected not only by primary structure of the enzyme, but also by secondary tertiary and quaternary structures.

A mathematical model of a single substrate-enzyme catalysed reactions was first developed by V.C.R Henri (1902), and also by L. Michaelis and M.L. Menten (1913). The kinetics of such enzyme-catalyzed reactions are known as saturation kinetics. It is to be mentioned that more complicated enzyme substrate interactions such as multisubstrates multienzyme reactions take place in biological systems.

However, saturation kinetics assumes a simple reaction scheme involving enzyme (E) substrate(S), complex (ES) formation by a reversible path followed by the irreversible dissociation of the complex to product(P) and the enzyme(E).

$$E + S \underset{k_{-1}}{\overset{k_1}{\rightleftharpoons}} ES \overset{k_2}{\text{-------}} E + P \tag{4.1}$$

When k_1 is the forward rate constant and k_{-1} is the backward rate constant for the reversible step and k_2 is the rate constant for the irreversible reaction.

Now two major approaches are used in developing a rate expression for enzyme catalysed reaction such as(4. 1): (1) rapid equilibrium postulated by Michaelis & Menten (1913) and(2) quasi steady state by G.E. Briggs and J.B.S. Haldane (1925).

Both the models assume the same initial steps as given below:

$$v = - dS / dt = dP / dt = k_2 [ES] \tag{4.2}$$

where v is the rate of decomposition of substrate or the rate of formation of product in kmol/(m^3) (s).

Now the material balance for ES and its rate of formation can be given as,

$$d[ES] / dt = k_1 [E] [S] - k_{-1} [ES] - k_2[ES] \tag{4.3}$$

And the material balance for enzyme can be given as

$$[E] = [E_0] - [ES] \tag{4.4}$$

Where [E] is the free enzyme concentration, $[E_0]$ is the initial concentration and [ES] is the concentration of enzyme that has combined with the substrate.

According to rapid equilibrium assumption of Michaelis and Menten, the equilibrium constant, K_M for [ES] complex formation can be expressed as

$$K_M = k_{-1} / k_1 = [E][S] / [ES] \tag{4.5}$$

Using the equation (4.4), [ES] is given as

$$[ES] = [E_0][S] / (KM' + [S]) \tag{4.6}$$

So the equation (4.2) reduces to

$$v = -dS / dt = dP / dt = k_2[E_0][S] / (K_M + [S]) = v_{max} S / (K_M + S) \tag{4.7}$$

where $v_{max} = k_2 [E_0]$ is the maximum forward velocity of the reaction, K_M is called the Michaelis Menten constant. A low value of K_M indicates a high affinity of the enzyme for the substrate.

The quasi-steady state was first proposed by Briggs and Haldane. In batch experiments initial concentration of substrate, S_0, greatly exceeds the initial concentration of enzyme, E_0 and they assumed that $d[ES] / dt = 0$ and it was claimed to be valid after brief transient if $(S_0) >> [E_0]$. By applying the above assumption,

We get,

$$[ES] = k_1 [E] [S] / (k_{-1} + k_2) \tag{4.8}$$

Using the equation (4.4), the equation (4.8) reduces to

$$[ES] = k_1 [(E_0) - (ES)] S / (k_{-1} + k_2) \tag{4.9}$$

Solving for [ES]

$$[ES] = [E0] [S] / [(k_{-1} + k_2) / k_1 + S] \tag{4.10}$$

Substituting [ES] into equation (4.2) we obtain,

$$v = -dS / dt = dP / dt = k_2[Eo][S] / (k_{-1} + k_2) / k_1) + [S] \tag{4.11}$$

or

$$v = v_{max} [S] / (K_M + [S]) \tag{4.12}$$

where K_M is $(k_{-1} + k_2) / k_1$ and $v_{max} = k_2[E_0]$

4.2 Evaluation of Kinetic Parameters of M-M Equation

The values of kinetic parameters K_M and v_{max} can be determined by several linearization techniques, as mentioned below:

1. Double – reciprocal plot or Lineweaver – Burk plot
2. Eadie Hofstee plot
3. Hans – Woolf plot
4. Integration method using batch kinetics

The first technique, **Lineweaver - Burk plot** is most popular one. The M-M equation is linearized as

$$1 / v = 1 / .v_{max} + K_M / v_{max} .1 / [S] \tag{4.13}$$

The plot of 1 / v vs 1 / S gives rise to a linear plot with a slope of K_M / v_{max} and the intercept of $1 / v_{maxM}$.

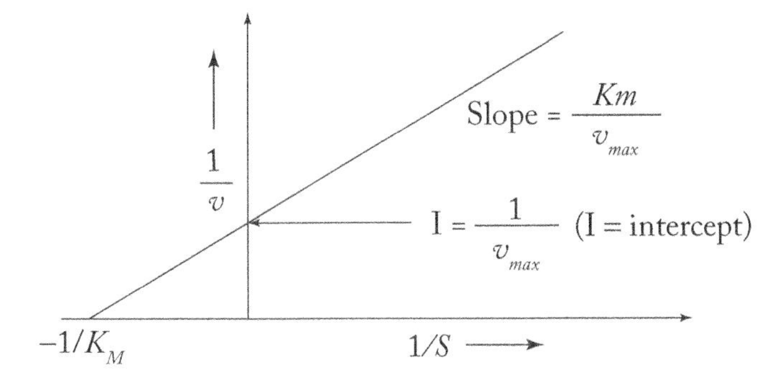

Fig. 4.1 Lineweaver Burk Plot

In the Eadie – Hostee plot, the linearized form of the equation is

$$v = v_{max} - K_M \, v / s \tag{4.14}$$

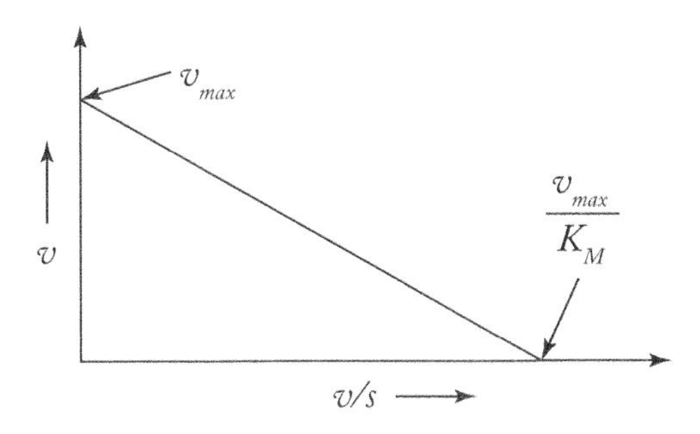

Fig. 4.2 Eadie - Hofste Plot

and is plotted in fig. 4.2

Hanes-Woolf Plot linearizes M-M equation in the following form:

$$S / v = K_M / v_{max} + S / v_{max} \tag{4.15}$$

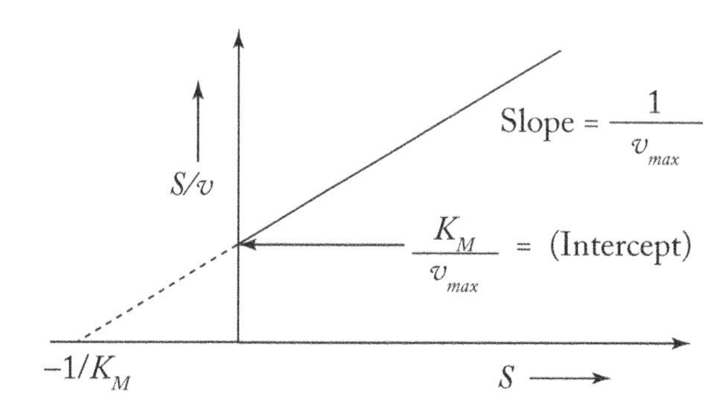

Fig. 4.3 Hanes - Wolf Plot

and is plotted in fig. 4.3

The Integration method involves the integration of the batch equation after separating the variable and then linearizing the equation,

$$v = -ds / dt = v_{max} S / (K_M + S)$$

after integration at $t = 0$, $S = S_0$, we get

$$v_{max} t = (S_0 - S) + K_M \ln S_0 / S \tag{4.16}$$

Linearizing and rearranging

$$(\ln So / S) / t = v_{max} / K_M - (So-S) / tK_M \tag{4.17}$$

The linearized form of equation (4.17) is plotted in fig. 4.4

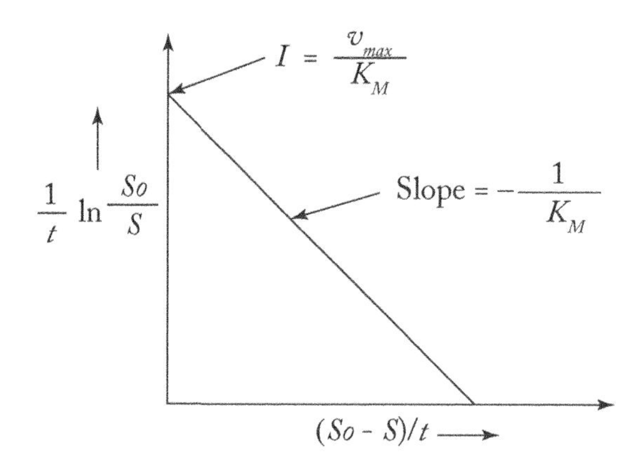

Fig. 4.4 Plot of Integration method

4.3 Significance of the Parameters k_M and v_{max}

K_M is an intrinsic kinetic parameter and function of temperatur and pH. v_{max} depends on the rate parameter, k_2 and initial enzyme concentration, E_0.

In many enzymatic reactions, crude enzymes are used. They are characterized by their specific activity which is defined as the number of units of activity per unit weight of total protein. One unit is the formation of one micromole product per minute at a specific pH and temperature with substrate concentration greater than the value of K_M.

4.4 Kinetics for Allosteric Enzymes

Some enzymes have more than one substrate binding site. This tendency of enzyme is known as allostery or co-operative binding. The rate model of the system has been presented in the form of Michaelis-Menten Kinetics.

$$v = -dS / dt = v_{max} S^n / (K_M' + S^n) \tag{4.18}$$

Where n is the co-operativity coefficient or Hill Constant, which may be positive or negative. K_M' is the overall binding constant.

Compared to Michaelis – Menten kinetics where v vs. S plot is asymptotic or hyperbolic, where as v vs S plot for allosteric enzyme is sigmoidal in nature as shown figure 4.5

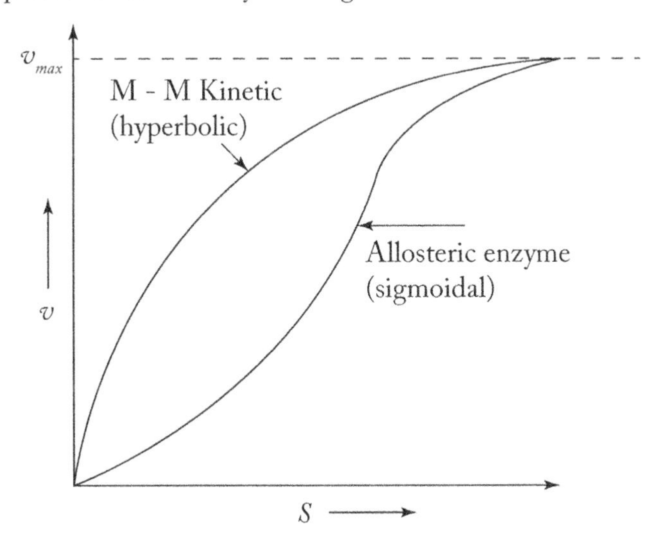

Fig. 4.5 Behaviour of allosteric enzymes compared to M-M kinetics

The non-linear equation (4.18) can be linearized as

$$\ln v / (v_{max} - v) = n \ln s - \ln K_M' \tag{4.19}$$

and by plotting $\ln [v / (v_{max} - v)]$ versus ln s, we can find out the parameters, n and K_M' from the slope and intercept respectively.

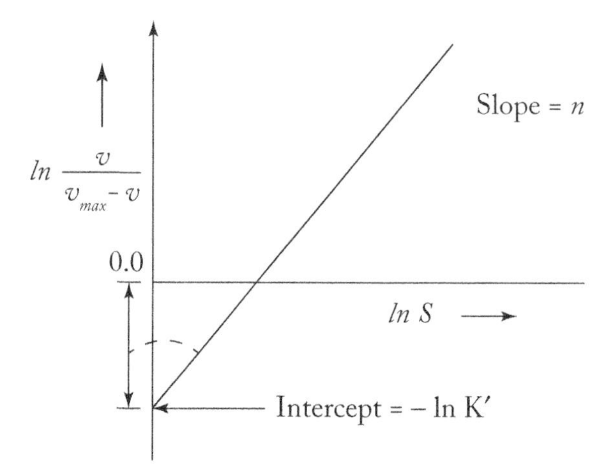

Fig. 4.6 The linearized plot of the rate model for allosteric enzymes

The Hill constant, n, is a measure of the cooperativity between the sites. If n = 1, binding is non-cooperative and usual Michaelis – Menten Kinetics exist. The Hill constant, n, may not be an integer, indicating incomplete cooperativity between the different binding sites. For example, n for the binding oxygen to haemoglobin (for which the number of binding sites is known to be 4) is 2.6 instead of 4.

Below is given a problem involving batch kinetics for enzymatic reaction.

Problem 4.1

From a series of batch runs with a constant enzyme concentration, the following initial rate. data were obtained as a function of initial substrate concentration.

S_0, gmol / m^3	v_0, initial reaction rate, gmol / (m^3) (min)
1.0	0.20
2.0	0.22
3.0	0.30
5.0	0.45
7.0	0.41
10.0	0.50

(i) Evaluate the Michaelis – Menten kinetic parameters by (a). Lineweaver – Burk plot and (b) Eadie-Hofstee plot

(ii) Compare the predictions from each method by plotting v verses s with the data points and discuss the strengths and weaknesses of each method.

Solution:

Lineweaver – Burk Plot:

$$1 / v = 1 / v_{max} + K_M / v_{max(}(1 / S)$$

Eadie-Hostee Plot:

$v = v_{max} - K_M v / S$

Solution of Problem 4.1

To make the two plots we prepare the following table

v_0, mol / (m^3)(min)	0.20	0.22	0.30	0.45	0.41	0.5
S_0, mol / m^3	1.0	2.0	3.0	5.0	7.0	10.0
$1 / v_0$	5.0	4.54	3.33	2.22	2.44	2.0
$1 / S_0$	1.0	0.50	0.33	0.20	0.14	0.10
v_0 / S_0	0.20	0.11	0.10	0.096	0.058	0.50

From the Lineweaver – Burk plot (method a)
$v_{max} = 0.454$ mol / (m^3) (min) $K_M = 1.22$ mol / m^3,
From Eadie & Hostee plot (method b)
$v_{max} = 0.60$ mol / (m^3) (min), $K_M = 2.0$ mol / m^3

S_0, mol / m^3	1	2	3	5	7	10
Predicted v (from method 1)	0.2	0.28	0.32	0.364	0.386	0.40
Predicted v (from method 2)	0.2	0.30	0.36	0.428	0.466	0.50
Expt v	0.2	0.22	0.30	0.45	0.41	0.50

Poor results are obtained at higher substrate concentration from method a.

(a)

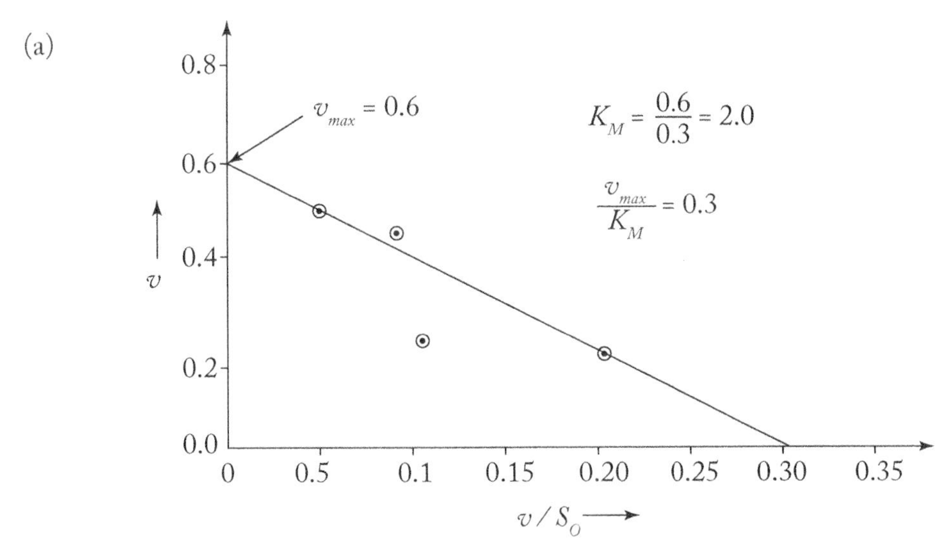

(b)

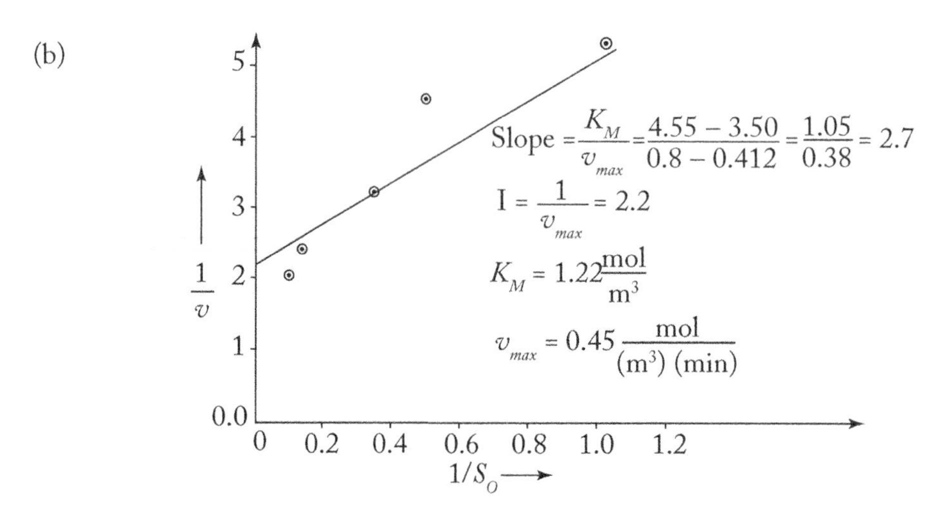

Fig. 4.7 (a) Eadie – Hofstee plot (b) Lineweaver – Burk Plot

(b) Rates predicted by the parameters evaluated by Lineweaver – Burk plot
$v_0 = v_{max}\ S_0 / (K_M + S_0) \quad = 0.454\ S_0 / (1.22 + S_0)$

S_0, mol / m^3	1	2	3	5	7	10
Predicted v_0 mol / m^3 (min)	0.2	0.28	0.322	0.365	0.38	0.40
Observed rate, v	0.2	0.22	0.30	0.45	0.41	0.50

Rates predicted by the parameters evaluated by Hofstee plot.
$v = 0.6\ S_0 / \ (2.0 + S_0)$

S_0	1	2	3	5	7	10
Predicted v_0	0.2	0.30	0.24	0.43	0.46	0.50

From the data it is evident that at low substrate concentration upto $S_0 = 3$ mol/m^3, the Lineweaver-Burk plot gives better results. But at higher substrate concentration, EadieHofste plot yields better results compared to the observed rate values predicted by the Lineweaver-Burk plot.

Problem 4.2

The K_M value of an enzyme is 0.01 kmol / m^3. Using the enzyme and the initial rate of reaction in a batch reactor it has been found that only 10 percent of the substrate has been consumed in 5 minutes. The initial substrate concentration was

$S_0 = 3.4 \times 10^{-4}$ kmole / m^3.

Assume Michaelis – Menten Kinetics,

(a) What is the maximum reaction rate (v_{max})

(b) What is the concentration of the substrate after 15 minutes?

Solution:

(a) The integrated form of Michaelis – Menten equation is
$K_M \ln 1/(1-\delta) + S_0 \delta = v_{max} t$
where δ = fractional conversion of the substrate, given as $(S_0 - S)/S_0$. Now putting the given values of K_M & δ, we have
$(.01) \ln 1/(1-0.1) + 3.4 \times 10^{-4} (0.1) = v_{max} (5)$
or
$1.053 \times 10^{-3} + 0.34 \times 10^{-4} / 5 = v_{max}$
Solving for v_{max}, we get
$v_{max} = 0.217 \times 10^{-3}$ kmol / (m³) (min)

(b) $(0.01) \ln 1/1 - \delta + (3.4 \times 10^{-4}) \delta = (0.217 \times 10^{-3}) \times 15 = 3.25 \times 10^{-3}$
From the above equation the conversion is obtained by trial and error, $\delta = 0.22$, when t = 15 minutes

4.5 Kinetics of Inhibited Enzyme Reactions

An inhibitor is a modulator which decreases the enzyme activity by binding with them. It is traditional to classify inhibitors by their influence on the Michaelis – menten kinetic parameters, v_{max} and K_M.

Reversible inhibitors are called competitive if their presence increases the value of K_M, but does not affect v_{max}. The effect of such inhibitors can be reversed by increasing the substrate concentration.

A non-competitive inhibitor decreases the v_{max} of the enzyme, but does not alter the K_M value. Consequently by increasing the substrate concentration to any level, the reaction rate possible with uninhibited enzyme cannot be achieved. Common non-competitive inhibitors are heavy metal ions which combine with the sulfhydryl (-SH group) of cysteine residue.

There are four major classes of inhibition kinetics, for example

1. Competitive
2. Non-competitive
3. Un-competitive
4. Substrate / Product inbition

Competitive Inhibition:

The competitive enzyme inhibition kinetics mechanisms can be shown as

$$E + S \underset{k_{-1}}{\overset{k_1}{\rightleftharpoons}} ES \xrightarrow{k_2} E + P \tag{4.20}$$

Where k_1 and k_{-1} are the forward and backward rate constants for the first reversible reaction and k_2 is the rate constant for the second step irreversible reaction.

$$E + I \rightleftharpoons EI\ldots, Ki \tag{4.21}$$

Ki is the equilibrium constant for the above reversible reaction.

Assuming rapid equilibrium

$K_M = [E][S] / [ES]$, $Ki = [E][I] / [EI]$

And $$E_o = E + ES + EI, \; v = k_2[ES] \tag{4.22}$$

The rate of enzymatic reaction for the above system has been deduced as,

$$v = v_{max} S / (Kmapp' + S) \tag{4.23}$$

when $$Kmapp' = K_M(1 + I / Ki) \tag{4.24}$$

where I = inhibitor concentration, Ki = inhibitor – enzyme equilibrium constant.

The competitive inhibition reduces the rate and the effect can be reduced by using increased amount of substrate.

- **Non-competitive Inhibition:**
  Here inhibitors bind on sites other than the active sites and reduce enzyme activity to the substrate. The mechanism for the system is,

$$E + S \underset{k_{-1}}{\overset{k_1}{\rightleftharpoons}} ES \xrightarrow{k_2} \text{-}E + P \tag{4.25}$$

Where k1 and k-1 are the forward and backward rate constants for the reversible. step and k2 is the rate constant for the irreversible step, K_M is the equilibrium constant. Other steps are:

$$E + I \underset{Ki}{\rightleftharpoons} EI + S \rightleftharpoons ESI \tag{4.26}$$

Where Ki is the equilibrium constant for EI complex and K_M' is the equilibrium constant for ESI complex

$$ES + I \underset{K_M'}{\rightleftharpoons} ESI \tag{4.27}$$

where

$K_M' = [E][S] / [ES] = [EI][S] / [ESI]$

$$Ki = [E][I] / [EI] = [ES][I] / [ESI] \tag{4.28}$$

The rate can be deduced as

$$v = vmappt / (1 + K_M' / [S]) \tag{4.29}$$

Where $$vmappt = v_{max} / (1 + I / Ki) \tag{4.30}$$

and vmappt = apparent v_{max}

The net effect of non-competitive inhibition is to reduce the value of v_{max}. Specific compounds are needed to block the binding of the inhibitor to the enzyme.

- **Un-competitive Inhibition:**
  In this case inhibitors bind to the ES only and have no affinity for the enzyme. The reaction scheme is
  $E + S \overset{K_M}{\rightleftharpoons} ES \overset{k_2}{----} E + P$

$$ES + I \rightleftharpoons ESI \tag{4.31}$$

Where $K_M' = [E][S] / [ES]$, $KI = [EI][I] / [ESI]$

Now the enzyme balance is

$$[E0] = [E] + [ES] + [ESI] \text{ and the rate } v = k_2[ES] \tag{4.32}$$

Further, the rate can be developed from the above equations as

$$v = vmappt\ S / (KMapp' + S) \tag{4.33}$$

where, $vmappt = v_{max}\ S / (KMappt' + S)$

$$KMappt' = K_M' / (1 + [I] / Ki). \tag{4.34}$$

The net effect of uncompetitive inhibition leads to reduction of both v_{max} and K_M parameters leading to overall reduction of the enzymatic rate.

- **Substrate Inhibition:**
  High substrate concentrations lead to inhibition to enzymatic reactions. The reaction mechanism is,
  $E + S \rightleftharpoons ES \overset{k_2}{---------\rightarrow} E + P$
  Where K_M' is the equilibrium constant for the reversible reaction and k_2 is the rate constant for the irreversible reaction. Another step is shown below:

$$ES + S \overset{Ksi}{\rightleftharpoons} ES^2 \tag{4.35}$$

Where Ksi is the equilibrium constant for the above reversible reaction

Now,

$$Ksi = [ES][S] / [ES^2]\ ,\ K_M = [E][S] / [ES] \tag{4.36}$$

On the basis of rapid equilibrium, the rate of enzymatic reaction can be given as

$$v = v_{max}\ S / (K_M + S + S^2 / Ksi) \tag{4.37}$$

At low substrate concentration, $[S]^2 / Ksi << 1$, there is no inhibition effect and the rate is

$v = v_{max}\ S / (K_M + S)$

At high substrate concentration $K_M / Ksi << 1$ and the inhibition is severe. The rate in this condition is,

$$v = v_{max} / (1 + S / Ksi) \tag{4.38}$$

and the linearized plot of the above equation (4.38) is

$$1 / v = 1 / v_{max} + [S] / Ksi\ v_{max} \tag{4.39}$$

and, will give the intercept

of $1 / v_{max}$ and the slope of $1 / K_{si}\ v_{max}$.

Further, the substrate concentration resulting to the maximum reaction rate,,

can be obtained by setting dV / dS at S = 0, we get

$$[S]_{max} = [K_M Ksi']^{1/2} \tag{4.40}$$

The double reciprocal plots of all inhibited enzyme reactions discussed above can be shown as follows:

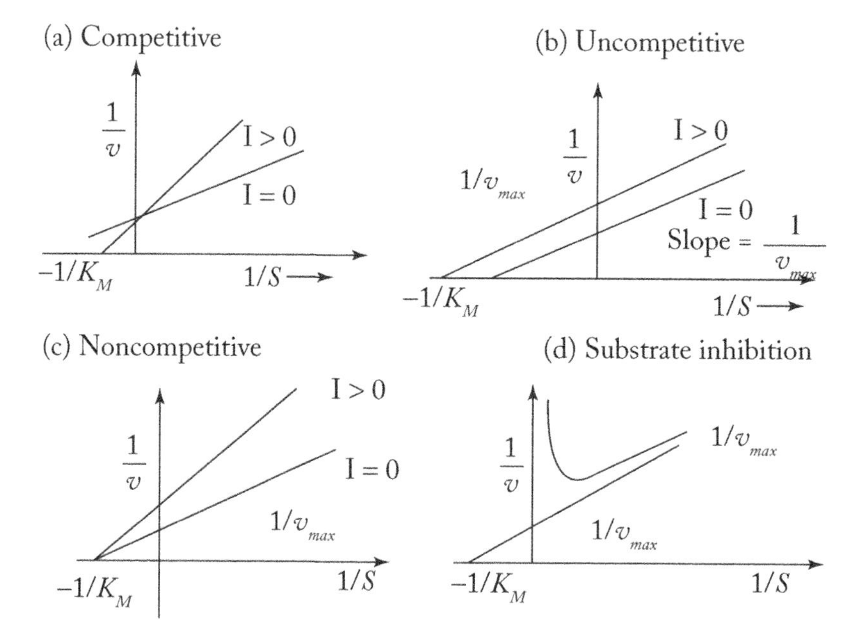

Fig. 4.8 Different forms of inhibition kinetics using Lineweaver – Burk plot

Several numerical problems are presented below along with their solutons.

Problem 4.3

The initial reaction rates of hydrolysis of acetylcholine (substrate) by the dog serum (crude enzyme) in the absence and presence of prostigmine (inhibitor) I = 1.5×10^{-7} kmol / m^3 were measured and the following data were obtained.

S, kmol / m^3	0.0032	0.0049	0.0062	0.0080	0.0095
v, kmol / (m^3) (min) without I	0.111	0.148	0.143	0.166	0.200
v, kmol / (m^3) (min) with the inhibitors	0.059	0.071	0.091	0.111	0.125

(a) Is prostigmine competitive or non competitive
(b) Estimate the Michaelis-Menten Kinetic parameters in the presence of inhibitor by employing the Langmuir plot.

Solution of 4.5

S, kmol / m^3	0.0032	0.0049	0.0062	0.0080	0.0095
v, kmol / (m^3) (min) without inhibitor	0.111	0.148	0.143	0.166	0.200
v' with I, inhibitor	0.059	0.071	0.091	0.111	0.125
1 / s	312.5	204.0	161.3	125	105
1 / v	9.00	6.75	6.99	6.02	5.0
1 / v'	16.95	14.08	10.98	9.00	8.0

From 1 / v vs 1 / S plot, it is evident that the inhibition is non-competitive.

- Without inhibition:

Slope, $S = 1 / v_{max} = 0.0266_2$

From the above, we get from the slope,

$v_{max} = 37.6$ kmol / (m^3)

- From the intercept, we obtain

$-1 / K_M = -1.0 \times 10^2$, or $K_M = 1 / 100 = 0.01$ kmol / m^3

with Inhibition, $1 / v_{max} = (13.4 - 8.0) / (2.0 - 0.8) \times 10^{-2} = 4.5 \times 10^{-2}$

$v_{max}' = 22.2 = v_{max} / (1 + I / Ki) = 37.6 / (1 + 1.5 \times 10^{-7} / Ki)$

Solving for Ki

$Ki = 2.17 \times 10^{-7}$ kmol / m^3

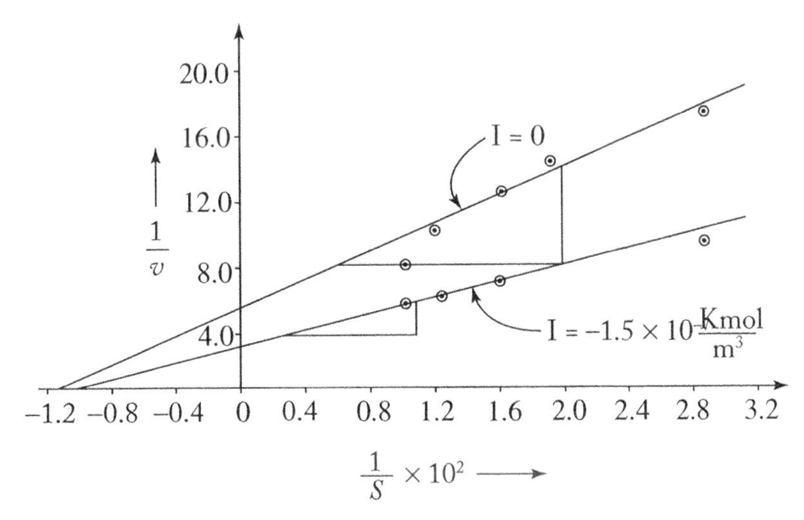

Fig. 4.9 Plot of 1 / v vs 1 / S with $I = 0$, $I = 1.5 \times 10^{-7}$ kmol / m^3

Problem 4.4

The experimental data on the enzymatic oxidation of phenol by phenol oxidase are given below:

$S, g / m^3$	10	20	30	50	80	90	110	130	140	150
v, g / m3hr	5	7.5	10	12.5	15	15	12.5	9.5	7.5	5.7

Solution: Data points from $S = 10\ g / m^3$ to $S = 80\ gm / m^3$ show the condition, not affected by substrate inhibition.

The rest are inhibited data.

The first set upto S = 80 gms / m3 will follow the Michaelis-Menten equation i.e.

$1 / v = 1 / v_{max} + K_M / v_{max} * 1 / S$

and the second set from the concentration range from 80 g / m3 to 150 gm / m^3, the rate is inhibited by substrate(phenol) and will follow the substrate inhibition model. The double reciprocal plot of the model is given for the substrate inhibition,

$1 / v = 1 / v_{max} + S / K_{si} v_{max}$

Solution

S	10	20	30	50	80
v	5	7.5	10	12.5	15
1 / s	0.1	0.05	0.033	0.02	0.0125
1 / v	0.2	0.133	0.10	0.08	0.067

Substrate (Phenol) inhibited data

S	90	110	130	140	150
v	15	12.5	9.5	7.5	5.7
1 / v	0.067	0.08	0.105	0.133	0.175

For the first set of data,

$1 / v$ vs $1 / S$ give the intercept $(1 / v_{max})$ and the slope (K_M / v_{max}) of 2.0

intercept $= 1 / v_{max} = 0.052$,

$v_{max} = 1 / 0.052 = 19.2\ gm / (m^3)(hr)$

slope $= K_M / v_{max} = 2.0$, then

$K_M = 2 \times v_{max} = 38.4\ gm / m^3$.

From the 2nd set of data with a substrate inhibition:

the plot of 1 / v vs S will give

Intercept = $1 / v_{max} = .052$

$v_{max} = 19.2$, slope = $1 / (Ksi' \; v_{max} = 0.000533$

$K_{si} \; v_{max} = 1917$, $K_{si} = 1917 / 19.2 \approx 99.84$ gm / m^3

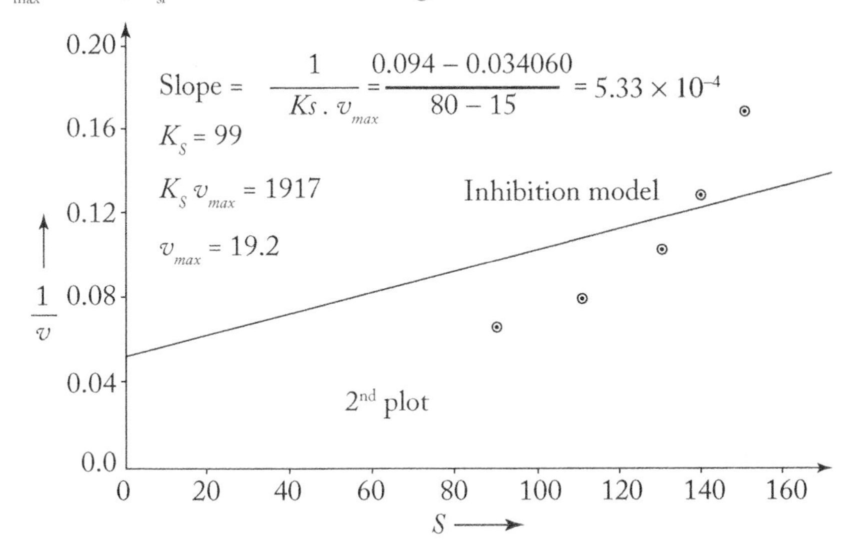

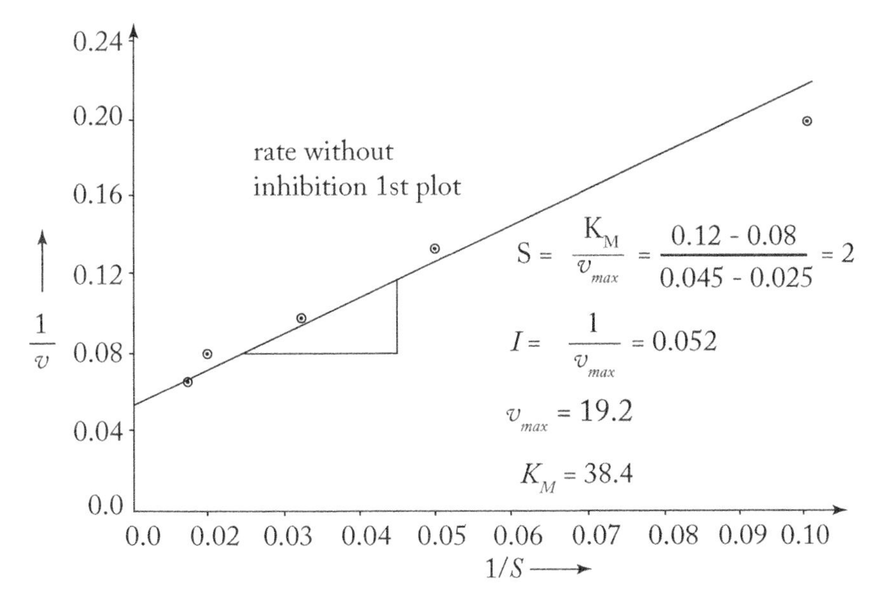

Fig. 4.10 Lineweaver-Burk Plots

4.6 Enzyme Reactors with Michaelis-Menten Kinetics

4.6.1 A batch and a steady state plug flow reactor

The simplest reactor configuration is a batch reactor which is provided with a stirrer and control devices for constant temperature and pH for enzymatic reactions. An ideal plug flow reactor is equivalent to a batch reactor where time is replaced by residence time which is the ratio of reactor–volume to volumetric flow rate. They are shown schematically as,

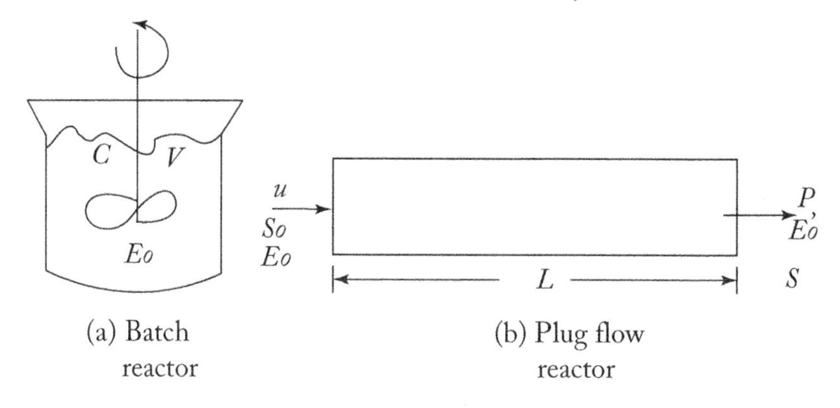

Fig. 4.11 Schematic sketch of a Batch reactor and a PFR

Batch reactor design equation can be obtained from the simple rate equation of enzymatic reaction.

$$-dS / dt = v_{max} S / (K_M + S) \tag{4.41}$$

Separating the variables, we get,

$$-[(K_M + S) / S] \, dS = v_{max} \, dt \tag{4.42}$$

After integration, we have

$$K_M \ln S_o / S + (S_o - S) = v_{max} t \tag{4.43}$$

With the known values of v_{max} & K_M, effluent concentration of the substrate, S, can be predicted with time.

If the batch data at a constant temperature and pH are obtained in the form of substrate vs time, the kinetic parameters, K_M & v_{max} can be evaluated from a linearized plot of equation (4.43) as given below by rearranging the equation (4.43)

$$(S_o - S) / \ln S_0 / S = -KM + v_{max} t / \ln (S_o / S) \tag{4.44}$$

By plotting$(S_o - S) / \ln S_o / S$ vs. $t / \ln S_o / S$ in the linear form from which the slope will give the v_{max} and the intercept gives K_M.

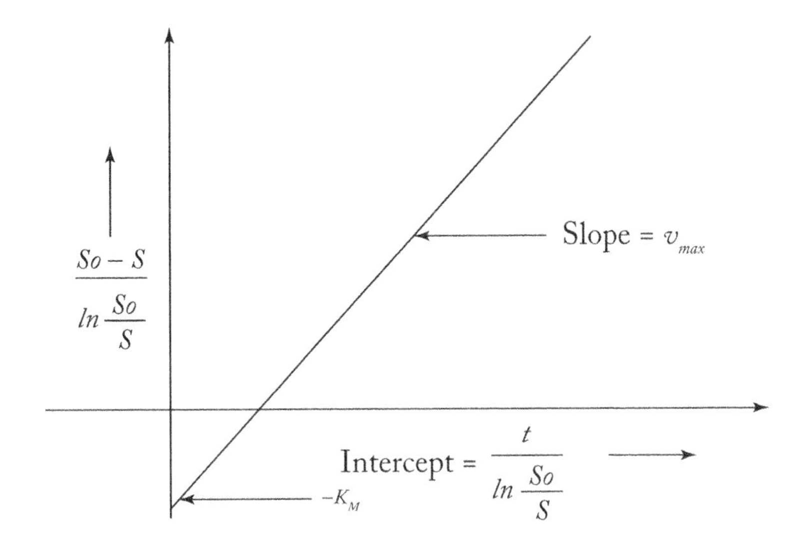

Fig. 4.12 The linear plot of a batch reactor with M-M kinetics

Problem 4.7

At room temperature, sucrose is hydrolysed by the enzyme as follows,

$$\text{Sucrose} \xrightarrow{\text{Enzyme}} \text{Products}$$

Starting with $S_o = 1.0$ mole / m^3 and sucrose, $E_o = 0.01$ mole / m^3.

The following data were obtained in a batch reactor by measuring the concentration of sucrose by optical rotation.

S, mol / m^3	1.0	0.68	0.16	0.006
t, hr	0	2.0	6	10

Apply Michaelis-Menten equation to find out the kinetics of enzymatic reaction.

Solution:

The Michaelis-Menten equation is given as

$v = ds / dt = v_{max} S / (K_M + S)$, where $v_{max} = K_2 E_o$

(a) by integration method:

Integrating the above equation and rearranging it in linear form, we get

$K_M \ln S_o / S + (S_o -)S = v_{max} t$

Rearranging in linear form,

$(S_o - S) / \ln S_o / S = -K_M + v_{max} t / \ln S_o / S$

(b) For differential method, the slopes, dS / dt are obtained by drawing tangents to the curve of S vs t

Table for integration method

t_1, hr	0	2	6	10
S mol / m^3	1.0	0.68	0.16	0.006
ln S_o / S	0	0.3856	1.8326	5.11
S_o – S / Ln S_o / S	0	0.8298	0.4583	0.1945
t / ln S_o / S	0	5.1817	3.274	1.95

S_o – S / ln S_o / S is plotted against t / ln S_o / S and from the Intercept,
$-K_M = -0.20$, $k_M = 0.20$ mol / m^3
Slope = $v_{max} = 0.18$ mol / (m^3) (hr)

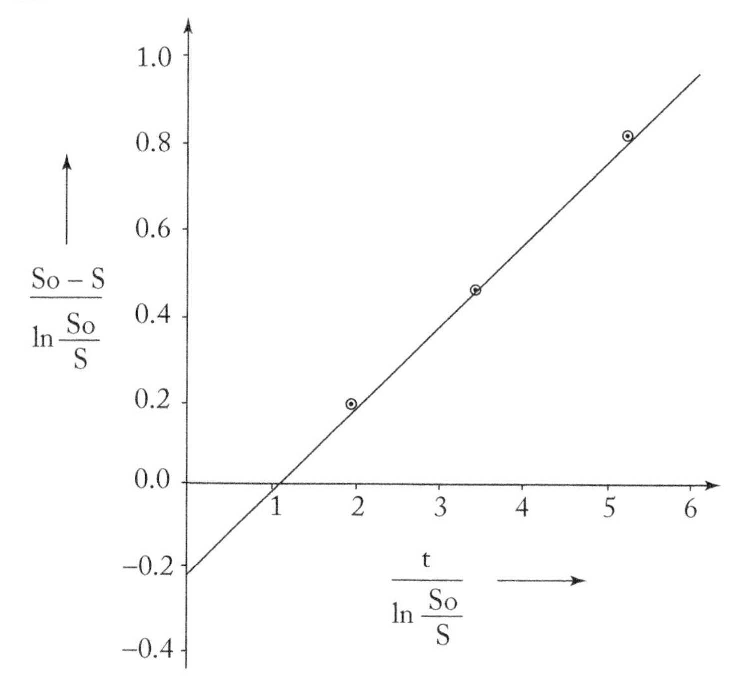

Fig. 4.13 Linearized Plot of (S_o – S) / ln S_o / S vs t / ln S_o / S

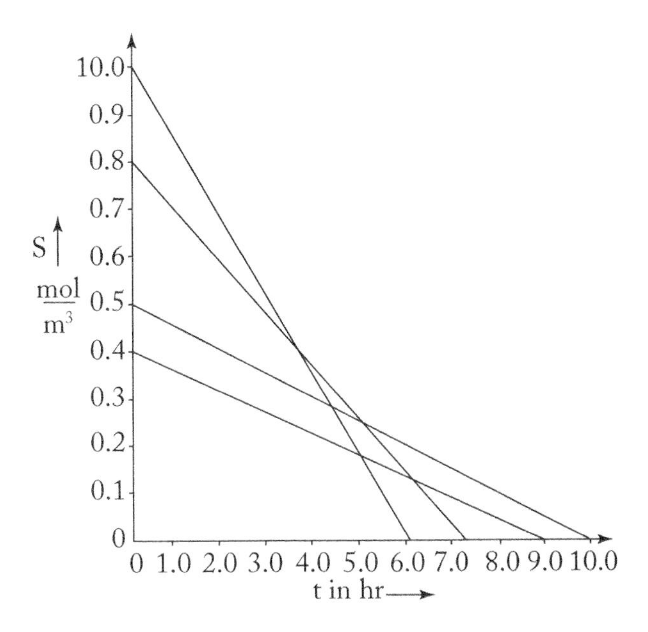

Fig. 4.14 Tangets to the curve of S vs t for dS/dt

From the figure 4.14 we get v = dS / dt, slopes of the curve at various values of S and the Lineweaver-burk plot is made in figure 4.15 and from the slope and intercept, we get $v_{max} = 0.2$ and $Km = 0.57 mol / m^3$

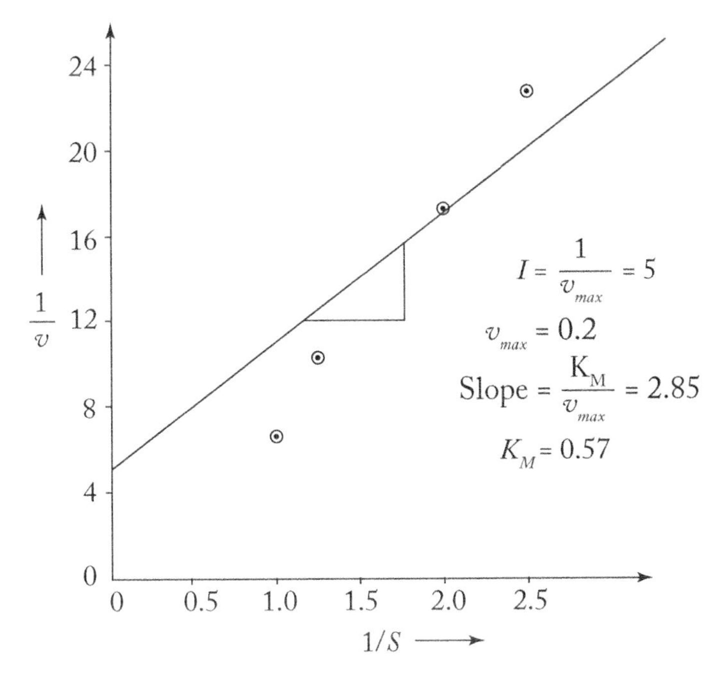

Fig. 4.15 Lineweaver – Burk plot

4.6.2 Plug flow reactor for enzymatic reactions

In plug flow reactor, the substrate enters at one end of a cylindrical tube along with the free enzymes dispersed in the substrate. The product stream leaves at the other end of the reactor. The properties of the flowing stream may vary with radial and axial directions. But the variation in the radial direction is small compared to that in the longitudinal direction and the system is called plug flow reactor. The ideal plug flow enzyme reactor can approximate the long tube, packed bed and hollow fiber reactor unit.

The basic equation for a plug flow reactor on a differential mass balance for an elemental length, dz is given as

$$v\, dS / dz = - v_{max}\, S / (K_M + S) \tag{4.45}$$

and integrating as before, we get

$$KMlnSo / S + (So - S) = v_{max} L / u = v_{max}\, t \tag{4.46}$$

where t is the residence, defined as $t = V / F = L / u$ (4.47)

4.6.3 Continuous stirred tank reactor with free enzymes

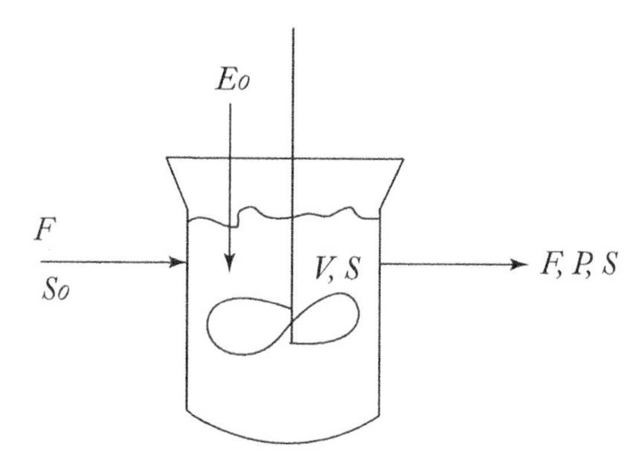

Fig. 4.16 CSTR with free enzymes

In the stirred tank reactor the reactor contents are well mixed. The concentration of the various components at the outlet stream are assumed to be the same as the concentration of those components in the reactor.

The substrate balance of a CSTR can be shown as
Input – Output + generation = rate of accumulation

$$F\,(S_o - S) - v(V_R) = V_R\, dS / dt \tag{4.48}$$

where V_R = reactor volume
For the steady state, $dS / dt = 0$

$$F / V_R = D = 1 / t = v_{max}\, S / (So - S)(K_M + S) \tag{4.49}$$

where $v = v_{max} S / (K_M + S)$ and v is the enzymatic reaction rate of MIchaelis- Menten type. and D is known as dilution rate and equal to reciprocal of residence time (t). Rearranging the equation (4.49) in the linearized form

where $t = \tau$, $$S = -K_M + v_{max} S t / (S_0 - S) \quad (4.50)$$

By plotting S vs $S t / (S_0 - S)$ we will get a straight, and from the slope and intercept' v_{max} and K_M can be evaluated respectively.

The Michaelis-Menten kinetic parameters are thus evaluated from a series of steady state CSTR runs with various flow rates..

Problem 4.6

A carbohydrate (S) decomposes in presence of an enzyme in a CSTR of volume of 0.5 m^3 with inlet flow rate of 0.1 m^3 / min and the substrate concentration,

$S_o = 0.30$ kmole / m^3. The kinetic parameters are $K_M = 0.2$ k mole / m^3, $v_{max} = 0.100$ kmol / (m^3) (min).

Calculate the substrate concentration at the outlet of the reactor.

Solution:

For a CSTR, $t = 0.5 / 0.1 = 5$ min

$V / F = t = (S_o - S) / v = (S_o - S) / [v_{max} S / (K_M + S)]$

or $t v_{max} S = (KM + S) (So - S)$

Putting the values of t, v_{max}, K_M, So

We have

$5 (0.1)S = (0.2 + S) (0.3 - S)$

or $0.5S = 0.6 + 0.1 S - S^2$

or $S^2 + 0.4S - 0.6 = 0$,

solving for S from the quadratic equation, we obtain

$S = (-0.4 \pm \sqrt{0.16 + 0.24}) / 2$

$= (-0.4 \pm 0.547) / 2 = -0.473, 0.073$

≈ 0.073 kmol / m^3

Problem 4.7

An enzymatic reaction with a substrate has the following Michaelis-Menten kinetic parameters: $K_M = 0.03$ kmole / m^3 and $v_{max} = 20$ kmole / (m^3) (min).

(a) If a CSTR is used to convert 95% of a substrate, with So = 10 kmol / m^3 a flow of 10 m^3 / hr, what is the volume of the reactor?

(b) If a plug flow reactor is used, what will be the volume of the reactor to achieve the same conversion?

Solution:

(a) $V / F = t$, $t = (S_o - S)(K_M + S) / (v_{max} S)$, in terms of conversion of substrate, $t = \delta [KM + So (1-\delta)] / v_{max} (1-\delta)$, putting the values of the parameters, we get $t = 0.95 [0.03 + 10 (1-0.95)] / 2.0 (1 - 0.95) = 5.03$ min
$V = Ft = 10 / 60 (m^3 / min) \times 5.03\ min = 0.84\ m^3$

(b) For plug flow reactor
$K_M \ln 1 / (1- \delta) + S_o \delta = v_{max} t$, putting the values
$0.03 \ln 1 / 1-0.95 + 10(.95) = 20 * t$
Or $t = 4.79$ min, $V = Ft = 10 / 60 \times 4.79 = 0.798$ m3

Problem 4.8

An enzymatic reaction of the Michaelis-Menten type is carried out in two one m^3 CSTRs in series at steady state.

(a) What will be the concentration of substrate leaving the 2nd reactor. The inlet substrate concentration, $S_o = 5\ kg / m^3$ and the flow rate is $0.5\ m^3 / min$. $K_M = 1.0\ kg / m^3$, $v_{max} = 2.0\ kg / (m^3) (min)$. The enzyme concentration in the two reactors is maintained at the same value all the time.

(b) Is the two-reactor system more efficient than one reactor whose volume is equal to the sum of two reactors?

Data given are:
$So = 5\ kg / m^3$, $F = 0.5\ m^3 / min$, $K_M = 1.0\ kg / m^3$, $v_{max} = 2.0\ kg / (m^3) / (min)$,
$t_1 = t_2 = 1 / .5 = 2$ min

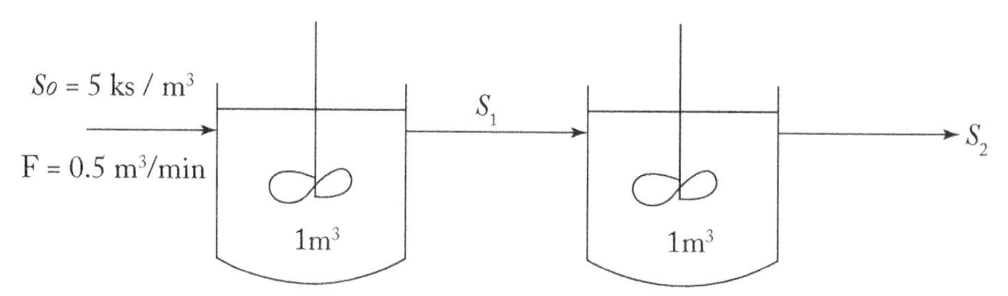

Solution of 4.8:

$\tau_1 = t_1 = (V/F) = 1/0.5 = 2$ min (residence time)

For the 1st CSTR, the outlet concentration is S1, the design equation is

$t1 = (S_o - S_1)(K_M + S_1) / v_{max} S_1$ and putting the values of the parameters

$2 (2) S_1 = (5 - S_1)(1 + S_1)$

Now solving for S_1

or $S_1^2 = 5$ or $S_1 = 2.23\ kg / m^3$, taking the positive value

For the 2nd CSTR, $S_1 = 2.23$ kg / m^3 is inlet concentration, the outlet concentration, S_2 is to be calculated. Now $t_2 = t_1 = V/F = 2$ min

$t_2 = (S_1 - S_2)(K_M + S_2) / v_{max} S_2$

Substitution of the values of known parameters in the above equation, we have

or $2 \times 2 S_2 = (2.23 - S_2)(1 + S_2)$

Solving for S_2

or $4 S_2 = 2.23 + 1.23 S_2 - S_{22}$

or $S_2^{\,2} + 2.77 S_2 - 2.23 = 0$

or $S_2 = [-2.77 \pm \sqrt{(2.77)_2 - 4(.1)(-2.23)}] / 2$

$= 0.65$ kg / m^3

b) $t = (5 - S_1)(1 + S_1) / (2) S_1$, here $t = V / F = 2 / 0.5 = 4$ min

Now writing the above equation in the form of a quadratic in S

$4 (2) S_1 = 5 + 4 S_1 - S_{12}$

or $S_1^{\,2} + 4 S_1 - 5 = 0$

solving for S_1

$S_1 = [(-4 \pm \sqrt{16 + 20})] / 2 = 1.0$ kg / m^3

So, the two reactor system is more efficient

Problem 4.9

A mixed flow reactor of volume, V = 6m^3 is used to carry out an enzymatic reaction with different concentration of substrate & enzyme using different flow rates. The inlet and outlet concentration of substrate at various flow rates are given.

Find out the kinetics of the reaction.

E_o, kmol / m^3	S_o, kmol / m^3	S (outlet) kmol / m^3	F, m^3 / hr
0.02	0.2	0.04	3.0
0.01	0.3	0.15	4.0
0.001	0.69	0.60	1.2

Solution

assuming the Michaelis-Menten equation, the mixed flow reactor design equation,

$V / F = t = (S_o - S) / k_2 E_o S / (k_M + S)$

Rearranging, $S = -K_M + k_2[tEoS / S_o - S] = f(s)$

where $f(s) = tE_0S / (S_0 - S)$

The following table is made for a plot, $\tau = t = V/F$ (residence time)

E_o	S_o	S	$V / F = \tau = t$, hr	... $E_oS / (S_o - S)$, f(S)
0.02	0.2	0.04	2	0.010
0.01	0.3	0.15	1.5	0.015
0.001	0.69	0.60	5	0.033

S is plotted against f(S) and from the linear plot, Slope = $k_2 = 25.2\ hr^{-1}$
from Intercept
$-K_M = -0.23$ or $K_M = 0.23\ kmol / m^3$

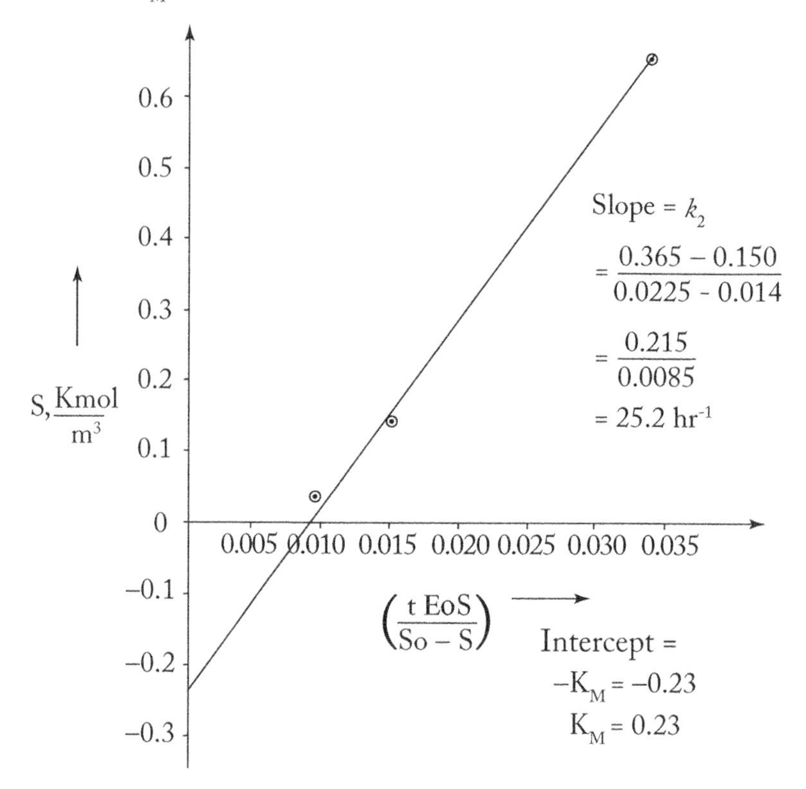

Fig. 4.17 Linear plot of S vs $tE_oS / (S_o - S)$ for CSTR

4.6.4 Kinetics for reversible reactions, two-substrate reactions, reaction with a co-factor (complex)

- **Reversible Reactions:**
  It has been observed that conversion of glucose to fructose by glucose-isomerase is considered to take place with reversible kinetics in the last step as shown below:

$$E + S \underset{k_{-1}}{\overset{k_1}{\rightleftharpoons}} ES$$

$$ES \underset{k_{-2}}{\overset{k_2}{\rightleftharpoons}} P + E \tag{4.51}$$

Considering the material balances on S, E_s and P in a well mixed vessel and using quasi steady state approximation, the rate can be given as

$$v = -\,dS/dt = dPdt = (vs\,/\,Ks)S - (vp\,/\,Kp)P\,/\,[1 + S\,/\,K_S + P\,/\,KP] \tag{4.52}$$

Where $vp = k_{-1}\,vs\,/\,k_2$, $Kp = k_1K_S\,/\,k_{-2}$ (4.53)
and v_s and K_S are identical to v_{max} and K_M respectively.

Enzyme Kinetics with two substrates
In this case it is presumed that a ternary complex is formed with two substrates (S1, S2) and Enzyme, E. The possible reaction scheme may be given as

Reaction	Dissociation Equilibrium constants
$E + S_1 \rightleftharpoons ES_1, K_1$	K_1, K_2, K_{12}, K_{21} are the equilibrium constants
$E + S_2 \rightleftharpoons ES_2, K_2$	
$ES_1 + S_2 \rightleftharpoons ES_1S_2, K_{12}$	
$ES_2 + S_1 \rightleftharpoons ES_1S_2, K_{21}$	

$$ES_1S_2 \xrightarrow{k} \text{-}P + E \tag{4.54}$$

and the rate, v can be given as, $v = k(ES_1S_2)$ (4.55)

k is the rate constant for product formation from ES_1S_2
Assuming equilibrium for the first four equations and constant mass of the enzyme, the rate can be given as

$$v = kE_0\,/\,[1 + K_{21}\,/\,S + K_{12}\,/\,S + 1\,/\,2(K_2K_{21} + K_1K_{12})\,/\,S_1S_2] \tag{4.56}$$

Further it is to be noted that $K_1K_{12} = K_2K_{21}$
The equation (4.56) may be simplified as

$$v = v_{max}{}^{*}\,S_1\,/\,(K_1{}^{*} + S_1) \tag{4.57}$$

$$\text{where, } v_{max}{}^{*} = kE_0S_2\,/\,(S_2 + K_{12}) \tag{4.58}$$

$$\text{and } K_1{}^{*} = K_{21}S_2 + K_1K_{12}/\,(S_2 + K_{12}) \tag{4.59}$$

If S_2 is kept constant, the reactions will follow Michaelis-Menten reaction kinetics.

- **Enzyme Kinetics with a cofactor**

  The cofactor may be a metal ion or a coenzyme like NAD or FAD and participates in one substrate- enzymetic reaction.

  The reaction scheme is

$$\underset{\substack{\text{(apo}\\\text{-enzyme)}}}{A} + \underset{\text{(cofacte)}}{C} \underset{}{\overset{Kc}{\rightleftharpoons}} \underset{\text{(enzyme)}}{E} + S \overset{K_S}{\rightleftharpoons} ES \xrightarrow{k} P + E \quad (4.60)$$

Where A is the apo-enzyme, and C is a cofactor, Kc is the equilibrium constant for the first reversible step and Ks is the equilibrium constant for the 2nd reversible step and k is the rate constant for the last irreversible step.

The overall rate has been given as,

$$v = k\,E_o\,CS / [CS + k_s\,(C + K_c)] \quad (4.61)$$

where C is the concentration of the co-factor.

If $C << K_c$ i.e. very little cofactor present, v is first order in C. Again, for $C >> K_c$, the rate is independent of C, but a function of S.

4.6.5 Kinetics of multiple substrates on a single enzyme

A polymeric substrate like starch does not behave as a single substrate when hydrolysed in presence of an enzyme.

The thinning of starch solution by amylases is an example of this type. The glucose production from starch by amyloglucosidase indicates the presence of many simultaneous reactions. The overall reaction considered is,

amyloglucosidase

Glucose polymers ------------------------------>❼ glucose

(maltose, maltotriose etc)

To develop a rate model for such systems, let us consider two substrate acting on a single enzyme. The reaction scheme may be presented as,

$$E + S_1 \rightleftharpoons ES_1,\ K_1$$

$$ES_1 + S_2 \rightleftharpoons ES_2\ K_2 \quad (4.62)$$

K_1 and K_2 are the equilibrium constants.

Slow steps:

$$ES_1 \overset{k_1}{\text{---->}} E + P1,\ K_1$$

$$ES_2 \overset{k_2}{\text{---->}} E + P2\ K_2 \quad (4.63)$$

Each substrate is assumed to bind certain fraction of the enzyme present. Following the earlier method and assuming quick equilibrium, the rates are given as,

$$dS_1 / dt = K_1EoS_1 / K_1 / [1 + S_1 / K_1 + S_2 / K_2] \quad (4.64)$$

$$dS_2 / dt = K_2EoS_2 / K_2 / [1 + S_1 / K_1 + S_2 / K_2] \quad (4.65)$$

The above simultaneous differential equations may be numerically solved for S_1 & S_2 with time.

4.7 Effects of pH and Temperature on Enzyme Kinetics

The enzyme as a protein may contain both positively or negatively charged groups at any given pH. Such ionisable groups are parts of the active sites. Acid and base type catalytic actions have been linked to several enzyme reaction mechanisms. However, the catalytically active enzyme may be a large or a small part of the total enzyme present depending upon the pH.

Let us consider the following simple model of the active site ionization state:

$$E \underset{\underset{K_1}{+H^+}}{\overset{-H^+}{\rightleftharpoons}} E^- \underset{\underset{K_2}{+H^+}}{\overset{-H^+}{\rightleftharpoons}} E^{2-} \quad (4.66)$$

Here E^- denotes the active enzyme, while E and E^{2-} are inactive forms by protonation and deprotonation of the active site, E^-, respectively and K_1 and K_2 are the equilibrium constants for the indicated reactions. The equilibrium relations are:

$$h^+ e^- / e = K_1, \quad h^+ + e^{2-} / e^- = K_2 \quad (4.67)$$

We have to find out the active part, e- as the fraction of the total enzyme. The total enzyme e_o can be given as

$$e_o = e + e^- - + e^{2-} \quad (4.68)$$

The active fraction, y-, is e- / eo and is given by

$$y- = 1 / [1 + h^+ / K_1 + K_2 / h^+] \quad (4.69)$$

There is a single maximum of y^- with respect to pH which occurs at the value,
H^+ optimum $= \sqrt{K_1K_2}$ or(PH)opt = ½ ($PK_1 + PK_2$)
Where PK_1 is defined as $- \log K_1, PK_2 = - \log K_2$ (4.70)
Protonation and deprotonation are very rapid processes compared to most reaction rates in solution.
The influence on the maximum reaction velocity, v_{max}' is thus obtained as,

$$v_{max}' = ke_oy_- = v_{max} / [1 + h^+/ K_1 + K_2 / h^+] \quad (4.71)$$

Further pH also has effect on $K_{M,}$ the Michaelis – Menten constant, if the substrate does not have different ionization steps.

For the case of ionizing substrate, the following mechanisms can be used to develop the rate with pH effect,

$$SH^+ + E \rightleftharpoons ESH \xrightarrow{k_2} E + HP^+$$

$$SH^+ \rightleftharpoons S + H^+ \quad (4.72)$$

The rate of enzymatic reaction can be developed as

$$v = v_{max}\, S / [k_M'\,(1 + K_1 / H^+) + S] \quad (4.73)$$

- **Effect of Temperature:**

  In chemical reactions, the temperature dependency of the rate constant, k is given by Arrhenius equation,

$$k = Ae^{-E/RT} \quad (4.74)$$

where A is the frequency factor and E is the activation energy of the reaction, R is the gas-law constant, T is the temperature in °K.

The above equation may be used to describe the temperature effect on the rate constant of many enzyme catalysed reactions. But it has been noted that the temperature range is small. At some higher temperature range, the rate constant decreases with temperature leading to deactivation of the enzyme. For many proteins, denaturation begins to occur at 45°C to 50°C. Arrhenius rate-temperature line breaks at high temperature as shown below:

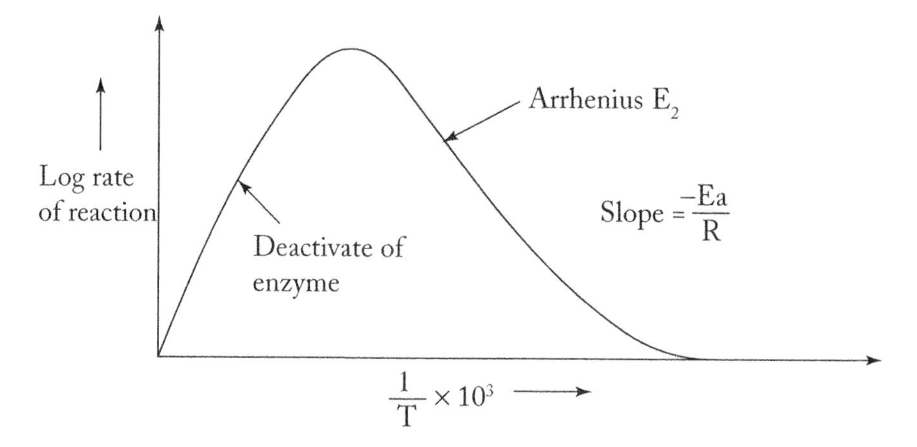

Fig. 4.18 Arrhenius plot of temperature dependency with active and deactivated enzyme at higher temperature

The rate parameter, v_{max} can be shown as

$$v_{max} = k_2[E_0] = A\exp(-Ea / RT)E_0 \quad (4.75)$$

where $k_2 = Ae^{-Ea/RT}$

where E_a is the activation energy of active enzyme. The ascending part of the Arrhenius plot is known as thermal denaturation. The kinetics of thermal denaturation is given as a first order decay,

$$-dE / dt = k_d E \quad (4.76)$$

or

$$E = E_o \text{ exponent}(-kdt) \quad (4.77)$$

where kd is the denaturation constant which also varies with temperature according to Arrhenius equation,

$$k_d = A \text{ exponent}(-Ed / RT) \quad (4.78)$$

The activation energies of enzme catalysed reaction, Ea, are in the range of 4 to 20 K.cals / g.mole and E_d, the activation energy for deactivation vary from 40 to 130 K.cals / gmole.

The overall effect of temperature on v_{max} (the maximum enzyme rate) is then expressed as,

$$v_{max} = A \exp(-Ea / RT) E_0 \quad (4.79)$$

temperature may also affect the other kinetic parameter of the enzymatic reaction, the Michaelis-Menten constant, K_M which may increase with temperature reducing the overall rate of reaction. If K_M is interpreted as equilibrium constant, the temperature dependency can be given as

$$K_M = \exp(-\Delta G / RT) = \exp(-\Delta Hr / RT)\exp(\Delta S / R) \quad (4.80)$$

and the plot of ln K_M vs 1 / T will give a straight line, with a slope of $-\Delta Hr/R$.

4.8 Enzyme Deactivation

It has been observed that a multitude of physical and chemical parameters cause changes in native protein geometries and chemical structure, along with concomitant reduced activity. Some factors of denaturation of protein are (i) elevated temperature, (ii) solution pH, (iii) mechanical forces etc. The other denaturing environments are solubility, tendency for crystallisation or gel formation, viscosity, antibody binding of protein solution.

4.8.1 Simplest deactivation model and the enzyme kinetics.

A simple deactivation model assumes that an active enzyme molecules, E_a, undergo an irreversible structural or chemical change to an inactive form:

E_a ------>❼ Ei

With a rate proportional to the active enzyme concentration, E_a

$$dE_a / dt = -k_d E_a \tag{4.81}$$

integrating

$$E_a = E_{ao} e^{-kdt} \tag{4.82}$$

Let us combine the deactivation model with simple catalytic reaction of the Michaelis-Menten type:

$$E_a + S \underset{k_{-1}}{\overset{k_1}{\rightleftharpoons}} E_aS \xrightarrow{k_2} E_a + P \tag{4.83}$$

$$E_a \xrightarrow{k_d} E_i \tag{4.84}$$

Assuming that the deactivation process is much slower than the reaction in equation (4.82) and using the quasi steady state approximation for (E_aS) gives the rate, v

$$v = dS / dt = K_2 E_{total} S / (K_M + S) \tag{4.85}$$

where E_{total} is the total concentration of active enzyme both in free and complexed forms. The rate of change of E_{total} is expressed as

$$dE_{total} / dt = -k_d\ E_a \tag{4.86}$$

using the Pseudo steady state assumption of E_aS, we have,
$d(E_aS)dt = k_1(E_a)(S)\text{-}k_{-1}(E_aS) - k_2(E_aS) = 0$

Or

$$(E_aS) = K_1(E_a)(S)(K_{-1} + K_2) = E_a(S) / K_M \tag{4.87}$$

Or $(E_{total} - E_a) = E_aS / K_M$
Solving for E_a

$$E_a = E_{total} / (1 + S / K_M) \tag{4.88}$$

Substituting E_a in equation (4.85)
we have,

$$dE_{total} / dt = -k_d\ E_{total} / (1 + S / k_M) \tag{4.89}$$

So using the equations (4.85) and (4.89), the substrate conversion and Enzyme deactivation rates are coupled, showing that the enzyme deactivation also depends on the substrate concentration.

4.9 Effect of Shear, Surface Tension

Mechanical forces in the form of a shear or shear rate may cause deactivation of enzyme, as the structures of the protein molecules are distorted. The activity of an enzyme has been found to decrease with increase of the product of shear rate (γ) and exposure time (θ).

Surface tension often causes denaturation of proteins. Foaming or frothing in protein solution commonly leads to denaturation of protein decreasing the activity of the enzyme.

4.10 Enzyme Kinetics for Insoluble Substrate

There are many reactions when the enzyme in solution acts on an insoluble substrates. Examples are enzymatic hydrolysis of cellulose in presence of cellulase enzymes, hydrolysis of newsprint or saw dust, food spoilage, solid waste treatment etc.

The usual mechanisms of such reactions are based on the assumption that enzyme in solution adsorbs on the surface of the insoluble substrate. In this condition, the reaction rate is directly proportional to the total concentration of enzyme; a limiting rate is reached as the enzyme concentration is increased.

4.10.1 Kinetic model

For the derivation of a rate model of such systems, it is assumed that A, number of vacant sites of a substrate combine with enzyme by adsorportion. The mechanism can be presented as

$$E(\text{soluble}) + A(\text{Solid}) \underset{k_{des}}{\overset{k_{ads}}{\rightleftharpoons}} EA(\text{solid}) \text{ -----------> } EA(\text{solution}) \xrightarrow{k_3} P + E \quad (4.93)$$

Where kads is the forward rate constant for adsorption and kdes is the backward rate constant for the reversible desorption reaction and k_3 is the rate constant for irreversible product formation.

If ao is the total number of moles of adsorption sites on the solid substrate per unit volume of the reaction mixture, we have

$$a_o = a + (E_a)$$

Now the adsorption equilibrium KA is given as

$$K_A = k_{desorp} / k_{adsorp} = [E]\,[a] / [E_a] = [E]\,[a_o - E_a] / [E_a]$$

Solving for [Ea],

$$[E_a] = [E]\,[a_0] / (K_A + E) \quad (4.90)$$

Assuming the irreversible decomposition of Ea

$$E_a \xrightarrow{k_3} \text{❼ } P + E \quad (4.91)$$

From the enzyme balance, free enzyme, E is

$$E = E_0 - E_a$$

If the initial concentration of the enzyme, E_o, is much larger than the substrate ($E_o >> a_o$), we can assume free enzyme, $E \approx E_o$. For the equation (4.91), assuming irreversible decomposition of Ea, the rate equation can be derived as

$$v = k_3 (E_a) = k_3 a_o E_0 / (K_A + E_0) \quad (4.92)$$

The rate equation, can be linearized and the Lineweaver- Burk plot of $1 / v$ vs $1 / E_0$ will give a straight line with a slope of $K_A / k_3 a_o$ and the intercept of $1 / k_3 a_o$. This type of behaviour has been observed for degradation of solid substrate (thiogel) by the soluble enzyme, trypsin.

4.10.2 Enzyme kinetics of cellulose degradation

The enzymatic hydrolysis of cellulose is carried out by the enzyme, cellulase. Cellulose materials are assumed to be composed of 1) crystalline phase (Sc), 2) permeable amorphous phase (Sa) and 3) non-hydrolyzable inerts (Sx) and their rates are different.

The enzyme cellulose is composed of three components: i) endoglucanases, ii) cellobiohydrolases & iii) cellobiose. But these are assumed to act as a single enzyme.

- **Kinetic model of Ryu et al (1982)**

  The above model was developed on the basis of the following assumptions:

  1. The cellulase systems is represented as a single enzyme
  2. The cellulase enzyme is first adsorbed, E* on the surface of the cellulose, forms an enzyme substrate complex, E*S and hydrolysed to release products and enzyme.
  3. The products (glucose and cellobiose) inhibit the cellulose enzyme competitively.

  The reaction scheme is represented as

$$E \underset{k_{de}}{\overset{k_{ad}}{\rightleftharpoons}} E^* \quad (4.93)$$

$$E^* + Sa \underset{k_{2a}}{\overset{k_{1a}}{\rightleftharpoons}} E^*Sa \quad (4.94)$$

$$E^* + Sc \underset{k_{2c}}{\overset{k_{1c}}{\rightleftharpoons}} E^*Sc \quad (4.95)$$

$$E^* + Sx \underset{k_{2x}}{\overset{k_{1x}}{\rightleftharpoons}} E^*Sx \quad (4.96)$$

$$E^* + P \underset{k_{2P}}{\overset{k_{1P}}{\rightleftharpoons}} EP \quad (4.97)$$

$$E^*Sa \xrightarrow{k_{3a}} E + P \quad (4.98)$$

$$E^{*}Sc \xrightarrow{k_{3c}} E + P \tag{4.99}$$

on the basis of Langmuir type adsorption isotherm,

$$E^{*} / E^{*}_{max} = K_e E / (1 + K_e E) \tag{4.100}$$

where E^{*}_{max} is the concentration of the maximum adsorbed enzyme and K_e is the adsorption constant. At low enzyme concentration ($K_e E << 1$), (4.101)
The equation (4.100) becomes,

$$E^{*} = K_e\ E^{*}_{max},\ E = K_d\ E \tag{4.102}$$

The rate of product formation can be given as

$$dP / dt = k_{3a} E^{*}S_a + K_{3c}\ E^{*}S_c \tag{4.103}$$

Now the total enzyme content (Eo) is

$$E_o = E + E^{*} + E^{*}S_a + E^{*}S_c + E^{*}S_x + EP \tag{4.104}$$

Now to develop the rate equation, we follow the Briggs-Haidane approach of the pseudo steady state. That is
$d(E^{*}S_a) / dt = 0$, $d (E^{*}S_c) / dt = 0$, $d (E^{*}S_x) / dt = 0$, $d (EP) / dt = 0$
from those equations in terms of variables, we can estimate

$$E^{*} = K_a / S_a (E^{*}S_a) = (K_c\ E^{*}S_c) / Sc \tag{4.105}$$

where

$$K_a = (k_{2a} + k_{3a}) / k_{1a} \tag{4.106}$$

$$K_c = (k_{2c} + k_{3c}) / k_{1c} \tag{4.107}$$

Substituting equation (4.104) in equation (4.103)

$$dP / dt = (k_{3a} + k_{3c} KaSc / K_c Sa)(E^{*}Sa) \tag{4.108}$$

Now to obtain an equation for ($E^{*}Sa$), we use the equations for the rates of intermediate complex and the equations of pseudo steady state for four species, (E^{*}, $E^{*}Sa$, $E^{*}Sc$, $E^{*}Sx$, EP), (the equations are not shown), and we can obtain,

$$(E^{*}Sa) = E_0 / [K_c / Sa(1 + 1 / Kd + Sa / Ka + Sc / Kc + Sx / Kx + P / K_P Kd)] \tag{4.109}$$

Where $Ka = (k_2 a + k_3 a) / k_1 a$, $K_c = (k_2 c + k_3 c) / k_1 c$

$K_x = k_2 x / k_1 x$, $Kp = k_2 p / k_1 p$

---

* formations (E*Sa, E*Sc, E*Sx, EP)

Substituting (E^*S_a) from equation (4.108), into equation (4.107), we get,

$$dP / dt = (k_{3a}S_a / K_a + k_{3c}S_c / K_c) E_0 / [1 + 1 / K_d + S_a / K_a + S_c / K_c + S_x / K_x + P / K_pK_d] \quad (4.110)$$

Defining
$$\Phi = S_c / (S_a + S_c) \quad (4.111)$$

and
$$\gamma = S_x / (S_a + S_c + S_{x)} = S_x / S \quad (4.112)$$

where $S = S_a + S_c + S_x$

The equation (4.109) can be written in terms of Φ and γ

$$dP / dt = [k_{31}(1\text{-}\Phi) + k_{3c}\Phi K_a / K_c]E_0(1\text{-}\gamma)E_0(S / [K_a(1 + 1 / K_d) + \{([1\text{-}\gamma)[(1\text{-}\Phi)\Phi K_a / K_c] + \gamma K_a / K_x\}S_0] \quad (4.113)$$

The above equation may be simplified if we use the condition of initial rate, i.e. $S = S_o$, $P = 0$, $\Phi = \Phi$ and putting $\gamma = 0$ when pure cellulose is a substrate

$$\frac{dP}{dt} = \frac{[k_{3a}(1-\phi) + k_{3c}K_a / K_c]E_0(1-\gamma)S}{K_a\left(1+\dfrac{1}{K_d}\right)\left\{(1-\gamma)\left[(1-\phi)+\phi\dfrac{K_a}{K_c}\right]+\gamma\dfrac{K_a}{K_x}\right\}S_0} \text{ at } t = 0 \quad (4.114)$$

The equation (4.114) can be rearranged in the form of Michaelis-Menten equation

$$dp / dt = v_{max}'S_0 / (K_M' + S_0) \quad (4.115)$$

where v_{max}' and K_M' at t = 0 are equivalent to the apparent maximum reaction rate and the apparent Michaelis-Menten constant respectively, which can be given as

$$v_{max}' = [k_{3a}(1\text{-}\Phi) + k_{3c}\Phi K_a / K_c]E0 / [(1\text{-}\Phi) + \Phi K_a / K_c] \quad (4.116)$$

$$K_M' = K_a(1 + 1 / K_d) / [(1\text{-}\Phi) + \Phi K_a / K_c] \quad (4.117)$$

The v_{max}' and K_M' can be calculated from the Lineweaver-Burk plot.

4.11 Summary

The chapter deals with the classical enzyme kinetics along with various inhibition systems. Various techniques for the estimation of kinetic parameters of several kinetic models have been employed and their relative merits and demerits have been discussed. The effects of different environmental conditions on the enzymatic kinetics have been quantitatively analysed. Simple reactor design with various enzyme kinetics have been presented with specific reaction systems. Simple enzyme deactivation models have been developed and incorporated in the Michaelis-Menten equation. Kinetic models have been deduced for the systems of soluble enzyme and insoluble substrates.

Problems (Exercise)

4.1 An enzymatic reaction, S ---------> (enzyme) ❼ R takes place in a batch, reaction with the following kinetics:
$v = 0.2\ S / (2 + S)$ k mol / (m^3)(min)
If the substrate S_o = 1 kmole / m^3 is introduced, what time is necessary for 90% conversion of the substrate?

4.2 Initial rates of an enzyme catalysed reaction for various substrate concentrations are given below:

S, kmole / m^3	4.1×10^{-3}	5.2×10^{-4}	4.0×10^{-5}	1.06×10^{-5}	5.1×10^{-6}
v_0, kmol / (m^3) (min) $\times 10^6$	177	125	80	67	43

(a) Evaluate v_{max} and K_M by Lline-weaver Burk Plot
(b) Using Eadi-Hofstee plot, determine v_{max} & K_M
(c) Use Hanse-Wolf plot to evaluate v_{max} and K_M

4.3 The following data are available for an enzymatic reaction with no inhibitor and with inhibitor. The initial rate data with substrate are given below:
Concentration of the inhibitor, I, is
$I = 10^{-3}$ kmole / m^3

S, kmol / m^3	3.3×10^{-4}	5.0×10^{-4}	6.7×10^{-4}	1.65×10^{-3}	2.21×10^{-3}
$v_0 \times 10^6$ without inhibitor	56	71	88	129	149
v_0, with Inhibitor $\times 10^6$	37	47	61	103	125

Evaluate K_I, v_{max}, K_M'

4.4 An enzymatic reaction gives the following data:
The initial substrate concentration, S_o = 1.0 kmole / m^3

δ, Substrate Conversion	0.32	0.84	0.99
Time, t hrs	2	6	10

Use Michaelis-Menten equation and evaluate K_M and v_{max} by integration method.

4.5 A carbohydrate (S) decomposes in the presence of an enzyme E. The Michaelis-Menten parameters were found to be:
v_{max} = 0.1 kmol / (m^3) (min)
K_M = 0.2 kmol / m^3
Calculate substrate concentration with time and prepare S vs t curve when the initial substrate concentration, S_o = 0.4 kmole / m^3.

4.6 A chemostat is used to carryout the above reaction with the kinetic parameters: $v_{max} = 0.1$ kmole / m^3 min, $K_M = 0.2$ kmol / m^3 and the flow rate is 0.1 m^3 / min and the initial substrate concentration is $S_o = 0.5$kmol / m^3. If the reactor volume is 1m3, what is the steady state substrate concentration at the exit of the reactor?

4.7 An enzymatic reaction for a specific substrate has the kinetic parameters or Michaelis-Menten equation
$K_M = 0.03$ kmole / m^3,
$v_{max} = 12$ kmole / m^3 min

(a) What should be the size of a steady state CSTR to convert 95% of incoming substrate, $S_o = 10$ kmole / m^3 with a flow rate of 1.5 m^3 / min?

(b) What should be the size of the reactor, if a plug flow reactor is used in the place of CSTR?

4.8 Cane Sugar is hydrolysed into glucose and fructose by the enzyme, invertase. The following table gives the rate of sugar inversion for various initial substrate concentrations. The amount of invertase was maintained constant in each case

S_o, k_s / m^3	18.9	67.0	98.5	199	300	400
v, k_s / (m^3)(min)	0.19	0.21	0.24	0.27	0.25	0.23

A Lineweaver plot of the above set of data did not give a straight line when the substrate concentration was high. To take into account of sugar inhibition effect, the following reaction scheme is suggested

$$E + S \underset{}{\overset{K_M}{\rightleftharpoons}} ES \xrightarrow{k_2} E + P$$

$$ES + S \overset{Ksi}{\rightleftharpoons} ESS$$

$$v = v_{max} S / [K_M + S + \frac{S^2}{Ksi}]$$

(a) Derive the rate equation using Michaelis-Menten model.

(b) Derive three kinetic parameters using the above experimental data. The parameters are v_{max}, K_M & Ksi.

4.9 The initial rates of reaction for the enzymatic cleavage of deoxyguanosin to phospite were measured as a function of initial substrate concentration as follows:

S_o mol / m^3	6.7	3.5	1.7
v_o mol / (m^3) (min)	0.30	0.25	0.16

(a) Calculate the Michaelis-Menten constants of the above reaction.

(b) When the inhibitor was added, I = 1.46 mol / m^3, the initial reaction rate decreased as follows:

S_o, mol / m^3	6.7	3.5	1.7
v_o, mol / (m^3) (min)	0.11	0.08	0.06

Is this competitive inhibition or noncompetitive inhibition? Justify your answer by showing the effect of the inhibitor.

4.10 Substrate (S) and Enzyme (E) flow through a CSTR of V = 1m^3. From the entering and exiting concentration of substrate and varied flow rates, find the rate equation in the form of Michaelis-Menten equation.

E_o kmol / m^3	S_o, kmol / m^3	S, kmol / m^3	F, m^3 / hr
0.02	0.2	0.04	0.5
0.01	0.3	0.15	0.67
0.001	0.70	0.60	0.20

Hint: $S = -K_M + k_2(E_0St / (S_0\text{-}S))$, where t = V / F

4.11 Cellulose breaks down according to the following reaction in presence of cellolase

Cellolase

Cellulose --------------------------> ❼ Sugar (Celluboise + glucose)

Both Celluboise and glucose inhibit the breakdown. Several experiments were made in a CSTR at 50°C using a feed of finely shreeded cellulose (S_o = 25 kg / m^3) Enzyme conc, E_o = 0.01 kg / m^3 is same for all the runs. Assume non-inhibition

Exit S, kg / m^3	1.5	4.5	9.0	21.0
t, hr	9.8	4.65	2.85	0.6

Find a rate equation to represent the breakdown of cellulose by cellulase

References

Adams, M.W.W., and R.M. Kelly, Enzymes from microorganism in Extreme Environments, Chemical Engineering News (Dec. 12) 32–42, 1995

Bailey, J.E. and D.F. Ollis, Biochemical Engineering Fundamentals, 2nd ed. McGraw Hill Book Co., New York, 1986

Bernhard, A. "Structures and Functions of enzymes", W.A. Benjamin, New York, 1968

Blanch, H.W. and D.S. Clark, Biochemical Engineering, Marcel Dekker, Inc. New York, 1996

Bohinski, R.C., Modern Concepts in Biochemistry, P. 105, Boston, M.A. Allya and Bacon, Inc. 1976

Briggs, G.E. and J.B.S. Haldane, "A Note on the kinetics of Enzyme Action", Biochem.J. 19, 338–339, 1925

Brown, A.J., "Enzyme Action", J. Chem. Soc. 81, 373–388, 1902

Charm, S.E. and B.L. Wang, "Enzyme Inactivation with Shearing", Biotechnol. Bioeng., 12, 1103–1109, 1970

Crueger, W. and A. Crueger, Biotechnology: A text book of Industrial Microbiology, pp. 1613–162, Medison, W.I. Science Tech. Inc. 1984

Dixon, M. and E.C. Webb, Enzymes, 3rd ed Academic Press, 1979

Eadie, G.S., "the inhibition of cholisterage by Physostigmine and Prostigmine" J. Biol. Chem., 146, pp 85–93, 1981

Hofstee, B.H.J., "Specificity of Esterages: I identification of two pancreatic Alieesterases", J. Biol Chem. 199, 357–364, 1952

Jones, E.O. and J.M. Lee, "Kinetic Analysis of Bio Conversion of Cellulose in Attrition Bioreactors", Biotechnol. Bioeng. 31, 35–40, 1988

Laidler, K.J. and P.S. Bunting, "The Chemical Kinetics of Enzyme Action", 2nd ed, Oxford University Press, London, 197

Levenspiel, O. Chemical Reaction Engineering, 3rd ed, pp 612–622, John Wiley & Sons, 1999

Lineweaver, H. and D. Burk, "The determination of enzyme dissociation constants", J. Amer, Chem. Soc. 56, 658–666, 1934

Mc. Laren, A.D. and L. Parker, "Some Aspects of Enzyme Reactions in heterogeneous systems" Advan Enzymol 33, 245, 1970

Michaelis, L. and M.L. Menten, "The Kinetik der Invertin Wir Kung", Biochem. ZeitSchr, 49, 333-360, 1913

Ryu, D.D.Y., S.B. Lee, T. Tassinazi and C. Maey, "Effect of Compression Milling on Cellulose structure and on Enzymatic Hydrolysis Kinetics", Biotechnol. Bioeng. 26, 10473–1067, 1982

Ryu, S.K. and J.M. Lee, "Bioconversion of waste cellulose by using an Attrition Bioreactor" Biotechnol. Bioeng., 25, 53–65, 1983

Schmid, R.D., "Stabilized soluble Enzymes" in Advances in Biochemical Engineering, Vol – 12, Immobilized enzymes edited by T.K. Ghosh, A. Fiechter, N. Blakebrough.

Thomas, C.R. and P. Dunnill, "Action of shear on enzymes studies with catalase and urease, "Biotechnol. Bioeng, 21, 2279-2302, 1985.

Wieseman, A. and B.J. Gould, "Enzymes, their nature and Role" pp 70–75 London, Hutchinson Education Ltd., 1970

Yang, S.T. and M.R. Okos, "A new graphical method for determining parameters in Michaelis Menten types", Kinetics for enzymatic lactose hydrolysis", Biotechnol. Bioeng. 34, pp. 763–773, 1989.

CHAPTER 5

Kinetics of Immobilized Enzymes and Reactor Design

5.0 Introduction

Enzymes can be immobilized on the surface or inside of an insoluble matrix either by physical or chemical methods. The advantages of immobilization of enzymes include (i) the separation of the enzyme from the reaction solution and (ii) stability of the enzyme structure. Immobilized enzymes can be easily retained in a continuous flow reactors.

5.1 Immobilization Techniques

Various immobilization techniques are available:

Chemical method involves the covalent attachment through functional groups, copolymerization of the enzyme; cross-linking using multi-functional reagents (Glutaradehyde), microencapsulation. Major immobilization methods can be presented as

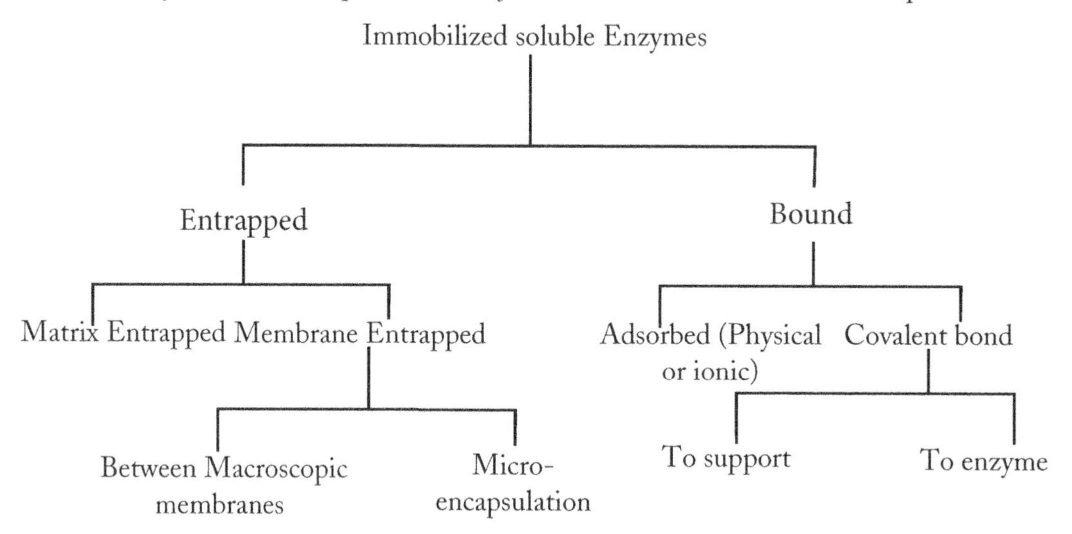

Membrane entrapment is possible with hollow fiber units. Membranes may be made of nylon, cellulose, polysulfone, polyacrylate. In microencapsulation, spherical membranes contain the enzyme solution.

Surface immobilization involves the immobilization on the surface of the solid support materials by physical adsorption or covalent bonding. Support materials may be alumina, silica, porous glass, ceramics, clay and other organic materials like cellulose (CMC, DEAE cellulose), starch, activated carbon ion-exchange resins like Amberlite, Sephadex, Dowex etc.

Covalent binding consists of the retention of enzymes on the surface by covalent bond formation through some functional groups.

Functional groups on support materials are activated by using chemical reagents like cyanogen bromide, glutaraldehyde etc.

5.1.1 Advantages of immobilized enymes or cells

Immobilized Enzymes or cells act as biocatalysts in packed bed reactors. There are some potential advantages of immobilized enzymes or cells over suspended systems with free enzymes or cells, and those may be listed as below:

1. High-Enzyme or cells concentration
2. Enzymes/ cells reuse and elimination of costly Enzymes/ cells recovery and recycle
3. No washout problems at high dilution rate for cells
4. Better performance of immobilized Enzymes or cells as biocatalysts due to good physical contact, reduced nutrient-product gradients or pH gradients
5. Improved structural stability for enzymes and genetic stability for microbial cells
6. Protection from shear damage of enzymes or cells

Methods of immobilization of Enzymes or cells

They may be briefly classified as

(a) Physical entrapment within porous matrices
(b) Gelation of polymers in solution
(c) Precipitation of polymers with Enzymes or cells. Such polymers are polystyrene, cellulose triacetate, collagen etc.
(d) Ion-exchange gelation: e.g. ca-alginate, carboxy- methyl cellulose, K-Carragenan and chitosan, polyphosphate etc.
(e) Polycondensation: e.g. epoxy-polyurethane, silicone gel, gelatin – glutaraldehyde etc
(f) Polymerization: e.g. Polyacrylamide beads, acrylamide, methacrylate amide
(g) Encapsulation: It is another method of entrapment. Microcapsules are hollow spherical particles, made of semipermeable membrane
(h) Entrapment: Entrapment of cells is found in macroscopic membrane based reactors, e.g. hollow fiber bioreactor
(i) Physical adsorption: Enzymes or cells are immobilized by physical adsorption
(j) Covalent bonding: Adsorption capacities of covalent bonding may vary. Adsorption capacity of porous silica is 2mg/gm. That of wood chips is 250 mg/gm, porous glass has the adsorption capacity of 10^9 cells/gm.

Cells or enzymes are also adsorbed on positively charged support surfaces like ion-exchange resin, gelatin. Cells are also attached to negatively changed surfaces by covalent bonding or H-banding.

Common materials for cell adsorption are: porous glass, porous silica, alumina, ceramics, gelatin, chitosan, activated carbon, wood chips, polypropylene, ion-exchange resins (DEA – Sephadex, CMC)

Covalent bonding is most widely used method for enzyme immobilization. Binding surfaces are to be specially treated with coupling agents like glutaraldehyde or carbodimide for covalent bonding.

A good support material should have rigidity, chemical inertness, high loading capacity for cells or enzymes For gel entrapment, gels should have porosity and small particle size to avoid intraparticle diffusion effect. Some immobilized cell or enzyme systems are illustrated in fig.5.1

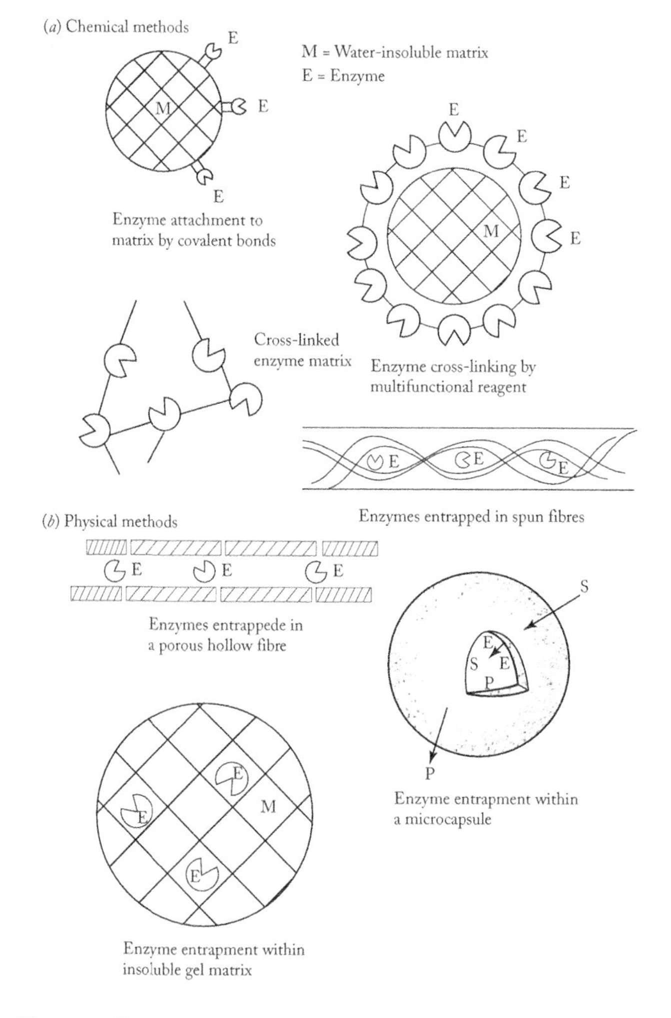

Fig. 5.1 Some immobilized cell systems (Baily & ollis, 1986)

Immobilized biological films

Biological films are generated by multi-layer growth of cells on solid support surfaces. These films have been utilized in many bioprocesses like biological waste water treatment, mold fermentation etc. In mixed culture microbial films, the presence of some polymer-producing organisms help in biofilm fermentation, enhancing the stability of the biofilm.

In stagnant biofilms in a packed bed system, nutrients diffuse into the biofilm and the products formed come out by diffusion into the liquid nutrient medium. The thickness of a biofilm affect the performance of the reactor. Thin films may cause low conversion due to less biomass concentration. Thick films may have diffusion limitation resulting lower yield. So there is optimal biomass thickness producing maximum rate of bioconversion. Mass transfer, kinetics and reactor design with these systems are discussed in following sections.

5.2 External Diffusion on the Nonporous Surface Immobilized Enzymes

For surface-bound enzymes, external diffusion from the bulk of solution can affect the rate of enzymatic reaction. Their significance can be characterized by a dimension less number known as Damköhler number, D_a defined as,

D_a = Maximum enzymatic rate of reaction / Maximum rate of diffusion rate

$$= v'_{max} / k_L S_b \qquad (5.1)$$

Where S_b is the substrate concentration in the bulk liquid (kmoles/m^3), k_L is the mass transfer coefficient (m/s). v'_{max} is the maximum enzymatic rate, kmoles/ (m^2) (s).

For non-porous solid support on which an enzyme is immobilized, a substrate diffuses from the bulk liquid through a thin liquid film surrounding the support surface as shown in figure 5.2

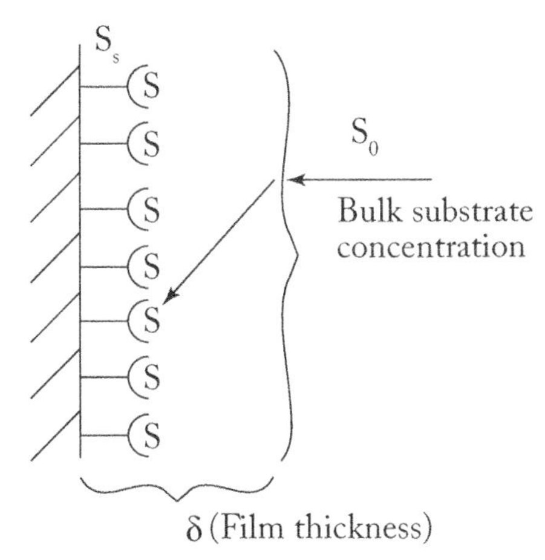

Fig. 5.2 Substrate concentration profile in a liquid film around adsorbed enzyme particle

At the steady state, the reaction rate, rp, is equal to the mass transfer rate,

$$r_p = k_L a\,(S_o - S_s) = v'_{max}\, S_s / (K_M + S_s) \text{ kmoles / (m}^3\text{)(sec)} \qquad (5.2)$$

Where v'_{max} is the maximum reaction rate per unit reactor liquid volume, k_L is the liquid phase mass transfer coefficient, (m/sec), S_s is the substrate's concentration on the solid matrix surface and So is the substrate concentration in the bulk, K_M is the Michaellis-Menten constant (Kmole/m^3). "a" is total surface area per unit volume of the reaction solution (m^2/m^3).

Equation (5.2) can be expressed as dimensionless form as

$$1 - S^* / D_a = \beta S^* / (1 + \beta S^*) \quad (5.3)$$

Where $S^* = S_s /So$, A dimensionless group called D_a

$$D_a = v_{max} / k_L a\ S_o \quad (5.4)$$

$$\& \beta = S_o / K_M$$

D_a is the dimensionless group known as Damkohler number given by equation (5.4).

Depending upon the magnitude of D_a, Equation (5.2) can be significant:

1. If $D_a << 1$, the mass-transfer rate is much faster than the reaction rate, then the overall rate of reaction is controlled by the enzyme reaction. The equation (5.2) is simplified to

$$r_p = v_{max} S_s / (K_M + S_s) \quad (5.5)$$

2. If $D_a >> 1$, the reaction rate is much greater than the mass transfer rate, the overall rate of reaction is controlled by the rate of mass transfer, leading to the overall rate of reaction as,

$$r_p = k_L a (S_o - S_s) \quad (5.6)$$

   Where S_o is the concentration of substrate in the bulk.

3. If $D_a = 1$, the diffusion and reaction resistances are both significant. The effect of mass transfer can be evaluated by the effectiveness factor of the immobilized enzyme defined as

   η = actual reaction rate based on S_s / rate not reduced by diffusion (based on S_o) (5.7)

$$\eta = v_{max} S_s / (K_M + S_s)/[v_{max} S_o / (K_M + S_o)]$$

$$= \beta S^* / (1 + \beta S^{*)} / [\beta / (1 + \beta)] \quad (5.8)$$

   If $S^* = S_s / S_o = 1$, η becomes l, that is there is no mass transfer effect. On the other hand, if $S^* \rightarrow 0$, where the mass transfer is very slow compared to the reaction rate.

4. An expression for overall rate of reaction involving mass transfer and enzyme kinetics parameters with some approximation, can be presented as

$$r_p = v = v_{max} S_o / (K_M \text{ appt} + S_o) \quad (5.9)$$

Where

$$K_M \text{ appt} = K_M [1 + v_{max} /\{ k_L (S_o + K_M)\}] \quad (5.10)$$

The apparent Michaelis-Menten constant, $K_M$appt becomes a function of stirring speed, as it contains the film mass transfer parameter, k_L,

Problem 5.1

The values of K_M and v_{max} for an enzymatic reaction at 25°C and pH = 7.0 are K_M = 0.004 kmole/m^3 and v_{max} = 10kmole / (m^3) (s). respectively. The enzyme was then immobilized by binding covalently to acrylamide gel in the form of a spherical particles of 1 mm. The effectiveness factor of the immobilized enzyme was found to be 70% of the rate with free enzyme when the substrate concentration was 0.5 kmole/m^3. The reaction was carried out in a stirred tank reactor agitated at 50 rpm.

(a) Estimate the concentration of the substrate at the surface of the immobilized enzyme particle.
(b) Estimate the mass transfer coefficient, k_L.

Solution:

(a) r_p (at S_s) = $\eta\, v_{max}\, S_o / (K_M + S_o)$
$= 0.7 (10) (0.5) / 0.004 + 0.5 = 6.94$, Thus
$r_p = 6.94 = v_{max}\, S_s / (K_M + S_s)$
putting the values of v_{max} and K_M
$6.94 = (10) (S_s)/ (0.004 + S_s)$, Now
solving for S_s (Concentration of substrate on the solid surface)
$S_s = 8.26 \times 10^{-3}$ kmole/(m^3) (s)

(b) We have the mass transfer equation,
$r_p (C_{Ss}) = k_L a (S_o - S_s)$,
Now the value of a is
$a = 4\pi r^2 / (^4/_3\, \pi r^3) = 3/r = 3/ 0.5 \times 10^{-3}\ m^{-1}$
$= 6 \times 10^{-3}\ m^{-1}$
Substituting the values of $r_p (C_{Ss})$, a, S_o, S_s
$6.94 = k_L (6 \times 10^{-3}) (0.5 - 8.26 \times 10^{-3})$
$k_L = 6.94 / 2.95 \times 10^{-3}$
$= 2.35 \times 10^{-3}$ m/s

Problem 5.2

Urea dissolved in aqueous solution is degraded to ammonia and CO_2 by the enzyme (urease) immobilized on the surface of non-porous polymeric beads. The reaction is controlled by transfer of urea to the surface of the beads through a liquid film and the reaction takes place on surface of the beads. The following data are given for the system:

$k_L = 2 \times 10^{-3}$ m/sec, $K_M = 0.2$ Kg/m^3
$v'_{max} = 10^{-3}$ kg urea/ (m^2 surface) (s), S_o (initial urea concentration)
$= 1.0$ kg/m^3

(a) Determine the surface concentration of urea
(b) Determine the rate of urea degradation under mass transfer controlled condition:

Solution:

We know, $k_L (S_o - S_s) = v'_{max} S_s / (K_M' + S_s)$

Now substituting the values of k_L, So, v'_{max} and K_M

$k_L (S_o - S_s) = v'_{max} S_s / K_M + S_s$

solving for S_s

$2 \times 10^{-3} (1 - S_s) = 10^{-3} S_s / 0.2 * S_s$

Or $(1 - S_s)(0.2 + S_s) = 0.5 S_s$

Or $S_s^2 - 0.3 S_s - 0.2 = 0$

Solving the quadratic

$S_s = \pm\sqrt{0.09 + 4.1\ 0.2 / 2}$

$= 0.621$ or $- 0.321$

Taking the positive value,

$S_s = 0.621$ kg/m^3.

5.3 Internal Diffusion for Enzymes Immobilized in Porous Solids

When enzymes are immobilized in porous solid matrix by copolymerization or microencapsulation, the intra-particle mass transfer significantly reduces the rate of an enzymatic reaction. The effect of the internal mass transfer can be quantitatively determined by effectiveness factor, η. To derive an expression for η, the following assumptions are made:

(i) The reaction occurs at every position within the immobilized enzyme
(ii) Mass transfer through the immobilized enzyme occurs via molecular diffusion
(iii) There is no mass transfer effect at the outer surface of the immobilized enzyme
(iv) The immobilized enzyme is spherical

The material balance for a spherical shell of thickness dr as shown in the figure 5.3

5.3.1 Effectiveness factor for a porous spherical pellet with miechalis-menten kinetics

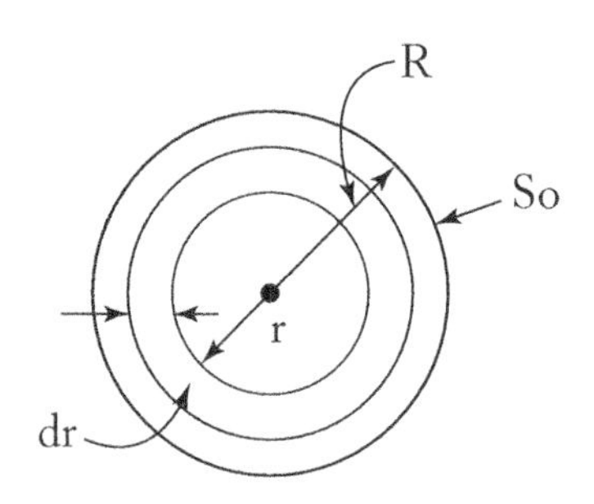

Fig. 5.3 Shell balance for a substrate in a spherical pellet

Shell balance for a substrate diffusing into the pellet can be given as

$4\pi (r+dr)^2 D_e [dS/dr + d/dr (dS/dr) dr]$	Output
$- 4\pi r^2 D_e dS/dr - (4\pi r^2 dr) r_{Ps}$	input Reaction
$= {}^4/_3 \pi r^2 dS/dt$	accumulation (5.11)

At the steady state, ds/dt =0

After simplifying by eliminating all the terms containing $(dr)^2$ and $(dr)^3$, a second order differential equation is obtained,

$$D_e (d^2S / dr^2 + 2/r\ dS/dr) = r_{Ps} \quad (5.12)$$

Where r_Ps is the enzymetic reaction rate in the pellet with first order kinetics

$$r_{Ps} = v_{max} S/(K_M + S) \approx (v_{max} S/K_M) \quad (5.13)$$

When S is small compared to K_M

De is the effective diffusivity of the substrate into the porous matrix and is given as

$$D_e = D_s € / \sigma \quad (5.14)$$

Where D_s is the molecular diffusivity, € is the peller porosity, σ is the tortuosity factor. to solve the equation (5.12), the boundary conditions are:

At r = 0, ds/dr = 0 (for central symmetry) (5.15)

We assume that substrate concentration at the external surface of the pellet is equal to the substrate concentration of the bulk liquid (C_{So})

Thus at r = R, S = So (5.16)

Now the observed overall rate of substrate utilization, r_{Po} is equal to the substrate diffusive flux into the pellet. Expressed in moles per pellet volume per unit time and can be given as

$$r_{Po} = A_p/V_p (D_e dS/dr),\ at\ r = R \quad (5.17)$$

Where V_p and the A_p are the particle volume and external surface respectively.

The effectiveness factor, η, for intra-particle diffusion is defined as

$$\eta = r_{Po} / r_p (S_o) = \text{observed rate / the rate without concentration gradient in the pellet} \quad (5.18)$$

Now the equation (5.12) can be transformed into a dimensionless form. Introducing dimension less variables,

$$\bar{S} = \frac{S}{S_o}, \bar{r} = \frac{r}{R}$$

We have,

$$\frac{d^2\bar{S}}{d\bar{r}2} + \frac{2}{\bar{r}}\frac{d\bar{S}}{d\bar{r}} = \frac{r_{p_o} R^2}{D_e S_o} = 9\varphi^2 \frac{\bar{S}}{1+\beta\bar{s}} \quad (5.19)$$

Where the dimension less parameters ϕ and β are defined by

$$\phi = R/3\sqrt{v_{max}/K_M/D_e},\ \beta = S_0/K_M \tag{5.20}$$

The dimensionless boundary conditions are

$$\text{at } \frac{\bar{S}}{\bar{r}=1} = 1 \text{ and } \frac{d\bar{S}}{d\bar{r}} = 0 \tag{5.21}$$

The square of the Thiele modulus, ϕ^2 has the physical significance of a first order reaction rate, $R^2 (v_{max} / K_M)$ So, divided by a diffusion rate RD_eS_o. The magnitude of the saturation parameter, β provides a measure of local rate deviations from first order kinetics. The large values of S indicate an approach to zero order kinetics.

In terms of the dimensionless variables, the effectiveness factor, η is

$$\eta = \frac{\left(\frac{d\bar{S}}{d\bar{r}}\right)_{\bar{r}=1}}{3\phi^2\left[\frac{1}{1+\beta}\right]} \tag{5.22}$$

From the above equation, the effectiveness factor, η is a function of ϕ and β or $\eta = f(\phi, \beta)$ 5.22 (a)

Now the general solution of the equation (5.19) for a first kinetics when $\beta = 0$ and introducing

$\propto' = \bar{r}\,\bar{s}$, we have

$$\frac{d^2 \propto'}{d\bar{r}2} - 9\phi^2 \propto' = 0 \tag{5.23}$$

The general solution of the above differential equation is

$$\propto' = C_1 \cos h\, 3\phi\bar{r} + C_2 \sin h\, 3\phi\bar{r} \tag{5.24}$$

Or

$$\bar{S} = \frac{1}{\bar{r}}(C_1 \cos h\, 3\phi\bar{r} + C_2 \sin h\, 3\phi\bar{r}) \tag{5.25}$$

Now $\bar{S}$ must be bounded as $\bar{r}^{p}$ approaches zero according to the first boundary condition. We must choose $C_1 = 0$. The second boundary condition requires that $C_2 = 1$ and we have,

$$\bar{S} = \frac{\sin h\, 3\phi\bar{r}}{\bar{r}\sin h\, 3\phi} \tag{5.26}$$

Now differentiating equation (5.26) at $\bar{r} = 1$, the effectiveness factor, η from equation (5.22) becomes

$$\eta = \frac{1}{\phi}\left[\frac{1}{\tanh 3\phi} - \frac{1}{3\phi}\right] \quad (5.27)$$

Where $\phi = R/3\sqrt{v_{max}/K_M/D_e}$
Where $\phi \leq 0.1$, η tends to 1.0 and

When $$\phi \geq 10,\ \eta \approx 1/\phi \quad (5.28)$$

The generalized plot of η vs ϕ at various values of β are shown:

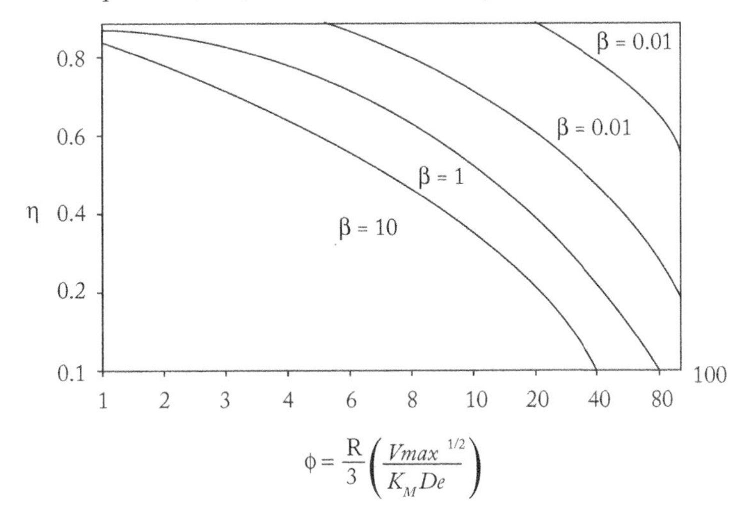

$$\phi = \frac{R}{3}\left(\frac{Vmax}{K_M De}\right)^{1/2}$$

Fig. 5.4 Effectiveness factor, η vs. Thiele modulus for first order reaction using different values of β

Problem 5.3

The enzyme, urease, was immobilized in Ca-alginate beads of 2 mm diameter. When the urea concentration was 0.5×10^{-3} kmoles/m^3. The rate of urea hydrolysis was $v = 10 \times 10^{-3}$ kmole/(m^3) (hr). Diffusivity of urea in the Ca-alginate beads is $D_e = 1.5 \times 10^{-9}$ m^2/sec and the Michaelis constant, $K_M = 0.2 \times 10^{-3}$ kmole/m^3. The liquid film resistance is neglected.

Determine the following

(a) Maximum rate of hydrolysis, v_{max}, Thiele modulus, ϕ and the effectiveness factor, η.
(b) What would be the v_{max}, ϕ and η values for a particle size of diameter, Dp = 4mm.

Solution

(a) To calculate v_{max}

$$v_{max} = \frac{v(K_M + S_o)}{S_o}$$

$$= \frac{10\times10^{-3}(0.2\times10^{-3} + 0.5\times10^{-3})}{0.5\times10^{-3}} = 14\times\frac{10^{-3}\,kmole}{(m^3)(hr)}$$

$$= 3.88 \times 10^{-6} \frac{kmole}{(m^3)(sec)}$$

R (radius of the spherical bead)

$$= \frac{2mm}{2} = 1mm = 10^{-3} m$$

$$\phi = \frac{R}{3}\sqrt{\frac{v_{max}}{K_M D_e}} = \frac{10^{-3}}{2}\left[\frac{3.88 \times 10^{-6}}{(0.2 \times 10^{-3})(1.5 \times 10^{-9})}\right]^{\frac{1}{2}} = 1.196$$

$$3\phi = 3.59$$

Now $\eta = \frac{1}{\phi}\left[\frac{1}{\tan h 3\phi} - \frac{1}{3\phi}\right] = \frac{1}{1.196}\left[\frac{1}{\tan h 3.59} - \frac{1}{3.59}\right] = 0.6$ Ans

(b) For immobilized enzyme system, the main variables are v_{max} and R, since the substrate conversion is constant, K_M and D_e are fixed.
Since the observed rate for the second case is not known, we assume v_{max} is the same as in the first case.

$$\phi = \frac{R}{3}\sqrt{\frac{v_{max}}{K_M D_e}}$$

Here $R = \frac{4mm}{2} 2mm = 2 \times 10^{-3} m$

$$\phi = \frac{2 \times 10^{-3}}{3}\sqrt{\frac{3.88 \times 10^{-6}}{(0.2 \times 10^{-3})(1.5 \times 10^{-9})}} = 2.39$$

$$3\phi = 7.17$$

$$\eta = \frac{1}{\phi}\left[\frac{1}{\tan h 3\phi} - \frac{1}{34}\right]$$

$$= \frac{1}{2.39}\left[\frac{1}{\tan h 7.17} - \frac{1}{7.17}\right] = 0.36$$ Ans.

5.3.2 Effectiveness factor in terms of Φ & β

Using equations (5.17) & (5.22), we obtain,

$$\phi = \left[\frac{R^2 r_{p_o}}{9\eta D_e S_o}(1 + \beta)\right]^{\frac{1}{2}} \tag{5.29}$$

Substitution of this into the right hand side of equation (5.22a) gives an implicit relationship between η, β and a new dimensionless observable modulus, Φ defined by

$$\Phi = \frac{r_{p_o}}{D_e S_o}\left(\frac{V_p}{A_P}\right)^{\frac{1}{2}} \tag{5.30}$$

Solution of this implicit equations yields η as a function of β and ϕ which has been plotted as below:

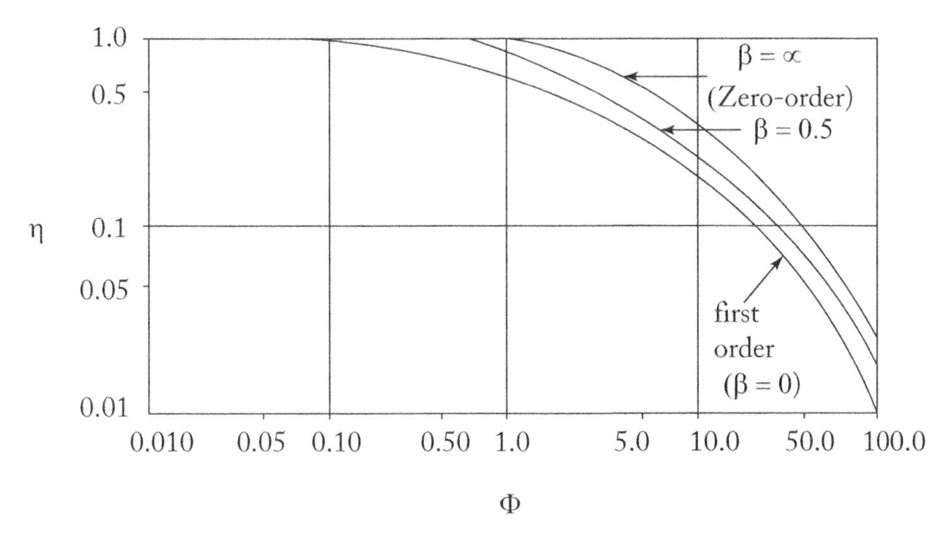

Fig. 5.5 Effectiveness factor for spherical particle with immobilized enzyme, ϕ : at various values of β

When the rate of diffusion is very slow compared to the rate of reaction, all substrate will be consumed in a thin layer near the exterior surface of the spherical particle. In this case we can assume that the layer has a flat geometry. Defining a control volume of thickness dr at a radial distance, r.

The material balance for the control volume at the steady state is given as,

$$4\pi r^2 D_e[\, ds/dr + d/dr\,(ds/dr)\,dr] - 4\pi r^2 D_e\, ds/dr + 4\pi r^2 D_e\, dr(rp) = 0 \tag{5.31}$$

Using Michaelis-Menteen Kinetics for rp and rearranging, we have

$$D_e\, d^2S/dr^2 = v_{max}\, S/(K_M + S_s) \tag{5.32}$$

We have defined the effectiveness factor, η

$$\eta = \frac{\dfrac{A_P}{V_p} D_e \dfrac{dS}{dr}}{\dfrac{v_{max} + S_o}{K_M + S_o}} \quad \text{at } r = R \tag{5.33}$$

Equation (5.32) may be rewritten as

$$D_e r\, dS/dr\, (dS/dr)^2 = v_{max}\, S / (K_M + S) \tag{5.34}$$

Integrating the above equation, we obtain

$$\left.\frac{dS}{dr}\right|_{r=R} = \left[\frac{2v_{max}}{D_e}\int_0^{S_0}\frac{S}{K_M + S}\right]$$

$$\left.\frac{dS}{dr}\right|_{r=R} = \frac{2v_{max}}{D_e}\int_0^{S_0}[S/(K_M + S)dS]^{1/2} \tag{5.35}$$

$$= \frac{2v_{max}K_M}{D_e}\left[\beta - ln(1 = \beta)\right]^{\frac{1}{2}} \tag{5.36}$$

Substituting (5.36) into equation (5.33) and simplifying, we get

$$\eta = \frac{\sqrt{2}(1+\beta)}{\phi\beta}\left[\beta - ln(1+\beta)\right]^{\frac{1}{2}}$$

Where
$$\phi = \frac{R}{3}\sqrt{\frac{v_{max}}{K_M D_e}}$$

$$\beta = \frac{S_o}{K_M} \tag{5.37}$$

5.3.3 Effectiveness factor with zero-order kinetics

In Michaelis-Menten equation, if $K_M << S$, the reaction becomes zero order, $r_p = v_{max} = k_o$

$$r_p = \begin{cases} -k_o & \text{if } S>0 \\ 0, & \text{otherwise} \end{cases} \tag{5.38}$$

The equation (5.13) becomes

$$\left(\frac{d^2S}{dr^2} + \frac{2}{r}\frac{dS}{dr}\right) - \frac{k_o}{D_e} = 0 \tag{5.39}$$

For $S > 0$

The boundary conditions for the above equation are

S is bounded at $r = 0$

$S = S_o$ (bulk concentration)

Defining by $\alpha = rS$ at $r = R$ (5.39a)

The equation (5.39) becomes,

$$\frac{d^2\propto}{dr^2} = \frac{k_{o_2}}{D_e} \tag{5.40}$$

Integrating with respect to r

$$\propto = \frac{1}{6}\frac{k_o}{D_e}r^2 + C_1 r + C_2 \quad (5.41)$$

Substituting the value of α in equation (5.39a)

$$S = \frac{1}{6}\frac{k_o}{D_e}r^2 + C_1 + \frac{C_2}{r} \quad (5.42)$$

Now, S must be bounded as r approaches zero according to the first boundary condition, then $C_2 = 0$. The second boundary condition gives

$$C_1 = S_o - \frac{1}{6}\frac{k_o}{D_e}R^2 \quad (5.43)$$

So the solution of equation (5.39) becomes,

$$\frac{S}{S_o} = \frac{1}{6}\frac{k_o}{S_o}\frac{R^2}{D_e}\left(\frac{r^2}{R^2} - 1\right) + 1 \quad (5.44)$$

Equation (5.39) is only valid when $S > 0$. The critical radius (Rc) can be calculated when S is zero, by solving,

$$\frac{1}{6}\frac{k_o}{S_o}\frac{R^2}{D_e}\left(\frac{R_c^2}{R} - 1\right) + 1 = 0 \quad (5.45)$$

Simplifying we get

$$\frac{R_c}{R} = \sqrt{1 - \frac{6S_oD_e}{k_oR^2}} \quad (5.46)$$

The actual reaction rate according to the distribution model is $\frac{4}{3}\pi(R^3 - R_c^3)k_oC_s$ and their rate without the diffusion is $\frac{4}{3}\pi R^3k_oC_s$.

Therefore the effectiveness factor, η, the ratio of actual rate to the rate without diffusion, is given as

$$\eta = \frac{\frac{4}{3}\pi(R^3 - R_c^3)k_oC_s}{\frac{4}{3}\pi R^3k_oC_s} = 1 - \left(\frac{R_c}{R}\right)^3 \quad (5.47)$$

Substitution of the equation (5.46) in equation (5.47) gives,

$$\eta = 1 - \left(1 - \frac{6D_eS_o}{R^2k_o}\right)^{\frac{3}{2}} \quad (5.48)$$

Thus the effectiveness factor increases with the increase of effective diffusivity. D_e and decreases with the increase of particle size (R) or k_o, where R is the radius of the reacting pellet., k_o is the zero order rate constant.

Problem 5.4

An enzyme which hydrolyzes the cellobiose to glucose and (3-glucosidase) is immobilized in a sodium alginate gel sphere of 2.5mm diameter. It is assumed that the zero order reaction occurs at every point within the sphere with $k_o = 0.0765$ kmol/(m^3) (s) and cellobiose moves through the sphere by molecular diffusion, $D_e = 0.6 \times 10^{-9}$ m^2/s.

Calculate the effectiveness factor, η of the immobilized enzyme when the cellobiose concentration in bulk solution is 10kmol/m^3.

Solution:

For zero order reaction immobilized enzyme, the effectiveness factor, η is

$$\eta = 1 - \left(1 - \frac{6D_e S_o}{R^2 k_o}\right)^{\frac{3}{2}}$$

$$\text{R (radius of the sphere)} = \frac{215}{2} \times 10^{-3}\,\text{m}$$

$$R^2 = 1.5625 \times 10^{-6}$$

$$\eta = 1 - \left[1 - \frac{6^{m^2}(0.6 \times 10^{-9})(10)}{(1.5625 \times 10^{-6})(0.0765)}\right]^{\frac{3}{2}}$$

$$= 1 - (0.6988)^{\frac{3}{2}} = 1 - 0.584 = 0.416$$

5.3.4 Effectiveness factor for first order reaction in a flat slab

The first order enzymatic reaction rate can be shown as

$$r_p = v_{max}\, S/K_M = k_1{'}S \tag{5.49}$$

Where $k_1 = v_{max}/K_M$ under the condition when $S << K_M$

The differential equation for the slab system can be developed as

$$D_e\, d^2S/dZ^2 = k_1 S \tag{5.50}$$

The boundary condition for the above system

At Z = L, S = S_o, L is the thickness of the slab, and at Z = 0, at the centre line symmetry, dS/dz = 0.

The solution of the equation (5.50) can be obtained by the method used for spherical particle and the effectiveness factor, η has been evaluated as

$$\eta = \tan h\ \phi_L / \phi_L \qquad (5.51)$$

Where $$\phi_L = L / 2\sqrt{v_{max} / K_M D_e} \qquad (5.52)$$

Where L is the thickness of the slab and reaction takes places on both sides of the slab.

The values of η for the slab coincides with the equation (5.27) for spherical particle to within 10 percent over the entire ϕ range. Differences are greatest for ϕ near unity and η diminishes rapidly for larger values of ϕ.

5.4 Limitations of Enzymatic Reactions in Porous Solids

In the formulation of inorganic catalysts, highly active supported metal is localized in a thin outer shall of the support for efficient use of the catalyst. Similarly, if enzyme catalyst is prepared by impregnating a porous support material with enzyme solution, it may lead to non-uniform loading which affects the apparent activity and deactivation properties of the immobilized enzyme catalyst.

Furthermore, immobilized enzyme kinetics may differ from the single substrate, irreversible Michaelis-Menteen Kinetics. Another complicating possibility is in the case of coupled diffusion and reaction system for substrate-inhibited form. If the external substrate concentration, S_o is greater than the substrate concentration corresponding to maximum rate as the reaction occurs in the pellet, the decreased substrate concentration will cause a large local reaction rate than that at the exterior surface of the catalyst, leading to effectiveness factor greater than one. Such a possibility may also occur in the case of autocatalytic reaction.

5.5 Effectiveness Factor for Simultaneous Film and Interparticle Mass Transfer

It is assumed that the substrate passes through the external film or boundary layer and then diffuses into the porous solid matrix with immobilized enzyme, where the reaction takes place.

Assuming a simple slab geometry for the immobilized particle and first order kinetics ($\beta \approx 0$), the steady state mass balance inside the particle for a substrate is given as

$$D_e\, d^2S/dx^2 - k\, S = 0 \qquad (5.53)$$

Where k (first order rate constant) = v_{max} / K_M

With center line symmetry

$$\frac{dS}{dx} = 0 \text{ at } x = 0 \qquad (5.54)$$

At the exterior surface of the slab (x = L), no substrate accumulates. Transport into the surface by the intraparticle diffusion is equal to the transport by the film diffusion giving the final boundary conditions as,

At x = L, $$-D_e \frac{dS}{dx} = k_L [S(L) - S_o] \quad (5.55)$$

Solving the equation (5.53) with the given boundary conditions, we have

$$\eta_o = \frac{\tan h\phi}{\phi\left[1 + \frac{\phi \tan h\phi}{\beta^i}\right]} \quad (5.56)$$

Where Thiele Modulus ϕ is defined as

$$\phi = L\sqrt{\frac{v_{max}}{K_M D_e}},$$

η_o = combined effectiveness factor and another important parameter, is the Biot number (Bi) defined as

$Bi = k_L L/D_e$ = characteristic film transport rate/characteristic intraparticle diffusion rate (5.57)

Rearranging the equation (5.56), we get,

$$\frac{1}{\eta_o} = \frac{1}{\eta} + \frac{\phi^2}{B_i} \quad (5.58)$$

Where $\eta = \frac{\tanh\phi}{\phi}$ denotes the effectiveness factor for the catalyst without film transport resistance.

η_o is the combined effectiveness factor consisting of intraparticle diffusion and external film diffusion.

The following cases for internal and external diffusion in immobilized pellet may be considered:

(i) $\eta\phi^2 / Bi = \phi / Bi = k'L/k_L << 1$ (5.59)
Where $k' = v_{max} / K_M$, k_L is film diffusion coefficient, L is the thickness of the sub,
The influence of external film diffusion resistance is negligible

(ii) If neither conditions is satisfied, both resistances must be considered.

(iii) If Bi is of the order of 100 or greater, the effects of external resistances are not significant. The above concepts apply equally to other geometries and intrinsic kinetics.

5.6 Effectiveness Factor for Cylindrical Particles with Immobilized Enzymes

The cylindrical particles are assumed to be infinite cylinders. The mass balance equation for the substrate in the system may be given as

$$D_e\left[\frac{d^2S}{dr^2} + \frac{1}{r}\frac{dS}{dr}\right] = \frac{v_{max}}{K_M} S = k' C_s \quad (5.60)$$

Where k (the first order rate constant)

$= v_{max}/K_M$

The equation (5.60) may be solved with the following boundary conditions:

For Central Symmetry at r = 0
dS / dr = 0, and at r = R, S = S_o.
The effectivenesss factor, η for the system can be developed as

$$\eta = \frac{2}{\phi}\frac{I_1(\phi_C)}{I_0(\phi_C)} \tag{5.61}$$

Where Thiele modulus for cylinder, ϕ_C is given as

$$\phi_C = \frac{R}{2}\sqrt{\frac{v_{max}}{K_M D_e}} \tag{5.62}$$

And I_0 and I_1 are the modified Bessel functions of zero order and first order.

5.7 Reactor Design for Immobilized Enzyme Systems

We have discussed different types of reactors for enzyme systems in soluble form and the system is assumed to be homogeneous. In the present case enzymes are immobilized in inert solid matrix. The same type of enzyme kinetics is applicable, but the magnitude of the kinetic parameters are different. Moreover intraparticle diffusion may be predominant if the ϕ(Thiele parameter) of immobilized particles are comparatively large . The effect of intraparticle diffusion may be predicted in terms of effectiveness factors. Mass balance equations for batch CSTR and plug flow are identical with those with free enzyme, but with reduced rates and diffusion effect in immobilised enzyme system.

5.7.1 CSTR for immobilized enzymes

For steady state CSTR with constant inlet and outlet flow rates, containing spherical immobilized enzyme particles, the reactor is gently stirred, so that the particles are not disintegrated. The mass balance for CSTR at the steady state can be given as

$$F(S_o - S) = \frac{\eta v_{max} S}{(K_M + S)V} \tag{5.63}$$

or

$$D(S_o - S) = \frac{\eta v_{max} S}{(K_M + S)} \tag{5.64}$$

Where D is the dilution rate given as D = F/V and F is the feed rate, m^3/hr, S0, S are the inlet and outlet substrate concentrations (kmole/m^3) respectively, η is the effectiveness factor which is a function of Thiele parameter, ϕ & β. v_{max} is the maximum substrate consumption rate in kmol/(m^3) of support (hr), K_M is the Michaelis-Menten constant in k.mol/m^3. The only difference with the free enzyme reaction is that the reaction rate is modified for mass transfer effect, which depends upon the particle size, R, the effective diffusivity D_e v_{max} & K_M are the intrinsic kinetic parameters.

5.7.2 Packed bed enzymatic reactor

Immobilized enzyme reaction are carried out in packed bed reactors which are designed on the basis of plug flow behaviour. With porosity, $\in$ of bed, and average fluid velocity, v in the axial direction, the change of S along the length of the reactor, Z is given as

$$\in u\frac{dS}{dZ} = -(1-\in)\eta v \tag{5.65}$$

Or

$$\in u\frac{dS}{dZ} = -(1-\in)\eta\frac{v_{max}S}{K_M + S} \tag{5.66}$$

which can be integrated by separating the variables.

Problem 5.5

A CSTR is used with 2mm beads with some immobilized enzyme to carry out an enzymatic reaction. The Michaelis-Menten parameters are:

$K_M = 2 \times 10^{-3}$ kmole/m^3, $v_{max} = 1.6 \times 10^{-5}$ kmole/(s)(m^3 of beads)

D_e = the effective diffusivity of the substrate (tyrosine)) with the matrix

$= 5.0 \times 10^{-10}$ m^2/s. External mass transfer is negligible. If the reactor is fed with a flow rate of 15m^3/day at the substrate concentration of S_o = 0.1 kmole/m^3 and 95% substrate is converted to product (DOPA), Calculate the volume of the CSTR. The immobilized catalyst is mushroom tyrosinase.

Solution:

$F = 15$ m^3/day $= 15/(24)(3600) = 1.736 \times 10^{-4}$ m^3/sec

The required Design equation of CSTR with internal diffusion effect

$D(S_c - S) = \eta\, v_{max}\, S/(K_M + S)$

Where D = (dilution ratio) = F/V

$S_o = 0.1$ kmole/m^3, at 95% conversion $S = 0.05 \times 0.1 = 0.005$ kmole/m^3.

Now, η is the effectiveness factor of the catalyst

ϕ (Thiele parameter) $= R/3\sqrt{v_{max}/K_M D_e}$

$\eta = 1/\phi\,[1/\tan h\, 3\phi - 1/3\phi]$

$\phi = 1 \times 10^{-3}/3\, \sqrt{(1.6 \times 10^{-5})/(2 \times 10^{-3})\,(5 \times 10^{-10})}$

$= 1.33$

$\eta = 1/1.33\ [1/\tan h\ 3.99 - 1/3.99]$

$= 0.56$

Now putting the values in the design equation of CSTR

$D(0.1 - 0.005) = 0.56\ (1.6 \times 10 - 5)\ (0.005)/\ (0.002 + 0.005)$

Or $D = 0.64 \times 10 - 5 / 9.5 \times 10^{-2}$

$= 6.7 \times 10^{-5}\ sec^{-1}$

$V = F/D = 1.736 \times 10^{-4} / 6.7 \times 10^{-5}$

$V = 2.6\ m^3$ Ans

Problem 5.6

Immobilized aspartase of 2mm diameter resin particles is used to convert ammonia solution of fumaric acid to aspartic acid in a packed bed of $0.5m^3$ with immobilized particles. The kinetic parameters of the system are:

at 32°C, $K_M = 4\ kg/m^3$, $v_{max} = 5.9\ kg/(m^3)(hr)$
$D_e = 5 \times 10^{-10}\ m^2/sec$

$S_o = 15\ kg/m^3$, at 90% conversion the exit substrate concentration, $S = 1.5\ kg/m^3$. Calculate the flow rate of the substrate to achieve this conversion.

$C_4H_4O_4$	+	NH_3	→	$C_4H_7O_4N$
(fumaric acid)		(ammonic)		(aspartic acid)

Solution: We use the plug flow design equation,

$$\frac{K_M l_n \dfrac{S}{S_o}}{\eta v_{max}} + \frac{(S_o - S)}{\eta v_{max}} = \frac{V}{F} = \tau$$

Solution:

To Calculate the effectiveness factor, η, we must calculate Thiele parameter, ϕ
$\phi = R/3 \sqrt{v_{max} / K_M D_e}$
$= 1 \times 10^{-3}/3 \sqrt{5.9/3600/(4)\ (5 \times 10^{-10})}$
$= 0.30$
$\eta = 1/0.30\ [1/\tanh 0.90 - 1/0/90]$
≈ 1.0

Now putting the values of the parameters in the packed bed or plug flow design equation

$$\frac{4 l_n \dfrac{15}{1.5}}{5.9} + \frac{(15 - 1.5)}{5.9} = \tau$$

Or $\tau = 3.84\ hrs.$

$$F = \frac{V}{\tau} = \frac{0.5\ m^3}{3.84\ hr} = 10.13\ m^3 / hr$$

The integrated form of the packed bed reactor design equation is,

$$K_M l_n \frac{C_{S_o}}{C_S} + C_{S_o} - C_S = \frac{(1-\epsilon)}{\epsilon} \eta v\max \frac{L}{u} \quad (5.67)$$

Where L is the length of the packed bed of immobilized enzyme particles and η is the effectiveness factor for intraparticle diffusion for the immobilized enzyme system.

5.8 Deactivation of Immobilized Enzyme Catalyst in a Packed Bed Reactor

Assuming first order decay of the active enzyme, v_{max} decreases with time as given below:

$$v'_{max} = k_3 E_a (t=0) e^{-kdt} \quad (5.68)$$

If deactivation starts in a packed bed immobilized enzyme reactor, the design equation for packed bed reactor may be presented as

$$\frac{1}{\tau}\left\{K_M l_n \frac{S_o}{S(L)} + S_o - S(L)\right\} = v_{max} e^{-kdt} \quad (5.69)$$

$$\tau = \frac{L(1-\epsilon)}{\epsilon u} \quad (5.70)$$

Where the exit substrate concentration, S(L) varies with time for the deactivation of enzyme. Under the steady state condition, for a particular value of τ, only one value of S(L) will be obtained. When deactivation starts, S(L) will decrease with time, t.

The equation (5.69) may be linearized as,

$l_n \left[\frac{1}{\tau}\left(K_M l_n \frac{C_{SO}}{C_S(L)} + C_{SO} - C_S(L)\right)\right] = l_n v_{max} - k\ dt$ and from the slope, the deactivation rate constant, k_d may be evaluated.

5.9 Summary

Immobilized enzyme kinetics have been discussed using Michaelis-Menten equation. Due to immobilization, the kinetic parameter, v_{max} remains unchanged, but K_M value is increased, so that overall reaction rate decreases. The effect external diffusion has been predicted by Damköhler number (D_a) or Biot number (Bi). The classical effectiveness factor, η, has been used to predict intra particle diffusion and η has been found to be a unique function of Thiele parameter, ϕ and saturation parameter β (S_o/K_M). For first order kinetics, η is a function of ϕ only.

Conventional reactor design equations have been used for CSTR & packed bed reactor. Effectiveness factor has been used to account for the intraparticle diffusion. Deactivation of enzymes has been considered in a packed bed reactor and deactivation kinetics have been introduced in the design equations of the packed reactor. The change of outlet concentration has been shown to be a function of time.

Exercise

Problem 5.1

Dextrose was produced from Corn starch using both soluble and immobilized (azo-glass beads) glucoamylase in a fully agitated batch reactor. The following set of data are given

For soluble enzyme

$S_o = 168\ k_g/m^3$, $E_o = 11{,}600$ units in $1 \times 10^{-3}\ m^3$, T = 60°C

t,min	0	15	30	45	60	75
P, K_g/m^3	0	28	64.5	82.3	108	123.5
S =(S_o-P) K_g/m^3		140	103.5	85.7	60	44.4

(a) Calculate K_M and v_{max} for the soluble enzyme using Michelis-Menten equation

(b) For immobilized enzyme, T = 60°C

$S_o = 336$ Kg/m^3, $E_o = 46{,}400$ units in $1 \times 10^{-3}\ m^3$

t,min	0	15	30	45	60	75
P, K_g/m^3	0	116.6	181.6	217.6	241.6	247.6
S =(S_o-P) kg/ m^3	0	219.4	154.4	118.4	94.4	88.4

Calculate K'_M and v'_{max} in proper units.

Hint: Use the linearized equation of integrated Michaelis-Mentin equation in both cases.

$$\frac{S_o - S}{ln\frac{S_o}{S}} = -K'_M + v_{max}\frac{t}{ln\frac{S_o}{S}}$$

Problem 5.2

An enzyme is immobilized in Ca-alginate beads and is used to carry out an enzymatic reaction which follows Michaelis-Menten Kinetics with the parameters

$K_M = 0.1$ kmole/m^3

$v_{max} = 10$ kmole/(min) (m^3 or solid-support)

D_e (effective diffusivity) = 4.5×10^{-9} m^2/s

Calculate the size of the particle for which the intraparticle diffusion is negligible.

Problem 5.3

An enzyme is immobilized on the surface of non porous glass particles and is used to carry out a reaction which follows Michaelis-Menten kinetics. It is suspected that there is film diffusion in the process. The following data are available for the process.

$K_M = 5$ gm/m^3,

$v_{max} = 4 \times 10^{-5}$ g/(m^2)(s)

$K_L = 6 \times 10^{-7}$ m^2/s

If the bulk concentration of substrate, $S_b = 100$ gm/m^3, Calculate surface concentration of the substrate, Ss.

Problem 5.4

Invertase from Aspergillus oryzae is immobilized in porous resin particles of diameter of 1.6 mm at a density of 0.1 μmol enzyme/gm of resin particle. The resin is sufficiently stirred so that external mass transfer effects are eliminated. Invertase catalyzes the reaction:

$$C_{12}H_{22}O_{11} + H_2 \rightarrow \underset{\text{(Glucose)}}{C_6H_{12}O_6} + \underset{\text{(Fructose)}}{C_6H_{12}O_6}$$

At a sucrose concentration of 0.85 kg/m^3, the observed rate of conversion is 1.25×10^{-3} kg/m^{3}s of resin. K_M for the immobilized enzyme is 3.5 kg/m^3. The effective diffusivity of sucrose in the resin is 1.3×10^{-11} m^2/s.

(a) Calculate the effectiveness factor for the system

(b) Determine the first order reaction rate constant for immobilized invertase

Problem 5.5

The resin particles of 1.6 mm immobilized with invertase are placed in a packed column of volume of 0.5 m^3 to convert sucrose to glucose and fructose.

The kinetic parameters are

$K'_M = 3.5$ kg/m^3,

$v'_{max} = 6.4 \times 10^{-3}$ kg/(s) (m^3), the effective diffusivity, D_e

$D_e = 1.3 \times 10^{-11}$ m^2/s

If a feed of initial sucrose concentration, $S_o = 1.0$ kgmole/m^3 is fed to the reactor and 90% conversion is to be achieved, find the flow rate of the substrate solution.

Problem 5.6

If the above reaction is carried out in a CSTR with good agitation to eliminate bulk diffusion, what will be the flow rate to achieve 90% conversion of the substrate $S_o = 1$ kg/m^3. The reactor volume is 0.5 m^3. The other kinetic parameters are the same as the previous problem.

References

Adams, M. W.W. and R.M. Kelly, "Enzymes from micro-organisms in extreme environments", Chemical and Engineering News, Dec 18, 32–42, 1995

Atkinson, B., Biochemical Reactors, Pion Ltd., London, 1974

Austrop, K., "Enzymes of Industrial interest: Traditional products" in D. Perlman (editor) Annual Reports in Fermentation Processes, Vol-2, Academic Press, N.Y. 1978

Bailey, J.E. and D.F. Ollis, Biochemical Engineering Fundamental, 2nd ed; McGrow Hill, 1986

Blanch, H.W. and D.F. Clark, Biochemical Engineering, Marcell Dekker, Inc. N.Y. 1996.

Chibata, I. (editor), Immobilized enzymes – Research and Development, Halsted Press, N.Y., 1978

Katchalski-Katzer, E., "Immobilized enzymes – Learning from past successes and failures", Trends in Biotechnology, 11, 471–478, 1993

Gutcho, S.J., Immobilized Enzymes: Preparation and Engineering Techniques p. 25, Park Ridge, N.J. Noyes Data Corporation, 1974

Itamunoala, G.F., "Limitations of methods of determining effective diffusivity coefficients in cell immobilization matrices", Biotechnol. Bioeng 31, 714–717, 1985

Messing, R.A. (editor), Immobilized enzymes for industrial reactors, Academic Press, N.Y. 1975.

Olson, A.C. and C.L. Cooney, Immobilized enzymes in Food and Microbial processes, Plenum Press, N.Y. 1978

Ryu, S.K., J.M. Lee, "Bioconversion of waste cellulose by using an attrition Bioreactor, "Biotechnol. Bioeng. 25, 53–65, 1983

Shuler, M. and F. Kargil, Bioprocess Engineering" Basic Concepts, Pearson, 2nd ed. 2004

Tanaka, H., M. Matsummura and E.A. Valiky, "Diffusion characteristics of substrates in Ca-alginate gel beads", Biotechnol. Bioeng., 26, 52–58, 1984

Zabrosky, O.R., Industrial Enzymes, CRC Press, Cleveland, Ohio, 1973.

CHAPTER 6

Cell Kinetics and Reactor Design

6.0 Introduction

The growth kinetics of microbial, animal and plant cells are important for the design and operation of bioreactors. In these systems, three rate processes are involved such as (i) growth rate of cells, (ii) substrate utilization and (iii) product formation rates. The cells usually grow in a liquid medium containing various nutrients, vitamins, trace elements, growth factors etc., under proper environmental conditions like pH, temperature, ionic strength, viscosity. One or two of the multiple nutrients which may control the growth rate of cells is/are called substrate(s). Cells grow in number or mass after consuming substrate(s) and release useful metabolic products in the medium. Cell reactions involve numerous complicated networks of biochemical and chemical reactions, transport processes involving multiple phases and multicomponent systems. Consequently accurate kinetic modeling on the basis of the above conditions is difficult and complicated and requires certain simplifications. Several models have been developed based on certain assumptions which consider cell components, population, environmental conditions.

The unstructured and distributed model is the simplest one having the basis of the following assumptions:

(i) Cells can be considered as a single component in the form of cell mass or cell number. The cell composition (proteins, DNA, RNA etc.,) is assumed to be uniform for the conditions of balanced growth.

(ii) The population of cellular mass is distributed uniformly throughout the culture.

In reality, the cells are heterogeneous in nature with respect to cell age, signifying young or old cells or muted cells.

The second assumption leads to uniform solution of cell suspension. The liquid medium is formulated with many components. But one or two components may be controlling the cell reaction rate and are called substrates. Other components may be present at sufficiently high concentration, but they do not affect the rate.

The environmental conditions of medium in which cells are in suspension are maintained constant during fermentation. They include pH, temperature, dissolved oxygen, ionic concentration, rheology etc. In the fermentation process, two interacting systems are present: (i) the biotic phase consists of cell population and (ii) the abiotic phase which comprises environmental factors of growth medium. Cells consume nutrients, convert substrate into products.

Several kinetic models such as structured, unstructured, segregated, deterministic or statistical have been characterized and developed. The cell reactions may generate heat. The medium temperature sets the temperature of the cells. Mechanical interactions occur through hydrostatic pressure, and flow effects the medium and the cells; medium viscosity changes due to accumulation of cells and metabolic products.

The cellular environment is often a multiphase system consisting of liquid with dispersed gas bubble, liquid-liquid system or three phase system containing cells, liquid medium and air bubbles.

Living cells are extremely small systems. In an extreme case there may be only one DNA molecular in a slowly growing bacterium. Stochastic or probabilistic models may be constructed for such systems. The popular deterministic model is often used for a population as small as 10^5 cells, compared to chemical system of 10^{23} molecules per gm-mole.

To evaluate kinetics of the cell growth several reactor models are to be considered. Two ideal type of continuous bioreactors will be discussed first: (i) batch or plug flow reactor (PFR) and (2) continuous stirred tank reactor (CSTR).

6.1 Cell Growth Kinetics

Cell growth is a result of both replication and change in cell size. Microorganisms grow under a variety of physical, chemical and nutritional conditions. In a liquid medium organisms consume nutrients and convert them into biological compounds. A part of the nutrients are used for energy production, parts are utilized for biosynthesis and product formation. Microbial growth process can be presented as

$$\text{Substrate} + \text{Cells} \rightarrow \text{cellular or metabolic products} + \text{more cells} \tag{6.1}$$

The rate of microbial growth is expressed by the specific growth rate, μ defined as

$$\mu = 1/X \; dX/dt \tag{6.2}$$

Where X is the cell mass concentration in kg/m^3, t is the time in hrs, μ is the specific growth rate, hr^{-1}. Microbial growth rate can also be expressed in terms of cell number, N and the specific replication rate, μ_R is defined as

$$\mu_R = 1/N \; dN/dt \tag{6.3}$$

6.2 Batch Growth Patterns

A seed culture known as inoculum is introduced into the liquid medium. The microorganisms slowly adjust to the environment, consume the nutrients and convert them into biomass. A typical batch growth curve may show the following phases: (1) lag phase, (2) logarithmic or exponential phase, (3) deceleration phase, (4) stationary phase, (5) death phase.

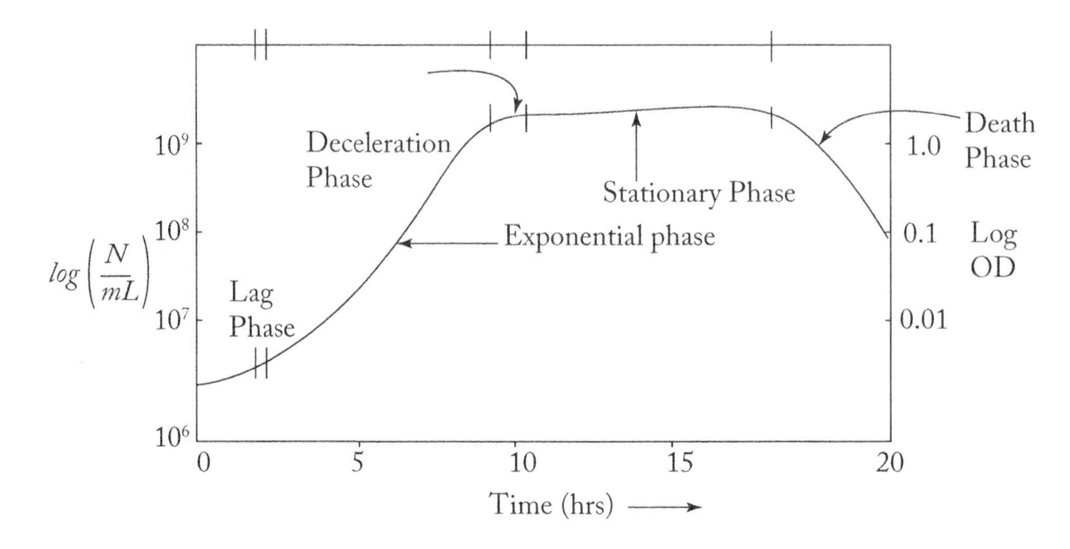

Fig. 6.1 Typical growth cycle of microorganism

The lag phase occurs immediately after inoculation, and cells adapt to the new environment during this period. Cell mass may increase little without an increase in the cell number density.

To reduce the lag phase, cells should be adopted from the exponential phase and the inoculum size should be large (5 – 10% of the reactor liquid volume).

Multiple lag phases may occur when the medium contains more than one carbon source. The phenomenon is known as diauxic growth, caused by a shift in the metabolic pathway in the growth cycle.

In the exponential or logarithmic phase, cells have adjusted to the new environment and multiply rapidly. Cell mass and cell number density increase exponentially with time. This is the period of balanced growth where all components of the cells grow at the same rate. The average composition of a single cell remains approximately constant during this phase. During this period, the cell growth rate,

dX/dt is given as $dX/dt = \mu x$ where $X = X_o$ at $t = 0$.

Integrating we have

$$\ln X/X_o = \mu t \text{ or } X = X_o\, e^{\mu t} \tag{6.4}$$

Where X and X_o are the cell concentration (kg/m^3) at the time t= t and t=0 and μ is called the specific growth rate The doubling time, td, of the cells is the time required to double the microbial mass and can be given as,

$$t_d = \ln 2/\mu = 0.693/\mu \tag{6.5}$$

Doubling time based on the cell number, specific rate of replication is,

$$t_d' = \ln 2/\mu_R \tag{6.6}$$

The deceleration growth phase after the exponential phase is characterized by the reduced growth due to either depletion of one or more essential nutrients or the accumulation of toxic products.

In the exponential growth phase cellular metabolic control system leads to maximum rates of reproduction. In the deceleration phase, the stresses induced by nutrient depletion or toxin accumulation restrict the cells to increase the prospects of survival in the hostile environment. The stationary phase follows the deceleration phase when the growth rate is zero and there is no cell division. The growth rate of cells is equal to death rate. But during this period, the cells are metabolically active and produce secondary metabolites. Primary metabolites are formed during the exponential phase and are called growth associated products (e.g. ethyl alcohol production). Secondary metabolites are non-growth associated (antibiotics and some hormones).

The stationary phase is characterized by the following factors:

(i) Total cell mass concentration remain constant, but the number of viable cell decrease

(ii) Cell lysis may occur and viable cell mass may be reduced. A second growth phase may occur when the cells consume the lysed cells.

(iii) Cells may not grow, but they are metabolically active and produce secondary metabolites.

During this phase, the cell catabolizes cellular reserves for new building blocs and for energy production for maintenance. This phenomenon is called Endogenous metabolism. The energy expenditure of the cells is required to maintain energized membrane transport of nutrients and for essential metabolic functions such as motility, repair of cell damage.

The loss of cell mass during the death phase can be described by the following rate equation,

$$dX/dt = -k_d X \text{ or } X = X_0 e^{-kdt} \tag{6.7}$$

Where X_o is the cell mass concentration at the beginning of the stationary phase, k_d is the first order decay rate constant.

At the end of the stationary phase, death phase follows due to either depletion of nutrients or accumulation of toxic products. Consequently cells in the stationary phase have a chemical composition different from that of cells in exponential phase. Cell, death is an exponential decay function:

$$dN/dt = -k_d'N \text{ or } N = N_o e^{-k'dt} \tag{6.8}$$

Where N_0 is the concentric of cells in number at the end of the stationary phase and kd' is the first order death constant.

During the stationary and death phase, cells are heterogeneous in nature as there is distribution of properties in the cell population. With a narrow distribution, cell death will occur simultaneously and with a broad distribution, a fraction of the cell population may survive for an extended period.

6.3 Yield Coefficients

To correlate the different rate processes for cell mass, substrate and product formation, yield coefficients are introduced. They are defined on the amount of consumption of another material. For example, yield coefficient of cell mass based on the substrate consumption is

$$Y_{X/S} = \Delta X/\Delta S = (X - X_o)/(S_o - S) \tag{6.9}$$

Yield coefficients based on the other substances or product formation may be defined.

$$Y_{X/O_2} = -\Delta X/\Delta O_2 = \text{Cell mass formed/mass of oxygen consumed} \tag{6.10}$$

$$Y_{P/S} = -\Delta P/\Delta S = \text{Product formed } (P - P_0)/\text{Substrate consumed } (S_o - S) \tag{6.11}$$

For organisms growing aerobically on substrate (glucose), $Y_{X/S}$ varies from 0.4 to 0.6 g/g for most yeast and bacteria.

Y_{X/O_2} values range from 0.9 to 1.4 g/g. For methane, $Y_{X/S}$ may be of the order of 0.6 to 1.0 g/g.

In most cases the yield of biomass on a carbon energy source is normally found to be 1.0 ± 0.4 gm of biomass per gm of carbon consumed.

A maintenance coefficient, m is defined as the specific rate of substrate utilization for the maintenance of cell activities (viz. repair of damaged cell, transfer of nutrients and products across the membrane, cells motility etc.).

$$m = -1/X\,(dS/dt)_m \tag{6.12}$$

Another yield coefficient, $Y_{P/X}$ is

$$Y_{P/X} = \text{mass of product formed/Increase in cell mass} \tag{6.13}$$

All these yield coefficients are assumed to be approximately constant.

6.4 Cellular Product Formation Kinetics

There are mainly three classes of product formation:

(i) Growth-associated: Products are formed simultaneously with microbial growth mainly during exponential growth phase of the cells. The specific rate of product formation, q_p can be presented as

$$q_p = 1/X\,dP/dt = \alpha\mu \tag{6.14}$$

Alcohol production is an example of Growth associated system.

(ii) Non-growth associated; product formation occurs during the stationary phase, when the cell growth rate is zero, but the cells are metabolically active, releasing a **secondary metabolite** product

$$q_p = 1/X\,dP/dt = \beta\ \text{(a constant)} \tag{6.15}$$

Antibiotics formation is an example of this category.

(iii) Mixed growth associated: Product formation involves slow growth during the stationary phase.

The specific rate of product formation may be expressed as follows:

$$q_p = 1/X\ dP/dt = \alpha\mu + \beta\ (\text{a constant}) \tag{6.16}$$

Fermentation of lactic acid and xanthan gum are the examples of this category which is characterized by both growth associated and non-growth associated characteristics. The equation (6.16) is popularly known as LEUDEKING-PIRET EQUATION. In this case, the first and second terms of equation (6.16) may be interpreted as energy used for growth and for maintenance respectively.

The product formation rate may be coupled with the substrate uptake rate as shown below:

$$dS/dt = -\ 1/Y_{X/S}\ dX/dt - 1/Y_{P/S}\ dP/dt - k_e X \tag{6.17}$$

Substrate utilization kinetics involve substrate conversion to cell mass and to product and a part of substrate is consumed for maintenance.

6.5 Environmental Conditions for Cell Mass Growth and Product Formation

Cell mass growth and product formation are extensively influenced by several environmental conditions viz Temperature, pH, dissolved oxygen (DO), redox potential and ionic concentrations of liquid medium. The optimum conditions of these parameters may be different for various organisms.

Temperature:

Temperature plays a significant role in the growth process. Microorganisms have been classified on the basis of temperature tolerance such as (i) Psychrophiles (Topt <20°C) (ii) Mesophiles (Temperature varies from 20°C to 50°C) (iii) Thermophiles (Topt > 50°C). As the temperature is increased towards optimum growth temperature (Topt) the growth rate approximately doubles for every 10°C increase in temperature. Above the optimal temperature the growth rate decreases to first order death rate. The net cell replication rate can be presented as

$$dN/dt = (\mu' - k_d')N \tag{6.18}$$

Where μ' is the specific replication rate constant and k_d' is the death rate constant. Their temperature dependency can be given according to Arrhenius equation

$$\mu' = Ae^{-Ea/RT}\ \&\ k_d' = A'e^{-Ed/RT} \tag{6.19}$$

Where Ea and E_d are the activation energy for cell growth and cell death respectively.

The activation energy for growth varies from 41.86 kJ/mole to 83.72 kJ/mole (10–20 kcals/mole) and E_d is in the range of 251 kJ/mole to 335 kJ/mole (60–80 kcals/mole). The maintenance coefficient, m defined as

$$m = (dS/dt)m/x \tag{6.20}$$

Which increases with increasing temperature with an activation energy of 62.8 kJ/mole to 83.7 kJ/mole (15–20 k.cals/mole).

Temperature may also affect the rate limiting step in a microbial process. With high temperature, the biochemical reaction rate may be very high and molecular diffusion becomes rate controlling. The activation energy for molecular diffusion is about 25.12 k.joules/mole (6 k.cals/mole).

pH:

The hydrogen ion concentration, pH, has strong effects on the enzyme activity. It affects the microbial growth reaction rate, since the latter is a multi-enzyme system. The optimal pH for growth varies for different microorganisms. For example, optimum pH for many bacteria ranges from pH 3 to 8, for yeast pH = 3 to 7, plant cells pH = 5 to 6. For animal cells: pH = 6.5 – 7.5. When pH differs from the optimum, the maintenance energy requirements increase pH of the medium and PH may be changed to select one organism over another.

The use of ammonia in the medium or its consumption by the microorganisms may alter the pH of the medium. The evolution or supply of CO_2 can also change pH significantly. The specific growth rate, μ, increases with pH, reaches a maxima and then may decrease with the increase of pH.

Dissolved Oxygen (DO)

Dissolved oxygen is an important parameter for aerobic fermentation. When the cells grow in the medium, the viscosity of the medium is increased, and transport of oxygen may be rate limiting. Under this condition, the specific growth rate, μ varies with D.O. and leads to saturation kinetics. A critical value of dissolved oxygen has been observed, above which the cell growth rate becomes independent of D.O. concentration. Oxygen is a growth limiting factor when D.O. level is below the critical D.O. concentration. The critical oxygen concentration is about 5 – 10 percent of the saturated D.O. concentration for bacteria and yeasts. For mold culture it may vary from 10 to 50% of the saturated D.O. Saturated D.O. (Solubility of oxygen in water) at 25°C and 1 atmosphere is about 7 ppm (7mg/litre or 7 gm/m^3).

Oxygen is usually supplied to fermentation broth by sparging air through the broth. Oxygen is transferred from gas bubbles to cells through a liquid film surrounding the gas bubbles. The oxygen transfer rate(OTR) from the gas to the liquid phase is usually given as

$$k_L a\,(C_L^* - C_L) = OTR \tag{6.21}$$

Where k_L is the liquid phase oxygen transfer coefficient (m/hr). a is the gas liquid interfacial area (m^2/m^3), $k_L a$ is the volumetric mass transfer coefficient (hr^{-1}). C_L^* is the saturated D.O. (kg/m^3) C_L is the actual DO concentration in the broth (kg/m^3).

Critical oxygen concentration is the DO concentration below which microorganisms can not survive.

The oxygen uptake rate by a microorganism, denoted as OUR is expressed as follows,

$$OUR = q_{O_2} X = \frac{\mu X}{Y_{X/O_2}} \tag{6.22}$$

Where q_{O_2} is the specific rate of oxygen consumption (kgO_2/(kg dry wt cells) (hr), Y_{X/O_2} is the yield coefficient of oxygen (kg dw cell/kg O_2 consumed), X is cell mass concentration (kg dry wt of cells/m^3).

When the oxygen transfer is the rate limiting step, the rate of oxygen consumption is equal to the rate of oxygen transfer,

$$\text{OUR} = \text{OTR} \tag{6.23}$$

$$\frac{\mu X}{Y_{X/O_2}} = k_{La}\left(C_L^* - C_L\right) \tag{6.24}$$

Or

$$\frac{dX}{dt} = Y_{X/O_2} k_{La}\left(C_L^* - C_L\right) \tag{6.25}$$

Thus the growth rate varies linearly with the D.O. concentration. DO limitation is eliminated by the use of O_2 enriched air or pure O_2 or operating under high oxygen pressure at 2–3 atmosphere.

Respiration quotient is the ratio of CO_2 formed to O_2 consumed.

Redox Potential:

An important environmental parameter in the fermentation broth is the redox potential which affects the cell growth and oxidation-reduction reactions. It is a complex function of DO, pH, and other ion concentrations of reducing and oxidizing agents. The electrochemical potential of fermentation medium is expressed by the following equation,

$$E_h = E_o^{'} + \frac{2.3RT}{F} logP_{O_2} + \frac{2.3RT}{F} log\left(H^+\right) \tag{6.26}$$

Where E_h is measured by milli volts by a pH meter or a voltameter. P_{O_2} is expressed in atmosphere units. The redox potential in a liquid medium can be minimized by passing N_2 gas or by the addition of reducing agent such as cysteine HCL or Na_2S. To enhance the redox potential, oxygen gas may be passed or some oxidising agents may be added to the medium. Ionic strength of fermentation medium may control the transport of certain nutrients in and out of the cells, the metabolic functions of cells and the solubility of some nutrients like dissolved oxygen.

The ionic strength is quantified by the following equation,

$$I = \frac{1}{2}\sum_{n=1}^{n} CiZ_i^2 \tag{6.27}$$

Where Ci is the concentration of an ion, Z is the charge, and I is the ionic strength of the medium, n is the number of ionic species.

Dissolved CO_2 (D_{CO_2})

The performance of organism is immensely affected by dissolved CO_2. High DCO_2 concentration may be found to be toxic to some cells. Some other cells may require more DCO_2 for their proper metabolic function.

Inhibition

The cell growth may be inhibited by the presence of excess substrate or excess product concentration. Some refractory compounds like phenol, toluene, methanol may act as inhibitor at even low concentration. (e.g. 1 kg/m^3).

In ethanol fermentation, the concentration of substrate, (glucose) above 200 kg/m^3 will inhibit the cell growth of yeasts. when the product ethanol concentration reaches about 50% (50 kg/m^3) the product inhibition takes place. Intermittent addition of substrate to the medium in a fed batch culture may eliminate the substrate inhibition.

Heat Generation by Microbial Growth

In aerobic fermentation, heat is released in the medium. For growing cells, energy requirement for maintenance is low. It has been estimated that 40–50% of the energy stored in a carbon compound is converted to the biological energy in the form of ATP during aerobic metabolism and the rest energy is liberated as heat. Use of cooling coil or jacket is used to eliminate excess heat and thus optimum temperature is maintained.

The heat generated during microbial growth depends on the heat of combustion of the substrate and of cellular material. The heat of combustion of the substrate is equal to the sum of the metabolic heat and the heat of combustion of the cellular material.

$$\frac{\Delta H_S}{Y_{X/S}} = \Delta H_c + \frac{1}{Y_H} \tag{6.27a}$$

Where ΔH_s is the heat of combustion of the substrate (KJ/g substrate), $Y_{X/S}$ is the yield coefficient (gm of cell mass per gm of substrate), ΔH_c is the heat of combustion of cells (KJ/gm cells) and $1/Y_H$ is the metabolic heat evolved per gm of cell mass produced. Typical ΔH_c value for bacterial cells is 23 KJ/g cells, Y_H for glucose = 0.42 g/k.cal.

The total rate of heat evolution in a batch fermenter, Q_{GR}, kcal/hr is expressed as,

$$Q_{GR} = V_L \mu X \frac{1}{Y_H} \tag{6.27b}$$

where V_L is the liquid volume (L) and X is the cell concentration (g/L)
In aerobic fermentation,

$$Q_{GR} = 0.12Q_{O_2} \tag{6.27c}$$

where Q_{O_2} is the oxygen uptake m mole/(hr)(L).

6.6 Monod Growth Kinetics (Unstructured Model)

Kinetic modelling of cell growth has been considered from several approaches like structured and unstructured modelling, segregated and unsegregated concept. Though the models with structured and segregation represent the more realistic behaviour of cell growth, they are very complex and difficult to handle.

The unstructured model is based on the assumption of balanced growth. The latter is defined as the condition where the average cell composition remains constant throughout the fermentation. This assumption is valid in a single stage, steady state continuous cultur, and during the exponential phase of the batch culture.

The basic assumption of Monod model are (i) balanced growth condition and (ii) unsegregated nature of the cells. If the concentration of one of the essential medium constituent is varied and the concentration of all other medium components are kept constant, the growth rate, μ changes with the controlling nutrient (substrate) hyperbolically. Monod (1942) proposed a functional relationship between μ and S in the following form,

$$\mu = \frac{\mu_{max} S}{K_S + S} \tag{6.28}$$

Where μ_{max} is the maximum growth rate when $S >> K_S$, S is the concentration of the substrate (one of the essential nutrients) (kg/m^3). K_S is called saturation constant, defined as the value of the substrate at which the specific growth rate, μ is half of μ_{max}.

The above equation (6.28) is in the form of Langnuir adsorpotion isotherm (1918) and the standard rate equation of enzyme catalysed reaction with a single substrate (Henri, 1902) and Michaelis-Menteen in 1913).

The latter equations are based on some mechanistic principles, where as the Monod model may be assumed to represent a multienzyme system empirical in nature Physical meaning is not known. In some special case, the model signifies the growth rate limitation by permease-mediated membrane transport rate.

In the equation (6.28), the term, $S/(K_S + S)$, has been regarded as a factor for calculating the deviation of μ from μ_{max} as the concentration of the substrate becomes smaller. It may be observed that for $S >> K_S$, μ is independent of S (zero order) and when $S << K_S$, μ is linearly dependent on S.

Problem 6.1

A batch aerobic fermentation of a bacterium growing on methanol gave the following data as shown in the table:

t,hr	0	2	4	8	10	12	14	16	18
X, kg/m^3	0.2	0.2	0.305	0.98	1.77	3.2	5.6	6.15	6.20
S, kg/m^3	9.23	9.21	9.07	8.03	6.8	4.6	0.92	0.77	0

Calculate

(a) Maximum growth rate, μ_{max}
(b) Saturation constant, K_S
(c) Mass doubling time, t_d at t = 10hrs
(d) Yield coefficient, $Y_{x/s}$
(e) Plot μ vs S and from the graph, evaluate approximately the value of μ_{max} and K_S.

Solution:

We know the growth kinetics,
$dX/dt = \mu X$
integrating between the limits
(t_1 and t_2) & (X_1 & X_2)
We have
$\ln X_2/X_1 = \mu (t_2 - t_1)$
or $\mu = (\ln X_2 - \ln X_1)/(t_2 - t_1)$

t,hr	0	2	4	8	10	12	14	16	18
X, kg/m^3	0.2	0.2	0.305	0.98	1.77	3.2	5.6	6.15	6.20
S, kg/m^3	9.23	9.21	9.07	8.03	6.8	4.6	0.92	0.77	0
ln X	−1.609	−1.556	−1.1874	0.0202	0.571	1.1631	1.7227	1.8164	1.8245
μ		0.265	0.291	0.275	0.296	0.280	0.0443		
S		9.22	8.55	7.415	5.7	2.76	0.4965		
1/μ		37.73	3.436	3.636	3.378	3.571	22.52		
1/S		0.108	0.117	0.134	0.175	0.362	2.010		

Taking the reciprocal of Monod Modal

$$\frac{1}{\mu} = \frac{1}{\mu_{max}} + \frac{K_S}{\mu_{max}} \frac{1}{S}$$

$\frac{1}{\mu} vs. \frac{1}{S}$ are plotted in fig. 6.2

From the Intercept = 3.0 = $1/\mu_{max}$
So $\mu_{max} = 1/3 = 0.33\ hr^{-1}$ (a) Ans
Slope = 2.5 = K_S/μ_{max} or
$K_S = 2.5\ (0.33) = 0.825\ kg/m^3$ (b) Ans

(c) $t_d = 0.693/\mu = 0.693/0.296 = 2.34$ hrs

(d) $Y_{x/s} = (x - x_o)/(s_o - s) = 6.20 - 0.2/9.23 - 0 = 0.65$

(e) μ vs s are plotted in fig. 6.3

From the graph, $\mu_{max} = 0.3\ hr^{-1}$.

Ks (at μ_{max}/s) = 1.2 kg/m^3

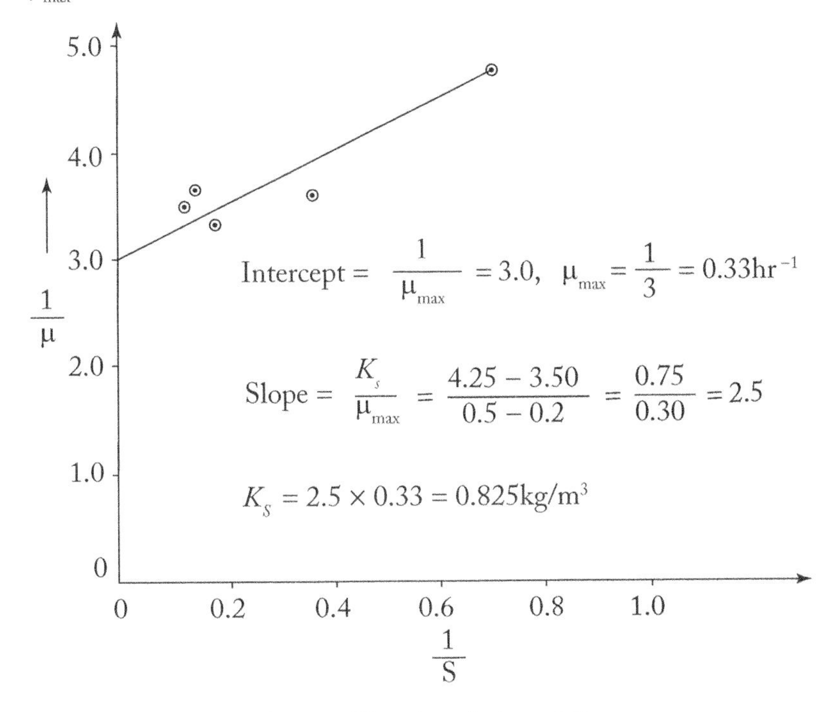

Fig. 6.2 The plot of $1/\mu$ vs. $1/S$

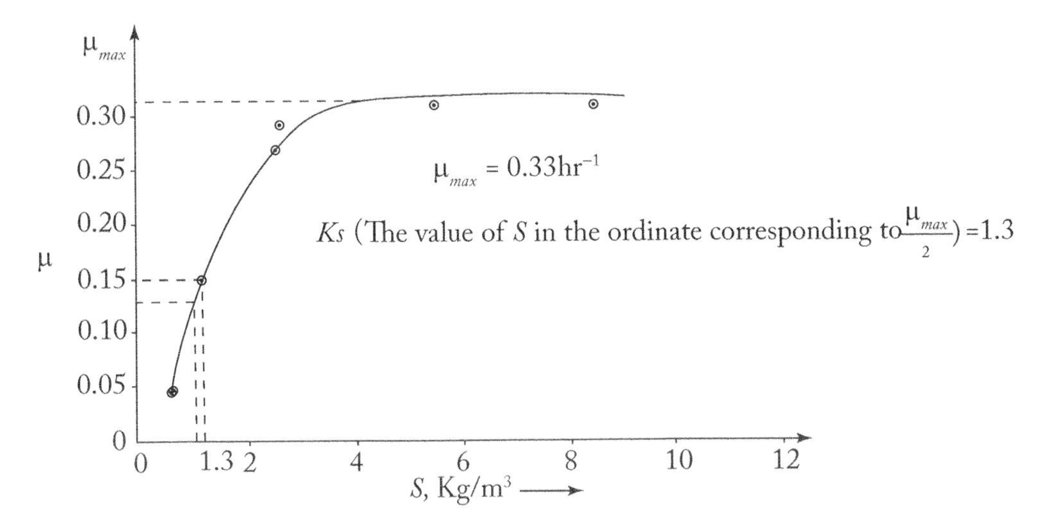

Fig. 6.3 The plot of μ vs. S

6.7 Other Forms of Cell Growth Kinetics

Though the other growth rate equations are much more complex than Monod Model, but they give better fit to the experimental data on some fermentations. The following models have been mentioned by Moser and Contois:

Tessier: $\mu = \mu_{max}\left(1 - e^{-S/K_S}\right)$

Moser: $\mu = \mu_{max}\left(1 + K_S S^{-\lambda}\right)^{-1}$

Contois: $\mu = \mu_{max}\dfrac{S}{Bx + S}$

The first two equation are complex algebraic equations, non-linear and it is difficult to find out the model parameters. The contois equation is equivalent to Monod equation, but contains an apparent Michaelis constant which is proportional to biomass concentration.

The specific growth rate, μ, may be inhibited by the medium constituents like substrate, product, metabolic toxic product etc.

Andrews Model for substrate inhibition (1968):

$$\mu = \mu_{max}\frac{S}{K_S + S + S^2 / K_I} \tag{6.29}$$

Product inhibition model of Aiba, Shoda and Nagatani (1968)

$$\mu = \mu_{max}\frac{S}{K_S + S}\cdot\frac{K_P}{K_i + P} \tag{6.30}$$

$$\mu = \mu_{max}\left(\frac{S}{K_S + S}\right)\left(1 - \frac{P}{P_{max}}\right)^{\alpha} \tag{6.31}$$

Where P_{max} is the product concentration at which growth stops and α is a constant.

Alcohol fermentation is an example of product inhibition for the anaerobic glucose fermentation by yeast.

Like Enzyme kinetics, we can develop non-competitive and competitive inhibition kinetic models for cells as given below:

1. **Non-competitive substrate inhibition**

$$\mu = \frac{\mu_{max}}{\left(1 + \frac{K_S}{S}\right)\left(1 + \frac{S}{K_I}\right)} \tag{6.32}$$

When $K_I >> K_s$, then

$$\mu = \frac{\mu_{max} S}{K_S + S + S^2 / K_I} \tag{6.33}$$

2. **For competitive substrate inhibition,**

$$\mu = \frac{\mu_{max} S}{K_S\left(1+\frac{S}{K_i}\right)+S} \tag{6.34}$$

Substrate inhibition may be reduced by slow, intermittent addition of the substrate to the growth medium

3. **Product inhibition**
   (a) Competitive product inhibition

$$\mu = \frac{\mu_{max} S}{K_S\left(1+P/K_P\right)+S} \tag{6.35}$$

   (b) Non-competitive product inhibition

$$\mu = \frac{\mu_{max}}{\left(1+\frac{K_S}{S}\right)\left(1+\frac{P}{K_I}\right)} \tag{6.36}$$

Ethanol fermentation of glucose by yeast is a non-competitive product inhibition, when ethanol concentration exceeds 50%.

Inhibition by toxic compounds

Competitive inhibition

$$\mu = \frac{\mu_{max} S}{K_S\left(1+\frac{I}{K_I}\right)+S} \tag{6.37}$$

Non-competitive inhibition

$$\mu = \frac{\mu_{max}}{\left(1+\frac{K_S}{S}\right)\left(1+\frac{I}{K_I}\right)} \tag{6.38}$$

Uncompetitive inhibition

$$\mu = \frac{\mu_{max} S}{\left[\frac{K_S}{\left(1+\frac{I}{K_I}\right)}+S\right]\left(1+\frac{1}{K_I}\right)} \tag{6.39}$$

If the toxic compounds in the medium causes the inactivity of the cells or death, the growth rate reduces to the form,

$$\mu = \frac{\mu_{max} S}{K_S + S} - kd \tag{6.40}$$

Where kd is the death rate constant (hr^{-1}).

With two substrate controlling Monod's growth model may be given as

$$\mu = \mu_{max} \frac{S_1}{K_1 + S_1} \cdot \frac{S_2}{K_2 + S_2} \tag{6.41}$$

Growth model with endogenous metabolism

$$\mu = \frac{\mu_{max} S}{K_S + S} - k_e \tag{6.42}$$

Endogenous metabolism involves reactions in cells which consume cell substance.

The additional term in the equation, (6.42), k_e is interpreted as death rate constant.

Using yield coefficient, $Y_{X/S}$, the rate of substrate decomposition r_S can be expressed as

$$-r_S = \frac{1}{Y_{X/S}} \frac{\mu_{max} SX}{K_S + S} \tag{6.43}$$

The substrate consumption may be used in parallel for cell growth and for maintenance requirement. The substrate consumption rate is then given as

$$-r_S = \frac{1}{Y_{X/S}} \mu X + mX \tag{6.44}$$

where m is the specific maintenance coefficient.

Another inhibition growth model was proposed by Verlhurst (1844) and Pearl and Reed (1920) which can be given as

$\frac{dX}{dt} = kX(1 - \beta X)$ with $X(O) = X_O$,(6.39), assuming that inhibition was proportional to X^2.

The equation (6.39), is integrated to obtain an expression known as **Riccati equation,**

$$X = \frac{X_O e^{kt}}{1 - \beta X_O \left(1 - e^{kt}\right)} \tag{6.45a}$$

The plot of the above equation is known as **logistic curve** shown in fig. 6.4 which is sigmoidal in nature, leading to stationary population size, $X_s = 1/\beta$

In another form, the equation can be given as

$$\frac{dX}{dt} = kX\left(1 - \frac{X}{X_S}\right) \tag{6.46}$$

Where X_S is the maximum biomass concentration

Integrating the equation(6.40), we get

$$X = \frac{X_O e^{kt}}{1 - \frac{X_O}{X_S}\left(1 - e^{kt}\right)} \tag{6.47}$$

And $$X_S = Y_{X/S} S_O + X_O \tag{6.48}$$

Further μ can be presented as

$$\mu = k\left(1 - \frac{X}{X_S}\right) \tag{6.48a}$$

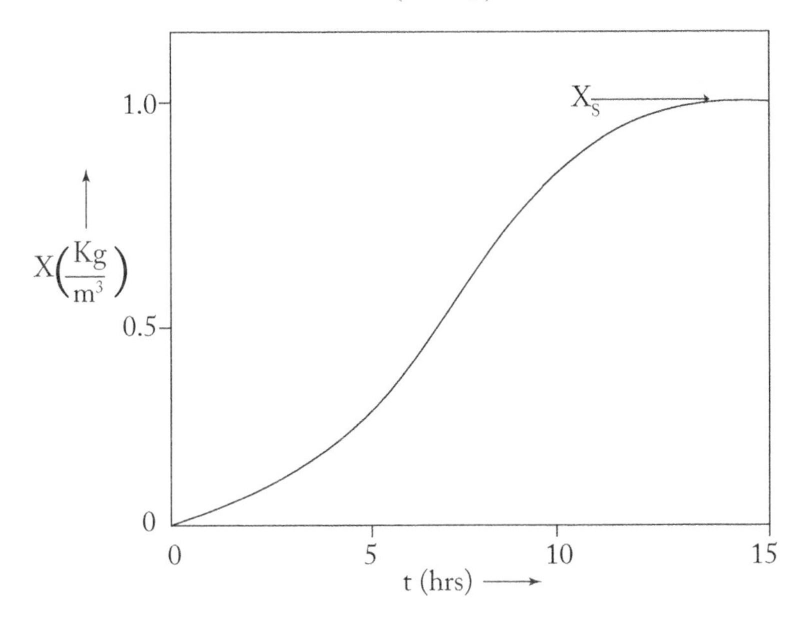

Fig. 6.4 Logistic growth curve

6.8 Growth Model of Filamentous Organisms

Filamentous organisms like molds form microbial pellets in the form of spheres at high cell densities. Growth models involve simultaneous diffusion and consumption of nutrients into the pellet and also formation of toxic products.

For isolated colony growth on solid surface or in sub-merged culture, some of the complexities of the process have been observed. Experimental studies of batch sub-merged culture have established that biomass increase takes place at a slower than the exponential rate with proportionality to $(\text{time})^3$.

It is presumed that a spherical pellet growing in a submerged culture follows a rate as shown below

$$\frac{dR}{dt} = k_O \text{ (a constant)} \tag{6.49}$$

Where R is the radius of the pellet at time, t. The biomass of the spherical pellet, M is then

$$M = \frac{4}{3}\pi R^3 \rho \tag{6.50}$$

Differentiating with respect to time. We obtain

$$\frac{dM}{dt} = \rho 4\pi R^2 \frac{dR}{dt} = k_O 4\pi R^2 \rho \tag{6.51}$$

Eliminating R from equation (6.51), using equation (6.50). we get

$$\frac{dM}{dt} = YM^{2/3} \tag{6.52}$$

Where $$Y = k_O \left[36\pi\rho\right]^{1/3} \tag{6.53}$$

Now integration of equation (6.52) gives

$$M = \left(Mo^{1/3} + \frac{Yt}{3}\right)^3 \tag{6.54}$$

where M_0 is the initial biomass which is quite small compared to M.

The equation (6.54) gives the cubic dependence of M on time t or M is proportional to t^3, which is supported by experimental data.

6.9 Continuous Stirred Tank Fermenters

There are two important basic types of stirred tank reactors for cell mass growth, for example (1) turbidostat and (2) chemostat.

6.9.1 Turbidostat

In the turbidostat, the cell concentration is maintained constant by monitoring the optical density of the culture and controlling the feed flow rate. A controller is used to maintain the set point value of turbidity by increasing or decreasing the flow rate through the pump. The culture volume is kept constant by removing an equal volume of culture from the reactor. The turbidostat is normally used in the laboratory for research purpose. This system is very useful for selecting subpopulation which can withstand environmental stress like excess substrate or excess product, because the cell concentration is maintained constant. It is a dynamic system with a sophisticated controller but more complicated than steady state chemostat.

6.9.2 Ideal chemostat

A perfectly mixed continuous flow reactor where cell reactions are carried out is called an Ideal chemostat. A schematic diagram of a chemostat is shown in figure 6.5 and material flow in the chemostat at the inlet and outlet is presented in figure 6.6

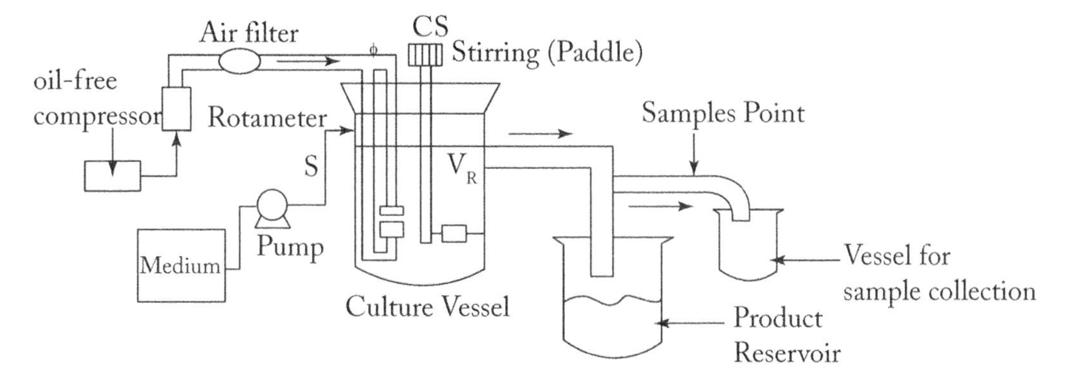

Fig. 6.5 A chemostat for continuous cultivation of cell population with components of a laboratory system

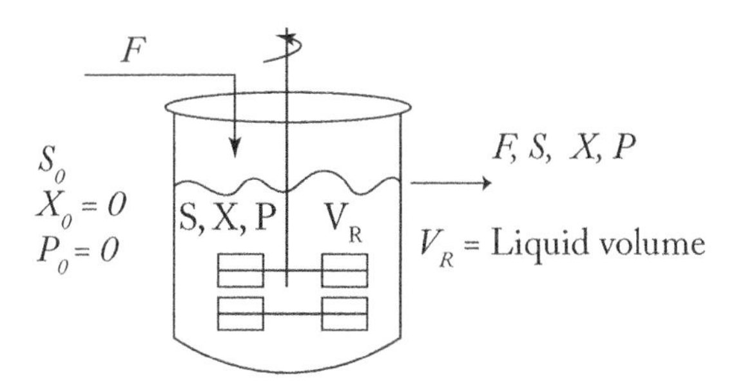

Fig. 6.6 Material balance in a chemostat

The reactor vessel of liquid volume V_R is agitated by a stirrer of impeller type, so that the concentrations of the cells(solid) and liquid phase are uniform throughout the reactor. For aerobic fermentation, air is introduced at the top through a sparger and dissolved oxygen concentration is the same at every region of the liquid phase. As shown in the figure 6.6, sterile feed culture is introduced into the reactor from the top and the exit stream is pumped out from the upper level of the culture so that the culture volume (V_R) in the vessel remains constant. The components inside the reactor are unreacted substrate (S), cells mass (X) and product (P). At the steady state, the concentrations of X, S, P in the outlet stream are the same as those inside the reactor due to efficient agitation by the stirrer and air bubbling. The optimum speed of the agitator is determined in such a way so that there is adequate mixing, causing no damage to the cell structure of microorganisms.

In the unsteady state, the mass balance for any component is based on the general mass balance principle

$$\begin{bmatrix}\text{Input of mass of a}\\ \text{reactant to the reactor}\end{bmatrix}-\begin{bmatrix}\text{Output of the mass of the}\\ \text{reactant from the reactor}\end{bmatrix}\pm\begin{bmatrix}\text{rate of generation or}\\ \text{consumption within the reactor}\end{bmatrix}$$
$$=\begin{bmatrix}\text{rate of accumulation}\\ \text{in the reactor}\end{bmatrix} \quad (6.55)$$

For the cell mass component balance (X) at the steady state

$$F\left(X_O - X\right) + V_R \mu X = V_R \frac{dX}{dt} = 0 \quad (6.56)$$

Dividing throughout by V_R and setting $D = F/V_R$
The equation (6.49) becomes

$$D\,(X_0 - X) + \mu X = 0 \quad (6.57)$$

where F is the volumetric feed rate V_R is the reactor-liquid volume, μ is the specific growth rate. D is called the dilution rate defined as $D = F/V_R$, which characterizes the holding time or processing rate of the CSTR. The dilution rate is equal to the number of tank liquid volumes which pass through the vessel per unit time. D is the reciprocal of mean holding time or mean residence time.

In the CSTR the cell populations adjust to a stead state environment and closely approach to a state of balanced growth condition. It is to be noted that achieving steady state condition in a chemostat may take hours or days. The long operation of CSTR may lead to contamination due generation of a mutant of useful type or wild type.

Now at the steady state ($dX/dt = 0$) and sterile condition ($X_0 = 0$), the equation (6.58) reduces to

$$D = \mu \quad (6.59)$$

Before the operation of the CSTR, an inoculum (X_0) of 5 – 10% of the reactor liquid volume is introduced, so that the growth of cells proceeds. The cells consume the substrate, grow in size or number, and release the metabolic product. The cell reactions may be presented as an autocatalytic reaction:

$$X_O + S \rightarrow X + P \quad (6.60)$$

Using Monod model with endogenous metabolism,

$$\mu = \frac{\mu_{max} S}{K_S + S} - kd \quad (6.61)$$

Putting $\mu = D$

$$\left(D + kd\right) = \frac{\mu_{max} S}{K_S + S} \quad (6.62)$$

Solving for S

$$S = \frac{K_S(D + k_d)}{\mu_{max} - (D + k_d)} \tag{6.63}$$

A material balance on the limiting substrate (S) in the absence of endogenous metabolism at the steady state gives

$$F(S_O - S) - V_R \mu X \frac{1}{Y_{X/S}} - V_R q_p \frac{1}{Y_{P/S}} = 0 \tag{6.64}$$

Where So and S are the feed and effluent concentration, (Kg/m^3), q_p is the specific rate of extracellular product formation in (Kg P/(Kg Cells) (hr)) and $Y_{X/S}$ and $Y_{P/S}$ are yield coefficient (Kg Cell X/Kg Substrate) and (Kg Product/Kg Substrate) respectively. For Cell mass and product with respect to substrate, the extra cellular product formation is negligible, the equation (6.64) reduces to

$$Y_{X/S} D(S_O - S) = \mu X \tag{6.65}$$

The Cell mass (X) is related to substrate (S) through yield coefficient, $Y_{X/S}$ as given below

$$X = X_O + Y_{X/S}(S_O - S) \tag{6.66}$$

Replacing S from the equation (6.55) and putting $k_d = 0$, we get

$$X = X_O + Y_{X/S}\left(S_O - \frac{DK_S}{\mu_{max} - D}\right) \tag{6.67}$$

The yield coefficient, $Y_{X/S}$ has been assumed to be constant. But it varies with S and μ. With endogenous metabolism ($D = \mu - k_d$), the steady state substrate mass balance with no. product formation may be given as

$$D(S_O - S) - \frac{1}{Y^M_{X/S}}(D + k_d)X = 0 \tag{6.68}$$

$Y_{X/S}$ is the maximum yield coefficient without endogenous metabolism and maintenance energy.

The equation (6.68) may be rearranged as follows:

$$D\left(\frac{S_O - S}{X}\right) - \frac{1}{Y^M_{X/S}}(D + kd) = 0 \tag{6.69}$$

or

$$D\left(\frac{1}{Y^{AP}_{X/S}}\right) - \frac{D}{Y^M_{X/S}} - \frac{kd}{Y^M_{X/S}} = 0 \tag{6.70}$$

Where $(S_0 - S)/X$ is $\frac{1}{Y^{AP}_{X/S}}$

Dividing throughout by D

Or
$$\frac{1}{Y_{X/S}^{M}}+\frac{kd}{Y_{X/S}^{M}D}=\frac{1}{Y_{X/S}^{AP}} \quad (6.71)$$

$$\frac{1}{Y_{X/S}^{AP}}=\frac{1}{Y_{X/S}^{M}}+\frac{m_S}{D} \quad (6.72)$$

Putting
$$m_S=\frac{kd}{Y_{X/S}^{M}} \quad (6.73)$$

Where m_s is the maintenance coefficient based on substrate $Y_{X/S}^{AP}$ is the apparent yield coefficient which varies with growth conditions and $Y_{X/S}^{M}$ is a constant. Values of $Y_{X/S}^{M}$ can be evaluated from the linear plot of $(S_0-S)/X$ vs $1/D$ from the equation (6.69)

- **Mass balance on the product**

For type I fermentation, $Y_{P/S}$ is constant, steady state mass balance on P is given as,

$$D(P_O-P)+Y_{P/X}\mu X=0 \quad (6.74)$$

or
$$P=P_O+Y_{P/X}\frac{\mu X}{D} \quad (6.75)$$

in terms of q_P, $\left(q_P=\frac{1}{X}\frac{dP}{dt}\right)$, equation (6.74) becomes, $D(P_O-P)+q_PX=0$ (6.76)

For non-growth association q_p is constant and equal to β; for growth association, q_p is a function of μ.

q_p = β (a constant) and qp = αμ

For both growth associated and non-growth associated system, qp= αμ + β, which is known as Luedeking-Piret Model.

- **Monod Chemostat Model**

The following mass balance equations on substrate and cell mass with non-sterile feed for the chemostat are known as Monod Chemostat Model.

The substrate balance:

$$D(S_O-S)-\frac{\mu_{max}SX}{Y_{X/S}(S+K_S)}=0 \quad (6.77)$$

Where
$$D=F/V_R,\ \mu=\frac{\mu_{max}S}{K_S+S}$$

and
$$\left(\frac{\mu_{max}}{K_S+S}-D\right)X+DX_O=0 \quad (6.78)$$

For sterile feed, X_o (inlet cell mass) = 0

$$X_{stinle} = Y_{X/S}\left(S_O - S_{steal}\right) = Y_{X/S}\left(S_O - \frac{DK_S}{\mu_{max} - D}\right) \tag{6.79}$$

And

$$S_{stenle} = \frac{DK_S}{\mu_{max} - D} \tag{6.80}$$

Equations (6.77) and (6.79) are known as Monod Chemostat Model. The dependence of *X*, *S*, and cell productivity (*XD*) on the dilution rate, *D* is graphically presented in figure 6.7 with parameters:

$\mu_{max} = 1.0\ hr^{-1}$, ks = 0.2 kg/m^3,

$Y_{X/S} = 0.5$, So (initial substrate concentration in the feed) = 10 kg/m^3.

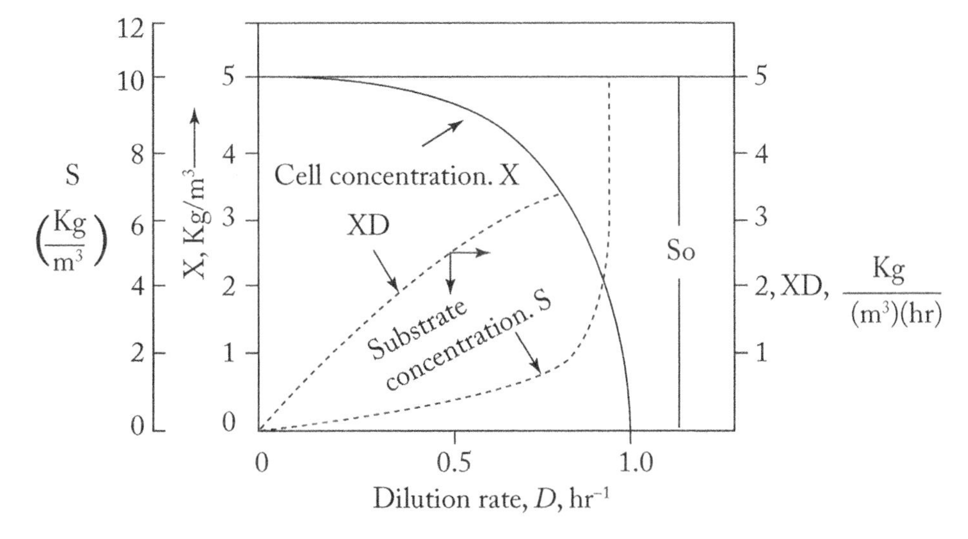

Fig. 6.7 Dependence of *X*, *S*, *XD* on *D* (the dilution rate)

For very slow rate ($D \rightarrow 0$), S tends to zero as all substrate is consumed, but the value of *X* tends to be maximum. From the figure (6.7) it is evident that *X* decreases with *D* but S increases with the increasing *D*. The productivity (*XD*) increases with *D*. But at a certain value of *D*, *XD* reaches the maximum indicating the maximum value of *D*, called D_{max}. Above certain value of D_{max}, the cell concentration reduces to zero and the corresponding *D* is known as $D_{washout}$. It is to be noted that near the washout, the reactor is very sensitive to variation in *D*, leading to unsteady state situation. D_{max} can be graphically evaluated as shown in the plot of *XD* vs *D* in figure 6.7

But mathematically, D_{max}, the dilution rate for maximum cell output can be evaluated by setting d(*XD*)/dD = 0 in the equation, $XD = DY_{X/S}\,(S_0 - Dks/(\mu_{max} - D))$.

After multiplying the equation (6.79) by D on both sides and taking the derivative with respect to *D* and setting the derivative,

$(dXD/dD = 0)$ to zero, we get

$$D_{max} = \mu_{max}\left(1 - \sqrt{\frac{K_S}{K_S + S_O}}\right) \tag{6.81}$$

Optimum cell concentration, Xopt can be derived using residence time, $\tau_m\left(1/D_{max}\right)$ for maximum cell productivity.

$$\frac{X}{\tau_m} = \frac{dX}{dt} = r_X = \frac{\mu_{max}SX}{K_S + S} \tag{6.82}$$

The productivity is maximum when $dr_X / dX = 0$ (6.83)

Now eliminating S from the equation (6.82) by the equation,

$$S = S_O - X / Y_{X/S} \tag{6.84}$$

We differentiate the equation (6.82) and setting the derivative to zero ($dr_X/dX = 0$), we obtain

$$X_{opt} = Y_{X/S}S_O\frac{\propto}{\propto + 1} \tag{6.85}$$

Where

$$\propto = \sqrt{\frac{K_S + S_O}{S_O}} \tag{6.86}$$

Problem 6.3

The following data were obtained in a steady state chemostat on the growth of a specific strain of backer's yeast (x) producing alcohol (P) from glucose (S)

D, hr-1	0.084	0.100	0.16	0.198	0.242
S_o, kg/m3	21.5	10.9	21.2	20.7	10.8
S, kg/m3	0.054	0.079	0.138	0.186	0.226
P, kg/m3	7.97	4.70	8.5	8.44	4.51
X, kg/m3	2.00	1.20	2.40	2.33	1.25

(a) Find the rate equation for cell growth
(b) Find out the dilution rate for maximum cell output (D_{max}) for the set I and the dilution rate for washout ($D_{washout}$) for So = 21.5 kg/m^3
(c) Find the rate for product (ethanol) formation from the 1st set of data.

Solution:

(a) Assume that the rate equation of cell growth is given by Monod model,
$\mu = \mu_{max}\ S/(K_S + S)$ ....(A)
For sterile feed ($X_o = 0$), $D = \mu$
Taking the reciprocal of equation (A)

We have $1/D = K_S/\mu_{max}\ 1/S + 1/\mu_{max}$

D, hr^{-1}	0.084	0.100	0.16	0.198	0.242
S,Kg/m3	0.054	0.079	0.138	0.186	0.226
1/D	11.9	10	6.25	5.050	4.132
1/S	18.51	12.66	7.24	5.376	4.424

1/D vs 1/S are plotted in the figure 6.8
From the linear plot
Intercept $= 1/\mu_{max} = 2.5$
or $\mu_{max} = 0.4 hr^{-1}$
Slope $= K_S/\mu_{max} = 0.5$
or $K_S = 0.2$ kg/m^3

(b)

$$D_{output}^{max} = \mu_{max}\left(1-\sqrt{\frac{K_S}{K_S+S_O}}\right)$$

$$= 0.4\left(1-\sqrt{\frac{0.2}{0.2+21.5}}\right) = 0.36 hr^{-1}$$

$$D_{washout} = \frac{\mu_{max} S_O}{K_S + S_O} = \frac{(0.4)(21.5)}{(0.2+21.5)} = 0.396 hr^{-1}$$

(c)

$$r_P = \frac{dP}{dt} = Y_{P/X}(\mu X) = Y_{P/X}(DX)$$

$$Y_{P/X} = \frac{P}{X} = \frac{7.97}{2} = 3.985 \quad (P \text{ and } X \text{ from the given table})$$

$$r_P = (3.985)(.084)(2) = 0.669 kg/m^3 hr$$

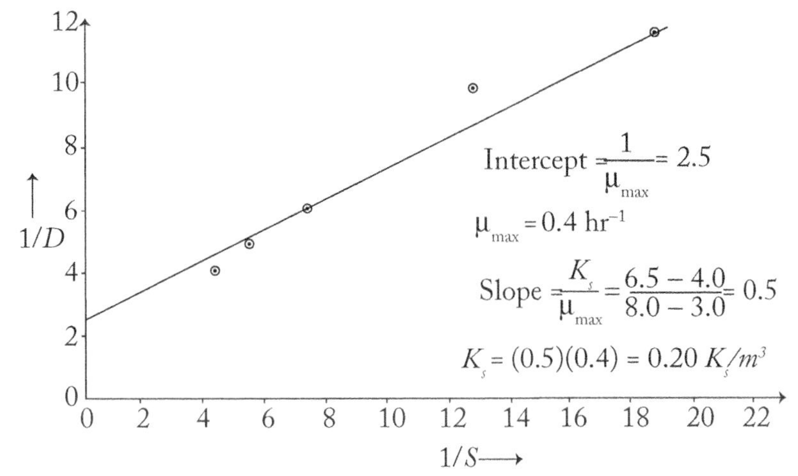

Fig. 6.8 Plot of $1/D$ vs $1/S$

Problem 6.4

Pseudomonas sp. has a doubling time, $t_d = 2.5$ hrs when grown on acetate. The saturation constant for this substrate is Ks = 1.2 kg/m^3 and the yield coefficient, $Y_{X/S} = 0.42$. For a chemostat operating with a feed of 35 kg/m^3 (So) of the acetate. Calculate the following

(a) A cell concentration when the dilution rate, D is one half of D_{max} output
(b) Substrate concentration when the dilution rate, D, is 0.75 D_{max}
(c) Maximum dilution rate for maximum output of cells
(d) Cell productivity, XD at D = 0.75 p_{max} output

Solution:

(a) td = 0.693/μ = 2.5
or μ = 0.693/2.5 = 0.277 hr^{-1}
$\mu = \mu_{max}\ S/(ks + S)$
$0.277 = \mu_{max}\ (35)/(1.2) + (35)$
Solving for μ_{max},
$\mu_{max} = 0.286$ hr^{-1}

$$D_{maxoutput} = \mu_{max}\left(1-\sqrt{\frac{K_S}{K_S+S_O}}\right)$$

$$= 0.286\left(1-\sqrt{\frac{1.2}{1.2+35}}\right) = 0.234 hr^1$$

Operating $D = 0.5\ D_{maxoutput} = 0.5 \times 0.234 = 0.117$ hr^{-1}

$$S = \frac{DK_S}{\mu_{max} - D} = \frac{(0.117)(1.2)}{0.286 - 0.117} = 0.83 Kg/m^3$$

$$X = Y_{X/S}(S_O - S) = 0.42(35 - 0.83) = 14.35 Kg/m^3$$

(b) $D = 0.75\ D_{max}$ output
= 0.75 (0.234) = 0.1755 hr^{-1}

$$S = \frac{DKs}{\mu_{max} - D} = \frac{(0.1755)(1.2)}{0.286 - .1755} = 1.90 K_g/m^3$$

(c) $D_{maxoutput} = 0.234$ hr^{-1}

(d) $DX = DY_{X/S}(S_O - S) = (0.1755)(0.42)(35 - 1.90) = 2.44 Kg/(m^3)(hr)$

Problem 6.5

The growth of a microorganism with a wide range of substrate concentration was studied in a chemostat of 1m^3 volume with sterile feed. The flow rate (*F*) and the inlet concentration of substrate were varied in the range of 30 kg/m^3 to 60 kg/m^3. The outlet concentration of the substrate at the steady state of fermenter was measured. The data obtained were:

F, m^3/hr	0.2	0.25	0.35	0.50	0.70	0.8	0.50	0.60	0.70
So, kg/m^3	30	30	30	30	30	30	60	60	60
S, kg/m^3	0.5	0.7	1.1	1.60	3.3	10	30	22	15

A substrate inhibition model was suggested in the following form.

$$\mu = \mu_{max} / \left[1 + \frac{K_S}{S} + \frac{S}{K_I}\right]$$

(a) Determine the kinetic parameters of the model (μ_{max}, K_s and K_I)

(b) If $Y_{X/S} = 0.46$ kg/kg, what was the steady state concentration of cell mass (X) where the flow rate is 0.20 m^3/hr.

(c) Explain the fact that the substrate inhibition in a continuous culture may lead to instability

(d) What will happen if the substrate concentration is suddenly increased from 30 kg/m^3 to 60 kg/m^3?

Solution:

(a) At low substrate concentrate concentration,
$S/K_I << 1$ so the model becomes

$$\mu = \frac{\mu_{max}}{1 + \frac{K_S}{S}} = \frac{\mu_{max} S}{K_S + S}$$

For a chemostat with sterile feed ($Xo = 0$), $D = \mu$,

$$D = \mu = \frac{\mu_{max} S}{K_S + S}$$

Taking reciprocal of both sides

$$\frac{1}{D} = \frac{K_S}{\mu_{max}} \frac{1}{S} + \frac{1}{\mu_{max}}$$

D (F/V) hr-1	0.2	0.25	0.35	0.50	0.70	0.8	0.50	0.60	0.70
S, kg/m^3	0.5	0.70	1.1	1.6	3.3	10	30	22	15
1/D	5	4	2.85	2.0	1.42	1.25	2.0	1.67	1.42
1/S	2	1.42	0.91	0.625	0.3	0.1	--	--	--

1/D vs 1/S is plotted.
From the linear plot
Intercept = $1/\mu_{max} = 0.85$
$\mu_{max} = 1.17$ hr^{-1}
From the slope, $K_S/\mu_{max} = 2.167$
$K_S = 2.167 (1.17) = 2.535$ kg/m^3

For the inlet substrate, $S_o = 60$ kg/m^3, the inhibition is prominent from the data. At high substrate concentration $K_S/S << 1$, so the given model reduces to

$$\mu = \frac{\mu_{max}}{1 + \frac{S}{K_I}}$$

Taking the reciprocal

$$\frac{1}{\mu} = \frac{1}{D} = \frac{1}{\mu_{max}} + \frac{S}{\mu_{max} K_I}$$

1/D is plotted again S, and from the linear plot

Intercept = $1/\mu_{max} = 0.85$

$\mu_{max} = 1.17$ hr^{-1}

Slope = $1/Ki\ \mu_{max} = 0.33$

Ki = 25.20 kg/m^3.

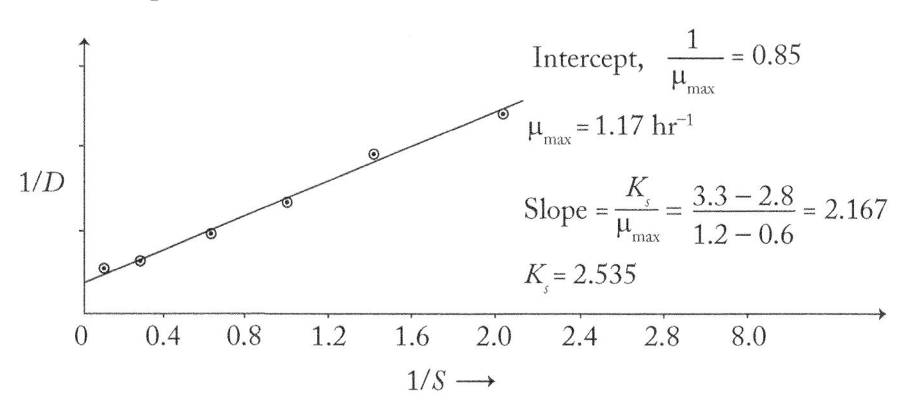

Fig. 6.9 Plot 1/D vs 1/S (no inhibition)

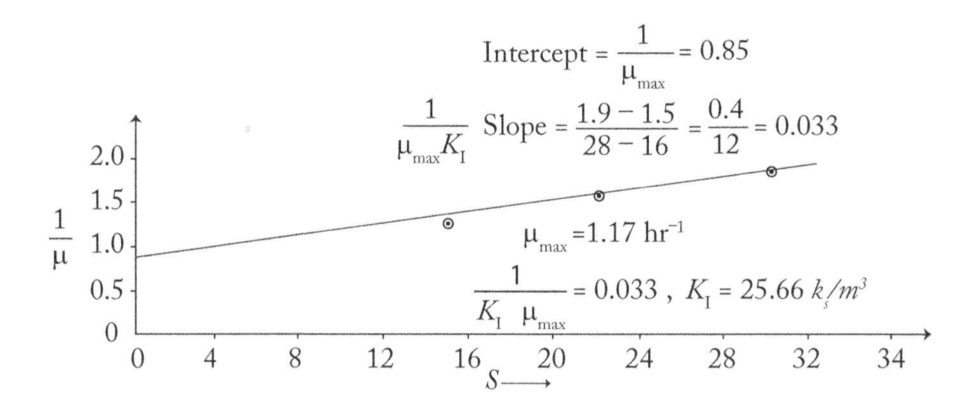

Fig. 6.10 Plot of 1/D vs 1/S

(b) $X = Y_{X/S}(S_O - S)$

$$D = \frac{F}{V} = \frac{0.2\frac{m^3}{hr}}{1.0m^3} = 0.2hr^{-1}$$

$$S = \frac{DK_S}{\mu_{max} - D} = \frac{0.2(2.535)}{1.17 - 0.2} = 0.522Kg/m^3$$

$$X = 0.46(30 - 0.522) = 13.56Kg/m^3$$

(c) We write down two equations for washout – one for low substrate and the other for high substrate with inhibit:

$$(D_{washout})_1 = \frac{\mu_{max}}{\left(1 + \frac{K_S}{S_O}\right)}$$

$$(D_{washout})_2 = \frac{\mu_{max}}{1 + \frac{K_S}{S_O} + \frac{S_O}{K_I}}$$

In the second case $(D_{washout})_2$ becomes smaller than the first one, leading instability with high substrate concentration

(d) With $S_O = 30K_g/m^3$

$$D_{washout} = \frac{1.17}{2.535 + 30} = 0.92hr^{-1}$$

$$D_{max} = 1.17\left(1 - \sqrt{\frac{2.535}{2.535 + 30}}\right) = 0.843hr^{-1}$$

With $S_O = 60K_g/m^3$

$$D_{washmout} = \frac{1.17}{1 + \frac{2.535}{60} + \frac{60}{25.66}} = 0.346hr^{-1}$$

$$D_{max} = 1.17\left(1 - \sqrt{\frac{2.535}{2.535 + 60}}\right) = 0.93hr^{-1}$$

So in the second case with high substrate, the system leads to instability as D_{max} is less than $D_{washout}$, where as with So = 30 kg/m^3, D_{max} is less than $D_{washout}$.

Problem 6.6

The growth of a microorganism is carried out in a steady state chemostat with a substrate (So = 10 kg/m^3) and an inhibitor (I = 0.05 kg/m^3). The inhibition kinetics are given below:

$$\mu = \frac{\mu_{max} S}{S + K_S\left(1 + I / K_I\right)}$$

The following kinetic parameters are available.

$X_O = 0,\ \mu_{max} = 0.5hr^{-1},\ K_S = 1Kg / m^3$

$K_I = 0.01\text{Kg} / m^3,\ Y_{X/S} = 1.0k_g / k_g$

(a) Plot on the graph *X* and *S* vs *D*
(b) Plot on the same graph *X* and *S* vs *D* when $I = 0$
(c) Plot *X* and *S* vs *D* if I increases with m taking $I = X/10$.

Solution:

(a) For sterile feed $\left(X_0 = 0\right)$, $D = \mu$

$$D = \frac{\mu_{max} S}{S + K_S\left(1 + {}^{I}\!/\!_{K_I}\right)} = \frac{0.5S}{S + 1\left(1 + \frac{.05}{.01}\right)} = \frac{0.5S}{6 + S} \qquad \text{(A)}$$

$D_{washout}$ takes place where $S = S_O = 10Kg / m^3$

$$D_{washout} = \frac{0.5(10)}{6 + 10} = 0.31hr^{-1}$$

Therefore *D* is varied from 0 to 0.3 at *D* = 0, from the equation (*A*), *S* = 0, and $X = Y_{X/S}$ (S0 – 0) = 0.10 (10 – 0) = 1.0 Kg/m^3

at *D* = 0.1, from Equation (*A*), 0.1 = 0.5*S*/(*S* + 6), and solving for *S*, *S* = 1.5, *X* = 1.0 (10 – 1.5) = 8.5 K_g/m^3

at *D* = 0.2, we have 0.2 = 0.5 *S*/6 + *S* or *S* = 4 k_g/m^3, *X* = 1.0 (10 – 4) = 6.0 k_g/m^3

at *D* = 0.3, 0.3 = 0.5*S*/6 + *S*, solving for *S*, *S* = 9 k_g/m^3 and *X* = 0.1 (10 – 9) = 0.1 K_g/m^3

(b) For the condition $i = 0$, we have

$D = \mu_{max} S / \left(K_S + S\right)$

$D_{washout} = \mu_{max}$ S0/(Ks + S0) = 0.5(10)/(1.0 + 10) = 0.45 hr^{-1} +

$D = 0.5S / (1 + S)$... (B), D_{max} is obtained from the equation given below

$$D_{max} = {}_{max}\left(1 - \sqrt{\frac{K_S}{K_S + S_O}}\right) = 0.5\left(1 - \sqrt{\frac{1}{1 + 18}}\right) = 0.35hr^{-1}$$

Table (1) of D, S, X at I = .05 Kg/m^3

D hr^{-1}	0	0.1	0.2	0.3	0.31
S, Kg/m^3	0	1.5	4.0	9.0	0
X, Kg/m^3	10	8.5	6.0	1.0	0

D hr^{-1}	0.4	0.45
S, Kg/m^3	4	9
X, Kg/m^3	0.6	.01

Table 2: at I=0

D,hr^{-1}	0.1	0.2	0.3	.35
S,Kg/m3	0.25	0.66	1.5	2.33
X,Kg/m3	9.75	9.34	8.5	7.62

X, S, VS, D are plotted in Fig. 6.11

At i=.05 and i=0

(c) $$D = \mu = \frac{\mu_{max} S}{S + K_S\left(1 + \dfrac{X}{10K_i}\right)} = \frac{\mu_{max} S}{S + K_S\left[1 + \dfrac{Y_{X/S}}{10K_i}\left(S_O - S\right)\right]}$$

where I= X/10 and X=Yx/s(S0–S)

Or $$D = \frac{0.5S}{S + 1\left[1 + \dfrac{1.0}{10 \times .01}\left(10 - S\right)\right]}$$

Or $$D = 0.55/100 - S, \quad D_{washout} = 0.5hr^{-1}, X = 0$$

D	0	0.1	0.2	0.3	0.4	0.5
S	0	7.142	8.69	9.375	9.75	10
X	10	2.856	1.31	0.625	0.25	0

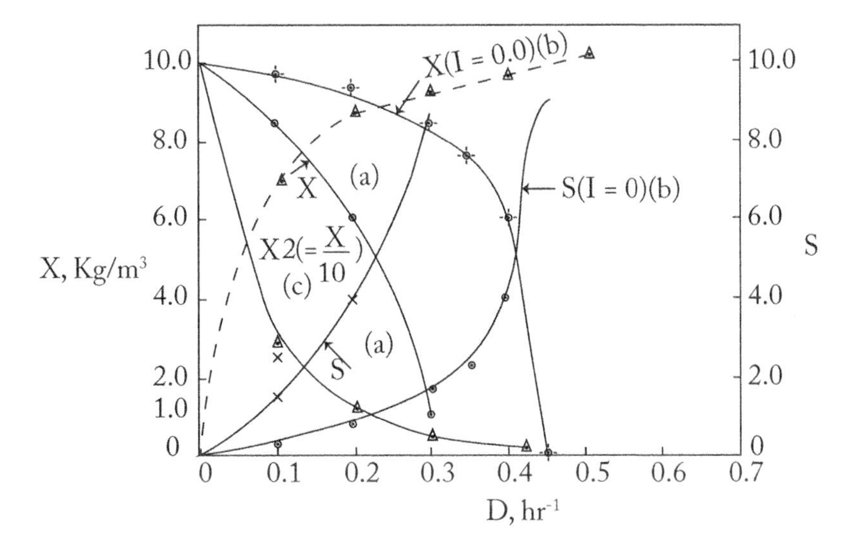

Fig. 6.11 Plot of X, S vs D at I = 0.5 & 0; X/10

Problem 6.7

E.Coli is cultivated in a synethetic medium in a chemostat of volume, V = 10m^3 with a sterile feed, flow rate, F = 5m^3/hr. The initial substrate concentration, So = 10 kg/m^3, $Y_{X/S}$ = 0.7, μ_{max} = 0.935hr^{-1}, Ks = 0.70 kg/m^3.

(a) What will be the doubling time and division rate of the cells in the chemostat?

(b) What will be the exit concentration of cells and substrate?

(c) If one more 10m^3 chemostat is connected for the first, what will be the exit concentrations of X and S in the second chemostat?

(d) If the flow rate is increased from 5m^3/hr to 8m^3/hr for the two fermenters, what will happen and why? What is the recommendation for the problem?

Solution:

(a) Doubling time, td = 0.693/μ

Dilution rate D, $\frac{F}{V} = \frac{5}{10} = 0.5hr^{-1} = \mu$

$$td = \frac{0.693}{0.5} = 1.386hrS$$

(b) $S_1 = \frac{DK_S}{\mu_{max} - D}$

$$= \frac{(0.5)(0.70)}{0.935 - 0.5}$$

$$= 0.80\ Kg / m^3$$

$$X_1 = Y_{X/S}(S_O - S_1)$$

$$= 0.6(10 - 0.8) = 5.76$$

(c) For sterile feed, for 2 CSTR in series, we have

$$S_2 = S_1 - \mu_2 X_2 / D_2 Y_{X/S},$$

Assume $X_2 = 6\ Kg/m^3$, X_1 = 5.76 then μ_2 is given by

$$\mu_2 = \mu_1\left(1 - \frac{X_1}{X_2}\right)$$

$$\mu_2 = 0.5\left(1 - \frac{5.76}{6}\right) = 0.04$$

$$S_2 = 0.8 - (.04)(6)/(.5)(0.6) = 0.2\ Kg/m^3$$

The exit cell mass concentration in $X_2 = 6\ Kg/m^3$

(d) Flow rate, F = 8m^3/hr, $D_2 = \frac{8}{10} = 0.8hr^{-1}$. $S_1 = \frac{D_1 Ks}{\mu_{max} - D_1}$

(i) Putting the values in the equation of S_1, we get

$$S_1 = (0.8)(0.7)/(.937 - .8) = 4.148 Kg/m^3,$$

$$X_1 = 0.6(10 - 4.148)$$

$$= 3.5112\ Kg/m^3$$

(ii) For the 2nd chemostel, assuming $X_2 = 5.0$ Kg/m^3 (by trial and error)

$$\mu_2 = D_2\left(1 - \frac{X1}{X2}\right), \text{ So } \mu_2 = 0.8(1 - 3.5/5) = 0.24 hr^{-1}$$

$$S_2 = 4.148 - (0.24)(5)/(0.8)(0.6) = 1.916 Kg/m^3$$

and $X_2 = 5\ Kg/m^3$

So increasing the flow rate, cell mass concentration in both the CSTRs have been reduced.

Therefore, the flow rate should not be increased.

Problem 6.8

The growth of a microorganism is given by the following kinetics

$\mu = \mu_{max}\left(1 - e^{-S/K_s}\right)$ (Tessier Equation)

Where $\mu_{max} = 0.4$hr^{-1}, Ks = 6.0 kg/m^3, $Y_{X/S} = 0.5$

(a) The microorganism is cultivated in 10m^3 chemostat with a flow rate of 3 m^3/hr. the substrate concentration of the feed stream, $S_o = 10$ Kg/m^3. The feed is sterile. What will be steady state concentration of X and S at the exit?

(b) Explain the difference between the above model and Monod Model from a μ vs s plot.

Solution:

D = F/V = 3/10 = 0.3 hr^{-1}

For a sterile feed in a chemostat

$$D = \mu = \mu_{max}\left(1 - e^{-S/K_S}\right) \tag{A}$$

Putting the values of D, μ_{max}, Ks in equation (A), we have

$$0.3 = 0.4\left(1 - e^{-S/6.0}\right)$$

Solving for S, by trial and error, we get S = 9.317 kg/m^3,

$$X = Y_{X/S}(S_O - S) = 0.5(10 - 8.317) = 0.84 Kg/m^3$$

μ values from Monod Model can be obtained by varying the values of S in the following equation,

$\mu = \mu_{max}$ S/ (Ks + S)

S, kg/m^3	1	2	4	8	10	16	20	30	40	50
μ hr^{-1}	0.57	0.10	0.16	0.228	0.25	0.29	0.30	0.33	0.347	0.357

μ values from the Tessier growth model $\mu = \mu_{max}\left(1 - e^{-S/K_S}\right)$ using different values of S, and the following table is made

S, kg/m^3	1	2	4	8	10	16	20	30
μ hr^{-1}	0.06	0.113	0.19	0.29	0.32	0.37	0.38	0.397

From the first plot of Monod model, the asymptotic value is reached after long time much beyond 40 hrs.

From the 2nd plot, the asymptotic value of μ (μ_{max} = 0.40) is reached at about 40 hrs.

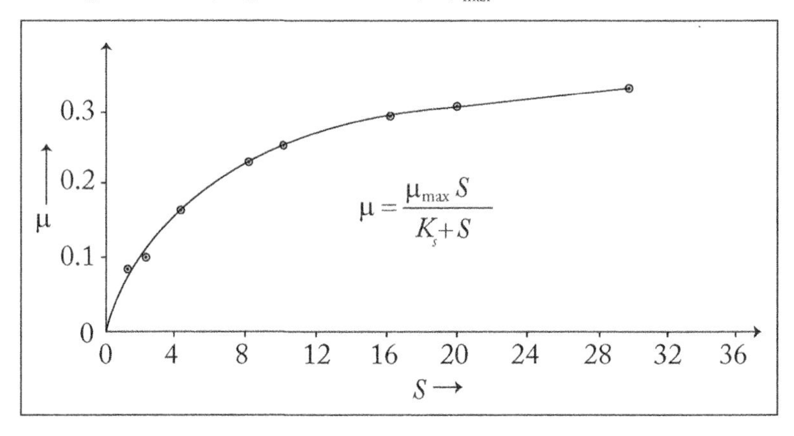

Fig. 6.12 The plot of μ vs S using Monod Model

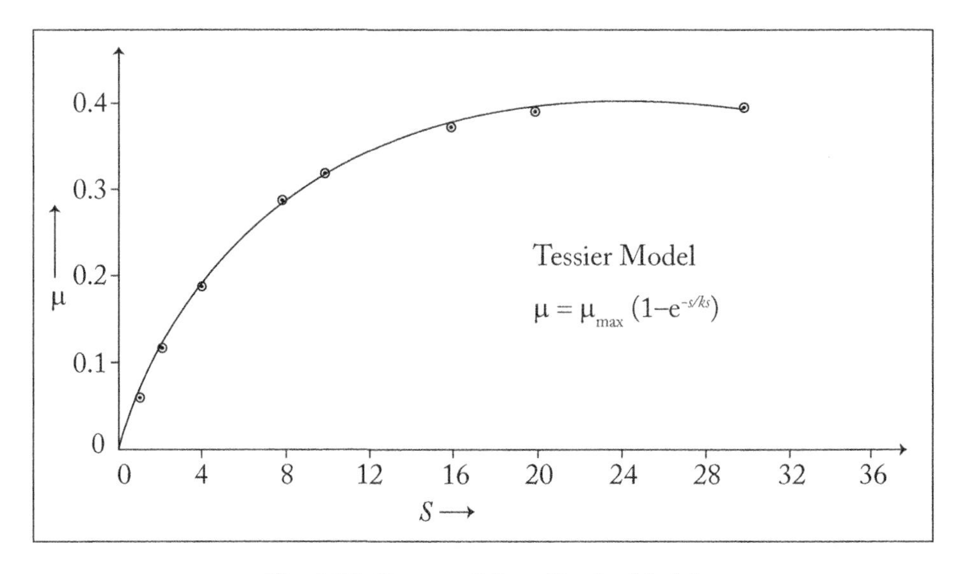

Fig. 6.13 Plot μ vs S from Tessier Model

6.9.3 Commercial chemostat (stirred tank fermentor)

Commercial chemostat or stirred tank fermenter is used for very large scale production of important biochemical products. For example a reactor of $100m^3$ capacity has been employed for penicillin production. Chemostat can be designed for both aerobic and anaerobic fermentation for wide range of cells including microbial, animal and plant cells. The mixing intensity may be varied by choosing a suitable impeller type and by varying agitator speed. The agitation and aeration are useful for suspension of cells, oxygenation, mixing of the culture and also for heat transfer. The stirred tank fermenter can also handle highly viscous media. Commercial fermenters are usually made of stainless steel. It can be operated under mild or vigorous conditions, depending upon the type of cells.

The disadvantage of the fermenter is its cost arising from the consumption of large amount of power beside its basic manufacturing cost.

Stirred tank fermenter has been modified for shear sensitive cells like animal or plant cells by modifying the stirrer design, reducing the agitator speed and removing baffles. Fluid Shear in mixing is produced by velocity gradients from tangential radial velocity components of fluids leaving the impeller region. The velocity profile of a regular flat bladed, disk turbine is parabolic in shape, with the highest observed velocity occurring at the centre line of each blade. The resultant velocity decreases away from the centre line. The reduction in velocity is about 85 percent at one blade width distance above or below the blade. As the blade width-to-tank diameter ratio increases, the velocity profile becomes less parabolic, yielding a lower amount of shear due to reduced velocity gradient. Thus by increasing the blade width, stirred tank vessel can be used for the cultivation of shear sensitive animal and plant cells.

The height to (tank) diameter ratio (H/D) varies from 2:1 to 3:1. The agitation is effected with two or three turbine impellers. The impeller shaft enters the reactor vessel from the top through a bearing house and mechanical seal assembly. The impeller diameters to tank diameter ratio (D_I/Dt) is usually in the range of 0.3 to 0.4. In a two-impeller system; the distance between the first impeller and the bottom of the vessel and between two impellers is 1.5 Dt. Normally four equally spaced baffles are provided to prevent vortex formation and to enhance mixing efficiency. The width of the baffle is one tenth of tank diameter (Dt). For aerobic fermentation, a single orifice sparger or a ring sparger with multiple holes is used to supply air to the fermenter. The sparger is placed between the bottom impeller and the bottom of the vessel. The pH of the medium is maintained by adding buffer solution by a pH controller. The temperature is controlled by cooling or heating through a coil or a jacket.

6.10 Batch Reactor

Many biochemical processes are carried out in batch reactors. *A* liquid medium with inoculum of living cells is placed in a vessel with a stirrer. The concentration of nutrients, cells and products vary with time. A material balance on a component, C_A, can be presented in a usual form as given below:

$$\frac{d}{dt}\left(V_R C_A\right) = V_R r_{fA} \tag{6.87}$$

For a constant culture volume, $V_{R,}$ we have $\frac{dC_A^{'}}{dt} = r_A$ (6.88)

For cell mass growth (X)

$$\frac{dX}{dt} = r_X = \mu X \tag{6.89}$$

Where μ is the specific growth rate.

Using the Monod module, $\mu = \mu_{max} S/(K_S + S)$, the equation (6.3) may be written as

$$\frac{dX}{dt} = \frac{\mu_{max} SX}{K_S + S} \tag{6.90}$$

Eliminating S from the equation (6.80) by the relation

$$S = S_O - \frac{\left(X - X_O\right)}{Y_{X/S}} \tag{6.91}$$

We have

$$\frac{dX}{dt} = \frac{\mu_{max}\left(Y_{X/S} S_O + X_O - X\right)X}{\left(K_S Y_{X/S} + Y_{X/S} S_O + X_O - X\right)} \tag{6.92}$$

Now integrating

$$\int_{X_O}^{X} \frac{\left(K_S Y_{X/S} + Y_{X/S} S_O + X_O - X\right)dX}{\left(Y_{X/S} S_O + X_O - X\right)X} = \left(\mu_{max}\right)t \tag{6.93}$$

After integration we get,

$$\frac{K_S Y_{X/S} + S_0 Y_{X/S} + X_0}{(Y_{X/S} S_0 + X_0)} \ln\frac{X}{X_0} - \frac{K_S Y_{X/S}}{(Y_{X/S} S_0 + X_0)} \times \frac{\ln\left[Y_{X/S} S_0 + X_0 - X\right]}{Y_{X/S} S_0} = \mu_{max} t \tag{6.94}$$

The X vs t curve from the equation (6.84) is sigmoidal in nature. The value of X asymptotically reaches to the value of $(Y_{X/S}\, S_0 + X_0)$.

The equation (6.90) may be integrated numerically for different values of t provided the parameters, v_{max}, K_S, X_0, S_0, $Y_{X/S}$ are known. Suitable numerical techniques like Euler or Rung-Kutta method may be used for integration.

The kinetic parameters of Monod Model (μ_{max} & K_S) can be evaluated from the batch reactor data as t vs X and S, and X vs t are plotted and the derivatives (dx/dt) are obtained at different values of X and dividing the slope by X, we get, $1/X$ dx/dt = μ. From a set of μ vs. S data and taking the reciprocal of the Monod Model, given as,

$$\frac{1}{\mu} = \frac{K_S}{\mu_{max}}\frac{1}{S} + \frac{1}{\mu_{max}} \tag{6.95}$$

$\frac{1}{\mu} vs. \frac{1}{S}$ is plotted in a linear form, the slope will give (K_S/μ_{max}) and the intercept $1/\mu_{max}$, from which K_S and μ_{max} are evaluated.

A simple numerical method has been used to solve the following problem

Problem 6.9

A batch fermenter is used to carry out a microbial reaction with the following kinetic and other parameters $\mu_{max} = 1.0\ hr^{-1}$, $K_S = 0.2\ kg/m^3$, $X_o = 0.1\ Kg/m^3$, $S_o = 10\ kg/m^3$, $Y_{X/S} = 0.5$

(a) Calculate at what time the substrate is completely consumed

(b) Make a plot of X, S vs t by using a numerical method (Euler method) which may be presented as

$$\frac{dy}{dx} = f(x, y)$$

$$\Delta y = f(xo, yo)\Delta x$$

$$y_1(x_1) = yo + \Delta y$$

$$= yo + f(xo, yo)\Delta x$$

$$y_2(x_2) = y_1 + f(x_1, y_1)\Delta x$$

$$y_n(x_n) = y_{n-1} + f(x_{n-1}, y_{n-1})\Delta x$$

Solution:

$$S = S_O - \frac{1}{Y_{X/S}}(X - X_o) = 10 - \frac{1}{0.5}(X - 0.1) = 10.2 - 2X$$

$$\Delta X = \left(\frac{\mu_{max} SX}{K_S + S}\right)\Delta t$$

Let us choose $\Delta t = 1.0 hr$

$$\Delta X = f(X_O, t_o)\Delta t = \left[\frac{(1)(0.1)(10)}{0.2 + 10}\right] \times 1 = 0.098 \times 1 = 0.098$$

$$X_1 = X_O + \Delta X = 0.1 + .098 = 0.198 Kg / m^3$$

$$t_1 = 0 + \Delta t = 1.0 hr$$

$$S_1 = 10.2 - 2(.198) = 9.804$$

$$X_2 = 0.198 + \left[\frac{(1)(.198)(9.804)}{(0.2 + 9.804)(10.004)}\right] = 0.392,\ t_2 = 1 + 1 = 2$$

$$S_2 = 10.2 - 2(.392) = 9.416$$

$$X_3 = 0.392 + \left[\frac{(1)(0.392)(9.416)}{0.2 + 9.416(9.616)}\right] \times 1$$

$$t = 2 + 1 = 3 = 0.776$$

$$S_3 = 10.2 - 2(.776) = 8.648$$

$$X_4 = 0.776 + \left[\frac{1 \times (.776)(8.648)}{0.2 + 7.14}\right] \times 1 = 1.53$$

$$S_4 = 10.2 - 2 \times 1.53 = 7.14$$

$$X_5 = 1.53 + \left[\frac{1 \times (1.53)(7.14)}{0.2 + 7.14}\right] = 3.02$$

$$X_5 = 10.2 - 2 \times 3.02 = 4.16$$

$$X_6 = 3.02 + \left[\frac{1 \times (3.02)(4.16)}{0.2 + 4.16}\right] \times 1 = 5.90$$

$$S_6 = 10.2 - 2 \times 5.9 = -1.6$$

Assume $\Delta t = 0.5$

$$X_6 = 4.46$$

$$S_6 = 10.2 - 2 \times 4.46 = 1.28$$

Assume $\Delta t = 0.75$

$$X_6 = 5.18$$

$$S_6 = 10.2 - 2 \times 5.18 = -0.16$$

Therefore substrate is exhausted at 5.75 hrs.
A table is prepared as given below:

(b)

t, hr	0	1	2	3	4	5	6
X, kg/m^3	0.1	0.198	0.392	0.776	1.53	3.02	5.18
S, kg/m^3	10	9.80	9.416	8.648	7.14	4.16	-0.16

(a) So after 5.75 hrs the substrate is completely consumed

6.11 Plug Flow Fermenter (Tower Fermenter)

The tubular flow fermenter is assumed to have plug flow behaviour, where there is no axial mixing. In such reactors, nutrients, seeded microorganisms enter at one end and the cells growth along the length. Since there is no mixing, the reactor is not suitable for cell growth reaction which is autocatalytic in nature. By certain amount of mixing, the efficiency of PFR may be enhanced. The performance of a plug flow reactor may be equivalent to a batch reactor. The design equation of PFR may be developed from that of a batch reactor:

$$\frac{dX}{dt} = \frac{\mu_{max} XS}{K_S + S} \tag{6.96}$$

Now replacing dt by dZ/u the above equation may be written as

$$u\frac{dX}{dZ} = \frac{\mu_{max} XS}{(K_S + S)} \tag{6.97}$$

Using substrate in the plug flow model we get

$$-u\frac{dS}{dZ} = \frac{1}{Y_{X/S}} \frac{\mu_{max} XS}{K_S + S} \tag{6.98}$$

Now eliminating X by the following stoichiometric relation

$$X = X_O + Y_{X/S}(S_O - S) \tag{6.99}$$

The equation (6.87) becomes

$$-u\frac{dS}{dZ} = \frac{\propto_{max}}{Y_{X/S}} \frac{\left[X_O + Y_{X/S}(S_O - S)\right]}{K_S + S} \tag{6.100}$$

Now integrating with the boundary condition at Z = 0, S = So

$$\int_{S=0}^{S} \frac{(K_S + S)dS}{X_O + Y_{X/S}(S_O - S)S} \tag{6.101}$$

We get,

$$X_0 + Y_{X/S}(S_O + K_S) ln\frac{X_o + Y_{X/S}(S_o - S)}{X_O} - K_S Y_{X/S} ln\frac{S}{S_O} = \mu_{max}(X_O + Y_{X/S} S_O)\frac{Z}{u} \tag{6.102}$$

The result is the same as that of batch reactor when Z/u is replaced by residence time, τ

With sterile feed ($X_0 = 0$), the PFR will yield zero biomass at the exit. This problem is overcome if the incoming stream is inoculated before entering the vessel by a recycle stream containing biomass.

For microbial processes, the plug flow fermenter gives maximum product yield in the effluent. But the continuous inoculation and the difficulties with gas exchange in PFTR lead to the use of a batch fermenter for high final product concentration. For exponential microbial growth the CSTR is more efficient that the plug flow or batch reactor.

6.12 Batch Reactor and Chemostat

Many biochemical processes are carried out in Batch reactors. For large scale production, the batch cycle time, preparation time for a new batch including cleaning, sterilizing and filling, and the number of cycles for a desired productivity are important.

For a batch fermenter, the total batch cycle time, t_c can be expressed as

$$t_c = \frac{1}{\mu_{max}} ln \frac{Xm}{Xo} + t_l \tag{6.103}$$

Where X_m is the maximum achievable cell concentration. X_0 is the cell concentration in the inoculum, t_l is the preparation time for the batch.

The maximum cell mass, X_m can be given as

$$X_m = X_O + Y_{X/s} S_O \tag{6.104}$$

The rate of cell mass produced in one batch cycle, r_b assuming $\left(r_{Opt}\right)$

$$r_b = \frac{Y_{X/S} S_O}{\frac{1}{\mu_{max}} ln \frac{X_m}{X_O} + t_l} \tag{6.105}$$

Now we proceed to find out the rate of biomass production (r_{optt}) from a chemostat. The optimum cell productivity from the equation (6.85) is

$$X_{opt} = Y_{X/S} S_O \frac{\sqrt{K_S + S_O}}{1 + \sqrt{K_S + S_O}}$$

The optimum dilution rate, Dopt is given by the equation (6.81)

$$D_{opt} = \mu_{max} \left(1 - \sqrt{\frac{K_S}{K_S + S_O}}\right)$$

Thus the best productivity of cell mass is given by Dopt Xopt and expressed as follows

$$D_{opt} X_{opt} = \mu_{max} \left(1 - \sqrt{\frac{K_S}{K_S + S_O}}\right) Y_{X/S} S_O \frac{\sqrt{\frac{K_S + S_O}{S_O}}}{1 + \sqrt{\frac{K_S + S_O}{S_O}}} \tag{6.106}$$

Now under usual condition, So >> K_S, so the rate of biomass product in the chemostat is

$$r_{copt} = D_{opt} X_{opt} = \mu_{max} Y_{X/S} S_O \tag{6.107}$$

So the ratio of biomass production in a chemostat and a batch can be given as

$$\frac{r_{copt}}{r_b} = ln\frac{X_m}{X_O} + \mu_{max}t_l \quad (6.108)$$

Most commercial fermenters operate
$Xm/X_o = 10$ to 20
For E.coli fermentation, $Xm/X_o = 20$ and $t_l = 5$hrs
$\mu_{max} = 1.0\ hr^{-1}$
So the ratio (rcopt/rb) comes to 8.0. So for large scale production, chemostats are preferable. Yet the batch reactors are employed for many biochemical processes. There are several advantages of batch reactor over CSTR or PFTR as mentioned below. Many secondary metabolic products are non-growth associated and growth represses product formation. Low dilution rate is required for better yield of product and that is much less than the optimum dilution rate. The productivity in a batch reactor may be higher than that in a simple chemostat for secondary products.

Another factor is genetic instability. In recombinant technology a chemostat gives a rapid growth of plasmid free- cells over the plasmid containing cells which release desired protein. In a batch culture, the tendency of the plasmid-free cells to outgrow the most productive strain is less. This is the phenomenon of genetic instability which may be controlled in a better way in a batch culture. Other considerations include market economy, simple construction and operation, multiple products formation. Some of the welknown continuous systems are SCP production, ethanol production, lactic acid production, waste-water treatment etc.

6.13 Non-ideal CSTR Model

A non-ideal CSTR model may be assumed to consist of two interconnected ideal CSTRs. The material balance and the flow between the two reactors is shown in the following figure (6.14)

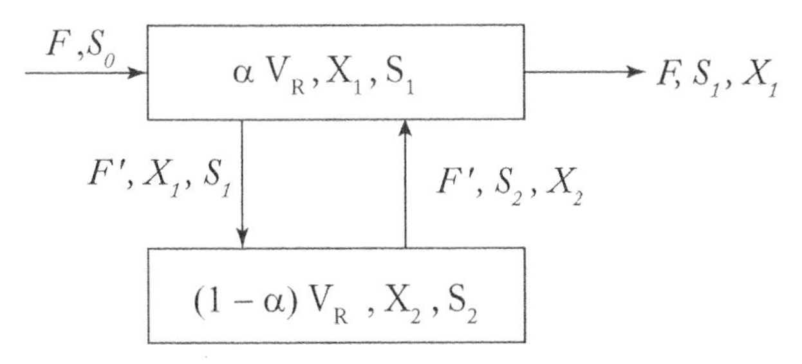

Fig. 6.14 Model for an incompletely mixed CSTR with a stagnant region

The reactor contents have been divided into two interconnected regions: one is a well mixed and the other is stagnant. The volume of each region is a fraction, of the total reactor volume, V_R. Monod growth kinetics are applicable. The mass balances for the two regions can be shown as follows.

Substrate balance:

Region 1: $X_1 = Y_{X/S}(S_0 - S_1)$ (6.109)

Region 2: $X_2 - X_1 = Y_{X/S}(S_1 - S_2)$ (6.110)

Cell mass balance for

Region 1:

$$X_2 + \alpha\gamma' \frac{\mu_{max} S_1}{K_S + S_1} X_1 = (1 + \gamma' D) X_1 \tag{6.111}$$

Where, $\gamma = V/F'$, $\alpha = V/V_R$, $D = F/V_R$

V = volume of the first CSTR, α is a fraction, $(1 - \alpha)V_R$ is the volume of the second reactor,

Region 2:

$$X_1 + (1 - \propto) Y' \frac{\mu_{max} S_2}{K_S + S_2} X_2 = X_2 \tag{6.112}$$

The dilution rate at washout, $D_{washout}$ is evaluated by putting $S_0 = S_1 = S_2$ and $X = 0$ in the equations (6.111 to 6.112) and we get

$$D_{washout} = \left(\frac{\mu_{max} S_O}{K_S + S_O}\right)\left[1 + \frac{(1-\propto)^2 \mu_{max} S_O}{K_S + S_O - (1-\propto) Y'(\mu_{max} S_O)}\right] \tag{6.113}$$

The expression on the right-hand side with the first bracket is the same obtained for $D_{washout}$ for an ideal chemostat (i.e. the perfectly mixed system). The part with the bracket on the right hand-side signifies the magnitude of non-mixing and the effect leads to the increased value of $D_{washout}$. When $\alpha = 1$, the expression of equation (6.113) reduces to $D_{washout}$ for ideal chemostat.

The variation of cell mass (X_1) with D with different values of α is shown in the figure 6.15, using kinetic parameter values such as $\mu_{max} = 1.0$ hr^{-1}, $K_S = 0.5$ Kg/m^3, $Y_{X/S} = 0.5$, So = 10 Kg/m^3

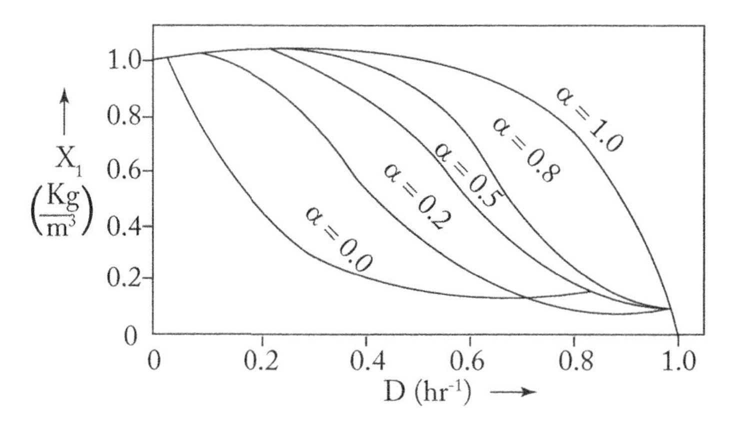

Fig. 6.15 Exit Cell concentration (X_1) as a function of D at different values of α

The factor, α. is a measure of dead zone. when $\alpha = 1$, there is complete mixing and the curve of X_1 vs D is similar to the theoretical plot of completely mixed reactor. As α decreases from 1 (perfect mixing) to 0 (fully stagnant zone), X_1 decreases with dilution rate D.

6.14 Multiple Fermenters Operated in Series

The best choice of fermenters combination depends on the shape of the curve obtained by plotting 1/rx vs x and the final conversion. If the desired conversion is less than optimum conversion, Xopt, a single CSTR is sufficient. For the conversion much larger than Xopt, a combination of two reactors is recommended and the sequence is a CSTR operated at Xopt and then followed by a PFR. A combination of two CSTRs is more preferred than one CSTR.

6.14.1 A chemostat and PFR in series

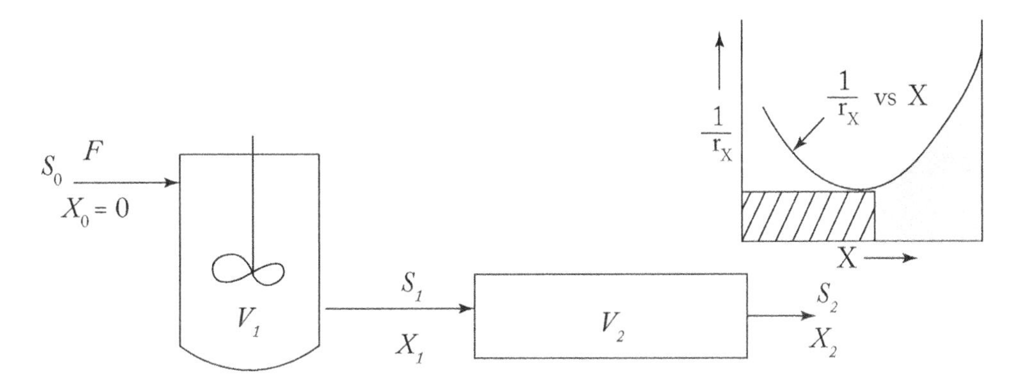

Fig. 6.16 Schematic diagram of CSTR and a PFR in series

For sterile feed (Xo = 0), the exit concentrations of S1, X1 and P1 for the CSTR can be given as

$$S_1 = \frac{K_S}{\tau_m \mu_{max} - 1} \tag{6.114}$$

Where $\tau_m = 1/D_{max}$

$$X_1 = Y_{X/S}(S_O - S_1) \tag{6.115}$$

$$P_1 = P_O + Y_{P/S}(S_O - S_1) \tag{6.116}$$

For the second reactor, PFR, the residence time, T_{P_2} can be evaluated from the following equation,

$$\tau_{P_2} = \int_{X_1}^{X_2} \frac{dX}{r_X} = \int_{X_1}^{X_2} \frac{(K_S + S)dX}{\mu_{max} SX} \tag{6.117}$$

Now
$$Y_{X/S} = \frac{X_2 - X_1}{S_1 - S_2} \tag{6.118}$$

The integration of the equation (6.117) can be given by the equation (6.119) by replacing X_0 and X_1

$$\tau_{P_2} \mu_{max} = \left(\frac{K_S Y_{X/S}}{X_1 + S_2 Y_{X/S}} + 1 \right) ln \frac{X_2}{X_1} + \frac{K_S Y_{X/S}}{X_1 + S_2 Y_{X/S}} ln \frac{S_1}{S_2} \tag{6.119}$$

Since X_2 is given, the final substrate concentration, S_2, can be evaluated from equation (6.118). The residence time of the second fermenter, τ_{P_2} is then calculated from equation (6.117) by numerical integration.

6.14.2 Chemostats in series

If the desired cell concentration, X is great than X_{opt}, multiple chemostats in series may be used. A schematic diagram for n number of chemostats in series is shown in figure. 6.17

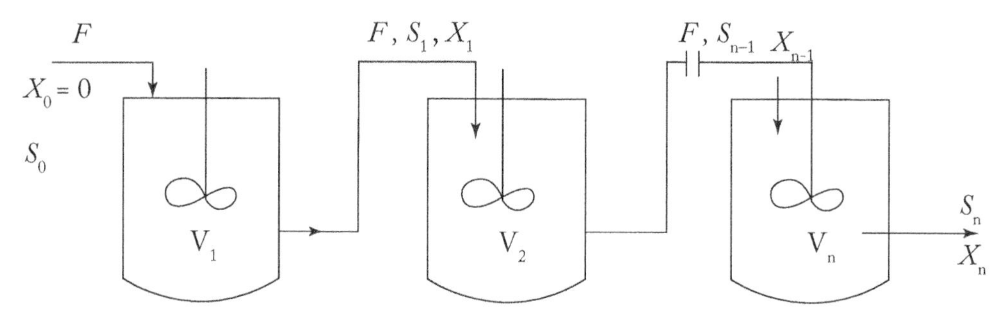

Fig. 6.17 Schematic diagram of multiple chemostats in series

For the nth steady state chemostat, the mass balance for the microbial cells can be written as

$$F(X_{n-1} - X_n) + V_n r_{Xn} = 0 \tag{6.120}$$

Where
$$r_{Xn} = \frac{\mu_{max} S_n X_n}{K_S + S_n} \tag{6.121}$$

Yield coefficient, $Y_{X/S}$ is given as

$$Y_{X/S} = \frac{X_n - X_{n-1}}{S_{n-1} - S_n} \tag{6.122}$$

The dilution rate for different CSTRs can be evaluated by solving the above three equations simultaneously if the final cell concentrations are known.

A graphical method can be used to calculate the outlet cell concentration for each stage and the number of CSTRs required to achieve final cell concentration. The dilution rate of the first CSTR for sterile feed (Xo=0) is given as

$$D_1 = \frac{F}{V_1} = \frac{r_{X1}}{X_1} \tag{6.123}$$

For the 2nd CSTR

$$D_2 = \frac{F}{V_2} = \frac{r_{X_2}}{X_2 - X_1} \tag{6.124}$$

And for nth reactor,

$$D_n = \frac{F}{V_n} = \frac{r_{Xn}}{X_n - X_{n-1}} \tag{6.125}$$

The growth rate, r_x based on Monod model is plotted against X, producing a parabolic curve as shown below:

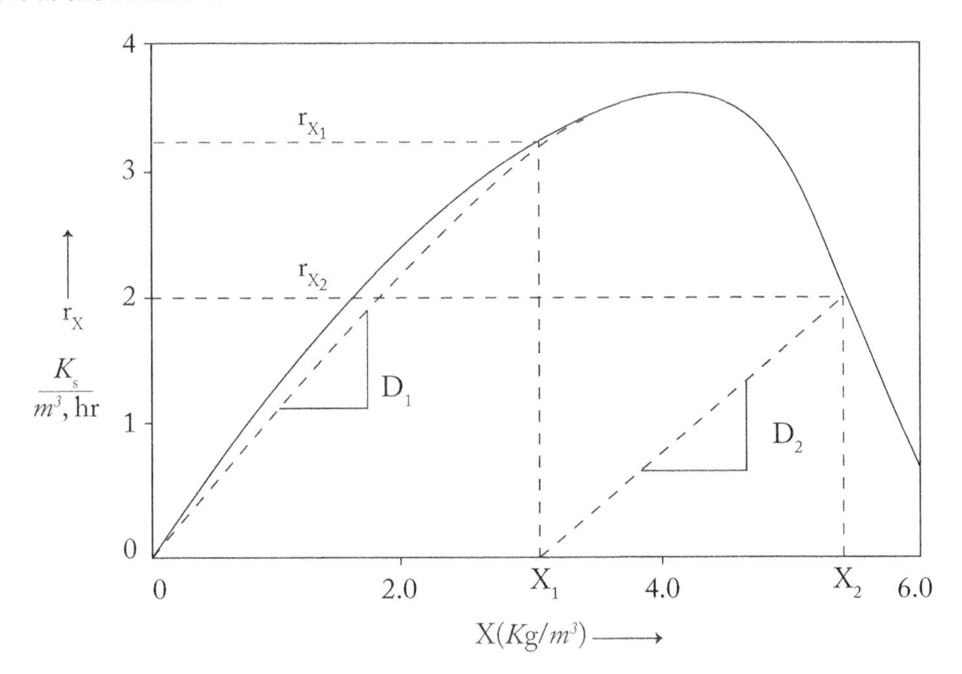

Fig. 6.18 Graphical solution of a two stage CSTR with $\mu_{max} = 0.94 hr^{-1}$, Ks = 0.7 kg/m^3, $Y_{X/S}$ = 0.6, So = 10 kg/m^3, Xo = 0

A line is drawn with a slope of F/V1=D1 passing through the origin, which intersects the curve at (X_1 and r_{x1}) and similarly for the 2nd CSTR (chemostat) with a slope (D_2) of the line connecting (X_1, 0) and (X_2, r_{x2}).

Thus knowing the dilution rate of each fermenter, the cell mass of each CSTR may be evaluated.

Problem 6.10

Two CSTRs in series are used to carry out a microbial reaction whose kinetic parameters are given by Mono Model.

$\mu_{max} = 0.8 hr^{-1}$, $Ks = 5\ kg/m^3$, $Y_{X/S} = 0.6$

If the inlet substrate concentration, So = 10 kg/m^3 and the outlet cell concentration in the second CSTR is 1.0 kg/m^3, calculate the outlet concentration of cells and substrate by a graphical method.

Solution:

$$D_2 = \frac{F}{V_2} = \frac{r_{X_2}}{X_2 - X_1} \text{ and } D_1 = \frac{F}{V_1} = \frac{r_{X_1}}{X_1 - X_0}$$

Now assuming different values of X, the values of S and r_x are calculated

$$X = Y_{X/S}\left(S_O - S\right) = 0.6\left(10 - S\right)$$

$$r_X = \frac{dX}{dt} = \frac{\mu_{max} SX}{K_S + S}$$

X, kg/m³	0	1	2	3	4	5	5.5	6
S, kg/m³	10	8.33	6.67	5.0	3.33	1.67	0.83	0
r_x, Kg/(m³) (hr)	0	0.5	0.914	1.2	1.28	1.0	0.62	0

r_x vs X is plotted with assumed values of X.

$X_2 = 0.6\ (10 - S) = 5.4\ kg/m^3$

X_2 is located on the x-axis and a perpendicular line is drawn to meet the curve at $r_{x2} = 0.6$. Now a straight line is drawn from ($r_{x2} = 0.6$, $D_2 = 0.3\ hr^{-1}$) to meet the X-axis at (0, X_1=3.7) and then a perpendicular line is drawn from ($r_x = 0$, $x_1 = 3.7$) to meet the curve at $r_{x1} = 1.24$, $x_1 = 3.7$). From that point, a line is drawn from that point to (0, 0) with slope of $D_1 = 0.33\ hr^{-1}$. For S_1, $X_1 = 0.6\ (10 - S_1)$

For $X_1 = 3.7$, $S_1 = 3.83\ kg/m^3$. Ans

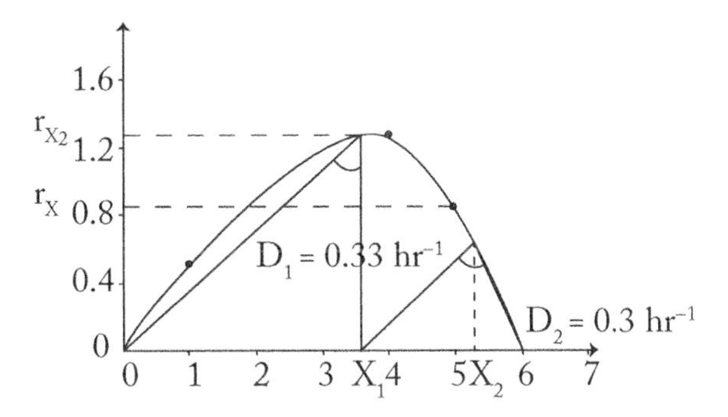

Fig. 6.19 Plot of r_x vs S for two CSTRs in series for Graphical method

An analytical method for a two-stage chemostat system can be presented by the steady state mass balances. For the first stage chemostat with sterile feed (Xo =0), we know,

$$S_1 = \frac{K_S D_1}{\mu_{max} - D_1}, \text{ and } X_1 = Y_{Y/S}\left(S_O - S_1\right)$$

For the second stage at the steady stage:

$$F\left(X_1 - X_2\right) + \mu_2 V_2 X_2 = 0 \tag{6.126}$$

Diving by V_2 and rearranging

$$\mu_2 = D_2\left(1 - \frac{X_1}{X_2}\right) \tag{6.127}$$

Where $X_1/X_2 < 1$ and $\mu_2 < D_2$.
The substrate balance in the second stage at the steady state is:

$$F\left(S_1 - S_2\right) - \frac{\mu_2 X_2}{Y_{X/S}} V_2 = 0 \tag{6.128}$$

Rearranging

$$S_2 = S_1 - \frac{\mu_2 X_2}{Y_{X/S} D_2} \tag{6.129}$$

Where $$D_2 = F / V_2, \; \mu_2 = \frac{\mu_{max} S_2}{K_S + S_2} \tag{6.130}$$

Equations (6.128) & (6.129) can be solved simultaneously for X_2 and S_2.

Problem 6.11

A fermentation is carried out with a microorganism which follows Monod's Kinetics with the following parameters:

$\mu_{max} = 0.8hr^{-1}$, $K_S = 5\ kg/m^3$, $Y_{X/S} = 0.7$

The microorganism will be cultivated in one fermenter or two in series. The flow rate and the substrate concentration of the inlet steam are

$F = 0.3m^3/hr$ and $So = 50\ kg/m^3$. The substrate concentration of the outlet stream is $5\ kg/m^3$.

(a) If we use one CSTR, what is the size of the fermenter and what is the cell ion-concentration in the outlet stream?

(b) If we use two CSTRs, what size of two fermenters will be most productive? What are the concentrations of cells and substrate at the exit of the first CSTR?

(c) What is the best combination of ferments types and volumes if we use two fermenters in series.

Solution:

(a) For a single steady stage CSTR with a sterilè feed, the dilution rate is equal to the specific growth rate, D

$$D = \frac{F}{V} = \frac{\mu_{max} S}{K_S + S} = \frac{0.8(5)}{5+(5)} = 0.4hr^{-1}$$

$$V = \frac{F}{D} = \frac{0.3m^3 / hr}{0.4^{-1}/_{hr}} = 0.75m^3$$

$X = 0.7\ (50 - 5) = 31.5\ kg/m^3$

(b) For two CSTRs in series, the first fermenter must be operated at X_{opt} and S_{opt}, which are evaluated as follows:

$$\propto = \sqrt{\frac{K_S + S_O}{K_S}}$$

$$X_1 = X_{opt} = Y_{X/S} S_O \frac{\propto}{\propto + 1}$$

$$\text{Or } \propto = \sqrt{\frac{5+50}{5}} = 3.31$$

$$X_1 = 0.7(50)\frac{3.31}{1+3.31} = 26.87Kg / m^3$$

$$S_1 = S_{opt} = \frac{S_O}{\propto + 1} = \frac{50}{1+3.31} = 11.50Kg / m^3$$

Now the residence of the first CSTR is τ_{m_1} given by

$$\tau_{m_1} = \tau_{opt} = \frac{\propto}{\mu_{max}(\propto - 1)} = \frac{3.31}{0.8(3.31-1)} = 1.79 hr$$

$$V_1 = T_{m_1} F = 1.79 \times 0.3 = 0.537 m^3$$

For the second fermenter, the mass balance equation can be given as:

$$F(X_2 - X_1) = V_2 \frac{\mu_{max} S_2 X_2}{K_S + S_2}$$

Substituting the values of the parameters

$$0.3(31.5 - 26.87) = V_2 \frac{0.8(5)(31.5)}{5+5}$$

Or $V_2 = 0.11 m^3$

So total volume, V= V_1+V_2 = 0.537+0.11 = 0.647 m^3

So in the 2nd case the total reactor volume is nearly 15.4% less than the value of single CSTR.

(c) The best combination is a CSTR operated at the maximum rate followed by a PFR. The volume of the first fermenter (CSTR) is 0.537 m^3 as calculated in the part (b)

For the second reactor (PFR) using the equation (6.107)

$$T_{P_2} = \frac{1}{\mu_{max}}\left[\left(\frac{K_S Y_{X/S}}{X_1 + S_1 Y_{X/S}} + 1\right) ln\frac{X_2}{X_1} + \frac{K_S Y_{X/S}}{X_1 + S_1 Y_{X/C}} ln\frac{S_1}{S_2}\right]$$

$$= \frac{1}{0.8}\left[\left(\frac{5 \times 0.7}{26.87 + 11.50 \times 0.7} + 1\right) ln\frac{31.5}{26.87} + \frac{5 \times 0.7}{26.87 + 11.50 \times .7} ln\frac{11.5}{5}\right]$$

$$= \frac{1}{0.8}[.1748 + .083] = 0.322 hrs.$$

$$V_2 = 0.3 m^3 / hr \times .322 = 0.097 m^3$$

Total volume V = 0.537 + 0.097 = 0.634 m^3. Ans

So the additional saving by using the second reactor (PFR) instead of a second CSTR is not significant in this case.

6.14.3 Chemostat with a recycle

To maintain a higher cell concentration inside the chemostat than the steady state value, some cells in the effluent can be recycled back to the reactor. The system has been used in Penicillin production, waste water treatment to improve the yield of product and to enhance the C.O.D. removal respectively. The performance equations may be developed for the chemostat with a recycle. A steady state cell mass and balance with a recycle stream of cells can be shown according to the schematic diagram in figure 6.20

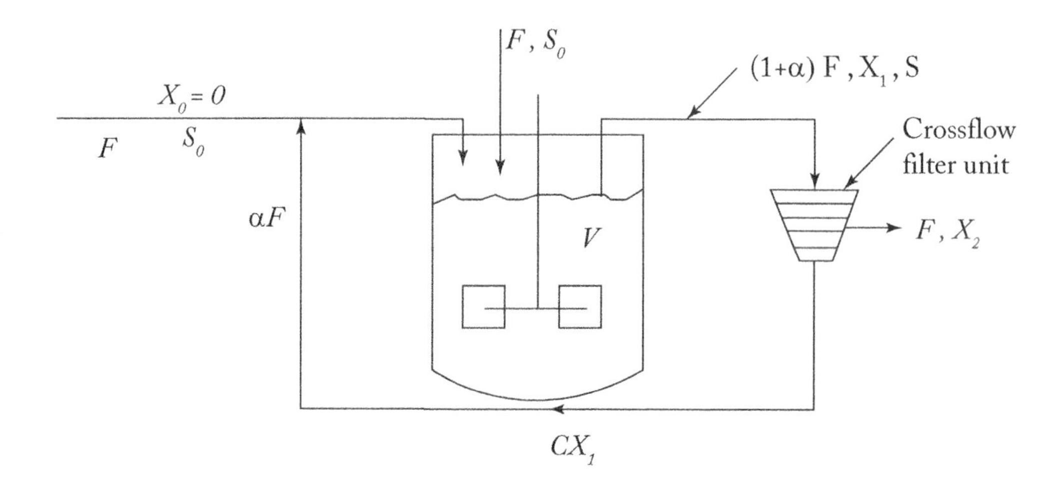

Fig. 6.20 Chemostat with a recycle

$$FX_0 + \propto FCX_1 - (1+\propto)FX_1 + V\mu X_1 = 0 \tag{6.131}$$

Where α is the recycle ratio based on the volumetric flow rate and C is the ratio of cell concentration in the recycle stream to the cell concentration in the reactor effluent, $C = (X_2/X_1)$

For sterile feed, $X_o = 0$ and dividing by V and setting $F/V = D$, μ can be evaluated from equation (6.131)

$$\mu = (1 + \alpha - \alpha c)D = [1 + \alpha(1 - C)]D \tag{6.132}$$

Since $C > 1$ and $\alpha\,(1 - C) < 0$, So $D > \mu$.

So, for a chemostat with a recycle, the dilution rate, D is higher than specific growth rate, μ even for sterile feed as evident from equation (6.132)

Since $C > 1$ and $\alpha\,(1 - C) < 0$, So $D > \mu$.

A mass balance for substrate (S) around the fermenter can be expressed as,

$$FS_0 + \propto FS - V\frac{\mu X_1}{Y_{X/S}} - (1+\propto)FS = 0 \tag{6.133}$$

Now at the steady state,

$$X_1 = \frac{D}{\mu} Y_{X/S}(S_0 - S) \tag{6.134}$$

replacing μ by equation (6.132), we have,

$$X_1 = \frac{Y_{X/S}(S_0 - S)}{(1 + \propto - \propto C)} \tag{6.135}$$

From the previous equation (6.135), it is evident that the cell mass concentration is increased by a factor, $1/(1 + \alpha - \alpha c)$. The substrate concentration in the effluent stream, S can be presented as

$$S = \frac{K_S D(1+\propto - \propto C)}{\mu_{max} - D(1+\propto - \propto C)} \tag{6.136}$$

Using equation (6.135), we can express the outlet concentration of cells X_1 as,

$$X_1 = \frac{Y_{X/S}}{(1+\propto - \propto C)}\left[S_O - \frac{K_S D(1+\propto - \propto C)}{\mu_{max} - D(1+\propto - \propto C)}\right] \tag{6.137}$$

It can be shown from the above equations that the cell concentration and cell productivities (XD) are higher in a chemostat with recycle. α, the ratio of recycle stream to the feed stream which is sometimes called the bleeding ratio. If α is reduced from 1 to 0.5, the cell concentration and cell productivity is doubled.

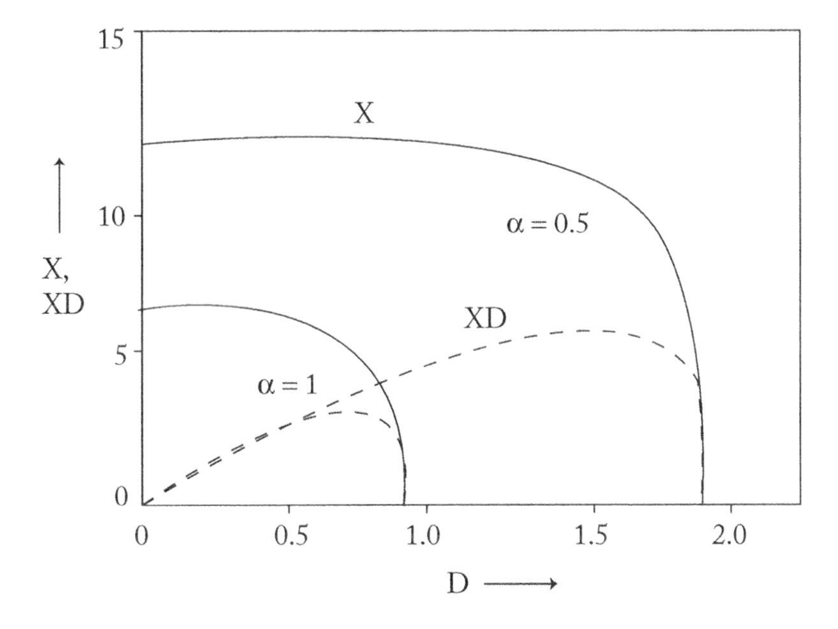

Fig. 6.21 The effect of bleeding ratio (α) on the cell (x) and the cell productivity

6.15 Fed-Batch Bioreactor

In fed-batch bioreactor, nutrients are continuously or semi continuously introduced into the reactor vessel and the effluent is removed discontinuously as shown in Fig. 6.22 with operational sequence.

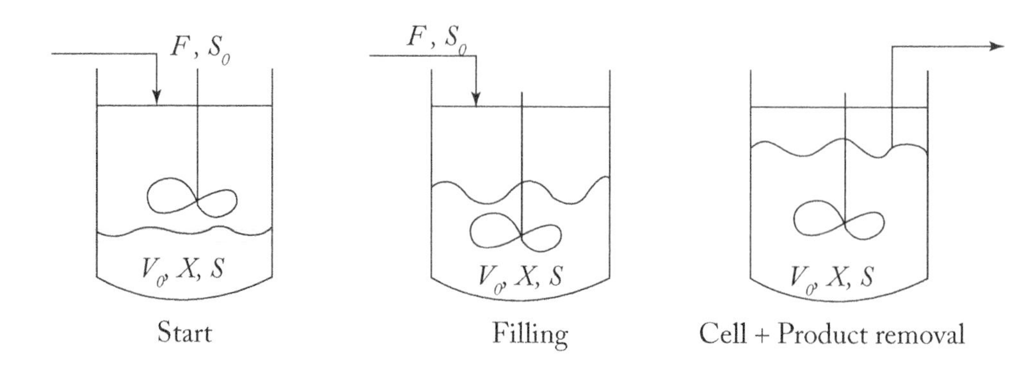

Fig. 6.22 Operation of a fed-batch bioreactor

The fed-batch system has been used in some specific cases such as substrate inhibition, penicillin production, toxic product removal. Secondary metabolic formation like lactic acid, plant cell & mammalian cell fermentation, protein production from recombinant DNA technology, high cell density fermentation. The system is characterized by intermittent feeding, volume change, intermittent product removal. To evaluate the characteristic features of the system, we consider first the batch operation and then continuous operation as a chemostat.

As a batch fermenter, the cell mass concentration, X at any time, t is

$$X(t) = X_O(0) + Y_{X/S}(S_O - S) \tag{6.138}$$

The biomass reaches maximum value, X_m when $S << S_0$ and $X_0 << X$

So, $$X_m = Y_{X/S}\, S_0 \tag{6.139}$$

The substrate is fed at a flow rate, F. The total amount of biomass in the reactor vessel is $X^t = VX$, where V is the culture volume at time t. The volume changes with time such as $dV/dt = F$ and integrating, we have

$$V = V_0 + Ft \tag{6.140}$$

The biomass concentration in the vessel at any time, t is

$$X = X^t V \tag{6.141}$$

Now differentiating (6.141) with respect to t,

$$dX/dt = [V(dX^t/dt) - X^t(dV/dt)] / V^2 \tag{6.142}$$

Now putting $dX^t/dt = \mu X^t$, $dV/dt = F$, $F/V = D$, the equation (6.142) becomes

$$dX/dt = (\mu - D)\, X \tag{6.143}$$

Under the condition when $S \rightarrow 0$ and $X_m = Y_{X/S}\ S_0$, the fed batch system operate at quasi steady state

$$(dX/dt = 0) \text{ and } \mu = D \tag{6.144}$$

Using Monod model

$$\mu = D = \frac{\mu_{max} S}{K_S + S} \tag{6.145}$$

and solving for S

$$S = \frac{DK_S}{\mu_{max} - D} \tag{6.146}$$

The mass balance for the substrate

$$\frac{dS^t}{dt} = FS_O - \frac{\mu X^t}{Y_{X/S}} \tag{6.147}$$

S^t is the total amount of substrate in the reactor and So is the concentration in the feed. At the quasi steady state ($dS^t/dt = 0$), $X^t = VX_m$ when all substrate has been consumed. Then we have

$$FS_O = \frac{\mu X^t}{Y_{X/S}} \tag{6.148}$$

From equation (6.142) at the steady state ($dX/dt = 0$), we get

$$\frac{dX^t}{dt} = X_m\left(\frac{dV}{dt}\right) = X_m F = FY_{X/S}S_O \tag{6.149}$$

Integrating equation (6.149) between the limits, $t = 0, X^t = 0, t = t; X^t = X^t$,

$$X^t = X_O^t + FY_{X/S}S_O t \tag{6.150}$$

From the equation (6.150), the total cell mass increases linearly with time in a fed batch fermenter. So at a particular time, cell mass X^t and the batch reactor volume ($V = V_0 + Ft$) can be calculated.

Product yield can also be evaluated as a function of time as presented below. Assuming the product yield coefficient, $Y_{P/S}$ to be constant and at the quasi steady state with $S << S_0$

$$P = Y_{P/S}\ S_0 \tag{6.151}$$

And the product output is

$$FP = Y_{X/S}\ S_0\ F \tag{6.152}$$

Now we assume that the specific rate of product formation, q_p is constant and then dP^t/dt can be given as

$$\frac{dP^t}{dt} = q_p X^t \tag{6.153}$$

where P^t is the total amount of product in the culture volume.

Substituting $X^t = VX_m = (V_O + Ft)X_m$

$$\frac{dP^t}{dt} = q_p X_m (V_O + Ft) \tag{6.154}$$

Integrating, we have

$$P^t = P_O^t + q_p X_m \left(V_O + \frac{Ft}{2} \right) t \tag{6.155}$$

The variation of product concentration (P) with time (t) can be given on the basis of the equation (6.155)

$$P = P_O \frac{V_O}{V} + q_p X_m \left(\frac{V_O}{V} + \frac{Ft}{2} \right) t \tag{6.156}$$

6.15.1 Repeated fed-batch culture

In the repeated fed-batch culture, a part of the culture volume is removed at certain interval, when the reactor-volume is full. The culture volume and μ undergo cyclic variation in this system.

If t_c is the cycle time and the reactor is always at steady state, the product concentration at the end of each cycle is expressed as

$$P_C = \gamma P_0 + q_p X_m \left(\gamma + \frac{D_C t_c}{2} \right) t_c \tag{6.157}$$

With Dc = F/Vc, where Vc is the culture volume at the end of each cycle, Vo is the residual culture volume after removal, γ is the fraction of culture volume remaining at each cycle, i.e. $\gamma = \frac{V_O}{V_C}$

Now
$$t_c = \frac{V_c - V_O}{F} = \frac{V_C - \gamma V_C}{F} = \frac{(1-\gamma)}{D_C} \tag{6.158}$$

Eliminating t_c from equation (6.157) by using equation (6.158), we have

$$P_C = \gamma P_O + \frac{q_p X_m}{2D_C}(1-\gamma^2) \tag{6.159}$$

6.16 Perfusion Systems in a CSTR

Perfusion Systems are modified CSTRs and used in animal cell culture. The characteristic features of perfusion systems are constant medium flow, viable cells retention, selective removal of dead cells, purified recycled medium, continuous product recovery. A schematic diagram of the system is shown in figure 6.23.

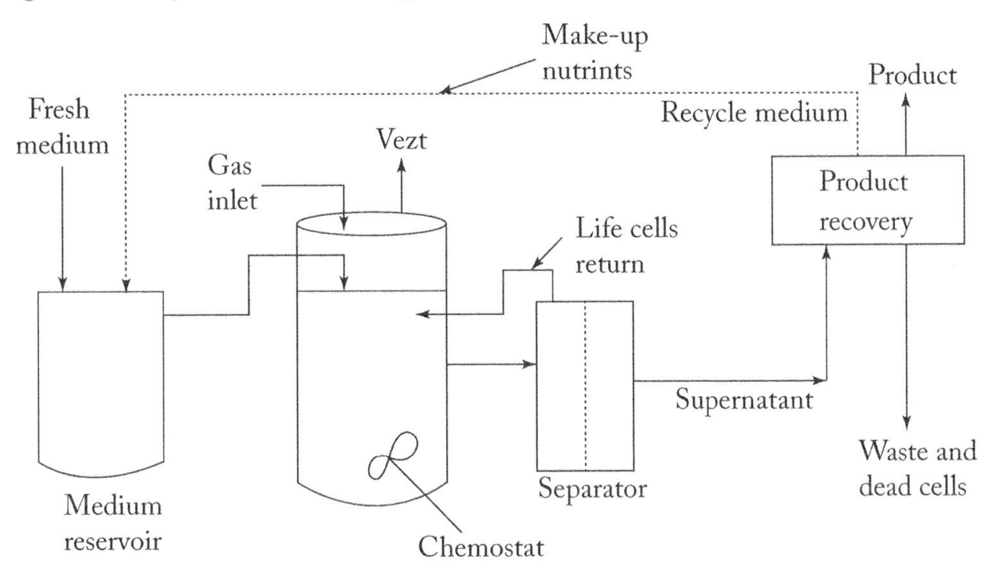

Fig. 6.23 Schematic diagram of a perfusion system in a chemostat

Cell retention is achieved by a membrane or a screen or by a centrifuge. Some of the reactor medium is removed to a separator from which live cells are returned. The supernetant is purified in the product recovery section. Waste and dead cells are rejected and purified medium (containing serum in animal cell culture) is recycled back to the reactor's fresh feed-stream.

The advantages of the system is the removal of dead-cells and inhibitory products like ammonia and lactic acid, high cell density, high volumetric productivity.

The disadvantages are the use of large amount of medium and incomplete utilization of medium compared to batch and fed batch systems.

High cost is involved in the complex reactor system containing membranes, pumps, centrifugal separator, use of large medium. Product cost should be based on the extra cost associated with improved product quality and high reactor productivity with increased cost of production.

Problems (Exercise)

6.1 An aerobic bacterium grown on methanol in a batch reactor and the following data are obtained.

t hr	0	2	4	6	8	10	12
X, kg/m^3	0.2	0.22	0.31	0.6	1.0	1.8	3.2
S, kg/m^3	10	9.5	9.0	8.5	8.2	6.8	4.8

Calculate the following

(a) Maximum growth rate, μ_{max}

(b) Yield coefficient, $Y_{X/S}$

(c) Mass doubling time, td

(d) Saturation constant, K_S

Hint $\mu = (\ln X_2 - \ln X_1)/(t_2 - t_1)$

6.2 Pseudomonas Putida is cultivated in a chemostat, using lactose as the carbon and energy source. The following data are given:

$\mu_{max} = 0.6\ hr^{-1}$, $D = 0.35\ hr^{-1}$, $S_0 = 5\ Kg/m^3$. The effluent lactose concentration, $S = 0.5\ Kg/m^3$. The growth rate is limited by the oxygen transfer and the following data are also given:

$$Y_{X/S} = 0.5\frac{K_S}{P_S},\ Y_{X/O_2} = 0.3\frac{K_S}{P_S},\ C_L^X = 8.0\frac{g_m}{m^3},\ C_L^X \text{ solubility of } O_2 \text{ in water}$$

(a) Calculate the steady state biomass concentration (X) and the rate of oxygen consumption, qo_2 for the microorganism

(b) What should be the volumetric mass transfer coefficient, k_{La} in order to overcome oxygen transfer limitation, when $C_L = 2qo_2/m^3$

Hint for (b) $k_{La}\left(C_L^X - C_L\right) = q_{O_2}X = \dfrac{\mu X}{Y_{X/O_2}}$

6.3 Single Cell Protein (SCP) is produced from a waste stream containing ethanol in a chemostat with ethanol concentration in the inlet So = 25 Kg/m^3. Oxygen can be transferred at the rate of 10 Kg/m^3 of liquid per hr. the micro organism used follows Monod kinetic with following parameter

$\mu_{max} = 0.6 hr^{-1}$, $K_S = 0.03\ Kg/m^3$

$Y_{X/S} = 0.5$ Kg Cells/Kg substrate

$Y_{X/O_2} = 2KgO_2 / KgEt_{opt}$

It is required to maximize biomass productivity and minimize the loss of unused ethanol in the effluent. Determine the required dilution rate and whether the O_2 supply is sufficient.

$$\text{Hint } D_{max} = \mu_{max}\left[1 - \sqrt{\frac{K_S}{K_S + S_O}}\right]$$

$$O_2 \text{ required} = D_{max} Y_{O_2/S}\left(S_O - S\right) Kg/\left(m^3\right)\left(hr\right)$$

$$S = DK_S / (\mu_{max} - D) Kg/m^3$$

6.4 A fermentation is carried out in a chemostat where product formation kinetics are given by Leudeking-Piert equation. Steady state is assumed. The following kinetic parameters are given

$D = 0.6hr^{-1}$, $\mu_{max} = 0.8hr^{-1}$

$K_s = 0.01$ kg/m^3, $Y_{X/S} = 0.5$ ks X/Ks S

$S_0 = 1$ Kg/m^3, $\alpha = 0.004$ kg P/ kg X

$\beta = 0.005$ hr^{-1}

Calculate the productivity DP of the chemostat

Hint: $qP = \frac{1}{X}\frac{dP}{dt} = Y_{P/X}\mu$

Or $qP = \propto \mu + \beta$

$$PD = P_O D + Y_{P/X}\mu X$$

6.5 E.Coli is to be cultivated in a steady state CSTR of volume, $V_R = 0.8m^3$ with a flow rate of 0.3 m^3/hr. The limiting substrate used is glucose, fed with initial concentration, So = 10 kg/m^3. The following kinetic data are given $\mu_{max} = 0.8hr^{-1}$, $K_S = 0.7$ kg/m^3, $Y_{X/S} = 0.6$

(a) What will be the doubling time, *td*

(b) What will be the cell and substrate concentration of the outlet stream?

(c) If one more CSTR of 0.5m^3 is added, what will be the cell and substrate concentration at the exit of the second reactor?

6.6 A chemostat of volume 1m^3 was used to study the kinetics of cell growth of a microorganism. The inlet stream is sterile. The flow rate and inlet substrate concentration were varied and the steady state outlet concentration was measured. The following data were obtained:

F m^3/hr	0.2	0.35	0.50	0.70	0.80
So kg/m^3	30	30	30	30	30
S kg/m^3	0.5	1.1	1.6	3.3	10

Use Monod model and find out the parameters.

6.7 The growth rate of E.Coli can be expressed by Monod model with the parameters:

$\mu_{max} = 0.9$ hr^{-1}, ks = 0.7 kg/m^3

$Y_{X/S} = 0.6$ kg X/kg S, $X_0 = 1$ kg/m^3

If the initial concentration of the substrate is So = 10 ks/m^3 and the cell grow exponentially at t = 0 in a batch reactor, what is the time required for the substrate to be reduced to S = 2 kg/m^3?

Hint $(t - t_O) = \frac{1}{\mu_{max}}\left[(\propto +1)\ln\frac{X}{X_O} + \propto \ln\frac{S_O}{S}\right]$

Where $\propto = \frac{K_S Y_{X/S}}{X_O + S_O Y_{X/S}}$

6.8 A fed batch culture operating with intermittent addition of glucose solution, is used to carry out a microbial reaction whose parameters are given when the system is in quasisteady state after time, t = 2 hrs

$V = 1\ m^3$, $F = dV/dt = 0.2\ m^3/hr$

$So = 1\ kg/m^3$, $\mu_{max} = 0.5\ hr^{-1}$

$K_S = 0.2$ kg glucose/m^3

Initial total cell mass at t = 0, $X_O^t = 0.20 \frac{Kg}{m^3}$

$$Y_{X/S} = 0.5 \frac{Kg\ cells}{Kg\ glucose}$$

(a) Find V_0 (the initial volume of the culture)

(b) Determine the concentration of the substrate in the vessel at the quasi steady state

(c) Determine the total concentration of biomass in the vessel at t = 2 hrs in the quasi steady state

(d) If $q_p\left(\frac{1}{X}\frac{dP}{dt}\right)$ = 0.3 Kg product/Kg Cells(hr) and P_0 = 0, determine the concentration of P in the vessel at t = 2 hrs

Hint for (C) $X^t = X_O^t + FY_{X/S}S_O t$

For (d): $P = q_P X_m \left(\frac{V_O}{V} + \frac{Dt}{2}\right)t$

And $X_m = Y_{X/S}S_O$

6.9 A chemostat with cell recycle is used for a fermentation process. The feed flow rate and culture volume are $F = 0.2 m^3/hr$ and $V_R = 1.0\ m^3$ respectively. Glucose is the rate limiting component and the inlet concentration, $S_0 = 2.5\ Kg/m^3$,

$Y_{X/S} = 0.5$ kg X/kg S

The kinetic constants of the microorganism are $\mu_{max} = 0.4\ hr^{-1}$, $K_S = 1.0\ kg/m^3$. The value of $C = X_2/X_1 = 1.5$, where X_1 is the cell mass at the outlet of the reactor and X_2 cell mass after cell concentration is returned to the inlet of the reactor and the recycle ratio, $\alpha = F_R/F = 0.7$ where F_R is the recycle flow rate and F is the initial flow rate.

(a) Calculate substrate concentration in the recycle stream (S)

(b) Calculate the specific growth rate, μ of the organism

(c) Find the cell mass concentration in the recycle stream X_1

(d) Find the cell concentration in the centrifuge effluent

Hint $\mu = \left[1 + \propto (1 - C)D\right]$,

$$S = \frac{\mu D}{\mu_{max} - D}$$

$$X_1 = \frac{D(S_O - S)Y_{X/C}}{\mu_{net}}$$

$$X_2 = (1+\propto)X_1 - \propto CX_1$$

6.10 A chemostat study on the specific strain of bakers year using glucose gave the following results

$D\ hr^{-1}$	0.084	0.10	0.16	0.2	0.242
So kg/m^3	21.5	10.0	21.2	20.7	10.8
S kg/m^3	0.054	0.08	0.138	0.186	0.226
P (ethanol) kg/m^3	7.97	4.70	8.57	8.44	4.51
X, kg/m^3	2.0	1.20	2.40	2.33	1.25

(a) Find the rate equation for cell growth
(b) Find the rate equation for the product (ethanol) formation

Hint for (a) $\frac{1}{D} = \frac{K_S}{\mu_{max}}\frac{1}{S} + \frac{1}{\mu_{max}}$

For (b) $\frac{dP}{dt} = \propto \mu X$

6.11 The rate equations for yeast (X) substrate (glucose) (S), product (alcohol) (P) in ethanol production are given below:

$$r_X = \frac{dX}{dt} = \mu_{max}\left(1 - \frac{P}{P_{max}}\right)^n \left(\frac{S}{K_S + S}\right)^X$$

$$r_S = \frac{dS}{dt} = -\frac{1}{Y_{X/S}}\frac{dX}{dt}$$

$$r_P = \frac{dP}{dt} = \frac{1}{Y_{X/P}}\frac{dX}{dt}$$

Where $K_S = 1.6\ kg/m^3$, $\mu_{max} = 0.4\ hr^{-1}$
$Y_{X/P} = 0.16$, $Y_{X/S} = 0.06$, $P_{max} = 100\ kg/m^3$
$P_o = 0$, $Xo = 0.1\ kg/m^3$, $n = 2$
Calculate the change of X, P and S as a function of time (0 – 10 hrs) when
$S_o = 100\ kg/m^3$
Hint Use a simple numerical method like Euler method

$$\Delta X = f(X,S,P)\Delta t$$

$$X_1 = X_O + f(X_O,S_O,P_O)\Delta t$$

Choose $\Delta t = 2hrs.$

$$X_2 = X_1 + f(X_1,S_1,P_1)\Delta t$$

References

Aiba, S.A., E. Humphrey and N.F. Millis, Biochemical Engineering, 2nd ed, Academic Press, 1973

Aiba, S.A., M. Shoda and M. Nagatani, "Kinetics of product inhibition on Alcohol fementation", Biotechnol. Bioeng. 10, 845–864, 1968

Andrews, J.F., "A Mathematical model for the continuous culture of Microorganisms, utilizing inhibitory substrates", Biotechnol. Bioeng, 19, 707–723, 1968

Atkinson, B., Biochemical Reactors, London, Pion Ltd., 1971

Atkinson, B. and F. Mavitona, Biochemical Engineering & Biotechnology, 2nd ed. Macmillan, 1991

Bailey, J.E. and D.F. Ollis, Biochemical Engineering Fundamentals, 2nd ed. Mc Graw Hill Book Co. 1986

Blanch, H.W. and D.S. Clark, Biochemical Engineering, Marcel Dekker, Inc. New York, 1996

Blanch, H.W., E.R. Papoutsakis & G. Stephanopoulos (editors) Foundations of Biochemical Engineering: Kinetics and Thermodynamics in Biological Systems: ACS Symposium Series, 207, American Chem. Soc. Washington D.C. 1983

Fredrickson A., G., R.D. Megee III, and H.M. Tsuchiya, "A Mathematical Model for Fermentation Processes", Adv. Appl. Microbiol 23, 417, 1970

Gaden, E.L. Jr., "Fermentation Kinetics and Productivity", Chem. Ind. Rev (London), 154, 1955

Herbert, D., R. Elleworth and R.E. Telling, "The continuous culture of Bacteria, A theoretical and experimental study" J.Gen. Microbiology, 14, 601, 1956

Lee, J.M. Biochemical Engineering, Prentice Hall, Englewood, Cliffs, New Jersey, 1988

Levenspiel, O., Chemical Reaction Engineering, 3rd ed. John Wiley & Sons, 1999

Leudeking, R. and E.L. Piret, A Kinetic study of the lactic acid fermentation, J. Biochem. Microbial Technol. Eng., 1:393, 1959.

Leudeking, R. "Fermentation Process Kinetics in Biochemical and Biological Engineering Sciences edited by N. Blakebrough, London, England, Academic Press, 181–243, 1967

Monod, J. The Growth of Bacterial Cultures, Am. Rev. Microbioals 3, 271–394, 1949

Pirt, S.J., Principles of Microbes and Cell Cultivations, Blackwel Scientific, Oxford, 1975

Roels, J.A., Energetics and Kinetics in Biotechnology, Elsevier, Amsterdam, 1983

Shuler M.L and F. Kargi, Bioprocess Engineering, Pearson Education Inc. 2003

Williams, F.M. "A Model of Cell Growth Dynamics", J. Theoret, Biol. 15, 190–267, 1967

CHAPTER 7

Transport Phenomena in Bioreactors

7.0 Outline

For aerobic fermentation, aerobic cells require oxygen in the dissolved state to consume nutrients for generating cell mass and product yield. This requires higher gas-liquid mass transfer of oxygen to the culture as the gaseous oxygen is sparingly soluble in aqueous solutions. The oxygen transfer rate depends upon local bulk-oxygen concentration (C_L), the diffusion coefficient (DO_2) and the local respiration rates in the aerobic regions : There are number of aerobic processes such as food spoilage in undesired oxidation, lake eutrophication due to inadequate aeration by natural oxygen supply or by excessive concentration of materials such as phosphate or nitrates. Other sparingly soluble gases used for fermentation as substrates are methane and light hydrocarbons.

The important aerobic processes are renewable resource bioconversion such as the cellulose, hemi cellulose, lignin fraction of agricultural and forests wastes as the fermentation feed stocks. Other mass transfer limited fermentations include large scale fermentation of Penicillin, extra cellular biopolymers (Xanthan gum) or the activated sludge waste treatment. Mass transfer coupled with momentum transfer takes place in the agitated stirred tank fermenter, aerated bubble columns and air-lift fermenters.

In some biochemical processes, heat transfer may be important, providing significant transport effect which may influence the bioprocess system behaviour. Such processes are relatively exothermic fermentations such as trickling filter for vine-vinegar production or waste water treatment, municipal waste composting and solid waste fermentations.

7.1 Gas-Liquid Mass Transfer in Microbial Cells

A sparingly soluble gas, mainly oxygen or light hydrocarbons is transferred from rising bubbles into a liquid phase containing microbial cells. Considering oxygen transfer, oxygen must pass through a series of transport resistances whose magnitude depends on solubility, bubble hydrodynamics, cellular activity, liquid density, interfacial phenomena and other factors. These resistances have been presented sequentially in the Fig.7.1 and the steps are:

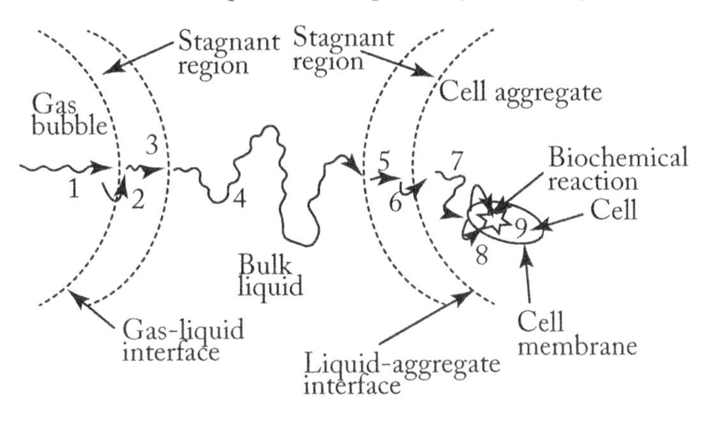

Fig. 7.1

Schematic diagram of steps involved in transport of O_2 from a gas bubble to the inside of the cell.

1. Diffusion from bulk gas to the gas liquid interface
2. Movement through the gas liquid interface
3. Diffusion of O_2 through the stagnant liquid film to the well mixed bulk liquid
4. Transport of O_2 (solute) through the bulk liquid to a second stagnant liquid film surrounding the cells
5. Transport through the second stagnant liquid film associated with cells
6. Diffusive transport through the cellular flocs
7. Transport across cell envelop and to the intracellular reaction site.

All the resistances listed above are not significant. Microbial cells preferably gather at the vicinity of gas-bubble liquid interface and only one stagnant liquid film is considered.

7.1.1 Basic mass transfer

For sparingly soluble solutes in water, there exists two equilibrated interfacial concentrations C_{gi} and C_{Li} on the gas and liquid sides respectively and they are related by a linear partition law known as Henry's Law,

$$C_{gi} = MC_{Li} \tag{7.1}$$

Where M is called the Henry's constant.

At the steady state, the oxygen transfer rate to the gas-liquid interface equals to the transfer rate to the liquid side film. Then the oxygen flux, NO_2' in kmoles/(m^2) (s) can be expressed as

$$NO_2 = kg\left(C_g - C_{gi}\right) \rightarrow \text{gas side} \tag{7.2}$$

$$NO_2 = k_L\left(C_{Li} - C_L\right) \rightarrow \text{liquid side} \tag{7.3}$$

Where kg(m/s) is the gas-side mass transfer coefficient and k_L(m/s) is the liquid phase mass transfer coefficient.

Now coupling those mass transfer processes, the mass transfer rate may be obtained in terms of overall mass transfer coefficient, K_L and the overall concentration driving force, $\left(C_{Li} - C_L\right)$ where C_L is the saturated concentration of the solute gas, which is in equilibrium with bulk gas phase, C_g, given by Henry's Law,

$$C_g = MC_L^* \tag{7.4}$$

Where M is called Henry's constant.

The solute flux $\left(N_A\right)$ is expressed as

$$N_A = K_L\left(C_L^* - C_L\right) \tag{7.5}$$

And the overall mass transfer coefficient, K_L is given by the equations (7.6) using the equations (7.1) to (7.5).

$$\frac{1}{K_L} = \frac{1}{k_L} + \frac{1}{Mk_g} \tag{7.6}$$

For sparingly soluble species, M is much larger than unity and k_g is also considerably larger than k_L. Therefore K_L is approximately equal to k_L.

Now the oxygen transfer rate per unit reactor volume is given by,

$$NO_2 = Oxygen\ absorption\ rate = \frac{(flux)(interfacial\ area)}{(reactor\ liquid - volume)}$$

$$= k_L\left(C_L^* - C_L\right)\frac{A}{V} \tag{7.7}$$

$$= k_L a'\left(C_L^* - C_L\right) \tag{7.8}$$

Where a' = A/V. and A is interfacial area , V is the reactor liquid volume

Sometime, the symbol, a is used as the ratio of interfacial area to reactor volume plus gas volume. The bubble volume per liquid- reactor volume is known as gas holdup, H., the fractional gas hold up, the value of a is given by

$$a = H\frac{6}{D} \tag{7.9}$$

Where D is the bubble diameter in mm. If a is used, it can be related to a' by the relation,

$$a'(1-H)=a \tag{7.10}$$

Where a' is the gas-liquid interfacial area per unit liquid volume

7.1.2 Metabolic oxygen uptake

In the design of aerobic biological reactors, it is important to know whether the oxygen transfer rate (OTR) or the oxygen uptake rate (OUR) is the controlling step.

The maximum possible oxygen consumption rate by the cells is given as $X\mu max/Y_{X/O_2}$. where Y_{X/O_2} is the ratio of kg cell mass formed per kg of oxygen consumed. Now if $k_L a' C_L^*$ is much larger than $\frac{X\mu max}{Y_{X/O_2}}$, reaction appears to be biochemical reaction controlled. Conversely the reverse inequality holds leading C_L nearly to zero, the reaction becomes mass transfer controlled system.

However, at the steady state, OTR = OUR (7.11)

Or
$$k_L a'\left(C_L^* - C_L\right) = \frac{X\mu max}{Y_{X/O_2}} = q_{o2}X \tag{7.12}$$

Where qo_2 is the specific oxygen consumption rate by a microorganism, X is the cell concentration (g dry wt of cells/L) , Y_{X/O_2} is yield coefficient on oxygen (g dry cells / g O_2).

If μ is given in term of C_L , μ can be given as

$$\mu = \frac{\mu \max C_L}{K_{O2} + C_L} \tag{7.13}$$

Combining (7.12) and (7.13), and rearranging we get

$$C_L = C_L^* \left[\frac{Y_{X/O_2} K_{O_2} a'/X \mu max}{1 - Y_{X/O_2} C_L^* / X \mu max} \right]$$

Provided
$$C_L \ll C_L^* \tag{7.14}$$

The resulting C_L must be maintained above the critical oxygen value, C_{Lcr} for a particular type of cells (about $3K_{O2}$). The critical oxygen requirement for some common organisms are in the range of 0.003 to 0.05 mml O_2/L. For higher critical oxygen values, such as for Penicillium molds, oxygen mass transfer may be controlling.

Many factors can influence the total microbial oxygen demand, $\left(x\mu / {}^{Yx}\!/_{O_2} \right)$ which determine the minimum values of $k_L a'$ needed for the process design (equation 7.12). The factors are cell species, culture growth phase, carbon nutrients, pH, X or P.

Problem 7.1

A fermenter has to attain $k_L a'$ = 25 hr-1. With its maximum Agitator speed air is sparged at 0.5m^{3}gas/((min). E. Coli with specific oxygen consumption rate, $q_{O2} = 10$ moles of O_2 / Kg drywt cells are to be cultured. The critical dissolved oxygen concentration is 0.2 g/m^3. The solubility of oxygen from air in the fermentation broth is 8.0 gm/m3 at 30°C.

(a) What maximum concentration of E. Coli can be sustained in the reactor?
(b) What concentration of E. Coli can be attained if pure oxygen is used?

Solution:

(a) We have the working equation OTR = OUR

$$k_L a\left(C_L^* - C_L\right) = q_{O_2} X \text{........ (A)}$$

$$C_L^* = 8.0 g / m^3 = \frac{8}{32} = 0.25 \frac{g\,moles}{m^3}$$

$$C_L = \frac{0.2}{32} = 0.00625 \frac{g\,moles}{m^3},$$

$$k_L a = 25 hr^{-1}, \; q_{O_2} = 10 \frac{g\,moles\,of\,O_2}{\left(Kg\,dry\,wt\right)\left(hr\right)}$$

Substituting the values in equation (A)

25 (0.25 - .00625) = 10 X

Or X = 0.61 Kg/m^3 Ans.

(b) $p_{O_2} = HC_L^* \quad C_L^* = 8 g / m^3 = \frac{8}{32} = 0.25\, gmole / m^3$

$$H = \frac{0.2}{0.25} = 0.8 \frac{atm - m^3}{gmole}$$

With pure O_2, $p_{O_2} = 1 atm$

$$C_L^* = \frac{1}{0.8} = 1.25 \frac{g\,mole}{m^3}$$

$$X = \frac{k_L a\left(C_L^* - C_L\right)}{q_{O_2}} = 25\left(1.25 - .00625\right) / 10 = 3.11 kg / m^3 \text{ Ans.}$$

7.2 Measurement of $k_L a'$

Four common methods are used to determine $k_L a'$ by experiments. Those include: (1) Unsteady State, (2) Steady State, (3) Dynamic, (4) Sulfite method.

In the first method of unsteady state, oxygen is removed from the liquid filled in a reactor by sparging with N2. Air is then introduced and the change in dissolved oxygen (DO) is measured at definite time intervals; the solution is nearly saturated. The O_2 absorption rate is given is

$$\frac{dC_L}{dt} = k_L a'\left(C_L^* - C_L\right) \tag{7.15}$$

Integrating at t = 0, $\left(C_L^* - C_L\right) = 0$ and t = t, $\left(C_L^* - C_L\right) = 0$, and t = t, $\left(C_L^* - C_L\right) > 0$, we get

$$l_n\left(C_L^* - C_L\right) = -k_{La}' t \tag{7.16}$$

The plot of $\ln\left(C_L^* - C_L\right) vs$ vs. t is linear with a negative slope of $-k_{La}'$.

2. **In the steady state method :** the whole reactor with cells in suspension is used as a respirator. The oxygen concentration is measured in the exit stream and also dissolved oxygen (C_L) is measured accurately. Oxygen uptake rate $\left(q_{O2}X\right)$ can be estimated with offline measurement of a sample of culture in the respirator. Determination of C_L^* in a large fermentor is a problem. We know that C_L^* is proportional to partial pressure of oxygen $\left(p_{O_2}\right)$. But p_{O_2} varies along the height of the reactor. In a bubble column, the log mean value of C_L^* based on p_{O_2} at the entrance and exit is to be used. In the perfectly mixed vessel, C_L^* would be based on p_{O_2} at the exit. The working equation used in the method is

$$k_{La}' = \frac{q_{O_2} X}{C_L^* - C_L} \tag{7.17}$$

Where q_{O_2} is the specific oxygen consumption rate of the cells in gm of O_2 consumed / (gm, dry. cells) (hr) X is the cell mass in gm/m³ and C_L^* and C_L in gm of O_2/m^3, $k_L a$ (the volumetric mass transfer coefficient) in hr^{-1}.

3. **The dynamic method** uses a fermenter with active cells, provided with a dissolved oxygen probe and a chart recorder. The air supply is shut off for a short period (less than 5 minutes) and then air sparging is continued.
The working equation is

$$\frac{dC_L}{dt} = k_{La}'\left(C_L^* - C_L\right) - q_{O_2 X} \tag{7.18}$$

The experimental results, CL vs. time (t) are plotted in figure 7.2.

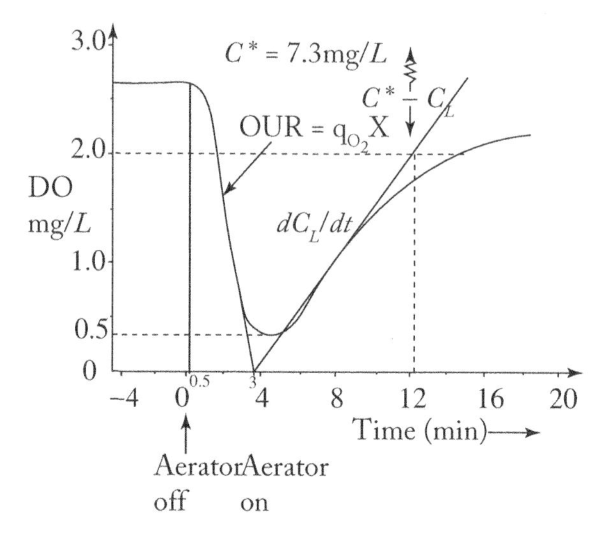

Fig. 7.2 plot of dissolved oxygen(DO) vs time

When the gas is off, there are no air bubbles in the culture and $k_{La}^{'} = 0$ and the slope of the descending curve will give $q_{O_2}X$. The lowest value of C_L obtained in the experiment must be above the critical oxygen concentration. Thus the slope of ascending curve is

$$\frac{dC_L}{dt} = -q_{O_2}X \tag{7.19}$$

When the air-sparging is started, the ascending curve can be used to determine $k_{La}^{'}$. We have,

$$k_{La}^{'} = \frac{dC_L / dt + OUR}{C_L^{*} - C_L} \tag{7.20}$$

The specific respiration rate, q_{O_2} is calculated from the equation,

$$q_{O_2} = \frac{OUR}{X} \text{ (the slope of the descending curve)} \qquad (7.21).$$

4th Method of sulphite reaction

The sulphite solution in presence of Cu^{++} is oxidized to sulphate $\left(SO_4^{2-}\right)$ in a zero order reaction.

The rate of sulphate formation is measured and it is proportional to the rate of oxygen consumption (½ mole of O_2 is consumed per 1 mole of SO_4^{2-}). Thus

$$\frac{1}{2}\frac{dC_{SO_4}}{dt} = k_L a' C_L^{*} \tag{7.22}$$

Where C_{SO_4} is the concentration of SO_4^{2-}, oxygen solubility, C_L^* is a constant dependent on medium composition, temperature and pressure. Thus, $k_L a' = \dfrac{\frac{1}{2} dC_{SO_4} / dt}{C_L^*}$ (7.23)

The sulphite method gives higher value of $k_L a'$

Problem 7.2

E. Coli have a maximum respiration rate of 250gm of O_2/(Kg dry wt. of cell) . The $k_L a$ is 100 hr^{-1}. in a 1 m^3 reactor. A gas enriched with oxygen which produces $C_L^* = 30 gm / m^3$ is passed into the reactor. The desired cell concentration is 20 kg dry wt of cells/m^3. If oxygen becomes growth limiting and respiration is slow, then the respiration rate is

$$q_{O_2} = \frac{q_{O_{2max}} C_L}{0.2 g / m^3 + C_L}$$

Where C_L is g/m^3

What dissolved oxygen concentration (C_L) is maintained when the cell mass concentration in the reactor is X = 20 Kg/m^3.

Solution:

We know

$$q_{O_2} X = k_{L_a} \left(C_L^* - C_L \right) \quad \text{..............} \quad (A)$$

Putting the q_{O_2} in the above equation (A)

$$\frac{q_{O_2 max} . C_L . X}{0.2 + C_L} = k_L a \left(C_L^* - C_L \right)$$

Putting the given values of $k_L a'$, X, CL* in the above equation, we get

$$\frac{250 C_L \times (20)}{0.2 + C_L} = 100 (30 - C_L)$$

$$\text{Or } \frac{50 C_L}{0.2 + C_L} = 30 - C_L$$

$$50 C_L = 6 - 0.2 C_L + 30 C_L - C_L^2$$

$$C_L^2 + 20.2 C_L - 6 = 0$$

$$C_L = \frac{-20.2 \pm \sqrt{(20.2)^2 + 24}}{2} = 0.29\,gm/m^3$$
Ans.

Problem 7.3

What value of k_La is to be maintained to sustain a population of animal cells of $1 \times 10^{13}\,cells/m^3$ when the oxygen consumption rate by the micro organism (animal cells) is $0.1 \times 10^{-15}\,kmol/(hr)(cell)$? Given, $p_{O_2} = HC_L^*$.

Where H (Henry's constant for O_2 in H_2O)

$$= 793 \frac{m^3 - bar}{kmol(O_2)}$$

Solution:

$$C_L^* = \frac{p_{O_2}}{H} = \frac{0.2\,bar}{793 \frac{m^3 - bar}{kmne}} = 2.52^{\times 10^{-4}} \frac{kmole}{m^3}$$

We know the relation,

$$q_{O_2}X = k_{L_a}(C_L^* - C_L)\,k_{L_a} = \frac{q_{O_2}X}{C_L^*}$$

Putting $C_L \approx 0$

$$K_{La} = \frac{0.1 \times 10^{-15}\left(\frac{kmole}{(hr)(cell)}\right)1 \times 10^{13}\,\frac{cells}{m^3}}{2.52 \times 10^{-4}\left(\frac{kmole}{m^3}\right)} = 3.96\,hr^{-1}$$

Problem 7.4

Lethal agents are added to a stirred and aerated tank to kill the organisms in the medium immediately. Increase in D. O. Concentration upon addition of lethal agents is followed with the aid of a DO analyzer and a recorder. The following data are obtained:

Time, t mins	1	2	2.5	3	4	5
DO (g/m^3)	1	3	4	5	6.5	7.2

C_L^* (Saturation concentrate of oxygen) = 9 gm/m^3. Calculate the value of K'_{La}.

Solution:

$$\frac{dC_L}{dt} = k_L a\left(C_L^* - C_L\right)$$

$$-\int_0^{C_L^*-C_L} \frac{d\left(C_L^* - C_L\right)}{C_L^* - C_L} = k_{La}\int_0^t dt$$

$$\ln\left(C_L^* - C_L\right) = -k_{La}t$$

t min	1	2	2.5	3	4	5
$ln\left(C_L{}^* - C_L\right)$	2.07	1.79	1.61	1.38	0.91	0.53

$ln\left(C_L{}^* - C_L\right)$ is plotted against time, t, and , from the slope,

$k_{La} = 0.385\min^{-1} = 23.13hr^{-1}$ Ans.

7.3 Gas — Liquid — Solid Reactors

The following four classes of reactors are commonly used for aerobic fermentation.

(a) Reactors with internal mechanical agitation (Stirred tank)

(b) Bubble columns which are agitated by gas sparging

(c) Loop Reactors or Air lift fermenters where mixing and agitation are carried out by liquid circulation and by the motion of an induced gas or by a mechanical pump or by a combination of pumping and mechanical agitation.

(d) Trickle bed reactors where agitation and mass transfer occur by spraying liquid and passing gas from the bottom. These reactors are complex reactors coupled with mass transfer.

7.3.1 Momentum and heat transfer (Continuous Stirred Tank Reactor [CSTR])

The traditional stirred tank reactors are shown in figure 7.3 (a). These reactors have mechanical agitation, high $k_L a'$ values for mass transfer, coupled with (four) baffles.

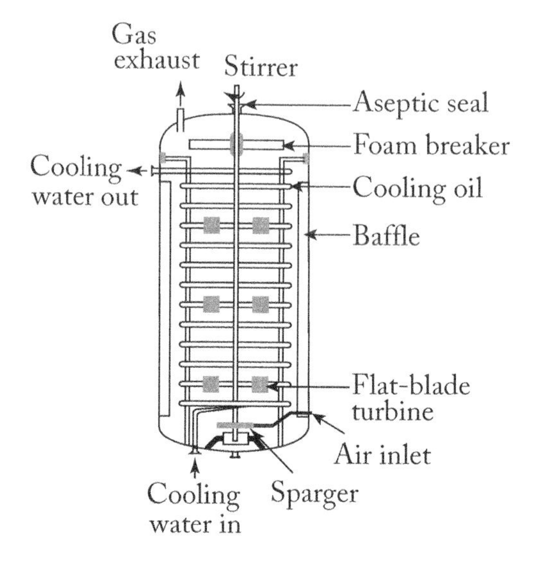

Fig. 7.3 (a) Stirred Tank Reactor

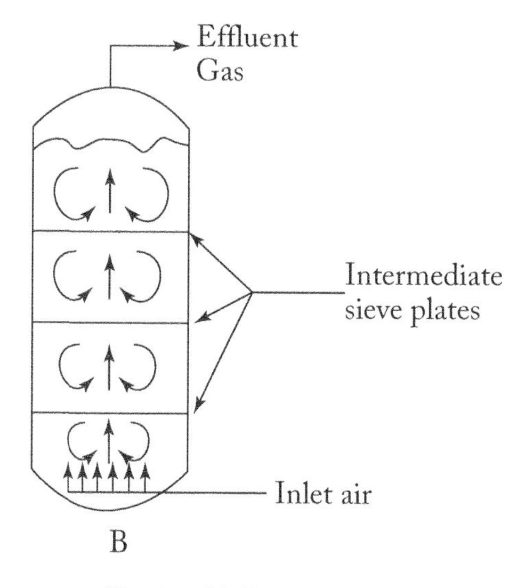

Fig. 7.3 (b) Bubble Column

Operation of the stirred tank reactor

Compressed gas is fed to the stirred vessel using a sparger in the form of a ring with holes or a tube with a single orifice from the bottom introduced from the top of the reactor. Lquid-dispersion is mainly effected by the use of impellers which provide rapid mixing to disperse bubbles throughout the vessel The stirrer speed is controlled for the shear sensitive cells such as animal or plant cells. The stratification of the reactor contents may be reduced by multiple impeller systems. Mostly disc and turbine type impeller are preferred. Marine and paddle impellers are specifically useful for cells of high shear sensitivity. The Rushton impeller (Fig.7.4) consists of a disc with 6 – 8 blades to pump a fluid in a radial direction.

Recently axial flow hydrofoil impellers are popular and they pump liquid either up or down and give superior performance compared to Rushton impeller with respect to low power consumption for the same level of oxygen transfer.

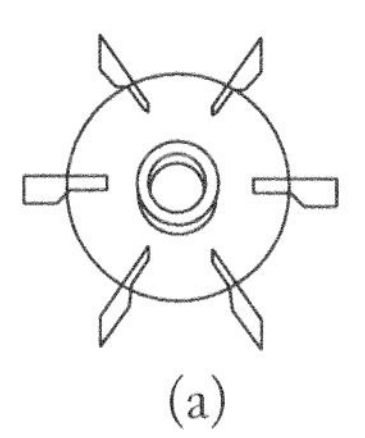

(a)

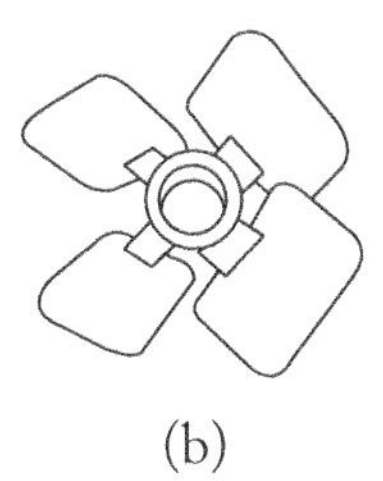

(b)

Fig. 7.4 Types of impellers: (a) Rushton radial flow impeller (b) Axial flow hydrofoil impellers

Most stirred tank fermenters have a height to diameter ratio of 2 to 3. For animal cells H/D ratio is close to 1.0. The Rushton impeller diameter is about 30% to 40% of the tank diameter and for the axial flow impeller, the impeller diameter may be 50% of the tank diameter. For large reactors, copper or stainless steel coils may be used for heat transfer instead of heat transfer jacket outside the reactor vessel.

In CSTRS, fluid shear in mixing is produced by velocity gradients from tangential and radial velocity components of fluids leaving the impeller region. The velocity profile of a regular flat bladed disc turbine is parabolic in shape. Moving away from the centre line, the resultant velocity decreases by 85% at one blade width-to-diameter ratio, above or below, creating a high shear region. As the blade width to diameter ratio increases, the velocity profile becomes more blunt (less parabolic shape) which yields a lower amount of shear and thus increasing the blade width, a CSTR can be used for cultivating on shear sensitive animal or plant cells.

The impeller shaft enters the vessel from the top or bottom of the vessel through a bearing housing and mechanical seal assembly. For two impeller systems, the distance between the first impeller and the bottom of the vessel and between the two impellers is 1.5 impeller diameters. For 3 impeller system, the gap between two impellers in one impeller diameter.

Four equally spread baffles are usually installed to prevent vortex formation which reduces the intensity of mixing. The pH in a fermenter is maintained by employing a buffer or a pH controller. The temperature is controlled by heating or cooling as required by the system.

The stirred tank fermentors are normally made of stainless steel of type 316 (mainly welded parts), type 304 used for covers and jackets. For plant and animal cell tissue culture, a low carbon steel (type 316L) is often used. Many bench scale reactors are made of glass with stainless steel cover plates. These materials are chosen for easy sterilization and handling corrosive culture medium. Stirred reactors of size up to 400 m^3 are used in antibiotic fermentation.

The problem presented below should have been placed much later in the text, since many factors have not yet been discussed.

Problem 7.5

A cylindrical tank of 1.22 m diameter is filled with water to an operating level equal to the tank diameter. The tank is equipped with four equally spaced with four equally spaced baffles whose width is one tenth of the tank diameter. The tank is agitated with 0.36 m diameter flat six blade turbine (impeller). The impeller rotation speed is 170 rpm. The air enters through an open ended tube situated below the impeller and the volumetric flow rate of air is 0.00416 m^3/sec at 1.08 atm and 25°C.

Data given:

Sauter mean diameter of bubble, D32 = 4 mm,

$$\rho H_2O = 997\frac{kg}{m^3},\quad \mu H_2O = 9\times\frac{10^{-4}\,kg}{m^{3s}},$$

Calculate the following.:

(a) Power requirement in watt
(b) Gas hold up
(c) Inter facial area (m^2/m^3)
(d) Volumetric mass transfer coefficient, k_La
(e) Mass transfer coefficient, k_L from correlation.

Solution:

(a) $R_{eI} = \rho N D_1^2 / \mu$ where R_{eI} = Impeller base diameter

$$R_{eI} = \frac{(997)\left(170/60\right)(0.36)^2}{8.9}\times 10^{+4} = 4.1\times 10^5$$

Since the flow is turbulent and the power number, $PN_O = 6$ for six blade turbine.

$$6 = PN_O = \frac{P}{\rho N^3 D_I^5},$$ PN_o is the dimensionless power numbers

$$P = 6\rho N^3 D_I^5$$

$$= 6(997)\left(\frac{170}{60}\right)^3(0.36)^5 = 819.8\,Watt$$ Ans. (a)

(b) Cross section of the reactor

$$= N/_4 Dt^2 = 0.7854(1.22)^2 = 1.169m^2$$

U_{gs} (Superficial gas velocity)

$$= \frac{0.00416 m^3 / S}{1.169 m^2} = 3.558 \times 10^{-3} m / s$$

V_t = bubble rise velocity

$$= 0.711\sqrt{gD_{32}} = 0.711\sqrt{(9.8)(4 \times 10^{-3})} = 0.1407 m / s$$

Approx Gas holdup, H

$$H = \frac{U_{gs}}{U_{gs} + V_t} = \frac{3.558 \times 10^{-3}}{3.558 \times 10^{-3} + 0.1407} = 0.024 \text{ Ans. (b)}$$

(c) Interfacial area, a

$$a = \frac{6H}{D_{32}} = \frac{6(.024)}{4 \times 10^{-3}(m)} = 36 m^{-1} \text{ Ans.}$$

(d) Volume of the reactor, V

$$V = \pi/_4 D^2 L = 0.7554(1.22)^2 (1.22) = 2.426m^3$$

$$P/_V = \frac{819.8W}{1.426 m^3} = 575W / m^3 \text{, where } P / V \text{ is the power per unit volume}$$

$$k_{La} = 0.026\left(P/_V\right)^{0.4} u_g^{0.5}$$

$$k_{La} = .026(575)^{0.4}\left(3.558 \times 10^{-3}\right)^{0.5} = 0.0197 sec^{-1}$$

$$k = \frac{0.0197}{36} \frac{sec^{-1}}{(m^{-1})} = 5.47 \times 10^{-4} m / sec$$

(e) For turbulent flow, mass transfer correlation:

$$Sh = \frac{k_L D_{32}}{D_{O_2}} = 0.13 Sc^{1/3} Re^{3/4}$$

Sh is the shearwood number D_{32} is volume - surface mean diameter of bubble.

$$Sc = \frac{\mu}{\rho D_{O_2}} = \frac{8.9 \times 10^{-4}}{(997)(2.5 \times 10^{-9})} = 357, \text{ where } Sc \text{ is the schmidt number}$$

Rei (already calculated) = 4.1×10^{5}, for turbulent flow, we can use the following relation for mass transfer correlation,

$$Sh = 0.13(357)^{1/3}\left(4.1 \times 10^{5}\right)^{3/4} = 1.5 \times 10^{4}$$

we know, $k_L = (Sh)(D_{O_2}) / D_{32}$, putting the values of the parameters

$$k_L = \frac{\left(1.5 \times 10^{4}\right) \times \left(2.5 \times 10^{-9}\right)}{4 \times 10^{-3}} = 0.93 \times 10^{-2}$$

$$= 9.3 \times 10^{-3} \text{ m/s}$$

7.3.2 Bubble columns

These reactors are tall columns containing low viscosity broths. Mixing is not vigorous. They are found to have higher energy efficiency than stirred reactors with respect to the amount of oxygen transfer per unit of power input. The other advantage is the generation of low shear suitable for fermentation of shear sensitive cells. Though bubble coalescence is a problem which reduces mass transfer, but it can be eliminated by using multiple stages and each stage may contain a perforated plate as shown in Fig.7.3B, for redistribution of gas flow. Higher concentration of cells exist is the lower part of reactor, which is not desirable to maintain homogeneity of the bed. However, the cells are lifted up from the lower region by higher flow rate of air. Bubble columns often work with a narrow range of gas flow rates. The operation of these columns becomes troublesome for foaming and bubble coalescence. The volumetric mass transfer coefficient, k_La is intermediate between the values of k_La obtained from CSTR and air lift fermenter. Bubble columns are more flexible than stirred fermenters. Higher concentration of cells in the lowest part of the column leads to the rapid initial fermentation followed by a slower one involving less desirable substrates. As the cell concentration increases in the fermenter, high air flow rates are required to maintain the cell suspension and mixing. But increased air flow rates can cause excessive foaming and high retention of air bubbles in the column.reducing the productivity of the column. As bubble rise in the column, they may coalesce rapidly leading to decrease in k_La value. To overcome the disadvantage of the column fermenter, several alternation designs have been proposed. A tapered column fermenter can maintain a high flow rate per unit area at the lower section of the fermenter. To enhance mixing, the fermentation broth can be pumped and recirculated by using an external pump.

7.3.3 Air lift fermenters

A loop or airlift fermenter is a vertical column with a liquid circulation loop through a central draft tube (riser) or an external loop. On the basis of liquid circulation it may be classified into three different types: air-lift, stirred loop and jet loop and have been illustrated in figure 7.5.

In air lift system, upward gas flow carries the fluid and cells up through central draft tube known as Riser. At the top, gas is separated from the liquid and the degassed liquid which is denser than the gassed liquid, descends through the annular space outside the draft tube, known as Down-comer. At the bottom of the reactor, the descending liquid meets the gas stream and it carries back to the draft tube. Another design with external loop as shown in figure 7.5 (b) is ICI airlift system used for the production of single cell protein (SCP) from methanol, provided with a heat exchanger and the fermenter capacity was 1500 m^3. Another design shown in Figure 7.3 (C) is coupled with a mechanical stirrer in the riser part to enhance liquid circulation and mixing. In the jet mode, the liquid is circulated externally using a pump and air is introduced through a nozzle and the mixed liquid and air come out as a jet at the bottom of the riser.

The volumetric mass transfer coefficient, $k_L a'$ is of intermediate value between stirred tank fermenter and bubble column. Air lift fermenter can handle moderately highly viscous broth. These reactors are also suitable for shear sensitive animal cells and plant cells, providing adequate oxygen transfer and mixing.

7.3.4 Trickling biological filter

Trickling Biological Filter is a packed bed reactor where biological oxidation of substrates is carried out by microbial films deposited on inert support particles like sand or gravels in presence of dissolved oxygen. It is a shallow packed bed reactor with diameter to height ratio of D/H = 3.0 to avoid clogging problems and axial variations. Waste liquid is fed from the top of the bed using rotary liquid distributor. Low flow rate is used to avoid shear forces so that biofilm is not removed from the support. Air enters the bed from the bottom and moves upward by natural convection. The driving force for air circulation is the temperature difference created by exothermic biological oxidation. The reaction zone is highly heterogeneous with respect to temperature, pH, D.O and nutrient profile. Dissolved oxygen limitation may be significant due to high cell density, complex hydrodynamic condition and diffusion barrier within the film. Oxygen transfer problem is reduced by preaeration of waste-water feed and high liquid circulation rate. As the liquid trickles downward in the column on the surface of the microbial film organic component consumed by the micro organics of the biofilm of the order of 0.01 mm to 0.25 mm, reducing the BOD value of the waste stream. A schematic diagram of trickle bed reactor is given in figure 7.6.

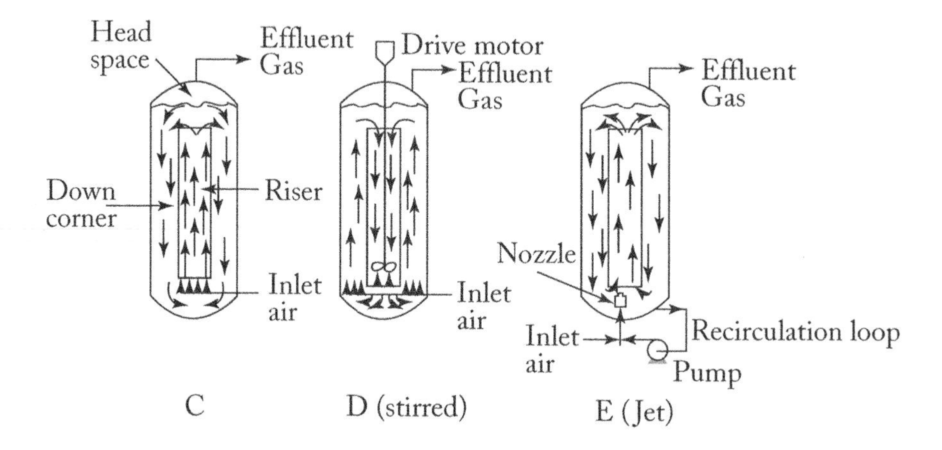

Fig. 7.5 Airlift Fermenters

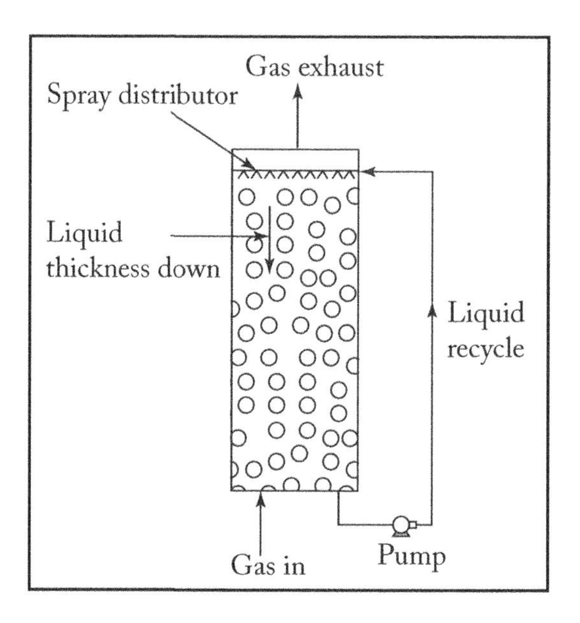

Fig. 7.6 Trickle bed Reactor

7.3.5 Relative merits and demerits of three basic bioreactors

Fermenters are often classified based on their vessel types as tank, column or loop fermenters. The tank and column fermenters are both constructed as cylindrical vessels. The height to diameter ratio (H/D) gives some distinguishing features,

$H/D < 3$ for the tank (CSTR)

$H/D > 3$ for the column fermenter.

The merits and demerits of three basic reactors may be tabulated.

Table 7.1: Merits and demerits of 3 basic reactors.

	Merits	Demerits
Stirred tank	1. Flexible and adaptable 2. Wide range of mixing intensity 3. Can handle highly viscous liquid	1. High power consumption 2. Damage shear sensitive cells 3. High equipment cost
Bubble column	1. No moving parts 2. Simple 3. Low equipment cost 4. High cell concentration	1. Poor mixing 2. Limited to low viscosity systems 3. Excessive foaming
Air lift	1. No moving part 2. Simple 3. High gas absorption efficiency 4. Good heat transfer	1. Poor mixing 2. Limited to low viscosity system 3. Excessive foaming

7.4 Estimation of Overall Mass Transfer Coefficient, k_La

Several correlations for k_La are given in terms of gas sparging and agitation parameters.

For stirred tank with water and coalescing bubbles,

k_La correlations are:

$$k_La = 2.6\times10^{-2}\left(P/V\right)^{0.4}\left(Ugs\right)^{0.5}\ Sec^{-1} \tag{7.24}$$

Where $V \le 2.6m^3$, $500\angle P/V < 10000\dfrac{w}{m^3}$

with non-coalescing bubbles,

$$k_La = 2.0\times10^{-3}\left(P/V\right)^{0.7}\left(Ugs\right)^{0.2}\ Sec^{-1} \tag{7.25}$$

Where $2\times10^{-3}m^3 < V < 4.4m^3$, $500\angle P/V < 10000\,w/m^3$.

Here Ugs is the superficial velocity (m/s) which is equal to the volumetric flow rate divided by the vessel cross section area and multiplied by the gas holdup (H).

The correlations are valid within 20–40% regardless of the type of stirrers (turbine, paddle, propellers, rods etc.).

Correlation for Bubble Column

Water and coalescing bubbles,

k_La is given for bubble column

$$k_La = 0.32\left(Ugs\right)^{0.7}\ Sec^{-1} \tag{7.26}$$

Where Ugs is in m/sec.

- **Correlation of Akita and Yoshida (1974)**

  k_L $(adt^2)/D$ = $0.6(Sc)^{1/2}(B_O)0.62$ $(G_a)^{0.31}H^{1.3}$........(7.27). For bubble columns, $O<V_{gs}<0.15$ m/sec. $100<\left(Pg/V\right)<1100\,w/m^3$ and Sc (Schnidt No.), $=\mu_L/\rho_L DO_2$

  Bond number, $B_o=gdt^2\rho_c/\sigma$, Galilio number, $G_a=gdt^3/U_L^2$, refer to tower diameter, dt, c refers to continuous phase.

  The limitations are: $dt\le 0.6m$, For $dt>0.6m$, the value of 0.6m should be used for dt in the equation (7.27).

  Correlation of Bottom etal (1980)for kla

-
$$k_L a/0.08=\left(\frac{Pg/V}{800}\right)^{0.75} \tag{7.27a}$$

  Where Pg is the gas power input, which can be calculated as $\left(P_g/V\right)=800\left(\frac{Vgs}{0.1}\right)^{0.75}$

  (7.27b)

  Where Vgs is the superficial gas velocity.

- Another Correlation for bubble columns is given by Wang and Fan (1978)

  $$k_L a=0.12U_L^{0.624}\left(\frac{U_g}{1.99U_g+47.1}\right)^{sec^{-1}}$$ where Ug and UL are gas and liquid superficial velocities (m/s) respectively. (7.28)

- **Air-lift Fermenter**

  For the above system, $k_L a$ is given by the equation **of Bello etal (1981)**

$$k_L a=\frac{0.0005(P/V)}{(1+Ad/Ar)}Sec^{-1} \tag{7.29}$$

  Where (Ad/Ar) is the ratio of areas of down comer and riser sections. P/V is the ratio of aeration power to reactor liquid volume.

  The dependence of $k_L a$ on P/V vanishes when $P/V\le 1.0$.

7.4.1 Power requirements for sparged and agitated vessels

For gas sparging into a column, the power used in compression to sparge a gas with volumetric flow rate, F_O at pressure p, is given as

$$P_g=\rho_g F_O\cdot\left[\frac{RT}{(MW)}l_n\frac{P_1}{P_2}+0.06\frac{U_0^2}{2}\right] \tag{7.30}$$

Where p_2 is the pressure at the top of the vessel and Uo is the gas velocity at sparger orifice. MW is the molecular at of air, P_1 is the pressure at the inlet.

- Power Consumption in a stirred tank

The power consumption for stirring non-aerated fluids depends on the following parameters: fluid properties, the stirrer speed, N_I and Impeller diameter, D_I drag coefficient of the impeller, C_D. The drag coefficient C_D varie with impeller Reynolds number, R_{eI} for each flow regime, Laminar, transition and turbulent.

Rushton, Costich and Everrett (1950) plotted power number, P_{NO} (a dimensional, power number) vs. Impeller Reynolds No., R_{eI} for three impeller geometries, (fig 7.7) where,

$$P_{NO} = \frac{P_g}{\rho_1 N_I^3 D_I^5} = \frac{Drag\ force\ on\ Impeller}{Intertitice\ force} \tag{7.31}$$

$$R_{eI} = \frac{\rho N_I D_I^2}{\mu} \tag{7.32}$$

In the turbulent region the power input is independent of R_{eI} (>20000

$$P \propto N_I^3 D_I^2 \quad PNO = \text{constant} \tag{7.33}$$

In the laminar flow region, $P \propto N_I^2 D_I^5$ Or, $P_{NO} \propto \frac{1}{R_{eI}}$ (7.34)

Power number, P_{NO} vs. R_{eI} are plotted for various impeller geometric in Fig.7.7

$P_{NO} = \propto \left(R_{eI}\right)^{\beta}$, and $\propto \& \beta$ are constants (7.35a).

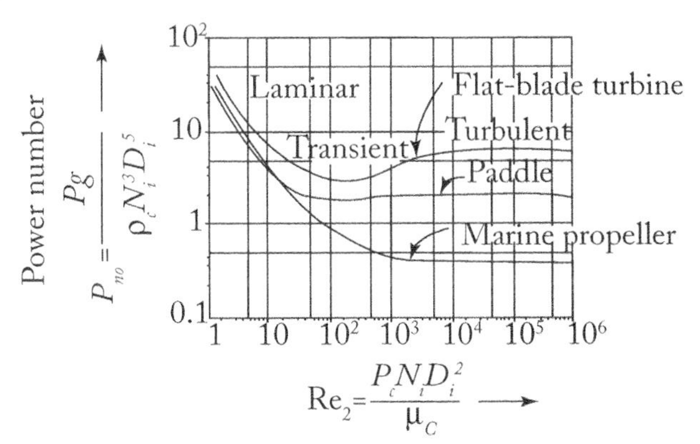

Fig. 7.7 Power number vs impeller based Reynolds number

For aerated and agitated stirred vessel, the power requirement for agitated vessel decreases with aeration. The ratio of power requirement in aerated vs. non-aerated vessel (Na) can be given as follows

$$N_a = \frac{F_g}{N_i D_I^3} \tag{7.35}$$

Where $N_a = P_a / P$ and F_g is the volumetric gas rate, Pa is the power of aeration and P is the power for agitation (mechanical).

- **For turbulent aeration of non-Newtonian fluids**

A correlation for P_a has been expressed as $P_a = m\left(\frac{P^2 NID_I^3}{F_g^{0.56}}\right)^{0.45} = m'\left(\frac{P^2\left(NID_I^3\right)^{0.44}}{Na^{0.56}}\right)^{0.45}$

where m′ = a constant (7.36)

and P is the non-aerated power input.

In some bioreactors, mixing is carried out by injection of a liquid jet into the vessel, the power dissipation, P is given as $P = \frac{8\rho_L F_L^3}{\pi^2 Dj^{0.4}}$ (7.37)

Where Dj is the jet diameter, F_L is the liquid flow rate.

7.5 Mass Transfer for Freely Rising or Falling Bodies

Based on the equation of change for description of conservation of mass of a species (such as oxygen) and the momentum balance, the following dimensionless groups are defined for mass transfer corrections:

$$Grash\,of\ number\,(Gr) = \frac{gravitational\ force}{iscons\ force} = \frac{gD^3\rho_L\left(\rho_L - \rho_g\right)}{\mu_L^2} \quad (7.38)$$

$$Sherwood\ number\,(Sh) = \frac{k_L D}{D_{O_2}} = \frac{total\ mass\ transfer}{mass\ diffusivily} \quad (7.39)$$

$$Schmidt\ number\,(Sc) = \frac{\mu_L}{\rho_L D_{O_2}} \quad (7.40)$$

Where D is the characteristics linear dimension (D, bubble diameteror tube diameter), μ_L is the viscosity of continuous phase.

7.5.1 Mass transfer coefficient, kL for single bubbles and bubble swarms

Near the gas-liquid interface, the dimensionless concentration, $\bar{c}\left(= c/_{co}\right)$ has a solution from the transport equation in the following form $\bar{c} = f\left(\bar{Z}, Sh, Sc, Gr\right)$ (7.41)

Where $\bar{Z}$ is the dimensionless distance.

It has been shown that the liquid phase mass transfer coefficient, k_L expressed in dimensionless group of Sherwood number (Sh) can be correlard as a function of Schmidt number (Sc) and Grashof number (Gr).

$$Sh = \frac{k_L D}{D_{O_2}} = g\left(Sc, Gr\right) \quad (7.42)$$

Where D is the characteristic bubble diameter.

Mass transfer from an isolated sphere with a rigid interface, applicable to small bubbles in a fermentation broth may be determined theoretically for the case, $Re = \rho_L DU / \mu_L << 1$ and Peclet no, $P_e = \frac{UD}{D_{O_2}} \gg 1$,

Where U is the velocity of the gas bubble relative to the liquid velocity. For Re(10^{-1} to 10^{-2}), the theoretical result gives equation.

$$Sh = 1.01 Pe^{1/3} = 1.01\left(UD / D_{O_2}\right)^{1/3} \ \& \ P_e = Re \cdot Sc \tag{7.43}$$

For small Reynolds number, the terminal velocity, U_t of a sphere is given as

$$Ut = gD^2(\Delta\rho) / 18\mu_L \tag{7.44}$$

Replacing V in equation (7.44) by equation (7.45), we have

$$Sh = 1.01\left(\frac{D^3 \Delta\rho g}{18\mu_L D_{O_2}}\right)^{1/3} = 1.01\left(\frac{D^3 \rho_L \Delta\rho g}{18\mu_L^2}\right)^{1/3} \times \left(\frac{\mu_L}{\rho_L D_{O_2}}\right)^{1/3} = 0.39 Gr^{1/3} Sc^{1/3}$$

Where $$\frac{D^3 \Delta\rho g}{18\mu_L D_{O_2}} = Rayleigh\,number\,(Ra) \tag{7.45}$$

For a large Reynolds number, the single bubble is equivalent to a non-circulating sphere in laminar flow, the correlation is given as,

$$Sh = 2.0 + 0.6 Re^{1/2} Sc^{1/3} \tag{7.46}$$

Where $Re \gg 1$

For many industrial sparged reactors, swarms or clusters of bubbles are produced. For this situation Calderbank and Moo-young (1961) reported two correlations for two distinct regimes of bubble swarms mass transfer. The division is based on a critical bubble diameter, D_C. In the absence of surfactants, $D_C \approx 2.5$ mm.

Bubbles larger than 2.5 mm are observed in agitated tanks and sieve plated columns. Smaller bubbles are found in sintered plate columns and in agitated vessels containing hydrophilic solutes in water. The correlations are:

For $D < D_C = 2.5 mm$

$$Sh = 0.31 Gr^{1/3} Sc^{1/3} = 0.31 Ra^{1/3} \tag{7.47}$$

For $D > D_c = 2.5 mm$

$$Sh = 0.42 Gr^{1/3} Sc^{1/2} \tag{7.48}$$

For individual cells, clumps and flocs as well as for Gas oil or other hydrocarbon dispersion, a more accurate correlation is: $Sh = k_L D / D_{O2} = 2.0 + 0.31(Ra)^{1/3}$ (7.49)

Rearranging equation (7.49) we have

$$\frac{k_L}{D_{O2}} = \frac{2.0}{D} + 0.3\left[\frac{\Delta \rho g}{\mu_L D_{O2}}\right]^{1/3} \tag{7.50}$$

The relative importance of the pure diffusion shows $\Delta\rho = 0, k_L = 2\frac{D_{O2}}{D}$. Large diameters due to flocs, films etc lead to greater contribution from the second term.

7.5.2 Mechanisms of mass transfer

Several mechanisms have been advanced to provide a theoretical basis for interphase mass transfer. Three best known theories are: **Two film theory, the Penetration Theory** and **surface Renewal Theory.** Two film theory assumes that the entire resistance to mass transfer exists in two fictitious films on the either side of the interface in which transfer occurs by molecular diffusion. According to the model, the mass transfer coefficient, k_L is proportional to the diffusivity, DAB and inversely proportional to the film thickness, Zf, thus

$$k_L = DAB/Z_f \tag{7.51}$$

Penetration theory (proposed by Higbie, 1935) is based on the turbulent eddies which travel from the bulk phase to the interface where they remain for a constant exposure time tc. The solute is assumed to penetrate into a given eddy during its stay at the interface by a process of unsteady state molecular diffusion. This model predicts that k_L is directly proportional to the square root of molecular diffusivity, DAB as

$$k_L = 2\left(\frac{DAB}{\pi t_c}\right)^{1/2} \tag{7.52}$$

Surface renewal theory proposed by Danckwerts, (1951), assumes that there is an infinite range of ages for the elements of surface and the surface age distribution function, $\phi(t)$, can be expressed as $\phi(t) = Se^{-3}$ (7.53)

where S is the fractional rate of surface renewal.

The theory predicts that k_L is proportional to the square root of molecular diffusivity,

$$k_L = (SDAB)^{1/2} \tag{7.54}$$

Last two theories require knowledge of unknown parameters such as exposure time, tc, the functional rate of surface renewal S, so their uses are not common and it is difficult to apply them. However, two film theory has been used extensively and k_L has been determined, using theoretical correlations.

7.5.3 Determination of interfacial area

7.5.3.1 Photographic technique

The interfacial area per unit liquid volume, 'a' can be determined from the Sautermean diameter, D_{32} and the gas hold-up, H (the volume fraction of the gas phase)

$$a = \frac{6H}{D_{32}} \tag{7.55}$$

The Sautermean diameter (a surface volume mean), D_{32} can be calculated by measuring drop sizes directly from the photographs of dispersion of bubbles by stroboscope according to the equation

$$D_{32} = \frac{\sum_{1}^{\infty} n_i D_i^3}{\sum_{2=1}^{n} n_i D_i^2} \tag{7.56}$$

7.5.3.2 Correlation for a and D_{32}

For gas sparging with no mechanical agitation (for bubble column and air-lift fermenter), we avoid the vicinity of a sparger where the bubbles may break or coalesce with others, an equilibrium size distribution may be obtained. For gas sparged through perforated plates and single orifice, the following correlation for D_{32} are available. These are

$$\frac{D_{32}}{D_C} = 20\left(\frac{gD_c^2\rho_c}{\sigma}\right)^{-0.50}\left(\frac{gD_C^3}{v_C^2}\right)^{-0.12}\left(\frac{V_{gs}}{\sqrt{gD_C}}\right)^{-0.32} \tag{7.57}$$

C refers to the continuous phase, σ is the surface tension

For the interfacial area, a

$$a\,D_C = \frac{1}{3}\left(\frac{gD_C^2\rho_L}{r}\right)^{0.50}\left(\frac{gD_C^3}{v_L^2}\right)^{0.3} H^{1.13} \tag{7.58}$$

Where D_c is the bubble column diameter, V_{gs} is the superficial gas velocity$\left(\leq 0.07\,m/s\right)$, $D_C \leq 0.3m$.

For gas-sparging with mechanical agitation:

Calderbank's correlation (1958)

Interfacial area, a_o for gas-liquid agitation in a flat blade disc turbine is given by,

$$a_o = 1.44\left[\frac{\left(P_m/V_R\right)^{0.4}\rho_C^{0.2}}{\sigma^{0.6}}\right]\left(\frac{V_{gs}}{V_L}\right)^{1/2} \tag{7.59}$$

For $V_{gs} < 0.02$ m/sc and

$$R_{eI}^{0.7}\left(\frac{ND_I}{V_{gs}}\right)^{0.3} < 20000$$

P_m is the mechanical power input.

For Vgs> 0.02 m/s

$$a_o = 1.44\left[\left(P_e / V_R\right)^{0.4} \rho_C^{0.2}\right]\left(\frac{V_{gs}}{V_L + V_{gs}}\right)^{1/2} \tag{7.60}$$

P_e = effective power consumption = $P_m + P_a$

Where P_a = power for aeration and Pm = power for mechanical agitation

Calderbank's Correlation (1955) for sauter mean diameter, D_{32} for dispersion of air in water,

$$D_{32} = 4.15\left[\frac{\sigma^{0.6}}{\left(P_m / V_R\right)^{0.4} \rho_c^{0.3}}\right] H^{0.5} + 9.0\times10^{-4} \tag{7.61}$$

7.5.4 Determination of gas hold up, H

Gas hold up depends mainly on the superficial gas velocity, power consumption and is very sensitive to the properties of the liquid. Gas-hold-up, H, can be determined by measuring the liquid height of aerated liquid, Z_F and that of the clear liquid, Z_I

$$H = \frac{Z_F - Z_I}{Z_F} \tag{7.62}$$

For bubble column, H is given as

$$H = \frac{V_{gs}}{V_{gs} + V_t} \tag{7.63}$$

Where Vg is the bubble rise velocity *and Vt is the terminal velocity*

Another, correlation for H for a bubble column is given by Akita and Yoshida (1973)

$$\frac{H}{\left(1-H\right)^4} = 0.2D\left(\frac{gD_C^2\rho_L}{\sigma}\right)^{1/8}\left(\frac{gD_C^3}{v_c^2}\right)^{1/12}\left(\frac{V_{gs}}{V_g D_c}\right) \tag{7.64}$$

Where D_c is the column diameter, D is the bubble diameter

Gas sparging with mechanical Agitation

For gas dispersion by agitation with a flat bade disc turbine impeller, Calderbank's correlation for gas hold-up, H is expressed as

$$H = \left(\frac{V_{gs} H}{V_L}\right)^{1/2} + 2.16 \times 10^{-4} \left[\left(P_m / V_R\right)^{0.4} \rho_c^{0.2} \right] \times \left(\frac{V_{gs}}{V_L}\right)^{1/2} \quad (7.65)$$

Where 2.16 x 10^{-4} has a unit (m) and V_L = 0.265 m/s. The bubble size, D_{32} is in the range of 2-5 mm diameter.

For higher superficial gas velocity V_{gs}> 0.02 m/s, P_m is to be replaced by P_e and V_L by $(V_I + V_g)$.

7.5.5 Power consumption in stirred reactors from dimensional analysis

Dimensional analysis gives the following relationship for power consumption in a stirred vessel.

$$\frac{P}{\rho N_i^3 D_I^5} = f\left(\frac{\rho N D_I^2}{\mu}, \frac{N^2 DI}{g}, \frac{D_T}{DI}, \frac{H}{D_I}, \frac{DW}{DI}\right) \quad (7.66)$$

Or

$$P_{NO} = f\left(\mathrm{R_{eI}}, \mathrm{F_r}, \text{all length ratios}\right)$$

Where Power Number,

$$P_{NO} = \frac{P}{\rho N_i^3 D_I^5}, \text{ Impeller based Reynolds number, } R_{eI} = \frac{\rho N D_I^2}{\mu}$$

Froude number, $F_r = \dfrac{N^2 D_I}{g}$

D_T, D_I, D_W, H are the tank diameters, Impeller diameters, Impeller width and Hold-up respectively.

Froude number (F_r) takes into account of the gravity forces which affects the power consumption due to vortex formation.

The vortex formation is reduced by the use of baffles.

For fully baffled reactor the effect of Froude number on the power consumption is negligible, and all length ratios are kept constant. The equation (7.66) simplifies to,

$$P_{No} = \propto \left(R_{eI}\right)^{\beta} \quad (7.67)$$

Where $\propto$ & β *are constants*

The plot of power no, P_{NO} vs $R_{e,I}$ (impeller based Reynolds number,) illustrated in fig 7.7 shows that for laminar flow P_{NO} decreases with an increase of R_{eI} and makes a constant value when the ReI is larger that 10,000. At this point, P_{NO} is independent of R_{eI} and become constant. For example, for a flat six-blade turbine impeller, at R_{eI} > 10,000 the power number (P_{NO}) is 6.

$$P_{NO} = \frac{P_m}{\rho N_i^3 D_I^5} = 6.0 \tag{7.68}$$

The power required by an impeller in a gas sparged system, Power is usually less than the power required by the impeller at the same speed in a gas free liquids.

7.5.6 Maximum stable bubble size

Theoretical correlations for the relationship between the maximum stable bubble size (D) and fluid properties employ a dimension less group, known as Weber number, we defined as,

$$W_e = \tau \frac{D}{\sigma} \tag{7.69}$$

where D is the bubble diameter, τ is the dynamic pressure and σ is the surface tension.

A suitable value of the dynamic pressures τ to calculate maximum bubble size for freely rising bubble is given as,

$$\tau = \frac{\rho_L U_t^2}{2} \tag{7.70}$$

Where U_t is the bubble terminal velocity for spherical bubbles and is expressed as

$$U_t = \frac{gD^2(\rho_L - \rho_g)}{18\mu_L} \tag{7.71}$$

The critical Weber number is that value of We for D = Dc, where Dc is the critical bubble size. From experiments with clean air and water system, the approximate value of critical Weber number, Wec = 1.05.

For complicated turbulent flow, estimation of. dynamic pressure, τ is difficult. But turbulence is found in bubble columns near the aeration nozzles. So bubble size determination under turbulent conditions is quite relevant.

7.5.7 Theory of turbulence by kolmogorov

From the statistical theory of turbulence by Kolmogorov, the turbulent flow field is assumed to be a collection of super eddies, characterized by fluctuating frequencies (on length scale and magnitude). The above theory predicts that smaller eddies are statistically independent of the primary eddies and are isotropic (spatially uniform). We know that the process of breaking up bulk flow is called dispersion. The degree of homogeneity as a result of dispersion is limited by the size of the smaller eddies which may be formed in a fluid. This size is given approximately by the Kolmogorov scale of mixing or scale of turbulence, .

$$\lambda = \left[\frac{\gamma^3}{\epsilon}\right]^{1/4} \tag{7.72}$$

Where λ is the characteristic dimension of the smallest eddies, γ is the kinematic viscosity $\left(\mu/\rho\right)$ of the liquid and ϵ is the rate of turbulent energy dissipation per unit mass of liquid. At the steady state, the rate of energy dissipation by turbulent eddies is equal to the power supplied by the impeller (P)

Or
$$P = \epsilon \rho D_I^3 \tag{7.73}$$

The fluid mass in the impeller zone is roughly equal to ρD_I^3, where ρ is the fluid density and D_I is the impeller diameter. Further P can be calculated from the plot of P_{NO} vs. Impeller based Reynolds number,Rei

$$P_{NO} = \frac{P}{\rho N_i^3 D_i^5} \tag{7.74}$$

$$R_{eI} = \rho N_i D_i^2 / \mu \tag{7.75}$$

At $R_{eI} > 10^3$, P_{NO} is constant. The value of the P_{NO} depends upon the type of impeller.

From the equation (7.72), the greater the power input to the fluid, the smaller are the size of the eddies. λ also depends upon viscosity at a given power input. Smaller eddies are produced in low viscosity fluids.

For low viscosity liquids such as water, δ is usually in the range of 30 – 100 μm. Within this range of eddies, homogeneity is achieved in about *one* second for low viscosity fluids. If the power input to a stirred vessel produces eddies of this dimension, mixing in a molecular scale can be achieved immediately.

To find out the critical bubble size for turbulent flow, a suitable dynamic pressure for the Weber number in turbulent flow can be obtained from the relation

$$\tau = \rho_L \left(U_{rms} / eddie\ of\ scale\ D_c\right)^2 \tag{7.76}$$

Now Dc (the critical bubble size) for turbulent shear stress

$$D_c = \propto \frac{\sigma^{0.6}}{\left(P/V_L\right)^{0.4} \rho_L^{0.2}} \tag{7.77}$$

Where $\propto$ is a constant. The maximum stable bubble size is reduced if the power dissipation per unit volume is increased. Now the bubble interfacial area, has been given as

$$a' = \frac{1}{Volume} n F_O t_b \cdot \frac{HD^2}{H/6D^3} \tag{7.78}$$

$$= \frac{nF_{Ot_b}}{V} \cdot \frac{6}{D} \tag{7.79}$$

Where t'_b is bubble residence time in the reactor, Fo is the volumetric flow rate per orifice, n is the number of orifices, D is the bubble diameter, V is the reactor liquid volume.

Now the t_b for large spherical cap-shaped bubbles (D) in Newtonian fluid is

$$t_b = \frac{hr}{\text{vt}} \tag{7.80}$$

Where hr is the height of the liquid column, vt is the terminal bubble rise velocity and is obtained by the relation,

$$v_t = 0.711(gD_{32})^{1/2} = 22.26\, D_{32}^{1/2} cm/sec \tag{7.81}$$

Where D32 is the Sautermean diameter..

For bubble clouds or swarms of bubble, v_t can be approximately calculated by the equation,

$$v_t\left(bubble\ clouds\right) = 0.5 v_t\left(single\ bubble\right) \tag{7.82}$$

The quantity, $n F_o t_b$ is equal to total bubble volume of the reactor. The bubble volume per reactor volume is known as the fractional gas-holdup (H). So the equation (2.79) reduces to

$$a' = \frac{6H}{D} \tag{7.83}$$

7.6 Problem on Turbulent Shear Damage in Animal Cell Culture

Microcarrier beads (120 μm in diameter) are used to culture recombinant (chinese hamster overy) cells for production of growth hormones. It is proposed to use a turbine impeller of a diameter, D_I = 0.06m to mix the culture in 0.0035 m^3 stirred reactor. Air and CO_2 are introduced through the reactor headspace. The microcarreir suspension has an approximate density of 1010 kg/m^3 and viscosity of 1.3 x 10^{-3} P_aS. It is known that shear due to the eddies may be avoided if the Kolmogorov scale of eddy size, $\lambda \geq \frac{2}{3}$ to ½ diameter of the beads.

The length criterion, λ is given as $\lambda = \left[\frac{\delta^3}{\epsilon}\right]^{1/4}$

Or $\lambda^4 = \frac{\delta^3}{\epsilon}$ ....

Where δ = Kinematic viscosity $\left(\mu/\rho\right)$

$\in$= the local rate of energy dissipation per unit mass of reactor fluid.

For turbulent flow in a CSTR, if R_{eI} (impeller based Reynolds number) $\geq 10^4$, the power number, PNo = 5 and the power, $P = \in\left(\rho D_I^3\right)$ where ρD_I^3 is approximately the mass of the fluid

Calculate the speed of the stirrer with turbine impeller, so that cell damage is avoided.

Solution:

Kolmogorov scale of turbulence is

$$\min \lambda^4 = \frac{\delta^3}{\in}$$

Now it is recommended that λ size should be such s that there is no cell damage.

$$\lambda = \frac{2}{3}(120\mu m) = 80\mu = 80\times10^{-6} m$$

$$\text{Kinematic viscosity, } \delta = \frac{\mu}{\rho} = \frac{1.3\times10^{-3} kg/ms}{1.01\times10^{3} kg/m^3} = 1.29\times10^{-6}\frac{m^2}{S}$$

$$\in = \frac{\delta^3}{\lambda^4} = \frac{\left(1.29\times10^{-6}\right)^3 m^6 S^{-3}}{\left(80\times10^{-6}\right)^4 m^4} = 0.052 m^2 S^{-3}$$

$$P = \epsilon\, \rho D_I^3 = .052\left(\frac{m^2}{S^3}\right)(1010)\frac{kg}{m^3}\left(6\times10^{-2}\right)^3 m^3$$

$$= 1.13\times10^{-2} J/S = 1.13\times10^{-2} \text{ Watt}$$

$$1 kg\, m^2/S^3 = 1J$$

$$P_{NO} = \frac{P}{\rho N^3 D_I^5}$$

$$\text{Now } R_{eI} = \frac{\rho N D_I^3}{\propto_4} = \frac{1010(N)\left(6\times10^{-2}\right)^2}{1.3\times10^{-3}} = 2.79\times10^3 \ldots\ldots$$

.. *Since the flow* is turbulent

Power number.,. $(P_{NO}) = 5$ from the fig 7.7

Now P=(Pno) ρ N3 Di5

$$N^3 = \frac{1.13\times10^{-2}}{(5)(1010)(6\times10^{-2})^5} = 2.87S^{-3}$$

$$N = 1.421S^{-1} = 85.26 \text{ .rpm........}$$

7.7 Forced Convective Mass Transfer

Mechanical mixing of air-liquid dispersion by different types of stirrers leads to evaluation of liquid phase mass transfer coefficient, k_L in terms of some important dimension less groups. Several important features in mechanical agitation in fermentation medium may be mentioned.

1. The high dynamic pressure near the impeller tip produces small bubbles, increasing interfacial area, a′.
2. The culture may contain suspensions of solids of microbial cells which may tend to rise or fall in the vessel. Mechanical mixing provides a more uniform volumetric dispersion of these phases in the bulk liquid.
3. The maximum size of aggregated mycelia, microbial slimes, mold pellets may be diminished by agitation leading to smaller particles. Product yield may be reduced due to cellular or extra cellular enzyme damage due to high agitation.
4. The liquid cell suspension may be highly viscous and mechanical agitation provides adequate gas-liquid mixing.

The following dimension less groups are important for predicting mass transfer under forced convection.

$$\text{Sherwood number, } Sh = \frac{k_L D}{D_{O_2}} \tag{7.84}$$

$$\text{Schmidt number, } S_c = \frac{\mu_L}{\rho_L D_{O_2}} \tag{7.85}$$

$$\text{Reynolds number based on impeller diamond, } R_{eI} = \frac{\rho_L N i D_I^2}{\mu_L} \tag{7.86}$$

$$\text{Froude number, } Fr = \frac{N_I^2 D_I^2}{gD} \tag{7.87}$$

7.7.1 Mass transfer for turbulent flow correlation

In a large stirred bioreactor provided with baffles, Frounde number becomes unimportant and vortex formation is eliminated.

From velocity and concentration fields, it has been shown

$$Sh/\overline{c} = f\left(\overline{Z}, R_e, Sc, Sh\right) \tag{7.88}$$

And Sherwood number, Sh is given as a function of $R_e \propto Sc$

$$Sh = \frac{k_L D}{D_{O_2}} = g\left(R_e, Sc\right) \tag{7.89}$$

Calderbank's Correlation (1961)

$$Sh\left(turbulent\ aeration\right) = 0.13\, Sc^{1/3} R_e^{3/4} \tag{7.90}$$

Using the power input per unit reactor volume, P/V, it can be shown with the variation of k_L in terms of Sh with P/V can be expressed as

$$Sh \propto \left(P/V\right)^{1/4} \tag{7.91}$$

Thus
$$k_L = 0.13\left(\frac{\propto^3 \mu_L (P/V)}{\rho_L D}\right)^{1/4} Sc^{-2/3} \tag{7.92}$$

Where $\propto$ is a constant in the equation of rms velocity with characteristic eddy size, (L or …λ..)

$$U_{rw} = \propto \left(P/V_L\right)^{1/3} \left(\frac{\mu L}{\rho_L}\right)^{1/3} \tag{7.93}$$

Another eddy of scale L correlation for stable bubble size, D_{32} for gas-liquid electrolyte

$$D_{32} = 2.25 \frac{\sigma^{0.6}}{\rho_L^{0.2}(P/V)^{0.4}} H^{0.4} \left(\frac{\mu_g}{\mu_L}\right)^{0.25}$$

Problem 7.7

A CSTR of volume 0.3 m^3 is stirred by a turbine impeller diameter, D_I = 0.1m and vessel diameter, D_t = 0.5m at rpm 200 with air flow rate of 2 x 10^{-3} m^3/min, density of the media = 1200 kg/m^3. Viscosity = 10^{-3} Pas, D diffusivity of air in H_2O = 1.5 x 10^{-9} m^2/s

Bubble diameter (mean) = 2.5 mm

(a) Determine k_L

(b) Determine k_{La}, a, H (hold up) and a′

Solution:

(a) We know

$$Sh\left(turbulent\ aeration\right) = 0.13\, S_c^{1/3}\, Re_I^{3/4}$$

$$Sc = \frac{\mu}{\rho D_{o2}} = \frac{10^{-3}}{(1200)(0.5\times10^{-9})} = 1660$$

Rel (impeller based Reynold number)

$$= \frac{\rho N D_I^2}{\mu} = \frac{1200(200/60)(0.1)^2}{10^{-3}} = 1.2\times10^3$$

$$Sh = \frac{k_L D}{D_2} = 0.13\, S_c^{1/3}\, Re_I^{3/4}$$

$$\text{Or } k_L = \frac{0.5\times10^{-9}}{2.5\times10^{-3}}\times0.13(1660)^{1/3}(1200)^{0.75}$$

$$= 0.628\times10^{-4}\, m/sec$$

$$= 0.628\times10^{-2}\, cm/sec$$

(b) We know the $k_L a$ correlation for a stirred vessel

$$k_L a = 2.6\times10^{-2}\left(P/V\right)^{0.4} Ug^{0.5}$$

Calculation of Ug, cross sectional area, $A = \pi/4\, D_T^2$

$$A = 0.7854(0.1)^2 = 7.854\times10^{-3} m^2$$

$$U_g = \frac{2\times10^{-3}}{7.854\times10^{-3} m^2} m^3/min = 0.254 m/min$$

$$= 4.23\times10^{-3} m/sec = 0.423\, cm/sec$$

At $R_{eI} = 1.2\times10^3$, from the plot P_{No} vs R_{eI}, we get P_{No} = 4.0 for turbine impeller

$$4 = \frac{P}{\rho_L N_I^3 D_I^5}$$

Or $P = 4(1200)(3.33)^3(0.1)^5 = 1.77 \times 10^{-3}$ W

$= 1.77$kW

$$P/_V = \frac{1.77W}{0.3m^3} = 5.9\text{kW} / m^3$$

$$k_{La} = 2.6 \times 10^{-2}(5.9)^{0.4}(0.423)^{0.5} = 3.43 \times 10^{-2}\, sec^{-1}$$

$$a = k_{La} / k_L = 3.43 \times 10^{-2} / .628 \times 10^{-2}$$

$$= 5.4\, cm^2 / cm^3$$

Now $a = \dfrac{6H}{D}$

Or $H = \dfrac{aD}{6} = \dfrac{5.4\frac{1}{cm} \times 0.25\, cm}{6} = 0.225$, where D (the bubble diameter) = 0.25 cm

Now a′ interfacial area per (liquid + gas)

$$a' = \frac{a}{1-H} = \frac{5.4}{1-.225} = 6.96\frac{cm^2}{cm^3}$$

7.8 Mass Transfer Across Liquid Surfaces

Mass transfer across free liquid surfaces has been observed in stream reaction, respiration of aerobic life near the surface, in lake communities. This transfer process is also important in trickle bed reactors, wine, vinegar manufacture and in waste water-treatment. Mass transfer into or out of the falling liquid film has been studied, but not under the conditions appropriate to microbial processes. However, we can make the above as the basis of calculating mass transfer rate on the free liquid surface. For a falling laminar liquid film of thickness, h, length L and width, W and with zero initial concentrations of dissolved gas the integrated absorption rate, (moles / unit time) $= WLC_L^* \left(\dfrac{4D_{O2}U_{max}}{\pi L}\right)^{1/2}$ (7.94)

Where U_{max} is the free surface velocity.

Reynolds number for the situation, $R_e = \dfrac{U_{max}R_h\rho_L}{\mu_L}$ (7.95)

Where R_h, the hydraulic radius used as the length parameter, defined

$$R_h = \frac{Wh}{2W + 2h} \approx \frac{h}{2} \text{ if } h \ll W \tag{7.96}$$

The mass transfer coefficient, k_L is used to obtain total mass transfer rate, Q

$$Q = k_L\left(C_L^* - C_L\right)WL \tag{7.97}$$

Assuming $C_L \rightarrow 0$ compared to C_L^*, using the previous equation, Sherwood number, Sh can be given as

$$Sh = \frac{k_L h}{D_{O2}} = 2b\left(Sc \cdot Re\right)^{1/2} \tag{7.98}$$

Where h is the length parameter for Sherwood number, and $b = (L / h)^{1/2}$

From the study of CO_2 absorption in an aqueous film moving down a known area, using water, algal suspension, nutrient media, k_L has been given as

$$k_L = 4 \times 10^{-5} Re^{2/3} cm / sec \tag{7.99}$$

For $2000 \leq Re \leq 8000$.

For turbulent flowing stream, we have to consider the depth where fluid carriers fresh, nearly saturated liquid from the surface to the bulk liquid. In this case k_L has been given in terms of Higbie's penetration theory as,

$$k_L = \left(\frac{D_{O2}}{\pi\tau}\right)^{1/2} \qquad 7.100)$$

Where τ is the residence time for fluid circulation. For flowing stream, the renewal time, τ, has been given by the relation

$$\tau_{stream} = \frac{h}{U_\omega} \tag{7.101}$$

Where h is the stream width and U_ω is the average stream velocity.

Then eliminating τ_{stream}, k_L can be expressed as

$$k_L = \left(\frac{D_{O2}}{\pi}\frac{U_\omega}{h}\right)^{1/2} \tag{7.102}$$

If the stream width is W, L is the length and h is the stream depth, the interfacial area, a per stream volume is

$$a = \frac{W(L)}{W(L)(h)} = \frac{1}{h} \tag{7.103}$$

Further $k_L a$ has been given by O'Conner – Dubbins (Bailey and Ollis, 1986)

$$k_L a = \left(\frac{D_{O2}}{\pi}\frac{Uw}{h^3}\right)^{1/2} \tag{7.104}$$

Diffusivity

The important parameter with the k_{La} expression is the diffusion coefficient of the solute in the solution. **The Wilke-Change correlation** has been widely used to calculate the diffusivity of a solute, A and is expressed by the relation,

$$D_{AB} = 7.4 \times 10^{-8} \frac{T(X_a M_A)}{\mu_L V_m^{0.6}} cm^2 / sec \tag{7.105}$$

Where M_A is the solute molecular weight, V_m = molecular volume of solute at boiling point, cm^3 / gmole, μ_L = liquid viscosity, Xa is the association parameter for H_2O, Xa = 2.6, 1.5 (ethanol), 1.00 (benzene, ether and heptane).

Skelland (1974) recommended the following equation

$$D_{AB} = \frac{1.12 \times 10^{-13}}{\mu^{1.1} V_m^{0.6}} m^2 / s \tag{7.105a}$$

Where μ is in kg/ms.
V_m = solute molecular volume at normal balling point, 0.0256 m^3/kmol for oxygen

7.9 Rheology: Newtonian and Non Newtonian Fluids

Rheology is the science of fluid deformation and fluid flow. Viscosity is an important aspect of Rheology. Viscosity is a property of the fluid and is determined by the ratio of the shear force causing the flow and velocity gradient in this flow. Two types of fluids have been found in fluids handling : (1) Newtonian and (2) Non-Newtonian. For the Newtonian fluid, viscosity is defined by Newton's law of viscosity,

$$\tau = -\mu \frac{dv}{dy} = \mu\dot{\gamma} \tag{7.106}$$

Where τ is the shear force which is proportional to the velocity gradient, dv/dy or shear rate $(\dot{\gamma})$ and μ is the proportionality constant, known as viscosity of the fluid. The negative sign is due to the negative velocity gradient and the direction of τ is considered positive. The velocity gradient, dv/dy is called the shear rate denoted by $(\dot{\gamma})$.

μ has the dimension of $L^{-1}MT^{-1}$. In SI units, the unit of viscosity is Pascal second (Pas), equal to 1 NS m^{-2} or kg m^{-1} s^{-1}. The viscosity of water at 20°c is 1 cp or 10^{-3} Pas. A modified form of viscosity is the kinematic viscosity, defined as $\delta = \mu/\rho$ where ρ is the density of the fluid. The fluids which obey Newton's Law of viscosity are known as Newtonian fluids. The plot of shear stress, τ vs shear rate, $\dot{\gamma}$ is a straight line with a slope equal to viscosity.

7.9.1 Non-newtonian fluids

Cell suspensions, slurries, solution of long chain polymers and other compounds like starch extra cellular polysaccharides like dextran, pollutan etc. are Non-Newtonian fluids which do not obey Newton's law of visocity.

Common types of Non-Newtonian fluids are (i) pseudoplastic, (ii) dilatant, (iii) Bingham plastic and (iv) casson plastics.

For non-Newtonian fluids, the ratio between the shear stress and the shear rate is not constant and the ratio is called the apparent viscosity, μ_a. It is not a property of the fluid, but it depends on the shear force exterted on the fluid.

7.9.2 Power law model of oswald – de- Waele

Pseudo plastics and dilatant fluids Obey the power law of Oswald-de-Waele,

$$\tau = K\dot{\gamma}^n \tag{7.107}$$

where τ is the shear stress, K is the consistency index, $\dot{\gamma}$ is the shear rate, n is the flow behaviour index. The parameters K and n characterize the rheology of Non-Newtonian fluids. Wher $n < 1$, the fluid is pseudo plastic, when $n = 1$, the fluid is Newtonian and when $n > 1$, the fluid is Casson fluids. From the power law model, the apparent viscosity, μ_a is given as

$$\mu_a = \frac{\tau}{\dot{\gamma}} = K\dot{\gamma}^{n-1} \tag{7.108}$$

For pseudoplastic fluids ($n < 1$), the apparent viscosity decreases with increasing shear rate and these fluids are called shear thinning. When the apparent viscosity increases with shear rate for dilatant fluids ($n>1$) and these are called shear thickening fluids.

Bingham plastic fluids do not produce motion until some finite yield stress has been applied and then they behave as Newtonian fluid. The model has been given as

$$\tau = \tau_O + K_P\dot{\gamma} \tag{7.109}$$

$$\mu_a = {}^{\tau_o}\!/_{\dot{\gamma}} + k_p \tag{7.110}$$

Casson fluids have been modelled as

$$\tau^{1/2} = \tau_O^{1/2} + k_p\dot{\gamma}^{1/2} \tag{7.111}$$

Once the yield stress is reached and thereafter the fluid behaves as a pseudo plastic. The apparent viscosity can be given as

$$\mu_a = \left[\left(\frac{\tau_O}{\dot{\gamma}}\right)^{1/2} + k_p\right]^2 \tag{7.112}$$

Bingham fluids may be visualized to have some internal structure that collapses at τ_O. Examples include slurries, pulp, ketchup. According to the equation, $\tau_S = -\eta_o \dot{\gamma}^n$ for a given polymer η_v (the absolute viscosity) is a decreasing function while, n is an increasing function of increasing temperature, which means that melt become more Newtonian (i.e. less shear thinning) with an increase in temperature.

At low shear rate, η_v is nearly independent of $\dot{\gamma}$ (Newtonian behaviour and approaches a limiting zero shear rate). η_v decreases with increasing $\dot{\gamma}$. Fluids that display this behaviour area called shear thinning.

The rheogram or the plot of shear stress, τ vs shear rate for Newtonian and non-Newtonian fluids has been shown

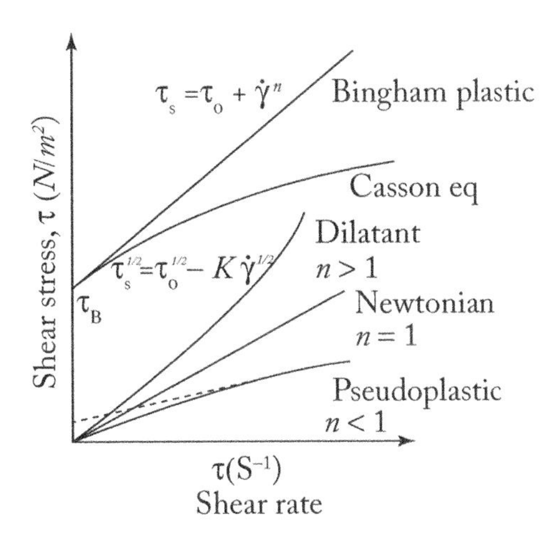

Fig. 7.8 Stress vs. Strain behaviour of Newtonian and Non-Newtonian fluids

Some of the non-Newtonian fluids are: Sewage sludge (Bingham plastic), Rubber latex (pseudoplastic), quick sand and some sand filled emulsion (dilatant), blood flow (Casson equation), Mycetial broth of Streptomyces grises (Bingham Plastic), Microbial flocs (Casson Equation)

7.9.3 Time dependent viscosity

If the apparent viscosity increases with time, the fluid is said to be rheopectic. Such fluids are rare in occurrence.

If the apparent viscosity decreases with time, the fluid is said to thixotropic. Fungal mycelia or extra cellular microbial polysaccharides show thixotropic behaviour. This behaviour may be attributed to reversible structural effects associated with the orientation of cells and macro molecules in the fluid.

Rheological properties vary during the application of shear force, because some time is necessary for the equilibrium to be established between the structure break down and re-orientation.

7.9.4 Kinematic viscosity

The ratio of absolute viscosity for the density of fluid, μ/ρ, is known as the Kinematic viscosity (δ). In the SI system, the unit for $\dot{\delta}$ is m²/sec. In the C.G.S. system, δ is called the stoke defined as 1 m²/s = 10^4 stokes. For liquids, δ varies with temp within a narrow range.

7.9.5 Viscosity measurement

Many viscosity measuring devices or viscometers are available. In any viscometer for measuring non-Newtonian fluid viscosity, the fluid is set with rotational motion and the parameters measured are torque, M and the angular velocity, Ω and these are used to calculate τ and $\dot{\gamma}$. Then τ and γ are used for the calculation of other viscosity parameters like, μ_a, K, n, τ_O.

Equations for different types of viscometers are available in literature. Three basic types of viscometers are normally used in bioprocess systems – (i) Cone-and-plate viscometer, (ii) Co-axial cylinder rotary viscometers, & (iii) Impeller viscometer.

The first two viscometers have number of problems for liquids containing solids e.g. solids setting, flocculation of deflocculating heterogeneous cell density near the rotating surface. So the third type i.e. impeller viscometer has been extensively used for the determination of viscosity.

7.9.5.1 Impeller viscometer

It consists of a small turbine impeller on a rotation shaft, which is used to shear the fluid sample. As the impeller rotates slowly in the fluid, the torque, M and the rotational speed, Ni are accurately measured. For turbine impeller under laminar flow conditions, the following equations are used.

$$\dot{\gamma} = kN_i \tag{7.113}$$

and

$$\tau = \frac{2HMk}{64D_I^3} \tag{7.114}$$

Where D_I is the impeller diameter, k is constant, for laminar flow and turbine impeller, $k \approx 10$ and Ni the number of rotations per second (rps).

7.9.6 Viscosity of suspensions

For dilute suspension of spherical particles, the effective viscosity of suspension is given by Vand equation,

$$\frac{\eta_{eff}}{\eta_{solvent}} = 1 + 2.5\phi + 7.20\phi^2 \tag{7.115}$$

where ϕ is the volume fraction of solids.

For higher volume fractions, solid suspension may behave as Bingham plastic fluids. The above equation has been applied to yeast and spore suspensions up to 14 vol% of solids. The cell concentrations strongly affect the apparent viscosity. A doubling in cell concentrating causes the apparent viscosity to increase by a factor of up to 90.

Similar results have been found for mould pellets in Antibiotic fermentations.

For dilute suspension, the viscosity of a suspension, η is given as

$$\eta = \eta_o \left(1 + k_E \phi\right) \tag{7.116}$$

where η_o is the viscosity of suspending liquid, K_E is called the Einstein Constant, ϕ is the volume fraction of the suspension. For spherical particle $K_E = 2.5$

Mooney's Equation

The viscosity of suspension is also given by the relation,

$$l_n \frac{\eta}{\eta_o} = \frac{k_E \phi}{1 - \dfrac{\phi}{\phi_m}} \tag{7.117}$$

where ϕ_m is the maximum packing volume fraction (ϕ_m = .065 for rod shaped particle, (ϕ_m = .074 for hexagonal close packing).

ϕ is the volume fraction of slid particle. k_E is a constant.

7.9.7 Viscosity of solutions containing macromolecules

For many biopolymers like X anthan gum, dextran,..the apparaent viscosity, μ_a is related by the equation,

$$\mu_a = A\left[P\right]^{2.5} \tag{7.118}$$

where P is the polymer concentration in gm/cm^3.

When the fermentation product is a polymer, continuous excretion in a batch culture raises the broth viscosity. In aurear bacterium pollulan, apparent viscosity measured as a shear rate of 1 sec^{-1} can reach as high as 24000 Cp.

Problem 7.8

The following rheological data were obtained for a blood sample:

$\dot{\gamma}$, Sec^{-1} Shear Rate	1.5	2.0	3.2	6.5	11.5	16.0	25	50	100
τ, Shear Stress, ...	12.5	16.0	25.2	40.0	62.0	80.5	120	140	475

Fit the data to the power law model of non-Newtonian fluid,

$\tau = \eta_o \left(\dot{\gamma}\right)^n$ Where η_o is the apparent viscosity and n is the flow index.

Solution:

$$\tau = \eta_o (\dot{\gamma})^n$$

Taking logarithm of both sides

$$l_n \tau = l_n \eta_o + n l_n \dot{\gamma}$$

$l_n \tau$ vs $l_n \dot{\gamma}$ are plotted and from the slope, η

$$n = \frac{5.1 - 4.0}{4.0 - 2.2} = 0.61$$

From the intercept $= l_n \eta_o = 2.0$

η_o (apparent viscosity) = 7.38 poise

7.10 Heat Transfer

Heat transfer takes place in many biological reaction systems. Some examples are:

1. Heat is required to sterilize a culture liquid at some desired temperature to kill some unactivated microorganisms or microbial spores in the total holding time.
2. An anaerobic sludge digester must be operated between 55 – 60°c and it requires heating.
3. The conversion of substrate generates excess heat for optimal reaction conditions to maintain viable cells and this excess heat must be removed.
4. The water content of a cell sludge is to be reduced by drying.

Heat Transfer equipments configurations may be a jacketed vessel, internal cooling or heating coils, circulation through heat exchanger, (SCP production using air lift fermenter), trickle bed reactor with phase change, natural temperature oscillation in lakes and rivers.

The fundamental steady state equation in convective heat transfer is related to the total rate of heat generation or heat removal through some heat transfer surface.

Thus net heat generation = removal of heat = $UA\Delta T$ (7.119)

Where ΔT = temperature difference between bioprocess fluid and heating or cooling fluid

A = heat transfer area, m^2

U = overall heat transfer coefficient, W/m^2°C

Let us consider a stirred bioreactor vessel where heat is generated from cell growth and maintenance, Qmet.

Now the cell growth is given as

$$\frac{dX}{dt} = \mu X \quad (7.120)$$

The corresponding instantaneous microbial heat generation rate, Qmet can be expressed as

$$Q_{met} = V_{reacter} \propto \times \frac{1}{Y_A} \quad (7.121)$$

Where Y_A is the heat generation coefficient (gm of cells/Kcal)

The corresponding equation for a continuous isothermal reactor at the steady state can be shown as

$$Y_A Q_{met} = V_{reacter} \mu X = (S_O - S) Y_{X/S} F - Xk_e \quad (7.122)$$

Or

$$\frac{Q_{met}}{V_{reacter}} = \left[(S_o - S) Y_{x/s} D - \frac{Xk_e}{V_{reacter}} \right] \frac{1}{Y_A} \quad (7.123)$$

Where D is the dilution rate (F/ V)

Problem 7.8

An aerobic fermentor of $100m^3$ with a cooking volume of $80m^3$ is used for microbial reactions using oxygen. The rate of oxygen consumption is 100 mole / (m^3) (hr). The desired operating temperature is 35°c. A cooling coil of I.D. 2.5cm is used to maintain the above temperature. The minimum allowable temperature difference between the cooling water and the broth is 5°c. Cooling water is available at 15°c.

Calculate the following:

(a) What is the required cooling water flow rate?

(b) Calculate the required length of cooling coil.

Data given: $C_{Pwater} = C_{Pbroth}$ = 4.18 kJ/Kg°c

Overall heat transfer coefficient, U = 1.420 KJ / sm^2 °C

The rate of metabolic heat evolution,

$Q_{GR} = 0.12\ Q_{o2}$ Kcal/hr

where Q2 is the rate of oxygen consumption in moles of O_2 / (m^3) (hr)

Solution: Temperature of the broth is 35°c which is constant. Since the minimum allowable temperature difference is 5°c, the outlet temperature of cooling wats is 30°c and the given initial temperature is 15°c.

(a) Arithmetic means temperature difference is Δtm

$$\Delta t_m = \frac{2(35) - (15 + 30)}{2} = 12.5°C$$

Now the total heat generated in the whole reactor is Q_t

$$Q_t = Q_{GR} \times V_R$$

$$= 0.12 \times 4.18 \frac{KJ}{mole} \times 100 \frac{mol}{(m^3)(hr)} \times 980 m^3$$

$$= 4013 KJ / hr$$

$$Q_t = m_w Cp_w \Delta T_m$$

$$4013 \frac{KJ}{hr} = m_w \frac{Kg}{hr} \cdot 4.18 \frac{KJ}{Kg °C} \cdot 12.5\ °C$$

Where, m_w (cooling water flow rate)

$$m_w = \frac{4013}{(4.18)(12.5)} = 76.8 kg / hr$$

(b) $Q_t = UA\Delta T = U(\pi DL)\Delta T$

$$U = 1.42 \, {KJ}/{m^2 s} \, °C = 1.42 \times 3600 = 5112 kJ / m^2 h\, °C$$

$$4013 = 5112(3.14 \times 2.5 \times 10^{-2} L)(12.5), \text{ solving for L}$$

$$\text{Or } L = \frac{4013 \, {kJ}/{hr} \, J}{5112 \times 3.14 \times 2.5 \times 10^2 \times 12.5} = 0.8 m$$

Where L is the length of the cooling coil.

7.10.1 Heat transfer correlations

The general equation of heat transfer design has already been given in the following form

$$Q_{net} = UA\Delta T,$$

For steady state heat transfer through a flat wall of thickness, L_W, separating the fermenter fluid at T_{bulk}, from heating or cooling fluid at T_{bulk2}, continuity of heat flux requires

$$h_{w1}\left(T_{bulk_1} - T_{wall_1}\right) = ks\left(\frac{T_{wall_1} - T_{wall_2}}{L_w}\right) - h_{w_2}\left(T_{wall_2} - T_{wall_1}\right), {w}/{m^2} \, °C \quad (7.124)$$

Where ks is the thermal conductivity of the wall in Watt / $m^{2o}c$. In terms of overall heat transfer coefficient, U

$$\frac{1}{U} = \frac{1}{h_{w_1}} + \frac{1}{k_s} + \frac{1}{h_{w_2}} \tag{7.125}$$

For heat transfer across a cylindrical tube wall in heating or cooling coils, the appropriate equation for U is,

$$\frac{1}{U_o d_o} = \frac{1}{h_o d_o} + \frac{l_n (do / di)}{2^{ks}} + \frac{1}{h_i d_i} \text{ (tube wall)} \tag{7.126}$$

Where di and do are the tube inside and outside diameters respectively.

U_o is the overall heat transfer coefficient based on outside diameter.

The individual heat transfer coefficient hw1 or hw2 or ho, hi are given in the above equation. Such individual heat transfer coefficients vary along the heat transfer surface, an overall local heat transfer coefficients is defined by the equation (7.124) and a detailed integration over the heat transfer area is required to calculate the total heat transfer. For fluid wall heat transfer the following dimension less groups are important:

Nusselt number, $$Nu = \frac{hd}{kf} \tag{7.127}$$

Prandtle number, $$Pr = \frac{C_p \mu f}{kf} \tag{7.128}$$

Brinkman number, $$Br = \frac{\mu u^2}{kf\left(T_{bulk} - T_{wall}\right)} \tag{7.129}$$

Froude no, $$Fr = \frac{U^2}{gD} \tag{7.130}$$

Reynolds number, $$Re = \frac{\rho u d}{\mu f} \tag{7.131}$$

An important correlation where viscosity at the bulk and the viscosity at the wall are significantly different,

$$Nu = f\left(Pr, Re, \frac{L}{d}, \mu b / \mu w\right) \tag{7.132}$$

For fluids of viscosity near that of water, a useful correlation for turbulent flow for cooling or heating:

$$Nu = \frac{hd}{k} = 0.023 Re^{0.8} Pr^{0.4} \text{ (Dittus-Boltus Equation)} \tag{7.133}$$

$10^4 \leq Re \leq 1.2 \times 10^3$

$0.7 \leq Pr \leq 120$

$L/D \geq 60$ (long tubes)

With viscosity correction, Seider-Tate modified the above equation:

$$Nu = \frac{hd}{k} = .027 Re^{0.8} Pr^{0.33} \left(\frac{\mu_b}{\mu_w}\right)^{0.14} \tag{7.134}$$

When natural convection is important due to the presence of non uniform fluid density, the Grashof number, (Gr) is used in the correlation :

$$Gr = \frac{gd^3 \Delta\rho \, \rho_{av}}{\mu^2} \tag{7.135}$$

And the correlation for the liquid flowing in horizontal tubes

$$Nu = 1.75\left[\frac{d}{L} \Pr Re + 0.04\left(\frac{d}{L} Gr\, Pr\right)^{0.7}\right]^{1/3} \left(\frac{\mu_s}{\mu_w}\right)^{0.14} \tag{7.136}$$

For vertical tubes, $\mu_b / \mu_w \approx 1.0$ and the constant 0.14 is replaced by 0.0723.

For non-Newtonian fluids, the correlations are changed. For pseudoplastic fluid, correlations have been given as

$$Nu = \frac{hd}{k} = 1.75\left(\frac{d}{L} RePr\right)^{1/3} \left(\frac{2n-1}{4n}\right)^{1/3} \tag{7.137}$$

where n is the flow behavior index.

7.10.2 Heat transfer in stirred liquids

The heat transfer coefficient in stirred vessel depends on the degree of agitation and the fluid properties. Heat transferred to and from a helical coil, h can be determined from the following relation

$$Nu = 0.87 R_{eI}^{0.62} Pr^{0.33} \left(\frac{\mu_p}{\mu w} \right)^{0.14}$$

Where R_{eI} (Impeller based Reynolds no.), given by

$$R_{eI} = \frac{\rho N D_I^2}{\mu}$$

and D in the Nu is the inside the diameter of the tank.

For heat transfer to or from a jacket,

$$Nu = 0.36 R_{eI}^{0.67} Pr^{0.33} \left(\frac{\mu b}{\mu w} \right)^{0.14} \quad (7.138)$$

Problem 7.9

A fermenter used for antibiotic production should be maintained at 28°c. Considering the oxygen demand or organism and heat dissipation from the stirrer, maximum heat transfer required is approximately 600 KW. Cooling water is available at 10°c and the exit temperature of the cooling water is estimated to be 25°c. The heat transfer coefficient of the cooling water is 14000 wm^{-2} oc^{-1}. An average fouling factor is 8500 wm^{-2} oc^{-1}. It is planned to install a helical cooling coil inside the fermenter whose dimensions are: do = 8 cm, pipe thickness, Δz = 5mm, thermal conductivity of steel is 60 w/m°c

1. Calculate heat transfer coefficient for the fermentation broth. Use the following equation:

$$Nu = 0.87 R_{eI}^{0.62} Pr^{0.33} \left(\frac{\mu b}{\mu w} \right)^{0.14}$$

Impeller diameter, D_I = 1.8m, D_t = 5 m (reactor diameter)

$$\rho = 1000 Kg/m^3, \; C_p = 4.2 kJ/Ks°C, \; kf = 0.70 \frac{w}{m} °C, \; \mu = 5 \times 10^{-3} kgS/(m)(s)$$

2. Calculate the overall heat transfer coefficient
3. Calculate the length of the cooling coil.

$$\text{Solution : (a) } R_{eI} = \frac{\rho N D_I^2}{\propto} = \frac{(1000)\left(\frac{60}{60}\right)(1.8)^2}{5 \times 10^{-3}} = 6.48 \times 10^5$$

$$P_r = \frac{C_p \mu}{k_f} = \frac{4.2\times10^3\ J/kg \times 5\times10^{-3}\ kg/(m)(s)}{0.7\,J/(s),(m)/(^\circ c)} = 30$$

$$Nu = \frac{hDt}{k_f} = 0.87 R_{eI}^{0.62} P_r^{0.33}\left(\frac{\mu s}{\mu w}\right)^{0.14}$$

$$Nu = 0.87\left(6.48\times10^5\right)^{0.62}\left(30\right)^{0.33} = 10718 \text{ (neglecting viscosity effects)}$$

$$h = \frac{(10718)(0.7)}{5} = 1500 w/m^2\,^\circ C$$

Over all U is calculated as

$$\frac{1}{U} = \frac{1}{h_a'} + \frac{\Delta Z}{k_{stu}} + \frac{1}{h_c} + \frac{1}{h_d}, \text{ where hd is the fouling factor}$$

$$= \frac{1}{1500} + \frac{5\times10^{-3}}{60} + \frac{1}{14000} + \frac{1}{8500} = 6.67\times10^{-4} + 0.833\times10^{-4} + 0.714\times10^{-4}$$

$$+1.1764\times10^{-4} = 9.39\times10^{-4} m^2 C/w$$

$$U = 1064\, w/m^2 C$$

Now ΔT is calculated as the arithmetic temperature difference. When one fluid in HE system, remains at a constant temperature such as in Fermenter, the arithmetic temperature difference ΔT is given as

$$\Delta T = \frac{2T_F - (T_{in} + T_{art})}{2} = \frac{2(28) - (10 + 25)}{2} = 10.5^\circ c, \text{ where } T_f \text{ is the feed temperature.}$$

$$q = UA\Delta T$$

$$A = \frac{q}{U\Delta T} = \frac{600\times10^3 W}{(1064)(10.5)} = 53.7\, m^2$$

$$A = 2\pi RL$$

$$\text{So } L = \frac{A}{2\pi R} = \frac{53.7}{2\times3.14\times4\times10^{-2}} = 213.77\, m$$

7.11 Scale-up

For the design of a large scale fermentation unit, it is necessary to translate the data of a small scale (model) to the large scale. The fundamental requirement for scale up is that the model and the prototype should be similar to each other.

Two approaches may be considered for scale up – A) Similarity criterion, B) Scale up criteria and other methods.

The similarity principle may be of two types:

1. Geometric similarity of the physical boundaries
2. Dynamic similarity of flow fields

In the case of geometric similarity, the model and the prototype must have the same shape and all the linear dimensions of the model must be related to the corresponding dimensions of the prototype by a constant scale factor.

In the dynamic similarity of the flow fields, the ratio of flow velocities of corresponding fluid particles is the same in the model and prototype as well as the ratio of all forces, acting on the corresponding fluid particles is the same in the model and prototype. To achieve dynamic similarity, the three dimension less numbers for the prototype and the model must be equal, as given below:

$$\left(\frac{P}{\rho N^3 D_I^5}\right)_P = \left(\frac{p}{\rho N^3 D_I^5}\right)_m \tag{7.139}$$

$$\left(\frac{\rho N D_I^2}{\mu}\right)_P = \left(\frac{\rho N D_I^2}{\mu}\right)_m \tag{7.140}$$

$$\left(\frac{N^2 DI}{g}\right)_P = \left(\frac{N_2^2 DI}{g}\right)_m \tag{7.141}$$

Where Power number, $P_{NO} = \dfrac{P}{\rho N^3 D_I^5}$,

Impeller Reynolds number, $R_{eI} = \dfrac{\rho N D_I^2}{\mu}$

Froude Number, $F_r = \dfrac{N^2 D_I}{g}$

If the fluid properties are the same for the model and the prototype, we have

$$(P)_P = (P)_m \left(\frac{N_P}{Nm}\right)^3 \left[\frac{(D_I)P}{(DI)_m}\right]^5 \tag{7.142}$$

If $(P)_P = (P)_m$

$$\frac{N_P}{N_m} = \left[\frac{(D_I)_m}{(D_I)_P}\right]^{5/2} \tag{7.143}$$

where N_p and Nm are the rpm of the prototype and that of model respectives.

For the equality of Reynolds number,

$$N_P = \left[\frac{(D_I)_m}{(D_I)_P}\right]^2 Nm \tag{7.144}$$

For the equality of Froude number

$$N_P = \left[\frac{(D_I)_M}{(D_I)_P}\right]^{1/2} Nm \tag{7.145}$$

So for the same fluid, we will get different results for the scale-up for the above models. However, there are many other methods of scale-up:

1. Fundamental method of making mass and momentum balance
2. Scale-up criteria
3. Dimensional Analysis
4. Time constants for conversion and transport processes.
5. Mixing time and Reynolds number
6. Maintenance of a constant substrate or product level (usually dissolved oxygen concentration).

The fundamental method involves setting of mass and momentum balances in the form of coupled differential equation and numerically solving them with given boundary condition. The method is complex and time consuming and requires machine computation.

7.11.1 Scale-up methods

The following scale-up methods are commonly used:

(i) Constant power input (P/V)

It involves constant oxygen transfer rate i.e. $k_L a$ is constant. It is to be noted that

$P \propto N^3 D_I^5$ and $V \propto D_I^3$

$Q \propto N D_I^3$, So $P/V \propto N^3 D_I^2$

So

$$N_P = N_m \left[\frac{(D_I)_m}{(D_I)_P}\right]^{2/3} \tag{7.146}$$

where p and m refers to prototype and model respectively.

(ii) Constant liquid circulation rate inside the vessel (pumping rate of impeller per unit volume)

$$Q/V\left(\frac{ND_I^3}{V}\right)_P = \left(\frac{ND_I^3}{V}\right)_m \tag{7.147}$$

If $V \propto D_I^3$

$$N_P = N_m\left(\frac{V_P}{V_m}\right)\left[\frac{(D_I)_m}{(D_I)_P}\right]^{2/3}, \tag{7.148}$$

(iii) Constant shear at impeller tip speed (ND_I)

$$N_P = N_m\frac{(D_I)_m}{(D_I)_P} \tag{7.149}$$

(iv) Constant Impeller Reynolds number, ReI

$$\left(\frac{\rho ND_I^2}{\mu}\right)_P = \left(\frac{\rho ND_I^2}{\mu}\right)_m \tag{7.150}$$

For the constant ρ and μ

$$N_P = N_m\left[\frac{(D_I)_m}{(D_I)_P}\right]^2 \tag{7.151}$$

(v) Constant mixing time, t_m:
Mixing time, t_m, is defined as the time it takes for the concentration of a compound to return to 95% of equilibrium value after a local perturbation in its concentration
Mixing times are experimentally determined by the step input of an electrolyte.
The mixing time t_m has been correlated empirically to reactor volume V, according to an expression,

$$t_m = T_b \cdot V^{0.3} \tag{7.152}$$

Another correlation for t_m is

$$t_m = 4V/1.5ND_I^3 \tag{7.153}$$

Where t_m is the mixing time and V is the reactor-vessel, N revolution per second, T_b is a constant which depends on impeller type and placement, vessel design, D_I is the impeller diameter.

(vi) DIMENSIONAL ANALYSIS
It is well known that power consumption by agitation is a function of physical properties, operating conditions, vessel and impeller geometry.
Dimensional analysis provides the following relationship

$$\frac{P}{\rho N^3 D_I^5} = f\left(\frac{\rho N D_I^2}{\mu}, \frac{N^2 D_I}{g}, \frac{D_T}{D_I}, \frac{H}{D_I}, \frac{D_w}{D_I}\right) \tag{7.154}$$

Where Power No. $\frac{P}{\rho N_i^3, D_i^5}$

Reynolds number based on impeller diameter, $D_I, R_{eI} = \frac{\rho N D_I^2}{\mu}$

Froude number, $F_r = \frac{N^2 D_I}{g}$

$\frac{D_T}{D_I}, \frac{H}{D_I}, \frac{D_w}{D_I}$ are the length ratios where D_T is the vessel diameter, H = the height of the reactor, D_I = impeller diameter, D_W = width of the impeller.

(vii) CHARACTERISTIC TIME CONSTANTS

The scale-up problems are related to transport processes. The system where the micro-kinetics control the system, may show transport controlled, when it is scaled up. In this situation, the above scale-up rules fail. So one approach is the use of characteristic time constants for conversion and transport processes. Some of the important time constants for different transport and conversion are shown in the table: 7.2.

Table 7.2 Some time constants with equations

Transport Process	Equation
Flow	V / Q
Diffusion	L^2 / D
Oxygen transfer	$1/k_L a$
Mixing	$t_m = \frac{4V}{1.5ND^3}$

Conversion process:

Growth	$1/\mu$
Chemical reaction	$1/k_t$ for n = 1, C_{A_0}/k_2 for n = 2
Substrate consumption	$C_s / r_{max}, C_5 \gg k_s$ $K_s / r_{max}, C_S \ll k_s$

Processes with small time constants compared to the main processes are assumed to be in equilibrium. For example, $1/k_{L_a} \ll t_{o_2}$ where t_{o_2} is the time constant for O_2 consumption, thus the broth would be statured with oxygen, because oxygen transfer is more rapid than conversion. If $1/k_{L_a} \approx t_{o_2}$, the dissolved oxygen concentration may be very low. D.O. measurement shows that dissolved oxygen concentration may pass through zero value. Many cells have regulatory circuits to respond to the changes from aerobic to anaerobic conditions, leading to constantly altering cellular metabolisms.

LIMITATIONS:

In practice, scale-up is highly empirical but produces sensible results if there is no change in the controlling regime during scale-up. Thumb rules for scale up are i) constant power to volume ratio (P/V), ii) constant kLa, iii) constant tip speed of the impeller, iv) combination of mixing time and Impeller Reynold number, v) constant substrate and product concentration. Each of these rules has both successful and unsuccessful cases. Newtonian fluids gave better results than Non-Newtonian liquids.

Problem 7.10

A stirred tank reactor is to be scaled down from 10m^3 to 0.1m^3. The dimensions of the large tank are Dt = 2m, D_I = 0.5m, N = 100 rpm

(a) Determine the dimensions of the small tank, (Dt, D_I& H) by the principle of geometric similarity

(b) What would be the rotational speed of the impeller in the small tank if the following criteria are used:

1. Constant tip speed
2. Constant impeller R_{eI} number
3. Constant P/V

Solution:

(a) For large tank: data given are :

Dt = 2m, D_I = 0.5m

$$V = \frac{3}{4} D_t^2 H \text{ Or } H = \frac{10\left(m^3\right)}{0.7854\left(2\right)^2 m^2} = 3.18m$$

$$\text{Scale down factor} = \left(\frac{0.1m^3}{10m^3}\right)^{1/3} = 0.2154$$

Dimension of the small tank Dt = 2 x 0.2154 = 0.43m, D_I = 0.5 x 0.2154 = 0.1077m, H = 3.18 x 0.2154 = 0.685 m

(b) Impeller speed of the small tank

1. Constant tip speed.

$$N_2 D_{I_2} = N_1 D_{I_1}$$

$$N_2 = N_1\left(\frac{D_{I_1}}{D_{I_2}}\right) = 100\left(\frac{0.50}{0.1077}\right) = 464 \text{ rpm}$$

2 refers to the small reactor, I refers to large reactor

2. Constant R_{eI}

$$N_2 = N_1\left(\frac{D_{I_1}}{D_{I_2}}\right)^2 = 100\left(\frac{0.50}{0.1077}\right) = 2155 \text{ rpm}$$

3. Constant P/V

$$N_2 = N_1\left(\frac{D_{I_1}}{D_{I_2}}\right)^{2/3} = 100\left(\frac{.5}{.1077}\right)^{2/3} = 278 \text{ rpm}$$

Problem 7.11

The optimum agitation speed for the cultivation of plant cells in a 0.5m^3 fermenter equipped with four baffles was found to be 150 rpm

(a) What should be the impeller speed of a geometrically similar 500m^3 fermenter if the scale up is based on the same power consumption per unit volume

(b) What should be the rpm for the larger fermenter if the scale-up is based on constant impeller tip speed?

(c) If we use the Impeller based Reynolds number is the scale up criterion, what is the rotational speed of the impeller?

Solution:

(a) For the smaller reactor the dimension are:

$$V = \pi/4\, D_t^2 \cdot D_t \text{ Or } D_t = \left(\frac{0.5}{0.7854}\right)^{1/3}$$

$$\text{Or } D_{t_1} = 0.86m \quad D_{I_1} = 0.33 \times .86 = 0.28m$$

$$\text{The scale up factor} = \left(\frac{50}{.50}\right)^{1/3} = 4.64$$

For large fermenter

$$D_{t_2} = 0.86 \times 4.64 = 3.99m$$

$$D_{I_2} = 0.33(3.99m) = 1.3167m$$

(a) For constant P/V

$$N_2 = N_1\left(\frac{D_{I_1}}{D_{I_2}}\right)^{2/3} = 100\left(\frac{0.28}{1.3167}\right)^{2/3} = 35.3$$

(b) $N_2 = 100\left(\frac{D_{I_1}}{D_{I_2}}\right) = 21.2$, for constant impeller speed

(c) $N_2 = 100\left(\frac{D_{I_1}}{D_{I_2}}\right)^2 = 4.49$, for equal impeller based Reynolds number

Problem 7.12

It is proposed to model a prototype process involving a long tube in which cells are cultured. Considering the nature of the process, the following parameters are thought to be relevant:

C_o = inlet concentration, kg/m³, C_L = concentration at the out of the reactor kg/m³, D_L = dispersion coefficient, $\frac{m^2}{s}$, u_L = Average velocity of the liquid inside the tube, m/s, L = length of the tube(m). It is assumed that the outlet concentration, C_L is the functions of Co, D, u, L.

By dimensional analysis, show how those variables are related in terms of dimensionless groups which can be used in the scaling process.

Solution:

It is assumed that

$C_L = f(Co, D, u, L)$

Applying Buckingham π theorem and using Rayleigh's method, we proceed as follows:

Number of variables = 5

Number of Fundamental units = 3 (M, L, T)

According to the Buckingham π theorem

Number of dimension less groups = 5 − 3 = 2

Rayleigh's method

$C_L = f(Co, D, u, L)$

Putting the units of each variable

$$ML^{-3} = (ML^{-3})^{\alpha} (L^2T^{-1})^{\beta} (LT^{-1})^{\gamma} (L)^{\delta}$$

Collecting coefficients of both sides for M, L, T

M : 1 = $\propto$ ....(1)

L: -3 = $-3\propto + 2\beta + \gamma + \delta$ .... (2)

T : 0 = $-\beta - \gamma$, $\beta = -\gamma$ .... (3)

From equations (1) & (3)

$$-3 = -3 - 2\gamma + \gamma + \delta$$

Or $\gamma = \delta$

Now, $\propto = 1, \beta = -y, \gamma = \gamma, \delta = \gamma$

$$C_L = \left(Co\,D^{-\gamma} u^{\gamma} L^{\gamma}\right)$$

$$\text{Or } \frac{C_L}{C_O} = \left(\frac{uL}{D}\right)^y$$

$$\text{So, } \frac{C_L}{C_O} = f\left(\frac{uL}{D}\right)$$

Where C_L/C_O = dimensionless concentration

uL/D = Peclet number.

7.12 Design of Bubble Column Reactor

Bubble columns are long cylindrical vessels with height to diameter ratio (H/D) >2. Mixing and aeration are carried out by gas sparing from the bottom. They require less energy than mechanical stirring. Other advantages are low capital cost, simple mechanical configuration, reduced operating costs based on lower energy requirement. Large gas sparging may cause foams which can be removed by mechanical means or by adding antifoam to the medium.

Bubble columns have been used for beer production, vinegar manufacture, for cultivation of microorganisms (single cell protein SCP) for use as animal feed.

Bubble column hydrodynamics and mass transfer characteristics depend on the nature of bubbles produced from the sparger. Homogeneous flow occurs at low gas flow rates. In homogeneous flow, bubble rises as plug flow with no back mixing of the gas phase. Liquid mixing in this flow regime has low intensity.

At higher gas velocities for lifting of the cells from the bottom to the upward regime, heterogeneous flow occurs when bubbles and liquid rise up to the centre of the column and there also occurs down-flow of some liquid near the walls. Liquid circulation entrains bubbles causing some backmixing. For homogenous flow, the following equation is used for upward liquid velocity at the centre of the column.

For 0.1m < D < 7.5 m and u_G < 0.4 m/s

$$u_G / u_L = 0.9\left(gDu_G\right)^{0.33} \tag{7.155}$$

where u_L is the linear liquid velocity, g is the gravitational constant, D is the column diameter, u_G is the gas superficial velocity (volumetric gas flow rate at atmospheric press and 25°c temp divided by the cross section of the reactor).

An equation for mixing time, t_m has also been given as

$$t_m = 11.0\frac{H}{D}\left(gu_G D^{-2}\right)^{-0.33} \tag{7.156}$$

where H is the height of the column.

The volumetric mass transfer coefficient for bubble column has been obtained as

For water and coalescing bubbles, non-viscous and heterogeneous flow.

$$k_{La} = 0.32\left(u_{gs}\right)^{0.7} sec^{-1} \tag{7.157}$$

where $u_g s$ is m/s

For gas transfer in bubble columns, Akita and Yoshida (1974) gave the following correlation for k_La:

$$\frac{k_L a d_t^2}{D} = 0.6\left(Sc\right)^{1/2} B_O^{0.62} G_a^{.31} H^{1.1} \tag{7.158}$$

Where Bond Number, $B_O = \dfrac{gd_t^2 \rho_c}{\sigma}$

Galileo number, $G_a = \frac{gd_t^3}{u_L^2}$, dt is the tower diameter, D is the bubble diameter, H is Gas holdup and dt < 60 cm.

The equation (7.158) is valid for bubbles with mean diameter about 6mm, 0.08m <dt< 11.6 m, 0.3 < L < 21m, 0 < u_G < 0.3 m/s.

If smaller bubbles are formed at the sparger and the medium is not coalescing, k_{La} will be larger than the value calculated by equation (7.158). In order to maintain interfacial area, a constant, the gas should remain in bubbling flow regime. Experimentally it has been found that the gas bubbles rising through the liquid will coalesce into slugs, if the gas hold-up (H) remains less than Hmax, which is roughly 0.3. The requirement that the gas holdup remains less than 0.3, the following design specification for the column diameter should be

$$F_G = u_G H \frac{\pi d_t^2}{4} \quad (7.159)$$

Where F_G and u_G are the gas volumetric flow rate and linear gas velocity respectively. It is reasonably assumed that u_G is the terminal velocity and V_g is the terminal velocity of a single gas bubble in a stagnant liquid. From the above equations (7.158) where H is less than 0.3, this condition will be satisfied if

$$d_t \geq 2\left(\frac{F_G}{u_G H_{max}}\right)^{1/2} \quad (7.160)$$

If the bubble growth is reduced due to hydrostatic head, u_t may decrease and in the extreme case a transition to the formation of cup-shaped bubbles will occur. Under these conditions, u_t or u_b is given as

$$u_b = 0.711\sqrt{gD_{32}}\ cm/sec \quad (7.161)$$

7.12.1 Plug flow model for bubble column

For the simplest modelling of bubble column reactor, it is assumed that bubbles rise in plug flow condition through the vessel, but maintain a perfect mixing in the liquid phase and C_L^*, the saturated dissolved oxygen concentration, will vary with position.

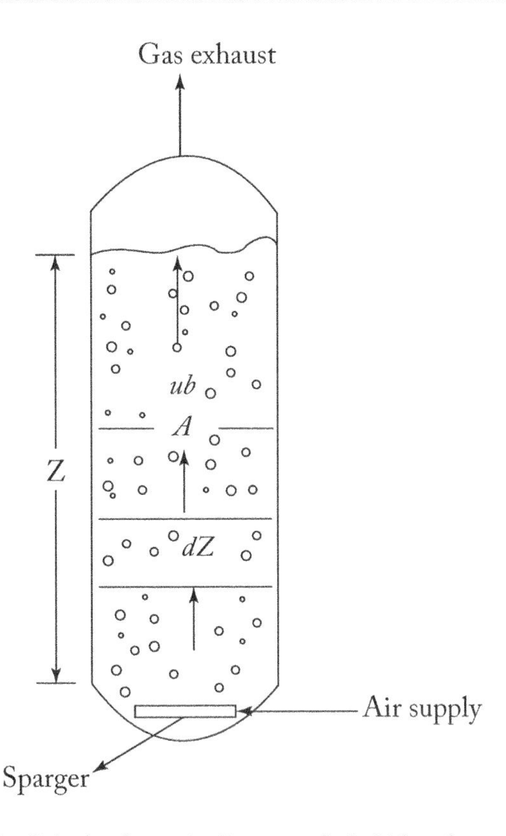

Fig. 7.9 A schematic diagram of a bubble column

Over a differential reactor height, dZ, the instantaneous loss of oxygen from the bubble can be expressed as

$$HA\,dZ\frac{dP_{O_2}}{dt}\cdot\frac{1}{RT} = -k_{La}\left(C_L^* - C_L\right)AdZ \tag{7.162}$$

Now replacing p_{o2} by Henry's law, $p_{o_2} = MC_L^*$, where M is the Hendry's constant, the equation (7.161) can be written as

$$\frac{H}{RT}\frac{dP_{O_2}}{dt} = \frac{HM}{RT}\frac{dC_L^*}{dt} = -k_{La}\left(C_L^* - C_L\right) \tag{7.163}$$

For a constant bubble rise velocity, u_b, $t = \frac{dZ}{u_b}$, the equation (7.162) reduces to

$$\frac{HM}{RT}u_b\frac{dC_L^*}{dz} = -k_{La}\left(C_L^* - C_L\right) \tag{7.163a}$$

Integrating from the bottom to the top,

z = Z, $\left(C_L^* - C_L\right)$ top, at the bottom, z = 0, $\left(C_L^* - C_L\right)$ bottom, we have,

$$\int \frac{dC_L^*}{C_L^* - C_L} = ln \frac{\left(C_L^* - C_L\right)_Z}{\left(C_L^* - C_L\right)_{bottom}} = -\frac{k_{La} RT}{HMu_b} Z \qquad (7.164)$$

The equation (7.164) gives the D.O. distribution along with reactor length, independent of biomass, X.

Problem 7.13

A bubble column reactor is used to cultivate E.coli whose kinetic parameters are:

$$\mu = 0.4 hr^{-1}, {}^{Y_X}\!/_{O_2} = 1.76 \frac{kg\ dry\ cell}{kg\ of\ O_2}$$

Air is introduced at the bottom of the column at the rate of 0.120 m^3/hr at 25°c and 1 atm pressure. The column diameter is 0.05m.

Hendry's constant $M = 793 \frac{m^3 - bar}{k \cdot mole}$

Bubble diameter, D_{32} = 3 x 10^{-3}m (3mm)

(a) Calculate the Gas-holdup, H

(b) Calculate k_{La}

(c) If the ratio of $\left(C_L^* - C_L\right)_Z / \left(C_L^* - C_L\right)_{bol} = \frac{1}{10}$, calculate the column height, Z

(d) If the molal flow rate of air, n_T is constant, mole fraction of O_2 in the inlet, y_{in} = 0.20 and that the outlet, y_{out} = 0.004, calculate the average cell mass concentration X (kg / m^3) and the total cell mass in the reactor.

Solution: (a) D_{32} = 0.003m, the bubble rise velocity,

$$V_t = 0.711(gD_{32})^{1/2} = .711\sqrt{(9.8)\left(3\times10^{-3}\right)} = 0.1219\, m / sec$$

Reactor diameter = 0.05m

Cross section of the reactor

$$= {}^{\pi}\!/_4 (.05)^2 = 1.9635\times10^{-3} m^2$$

ugs (superficial gas velocity)

$$= \frac{0.12 m^3 / hr}{1.9635\times10^{-3} m^2} = 61.1 m / hr = 0.01697 {}^{m}\!/_{sec}$$

$$H \text{ (Gas hold-up)} = \frac{u_g s}{u_g s + v_t} = \frac{0.01697}{0.01697 + 0.1219} = 0.122$$

(b) $k_L a = 0.32\left(u_g s\right)^{0.7} = 0.32\left(.01697\right)^{0.7} Sec^{-1}$

$= 0.043\, Sec^{-1} = 156.3\, hr^{-1}$

(c) We have the design equation of a bubble column reaction on the basis of upward plus flow of bubbles

$$-ln \frac{\left(C_L^* - C_L\right)_Z}{\left(C_L^* - C_L\right)_{bolton}} = \left(\frac{k_L aRT}{HMV_t}\right) Z$$

$$R = 0.0831 \frac{m^3 bar}{\left(k_{mole}\right)\left(K\right)}$$

$$T = 25°\text{C} = 298K$$

$$M\left(Henry's\, Const\right) = 793 \frac{m^3 - bar}{k_{mole}}$$

$$-ln\frac{1}{10} = \frac{(.043)(.0831)(298)}{(0.122)(793)(.1219)} Z$$

$$Z = \frac{2.302 \times 0.122 \times 793 \times 0.1219}{(.043)(.0831)(298)} = \frac{27.148}{1.0648} = 25.5\, m$$

(d) Volume of the reactor

$$= \pi/4\, D_t^2 Z = 0.7854(.05)^2 (25.5) = 0.05\, m^3$$

The oxygen balance for constar molal flow ratio of air is

$$\eta_T\left(y_{in} - y_{out}\right) = \frac{\mu \times V_L}{Y_X / O_2}$$

Flow rate of air at 0°c

$$= 0.12 \frac{m^3}{hr} \times \frac{273}{298} = 1.1 m^3 / hr$$

$$\eta_T = \frac{1.1 m^3 / hr}{22.4 m^3 / k_{mole}} = 0.049 kmoles / hr$$

$$Y_{X/O_2} = 1.76 \frac{kg\ dry\ wt\ cells}{kg\ of\ O_2} = 56.32 \frac{kg\ dry\ wt \cdot cells}{kg\ mole\ O_2}$$

$$\eta_T \left(y_{in} - y_{on} \right) = \frac{\mu X V_L}{Y_{X/O_2}}$$

$$\text{Or } X = \frac{0.049 \frac{K_{mole}}{hr} \left(0.2 - .004 \right) \times 56.32 \frac{kg\ dr.w\ cell}{K.mole}}{0.4 \frac{1}{hr} \times 0.05 m^3} = 27.04 \text{ kg/m}^3$$

$$\text{Total cell mass} = 27.04 \times 0.5 = 1.352\ kg$$

7.12.2 Decker's dispersion model for bubble column

To develop the above model, the following assumptions have been considered.

(i) The superficial gas velocity is selected to maintain the bubbles in two phase flow regime.

(ii) The pressure gradient due to liquid height should be considered

(iii) Axial dispersion is present

(iv) The effect of segregation may be important due to the presence of small and large bubbles having different residence times and interfacial areas.

(v) The gas phase molar flow rates may change with position due to gas absorption.

Decker (1985) has presented mathematical models, taking into account of axial dispersion and the pressure gradient. He correlated axial dispersion, D_L by the following equation,

$$D_L = 2.7 dt^{1.4} u_G^{0.3} cm^2 / sec \tag{7.165}$$

The pressure at the bottom of the column is more than 30% larger than the pressure at the top above the liquid column. For the design methodology, four independent variables maybe optimally selected, based on economic considerations. These are: specific growth rate,.Biomass concentration (X), volumetric mass transfer coefficient, k_{La}, and the dispersion coefficient, D_L. If the gas phase segregation is important, some bubbles will have a lower mole fraction of oxygen than the value predicted by complete mixing model. Assuming liquid phase superficial velocity is nearly zero, the dissolved oxygen balance may be given as

$$D_L \frac{d^2C_O}{dx^2} + k_L a'\left(C_O^* - C_O\right) - \frac{\mu X}{Y_{X/O_2}} = 0 \qquad (7.166)$$

Where C_O is the dissolved oxygen concentration, Co* is the saturated concentration given as

$$C_O^* = \frac{P_T y}{H_O}\left[1 + \propto\left(1 + Z\right)\right]$$

$$\propto = \rho_L g\left(1 - \epsilon_G\right)L / P_T$$

where x is the height above the bottom of the column, P_T is the pressure at the top of the column $\propto$ is the ratio of the liquid head to the pressure at the top of the column, X = cell mass concentration

H_o is the Henry's Constant

ρ_L is the liquid density, the boundary conditions are:

$$\frac{dC_O}{dx} = 0 \text{ at x = 0 and x = L} \qquad (7.167)$$

Now introducing the dimension less parameters,

$\overline{x} = x / L, \overline{C} = C_o / C_{oT}^*$ and $C_{oT}^* = P_T y / H_o$, is the liquid phase concentration of oxygen in equation with the gas phase at the top of the column,

$$B_1 = \frac{D_L}{L^2 k_L a'} \qquad (7.168)$$

$$B_2 = \frac{\mu X}{Y_{X/O_2} k_L a' C_T^*} \qquad (7.169)$$

The equation (7.166) can be expressed in terms of dimension less variables,

$$B_1 \frac{d^2\overline{C}}{d\overline{x}^2} - \overline{C} = B_2 - 1 - \propto + \propto \overline{x} \qquad (7.170)$$

which has the solution as,

$$\overline{C} = A_1 e^{-M_1\overline{x}} + A_2 e^{M_1\overline{x}} + \left(1 + b - B_2\right) - \propto \overline{x} \qquad (7.171)$$

where,

$$A_1 = \frac{\propto\left(e^{M_1} - 1\right)}{M_1\left(\overline{e}^{M_1} - e^{M_1}\right)} \qquad (7.172)$$

$$A_2 = \frac{\propto\left(e^{-M_1} - 1\right)}{M_1\left(\overline{e}^{M_1} - e^{M_1}\right)} \tag{7.173}$$

$$M_1 = \frac{\sqrt{L^2 k_L a'}}{D_L} = \frac{1}{\sqrt{B_1}} \tag{7.174}$$

$$B_1 = D_L / L^2 k_L a' \tag{7.174a}$$

$$B_2 = \mu X / \left(Y_X/_{O_2} \, k'_{L_a} \, C_T^* \right) \tag{7.174b}$$

Dividing the equation (7.170) by α , we get

$$\frac{\overline{c}}{\propto} - \frac{(1 - B_2)}{\propto} = \frac{A_1}{\propto} e^{-M_1 \overline{x}} + \frac{A_2}{\propto} e^{M_1 \overline{x}} \tag{7.175}$$

For all values of α and B_2 in which $\overline{e} > 0$ for $\overline{x}$=1.0

With the constant cell mass growth (μX), the oxygen balance for a constant molar flow rate, n_T is

$$n_T\left(y_{in} - y\right) = \mu X V_L / Y_{X10} \tag{7.176}$$

Where y_{in} is the inlet molefraction of dissolved oxygen and y is the outlet value.

Column diameter, dt is given by the following equation,

$$P_T u_g \pi d_t^2 = 4 n_T RT \tag{7.177}$$

The equation (7.175) may be graphically presented showing the effect of axial dispersion and pressure gradient on $\overline{c}_L$.

7.13 Design of Air-Lift Fermenters

An airlift fermenter with an inner draft consists of three sections: 1) riser, 2) downcomer and 3) the headspace. The liquid and air bubbles go up through the riser. The gas is separated at the headspace at the top of the column prior to entering the down flow sections (the downcomer, the annular section). The actual liquid velocity in the downcomer must be larger than the rise velocity of small bubbles and smaller than the rise velocity of large bubbles. Liquid velocity should be around 10-30 cm/sec. However the liquid velocity should increase with gas superficial velocity and they are correlated by the equation,

$$\frac{u_G}{\epsilon_G} = 1.03 u_M + V_D \tag{7.178}$$

where V_D = 33 cm/sec is approximately the gas velocity relative to the superficial velocity of the mixture (u_M).,ug is the gas velocity, €g is the gas hold-up

The axial dispersion coefficient D_L in air-lift upflow section has been measured and found to be constant at D_L = 58 cm^2/sec for low superficial gas velocities. The liquid phase may be modelled using a plug flow model for both upflow and downflow sections. A complete mixing model has been recommended for the headspace.

The design methodology for the airlift fermenter is analogous to bubble column described in the previous section.

The fraction of oxygen consumed, the feed gas superficial velocity in the draft tube, the ratio of upflow and downflow areas may be considered as independent variables together with μ and X

A trial and error calculation is required to find the operating pressure in an acceptable dissolved oxygen concentration.

Volumetric mass transfer coefficient, k_{La} for airlift fermenter has been given by the following equation (Bello etal, 1981):

$$k_{La} = \frac{0.0005\left(P/V\right)^{0.8}}{\left(1 + Ad/Ar\right)} Sec^{-1} \tag{7.179}$$

where Ad / Ar = ratio of areas of downcomer and riser sections and (P/V) is the power input per volume (W/m^3).

For gas sparging into a column, the power used in compression to sparged gas with volumetric flow rate, Fg at pressure, P_1 is given as,

$$\rho_g = \rho_g F_g \left[\frac{RT}{(MW)} ln \frac{P_1}{P_2} + 0.06 \frac{u_o^2}{2} \right] \tag{7.180}$$

Where u_0 is the gas velocity at the sparger orifice

Problems (Exercise)

7.1 Estimate the mass transfer coefficient for oxygen dissolution in water at 25°c in a mixing vessel equipped with flat blade disk turbine and sparger. Assume average bubble size is 3 mm. Density of water at 25°c ρ = 1000 kg/m^3, density of air at 25°c ρ_g = 1.18 kg/m^3, viscosity of water = 10^{-3} Pas, Do_2/H_2O = 2.5 x 10^{-9} m^3/s.
Hint for D> Dc = 2.5mm
Sh = 0.42 $Gr^{1/2}$ $Sc^{1/3}$
Where Sh = $k_L D/Do_2$ and D is the bubble diameter

$$G_r = \frac{D^3 \rho_L \Delta\rho g}{\mu_L^2}, S_c = \frac{\mu}{\rho D_{O_2}}$$

$$k_L = 4.20 \times 10^{-4} m / sec$$

7.2 A strain of Azoto bactor vinelandii is cultured in a $5m^3$ stirred fermenter for alginate production. Under given operating conditions, k_La = 0.1 sec^{-1}. The specific rate of

oxygen uptake, $q_{o_2} = 12.5 \times 10^{-3} \frac{k_{mol} O_2}{(k_g X)(hr)}$

The solubility of O_2, C_L^* is given by Hendry's Law, $po_2 = HC_L^*$ where H (Henry's

count) $= 800 \frac{bar - m^3}{Kmole}, p_{o_2} = 0.2 bar$

(a) What is the value of kLa ?. What is the maximum cell concentration in kg/m^3. If the dissolved oxygen concentration (C_L) in the fermenter is 10 percent of the saturated value (C_L*)?

(b) The bacterial growth is effected by an inhibitor by the addition of copper sulphate, causing the reduction of oxygen uptake by 50%. What maximum cell concentration is expected in the fermenter . If the other conditions remain same?

Hint for (a) $q_{o_2} X = k_{La}(C_L^* - C_L)$

For (b) $0.5 q_{o_2} X = k_{La}(C_L^* - C_L)$

7.3 A stirred tank of $1.0m^3$ volume is equipped with four equal spaced baffles and 80% of the tank is filled with a liquid. The tank is agitated with 0.3 m diameter flat six blade disk turbine.

The stirrer is rotated at 160 rpm. The air is introduced below the impeller at the rate of 0.05 m^3/s at 1.13 bar at 25°c. Tank diameter is 0.8m.

The properties of the liquid

$\rho_L = 1100 kg / m^3, \mu_L = 1.5 \times 10^{-3} \rho$ as

$D_{o_2} = 0.5 \times 10^{-9} m^2 / s, D_{32} = 3.0 mm$

Calculate the following

(a) Power requirement
(b) k_L
(c) k_{La}
(d) Gas hold up
(e) Interfacial area

7.4 The optimum agitation speed for the cultivation of plant cells in a $0.5m^3$ fermenter equipped with 4 baffles has been found to be 150 rpm

Given D_I=0.3Dt, vessel diameter, Dt= 1.0m

(a) What should be the impeller speed of a geometrically similar 20m^3 fermenter if the scale-up is based on the same (P/V)

$$\text{Hint}: N_2 = N_1\left(\frac{D_{I_1}}{D_{I_2}}\right)^{2/3}$$

(b) If the impeller diameter is 1/3 of vessel diameter of 0.8m, what should be the rpm of the prototype fermenter based on the equal impeller based Reynolds number?

$$\text{Hint } N_2 = N_1\left(\frac{D_{I_1}}{D_{I_2}}\right)^2$$

7.5 Power consumption by aeration is a function of physical properties and vessel and impeller geometry. Power can be expressed as function of the following variables, where other dimensions of the vessel are constant

$$P = f\left(\rho, \mu, N, D_I, g\right)$$

where p is the power whose dimension in fundaments units are L^2MT^{-3}.

Show that

$$\frac{P}{\rho N^3 D_I^5} = f\left(\frac{\rho N D_I^2}{\mu}, \frac{N^2 D_I}{g}\right)$$

where $P/\rho N^3 D_I^5$ = Power No. (P_{No})

$\frac{\rho N D_I^2}{\mu} = R_{eI}$ (impeller based Reynolds no), $N^2 D_I / g$ = Fr (Froude No.)

7.6 The fungus Aureobasidium pullulan is used to produce an extracellular polysaccharide by fermentation of sucrose. After 120 hrs fermentation, the following measurements of shear stress and shear rate were made with a rotating cylinder viscometer:

$\tau, dyne/crm^2$	44.0	235.3	357	457	638
$\dot{\gamma}, sec^{-1}$, shear rate	10.2	170	340	510	1020

(a) Plot the rheogram of the fluid

(b) Determine non-Newtonian fluid parameters

(c) What is the apparent viscosity at shear rate of (i) 15 sec^{-1}, (ii) 200 sec^{-1}

Hint for (b) $\tau = K\dot{\gamma}^n$ for c, $\mu_a = \frac{\tau}{\dot{Y}} = k\dot{\gamma}^{n-1}$

7.7 A small tank of 0.1m^3 volume is stirred with a helical ribbon stirrer with a diameter of D_I = 10cm at the rate of 100 rpm. The flow behaviour inside the stirrer is laminar. Under such condition, Power no. (P_{No}) is inversely proportional to impeller based

Reynolds number (R_{et}) and the proportionality constant is 1000.
Calculate the power required for the system in KW

7.8 A stirred bioreactor with L/D = 2 with a diameter of 1 meter is agitated by a Rushtine turbine impeller with a diameter D_I = 0.3Dt. The reactor contains Newtonian PaS culture broth with ρ_L = 1000 kg/m^3, $\mu = 4 \times 10^{-3}$ PaS

(a) If the specific power consumption should not exceed 1.5 KW/m^3, determine the maximum allowable stirrer speed? What is the mixing time under these conditions?

(b) In the aerated tank, the presence of bubbles gives the approximate relation between the ungassed power number P_{No} and gassed power number, P_{Ng} is P_{Ng} = 0.5 P_{No}.

What maximum stirrer speed is possible in the sparged reactor? Under such conditions what is the mixing time?

For (a) & (b) mixing time,

$$t_m = \frac{1.54V}{N_i D_I^3}$$

7.9 A bubble column of 31.4m^3 liquid volume with Dt = 2m is sued to produce saacharomyces cerevisiae at air flow rate of 0.5 m^3/sec at 25°c and atmospheric pressure. The other data for the process are

$\in_G$ (gas- hold up) = 0.1,

$C_L^* = 0.25 \times 10^{-3}\, kmol/m^3$,

H = 800 bar – m^3 / kmol, R (gas constant) = 0.0831 $\frac{bar\, m^3}{kmol\,°K}$, T = 298°K

L = 10m

(a) Calculate kLa for the bubble column

(b) Design equation of bubble column is

$$\frac{\left(C_L^* - C_L\right)_Z (top)}{\left(C_L^* - C_L\right)_{bottom}} = exp\left[-\frac{k_L aRTZ}{E_G\, H u_G}\right]$$

Evaluate the right hand side with the given values.

(c) If the average value of (C_L* - C_L) on the bottom and top values is given as 0.6, calculate the concentration of saccharomyces cerevisiae when the specific oxygen consumption rate of cells is

$$q_{O_2} = 8 \times 10^{-3} \frac{kmol\, O_2}{(kg\ cell)(hr)}$$

Calculate the cell mass concentration (X) in the reactor.

Hint :k_{La} 0.6 (C_L* - C_L) = qo_2 X and C_L = 0.1 C_L*

(d) Also calculate the liquid flow rate

7.10 An air-lift fermenter of length, L = 100 cm has the following data:
Superficial liquid velocity in the riser = 30 cm/sec.
Gas velocity in the riser = 10 m/sec,

$$P/V = 2\frac{watt}{m^3}$$

Dispersion coefficient, D_L = 58 cm^2/sec.
The ratio of Downcomer to riser areas = 1.25
(a) Calculate holdup in the riser
(b) Disperson no.
(c) k_{La}

Given $$k_{La} = \frac{5\times10^{-4}\left(P/V\right)^{0.8}}{1+(Ad/Ar)}\cdot sec^{-1}$$

Make a comment on the k_{La} value

References

Akita, K. and F. Yoshida, "Bubble size, interfacial area and liquid phase mass transfer coefficient in Bubble columns" I & EC Process Des. Develop. 13, 84, 1974

Andrew, S.P.S., "Gas liquid mass transfer in Microbial Reactors", Tran. IChE 60, 3–13, 1982.

Andrews, G.F., J.P. Fonte, E. Marrota and P. Stroeve, The effects of cells in oxygen transfer coefficients Chem. Eng. J. 29, B39, B47, 1984

Bailey, J.E. and D.F. Ollis, Biochemical Engineering Fundamentals, 2ned. McGraw Hill Book Co., New York, 1986

Bello, R.A., C.W. Robinson and M. Moo-Young, "Mass transfer and liquid mixing in External circulating Loop reactors", Adv. Biotechnol.1, 547, 1981

Blakebrough, N. and K. Sambamurthy, "Mass transfer and mixing rates in Fermentation Vessels", Biotechnol.Bioeng., 8, 25, 1966.

Blanch, H.W. and D.S. Clark, Biochemical Engineering, Marcel Dekker, Inc. New York, 1996.

Bull, D.N., R.W. Thomas and T.E. Stinnett, "Bioreactors for submerged culture", Adv. Biotehnol processes, I 1–30, 1983.

Blenke, H. "Loop Reactors" P.121 in Advances in Biochemical Engineering Vol. 13, edited by T.K. Ghosh, A. Fiechter, B. Blakebrough, Springer Verlag, 1979

Calderbank, P.H. and M. Moo Young "The continuous phase Heat and Mass transfer properties of dispersions", Chem. Eng. Sci. 16, 39, 1961

Calderbank, P.H., "Mass Transfer in fermentation equipment", p. 102 in B. Blaebrough's (editor) Biochemical and Biological Engineering Science, Vol. 1, Academic Press, New York, 1967

Chang, H.T. and D.F. Ollis'. "Generalized power law for polysaccharide solution" Biotechnol, Bioeng. 24, 2309, 1982.

Chisti, Y. and M. Moo-Young, Airlift Reactors: Characteristics, Applications and design considerations", Chemical Eng. Comm. 60, 195–242, 1987

Chisti, M.Y. Airlift Bioreactors, Elsevior, Applied Science, Barking, 1987

Chisti, Y. and M. Moo-Young, "Improve the performance of Airlift Reactors", Chem. Eng. Progr.89(6), 38–45, 1993.

Cooney, C.L., D.I.C. Wang and R.I. Mateles, "Measurement of heat evolution and correlation with oxygen consumption during Microbial Growth, "Biotechnol.Bioeng., 11, 269, 1968

Cooney, C.L., "Bioreactor, Design and Operation, Science 219, 728 – 733, 1983.

Deckwer, W.D., "Bubble column Reactors" in Biotechnology Vol-2, edited by H.J. Retm and G. Reed, P. 445–464, VcH, Weinheim, 1985

Doran, P.M., Bioprocess Engineering Principles, Academic Press, San Diego, 1995

Finn, R.K., "Agitation and Aeration, P. 69 in Black brough's (editor), Biochemical and Biological Engineering Science, Vol – 1, Academic Press, 1967.

Heijnen, J.J. and K. Van Riet, "Mass transfer, mixing Heat Transfer Phenomena in Low Viscosity Bubble column Reactors", Chem. Eng., J. 28, B21–B42, 1984

Hughmark, G.A., "Holdup and Mass Transfer in Bubble columns", Ind. Eng. Chem. Process Des. Develop. 6, 218, 1967

Jost, J. "Selected Bioengineering Problems in stirred tank Fermenters", in S.L. Sandler and B.A. Finalyson (editors), Engineering Education in a changing Environment", Engineering Foundation, New York, 1988

Kargi, F. and M. Moo-Young, "Transport Phenomena in Bioprocesses", in Comprehensive Biotechnology edited by M. MooYoung, Pergamon Press, Oxford, U.S.A., Vol. 2, 5–55, 1985

Leduy, A., A. Morson and B. Corpal, "A study of the Rheological properties of non-Newtonian Fermentation Broth (Pullulans)", Biotechnol.Bioeng., 16, 61, 1974

Mukhopadhyay, S.N. and T.K. Ghosh, "A simple Dynamic Method of k_{La} determination in a Laboratory Fermenter", J. Ferment Technology, 54, 406–419, 1976

Oldshue, J. "Fermentation Mixing Scale-up Techniques", Biotechnol. Bioeng., 8,3, 1966

Prave, P. and others, Fundamentals of Biotechnology, VCH, Verlagsgesellchaft, Weinheim, Germany, 1987

Rushton J.H., E.W. Gostich and H.J. Everett, "Power Characteristics of Mixing impellers" Part 2, Chem. Eng. Prog.46, 467, 1950.

Roels, J.A., J. Vander Berg and R. R.K. Voucken, "The Rheology of Mycelial broths (Penicillium)", Biotechnol.Bioeng., 16, 181, 1974

Shuler, M.L and F. Kargi Bioprocess Engineering: Basic concepts, Chapter 9, Prentice Hall, New Jersey, 1992

Shumpe, A and W.D. Decker, "Estimation of O_2 and CO_2 solubilities in Fermentative media", Biotechnol.Bioeng. 21, 1075–1078, 1979

Stephanopoulos, G., Biotechnoogy Vol-3 of Biotechnology edited by H.J. Rehm and G. Reed (editors)

Taguchi, H. and S. Miyamoto, "Power Requirement in Non-Newtonian Fermentation broth", Biotechnol.Bioeng. 8, 43, 1966

Thomson, N. and D.F. Ollis, "Evaluation of power law parameters for Xanthan and Pollulan Batch Fermentation, Biotechnol. Bioeng., 22, 875, 1980

Van Riet, K., "Review of Measuring Methods and results in non-viscous gas-liquid mass transfer in stirred vessels, "Ind. Eng. Chem. Process Des. Develop. 18, 357–364, 1979

Van Riet, K., "Mass Transfer in Fermentations", Trends in Biotechnology, 1 (4), 113, 1983

Van Riet K., and J. Tramper, Basic Bioreactor Design, Chapter 1, Marcel Dekker, New York, 1991

Wang, K.B. and L.T. Fan, Mass Transfer in Bubble columns packed with Motionless Mixers, Chem. Eng. Sci. 33, 945, 1978.

CHAPTER 8

Sterilization Reactors

8.0 Introduction

Sterilization of media or equipment is essential requirement before the starting of desired fermentation. It is necessary to kill the unwanted microorganisms like bacteria, mold by means of heat (moist or dry), chemical agents, ultraviolet light, x-rays and mechanical means (sonic or ultrasonic vibrations). The moist heat is more effective than the dry heat, as the death rate is much lower for the dry cells than for moist ones.

Laboratory autoclaves are commonly sterilized by using steam at a pressure of 30 Psia and at a temperature of 121°C. Chemical agents are employed for disinfection for pathogenic organisms. Common chemicals are phenol and phenolic compounds (cresol, orthophenyl phenol), ethyl and methyl alcohol, halogens, hypochlorites, detergents, dyes, ethylene oxide, formaldehyde etc. X-rays are lethal to microorganism and have penetration ability, but are not commonly used for their high expense and safety factors.

Filtration is effectively employed for the removal of microorganisms from air or other gases and also from liquid solutions containing thermally unstable compounds like serums and enzymes. In most cases, heat is used to sterilize medium, where microbial death kinetics by heat is important.

8.1 Thermal Death Kinetics

Thermal death of microorganisms at a particular temperature is expressed by first order decay kinetics,

$$\frac{dn}{dt} = -k_d n \tag{8.1}$$

Where k_d is the specific death rate which depends on the type of species and on temperature. For example k_d for bacterial species at 121°C is 1 min^{-1}. For vegetative cells, k_d may vary from 10 to 10^{10} min^{-1}.

Integrating the equation (8.1), we have

$$l_n \frac{n}{n_0} = -\int_0^t k_d dt \qquad (8.2)$$

where n_0 is the initial number of microorganisms and n is the number of viable microorganisms after thermal treatment at temperature, T.

$$k_d = k_{do} e^{-E_d/RT} \qquad (8.3)$$

Where E_d is the activation energy for the thermal death of a microorganisms. E_d for E.Coli is 127 K.cals/gmole and that of Bacilus Stearothermophilus is 68.7 k.cals/gmole.

8.2 Design Criterion, Del

The design criterion for sterilization, Δ is defined as

$$\Delta = ln\ N_0/N = k_d\ dt = kd_0\ exp(-\ E_d/RT)\ dt \qquad (8.4)$$

The Δ is known as the Del factor, which is a measure of the size of sterilization.

Problem 8.1

A fermenter tank of 50 m^3 is to be sterilized with the residence time of 2 hrs. The unsterilized medium contains spores of $10^7/m^3$. The value of kd has been determined to be 1 min^{-1} at 121°C and 61 min^{-1} at 140°C. For each temperature, determine required residence time in the holding section to 99% sterility level for the time of 4 weeks of continuous operation.

Solution:

Total initial number of spores = No

Flow rate = 50 m^3/ 2 hrs =25 m^3/hr,

$$No = 10^7 \frac{Spores}{m^3} \times 25 \frac{m^3}{hr} \times 24 \frac{hrs.}{day} \times 28\,day$$

$$= 1.68 \times 10^{11}\ Spores$$

From the required sterility level, $N = \frac{1}{100} = 0.01$

From the sterilization kinetics

$$-\frac{dN}{dt} = k_d N$$

Integrating from $t = 0, N = N_0,\qquad t = t, N = N$

$$\frac{N}{N_0} = e^{-k_d t}$$

$$ln\frac{N}{N_0} = -k_d t$$

$$t = \frac{ln\,{}^{No}\!/_{N}}{k_d}$$

$$= ln\left(\frac{1.68\times 10^{11}}{10^{-2}}\right)/\,1.0$$

$$= 5.18\ min$$

Where $k_d = 1\ min^{-1}$ at $121^{\circ}C$

At $140^{\circ}C$, $k_d = 61\ min^{-1}$

$$t = \frac{ln N_0 / N}{k_d} = \frac{30.44}{61} = 0.5\ mins\ \text{Ans.}$$

If the final number of cells surviving after sterilization is 1 in 1000, then the del factor to reduce the number of cells from 10^{10} viable organism to 0.001 is

$$\nabla = ln\frac{10^{10}}{0.001} = 30 \tag{8.5}$$

Based on this sterilization criterion the sterilization unit may be designed.

8.3 Batch Sterilization

In batch sterilization, sterilization cycles have three stages viz. (i) heating from the initial temp to the desired temperature, (ii) holding the temperature for a constant period and (iii) cooling from the desired temperature to the initial temperature.

The total del factor required should be equal to the sum of the del factors for heating. Holding and cooling:

$$\nabla_{total} = \nabla_{heat} + \nabla_{hold} + \nabla_{cool} \tag{8.6}$$

The values of ∇_{heat} and ∇_{cool} are determined by the methods used for the heating and cooling. The value of ∇_{heat} is determined by the controlled holding period.

To calculate Δ_{total}, it is required to measure the temperature versus time profile during heating, holding and cooling cycles of sterilization. The temperature vs time profile for batch sterilization is given in fig. 8.1 and the reduction in the number of cells with time, t is given in figure 8.2

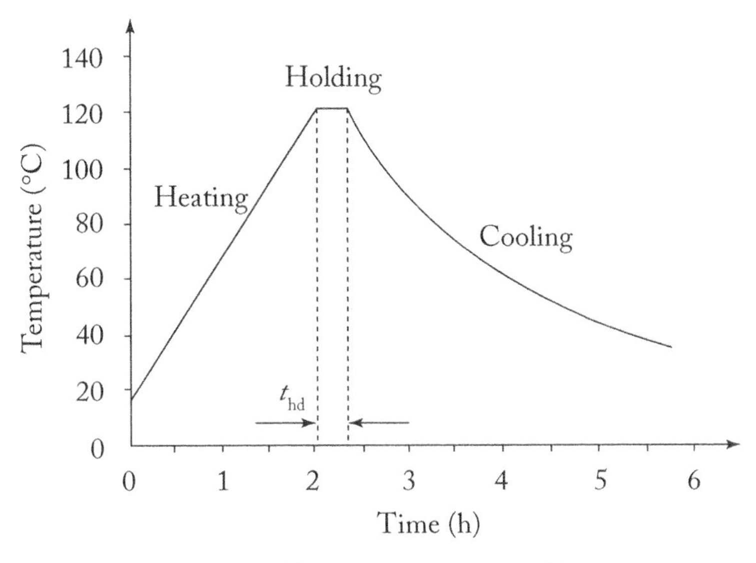

Fig. 8.1 Temperature vs time profile

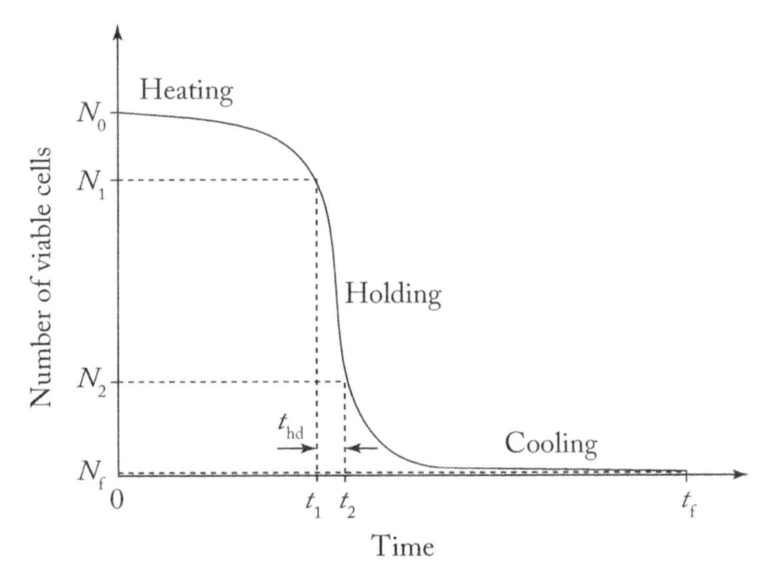

Fig. 8.2 Reduction of cells with time

For constant temperature sterilization during the holding time, the holding time t_{hd} can be given as,

$$t_{hd} = \frac{ln \, N_1 / N_2}{k_d} \tag{8.7}$$

where N_1 is the number of viable cells at the start of the holding time t_1 an N_2 is the number of viable cells at the end of holding at t_2.

Temperature versus time profile is measured during heating, holding and cooling cycles of sterilization. If the experimental measurements are not possible, theoretical equations for heating and cooling can be applied. These equations are linear, exponential or hyperbolic and have been presented in the table 8.1.

Table 8.1 General equations for temperature vs time during heating and cooling periods of batch sterilization

Heat Transfer Method	Temperature – Time profile	Equation type
1. Heating Direct sparging with time	$T = T_o\left(1 + \frac{hMst/(M_m C_p T_o)}{1 + \frac{Ms}{M_m}t}\right)$	Hyperbolic
2. Electrical heating	$T = T_o(1 + Qt/(MmC_pT_o)$	Linear
3. Heat transfer from isothermal steam	$T = T_s[1 + (T_o - T_s)/T_s\, exp(-UAt/MmC_p)$	Exponential
4. Cooling heat transfer to non-isothermal cooling water	$T = T_a\left\{1 + \frac{T_o - T_a}{T_a} exp\left[\frac{\left(-M_m C_{pw} t\right)}{M_m C_p}\left(1 - e\frac{[UA]}{MwC_p}\right)\right]\right\}$	

Where

A = Surface area for heat transfer, m^2

Cp = Specific heat capacity of medium, W/m°C

Cpw = Specific heat capacity of cooling water

h = Specific enthalpy difference between steam and raw medium, Kj/Kg

M_m = Initial mass of medium, Kg

M_s = Mass flow rate of steam, Kg/hr

M_w = Mass flow rate of cooling water, Kg/hr

Q = Rate of heat transfer, $W/m^{2}{}^{\circ}C$

T = Temperature, °C

T_o = Initial medium temperature, °C

T_a = Initial temperature of cooling water, °C

T_s = Steam temperature, °C

t = time

U = Overall heat transfer coefficient, W/m^2°C

Now k_d is plotted as a function of time. We integrate the area under k_d versus time curve for heating, that is

$$ln\frac{N_0}{N_1} = \int_0^{t_1} Ae^{-E_d/RT}\,dt \tag{8.8a}$$

and for cooling

$$ln\frac{N_2}{N_f} = \int_{t_2}^{t_f} Ae^{-E_d/RT}\,dt \tag{8.8b}$$

where t_1 is the time at the end of heating, t_2 is the time at the end of the holding and t_f is the time at the end of cooling.

By integration, ∇_{heat} and ∇_{cool} are calculated using the temperature profiles.

Now, the holding time, t_{hold}, can be estimated from the relation,

$$t_{hold} = \frac{\nabla_{hold}}{k_d} = \frac{\nabla_{total} - \nabla_{heat} - \nabla_{cool}}{k_d} \tag{8.9}$$

Our design requirement is the estimation of holding time, t_{hd}.

Problem 8.2 (Batch sterilization)

A fermenter containing 10m^3 of medium (25°C) is desired to be sterilized by passing saturated steam (500kPa gauge pressure, saturation temp 152°C) through the heating coil in the fermenter. The medium contains about $3\times10^{12}\,m^3$ microbial spores, which is to be reduced to the contamination level of 1 in 100. The fermenter will be heated until the medium reaches 115°C. During the holding time, the heat loss through the vessel is assumed to be negligible. After the proper holding time, the fermenter will be cooled by the flow of 20m^3/hr of water at 25°C through the coil in the fermenter until the medium reaches 40°C. The coil has a heat transfer area of 40m^2 and the average overall heat transfer coefficients (U) for heating and cooling are 5, 500 and 12,500 KJ/hr m^{2}k respectively. The heat resistant bacterial spores in the medium can be characterized by the Arrhenius Coefficient, (k_{do}) of 5.7×10^{39} hr^{-1} and an activation energy of $E_d = 2.834\times10^5\,kJ/mole$. The heat capacity and the density of the medium are 4.187 kJ / kg K and 1000 kg/m^3 respectively.

Estimate the required holding time, t_{hd}.

Solution:

The design criterion can be calculated form the equation:

$$\nabla = ln\frac{no}{n} = ln\frac{3\times 10^{12}\times 10}{0.01} = ln 3\times 10^{15} \approx 35.64$$

Saturation steam temp = 152°C at 500 kPa

For heating time – Temperature relation is given

$$T = T_H + \left(T_0 - T_H\right) exp\left(-\frac{UAt}{C_p M_m}\right)$$

Where T_H = 152°C, T_O = 25°C, C_p = specific heat of medium = 4.187 kJ/kgK

M = mass of medium = 10 × 1000 = 10000 kg

U = overall heat transfer coefficient = 5500 kJ/hrm^{2}K

A = heat transfer area = 40 m^2

$$T = 152 + (25 - 152) exp\left(-\frac{5500\times 40t}{(4.187)(10000)}\right) = 152 - 127 exp(-5.25t)$$

Time required to reach the maximum temperature of the medium, t_h in, hr

115 = 152 – 127 *exp* (–5.25t)

Solving t_h = 0.24 hrs

- Now ∇_{heat} is calculated from the relation

$$\nabla_{heat} = k_{do}\int_0^{t=0.24} e^{-Ed/RT} dt$$

$$= 5.7\times 10^{39}\int_0^{t=0.24} exp\left[-\frac{2.834\times 10^5}{8.314\mathrm{T}}\right] dt \tag{A}$$

Where T = 152 – 127 *exp* (–5.25 t)

The equation (A) is integrated numerically using Simpson's rule

$$I = \frac{h}{3}\left(y_0 + 4y_1 + y_2\right)$$

The function (A) is calculated with three values of t: 0, 0.12, 0.24 hrs.

(i) At t = 0 T = 152 – 127 = 25°C = 298 K

$$y_0 = 5.7\times 10^{39} exp\left[-\frac{2.834\times 10^5}{(8.314)(298)}\right] = 1.198\times 10^{-10}$$

(ii) At $t = 0.12$ *hr*

$T = 152 - 127\ exp\ [-5.25 \times 0.12] = 84°C = 357\ K$

$$y_1 = 5.7 \times 10^{39}\ exp\left(-\frac{2.834 \times 10^5}{8.314 \times 357}\right) = 1.94 \times 10^{-2}$$

(iii) At $t = 0.24$ *hrs*

$T = 152 - 127\ exp\ [-5.25 \times .24] = 115 = 388$

$$y_2 = 5.7 \times 10^{39}\ exp\left(-\frac{2.834 \times 10^5}{8.314 \times 388}\right) = 5.0$$

$$\nabla_{heat} = I = \frac{0.12}{3}\left[0 + 4(.0194 + 0.5)\right] = 0.20\ hrs.$$

- To calculate $\nabla_{cooling}$, the time – temperature relationship is given as

$$T = T_{co} + (T_o - T_{co}) \times exp\left\{\left[1 - exp\left(-\frac{UA}{\overline{m}_c C_{PC}}\right)\right]\frac{\overline{m}_c t}{M}\right\}$$

Where T_{co} = coolant temp.

The above equation is integrated by Simpsons rule,

(i) $t = 0$, $T = 115°C = 388\ K$

$$y_0 = 5.7 \times 10^{39}\ exp\left[-\frac{2.834 \times 10^5}{(8.314)(388)}\right] = 39.96$$

(ii) $t = 2.3$ *hrs*, $T = 25 + 90\ exp\ (-3.88 \times 2.3) = 62°C = 335\ K$

$$y_1 = 5.7 \times 10^{39}\ exp\left[-\frac{2.834 \times 10^5}{(8.314)(335)}\right] = 3.67 \times 10^{-5}$$

(iii) $t = 4.6$ *hrs*, $T = 25 + 90\ exp\ (-3.88 \times 4.5) = 40°C = 313\ OK$

$$y_2 = 5.7 \times 10^{39}\ exp\left[-\frac{2.834}{(8.314)(313)}\right] = 2.88 \times 10^{-8}$$

$$\nabla_{cool} = \frac{2.3}{3}\left[39.96 + 4(3.67 \times 10^{-5}) + 2.85 \times 10^{-5}\right] = 30.63$$

$$\nabla_{hold} = \nabla_{total} - \nabla_{heat} - \nabla_{cool}$$

T_o = temperature of the medium before cooling starts (115°C)

M_c = mass of the coolant

M = mass of the medium

U = heat transfer coefficient for the cooling ..., 2500 kJ / m^2 hr K

A = heat transfer area of the cooling coil, 40 m^2

$$T = 25°C + (115 - 25) exp\left\{\left[1 - exp\left(-\frac{2500 \times 40}{(20 \times 1000)4.187}\right) \times \frac{20 \times 1000t}{10 \times 1000}\right]\right\}$$

Or $T = 25 + 90\ exp\ [-0.388\ t]$

When $T = 40°C$

$$\frac{40 - 25}{90} = exp(-0.388\, t)$$

Or $t = 4.60\ hrs.$

$$\nabla_{cool} = 5.7 \times 10^{39} \int_0^{4.60\ hrs} -exp\left(-\frac{2.834 \times 10^5}{8.314\ T}\right) dt \quad \text{(B)}$$

where $T = 25 + 90\ exp(-U \cdot 388\ t)$

The equation (B) is numerically integrated as before, ∇_{cool} is, ∇_{cool} = 30.63.

$$\nabla_{total} = ln\frac{no}{n} = ln\left[\frac{3 \times 10^{12} \times 10}{0.01}\right] = 35.64$$

$$\nabla_{hold} = 35.64 - 0.20 - 30.63 = 4.81$$

Now k_d at 115°C (388 K)

$$k_d = 5.7 \times 10^{39} exp\left(-\frac{2.834 \times 10^5}{8.314 \times 388}\right) = 39.96\ hr^{-1}$$

$$t_{hold} = \frac{\nabla_{hold}}{k_d} = \frac{4.81}{39.96} = 0.12\ hrs = 7.2\ minutes \quad \text{Ans.}$$

8.4 Continuous Sterilization

Continuous sterilization has several advantages over batch sterilization and may be listed as below:

(i) It provides simplified production planning, maximum plant utilization and minimum delays.

(ii) It can be operated at high temperature (140°C) compared to 121°C in batch process and sterilization time of 1–2 minutes are required in the continuous process.

(iii) It requires less steam by recovering heat from the sterilized medium. It also requires less cooling water.
(iv) The process is run with automatic control, requires less labour power
(v) It provides reproducible conditions.

Heating Section:

Method of heating may be of two types – i) direct stream injection or ii) indirect heating in shell-and-tube or plate and frame type heat exchangers. The steam injection raises the temperature of the medium to the peak value(140°C) Sterilization during this heating period is not significant.

For indirect heating, the plate-and-frame heat exchange (fig. 8.3b) is more effective than the shell-and-tube type heat exchange (fig. 8.3a) as the latter requires more heat transfer area.

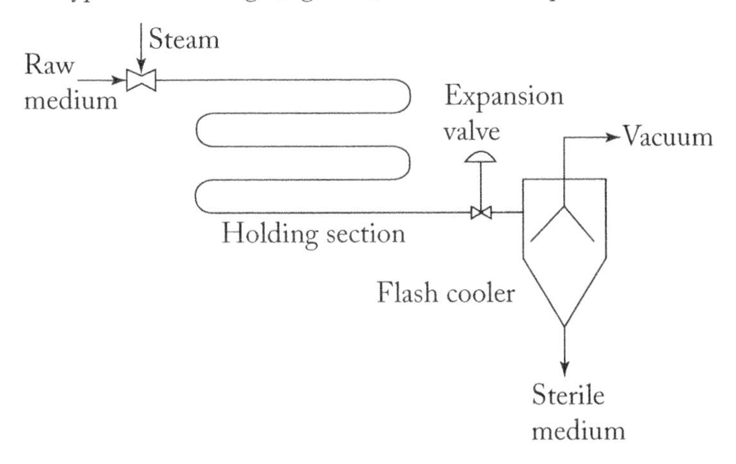

(a) Continuous injection type

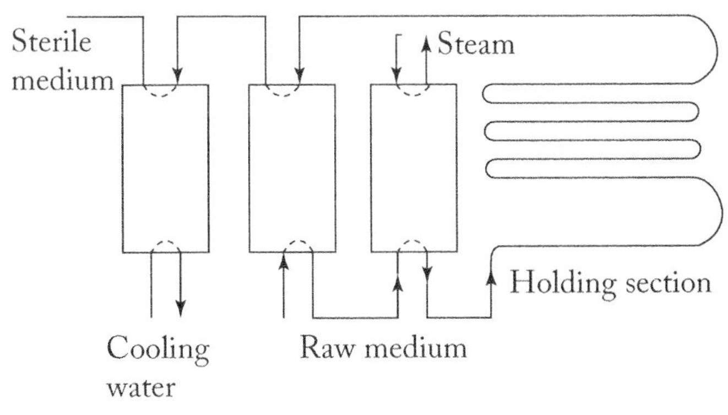

(b) Continuous plate exchanger type

Fig. 8.3 (a) Continuous steam injection & (b) continuous plate and frame heat exchanger

The temperature change with respect to residence time $\bar{\tau}_{heat}$ as the medium passes through an isothermal heat source, can be given as

$$T_{C_2} = T_H - \left(T_u - T_{C_1}\right) exp\left(-\frac{UA\bar{\tau}_{heat}}{C_p W}\right) \tag{8.10}$$

For heating, using a counter current heat source of equal flow rate and heat capacity,

$$T_{C_2} = T_{C_1} + \frac{\Delta TUA\bar{\tau}_{heat}}{C_p W} \tag{8.11}$$

Where $$\bar{\tau}_{heat} = \frac{L}{U} \tag{8.12}$$

Where T_{C1} & T_{C2} are the temperatures of inlet water and outlet water respectively. T_s is the temperature of the steam. A is heat surface area, C_p is the specific heat of medium, $\dot{W}$ is the flow rate of cooling water.

8.4.1 Holding section: plug flow model

The heated medium passes through a holding section in the form of a long tube. Adiabatic conditions are maintained in the holding section. If the heat loss is minimum in this section, the temperature of the holding section may be assumed constant.

Defining $\bar{\tau}_{hold} = \frac{L}{U}$

Using the above relation, the delfactor can be calculated as

$$\nabla_{hold} = ln\, n / n_0 = k_{d0}\, exp(-E_d/RT) \tag{8.13}$$

The calculation of $\bar{\tau}_{hold}$ is based on the assumption of plug flow in the holding section. But the slippage due to the viscous nature of the fluid, friction of the pipe wall, and the turbulent eddies of the flowing fluid cause the flow to deviate from plug flow. For laminar flow of Newtonian fluid, the average velocity, $\bar{U} = 0.5\ V_{max}$, for turbulent flow with $R_e = 10^6$, $\frac{\bar{u}}{V_{max}} = 0.88$ where V_{max} is the maximum velocity at the centre of the pipe. If we use the mean velocity for calculating the residence time for sterilization, some portion of the liquid medium may remain unsterilized causing serious contamination problem for the medium. The deviation from plug flow due to axial mixing is given by dispersion model. So the sterilization of medium can be calculated by dispersion model.

8.4.2 Dispersion model for sterilization

We consider a differential length, dz in the holding tube. The basic material balance for microorganism suspended in the medium may be given in terms of axial dispersion and first order microbial death kinetics:

$$input\ of\ cells - output\ of\ cells - cells\ killed\ by\ sterilization = Accumlation \tag{8.14}$$

At the steady state, the accumulation term is zero. The x-directional flux of microorganisms suspended in medium due to axial mixing can be presented for microbial flux, Jn can be presented

$$Jn = D_L \, dn/dx \tag{8.15}$$

Where D_L is the axial dispersion coefficient, characterized by back mixing. If $D_L = 0$, the flow is plug flow and if D_L =(infinity), the fluid in the pipe is well mixed.

For turbulent flow, the dispersion coefficient in the form of dimension less group (dispersion number), $\frac{D_L}{\bar{u}d_t}$ is a function of Reynolds number as shown in the figure 8.4

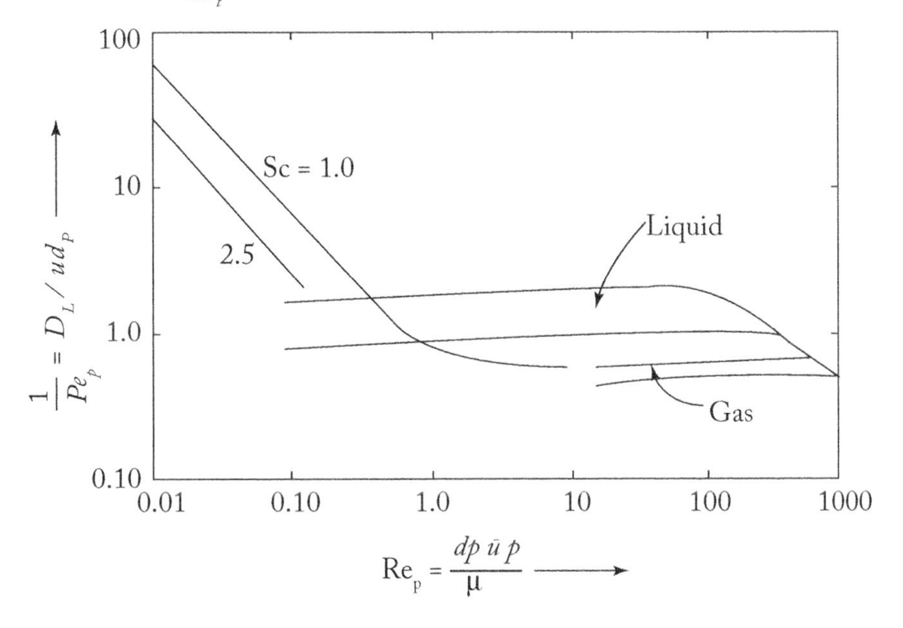

Fig. 8.4 Correlations of $\frac{D_L}{\bar{u}d_t}$ as function of Reynolds number

Substituting the values of input and output of cells and thermal death of microorganisms by first order Kinetics and simplifying we have,

$$\frac{d}{dx}\left(D_L \frac{dn}{dx}\right) - \frac{d(\bar{u}n)}{dx} - k_d n = 0 \tag{8.16}$$

For constant D_L and $\bar{u}$, we have,

$$D_L \frac{d^2 n}{dx^2} - \bar{u}\frac{dn}{dx} - k_d n = 0 \tag{8.17}$$

Introducing dimension less groups:

$$n' = \frac{n}{n_0},\ x' = \frac{x}{L},\ P_e = \frac{\bar{u}L}{D_L{}^{\circ}} \tag{8.18}$$

Where P_e is the peclet number

$$\frac{d^2n'}{dx'^2} - P_e\frac{dn'}{dx'} - P_e Dan = 0 \tag{8.19}$$

Where Pe_e (peclet no, $\frac{\bar{u}L}{D_L}$) $\rightarrow \infty$, the flow behaves as a plug flow. Da (Damköhler number) is $k_d L/u$

The boundary conditions for the solution of this ordinary second order differential equation,

$$\frac{dn'}{dx'} - P_e(1-n') = 0 \text{ at } x' = 0 \tag{8.20}$$

$$\frac{dn'}{dx'} = 0 \text{ at } x' = 1 \tag{8.21}$$

Using the above two boundary conditions, the solution of equation (8.19) can be presented

$$\frac{n(L)}{n_0} = \frac{4y\,exp\left[\frac{P_e}{2}\right]}{(1+y)^2 exp\left[\frac{(P_e)(y)}{2}\right] - (1-y)^2 exp\left[-\frac{(P_e)(y)}{2}\right]} \tag{8.22}$$

Where,

$$y = \left(1 + \frac{4Da}{P_e}\right)^{\frac{1}{2}} \tag{8.23}$$

Where Da (Dam Kohler No) $k_d L/u$

For small deviation from plug flow ($\frac{1}{P_e}$ is mall), the equation (8.25) reduces to the following form;

$$\frac{n(L)}{no} = exp\left(-Da + \frac{Da^2}{P_e}\right) \tag{8.24}$$

The results of equation (8.22) are displayed as a plot of remaining viable fraction, $\frac{n_L}{n^0}$ Vs the dimension less group, Da for various values of Peclet number (P_e) in figure 8.5.

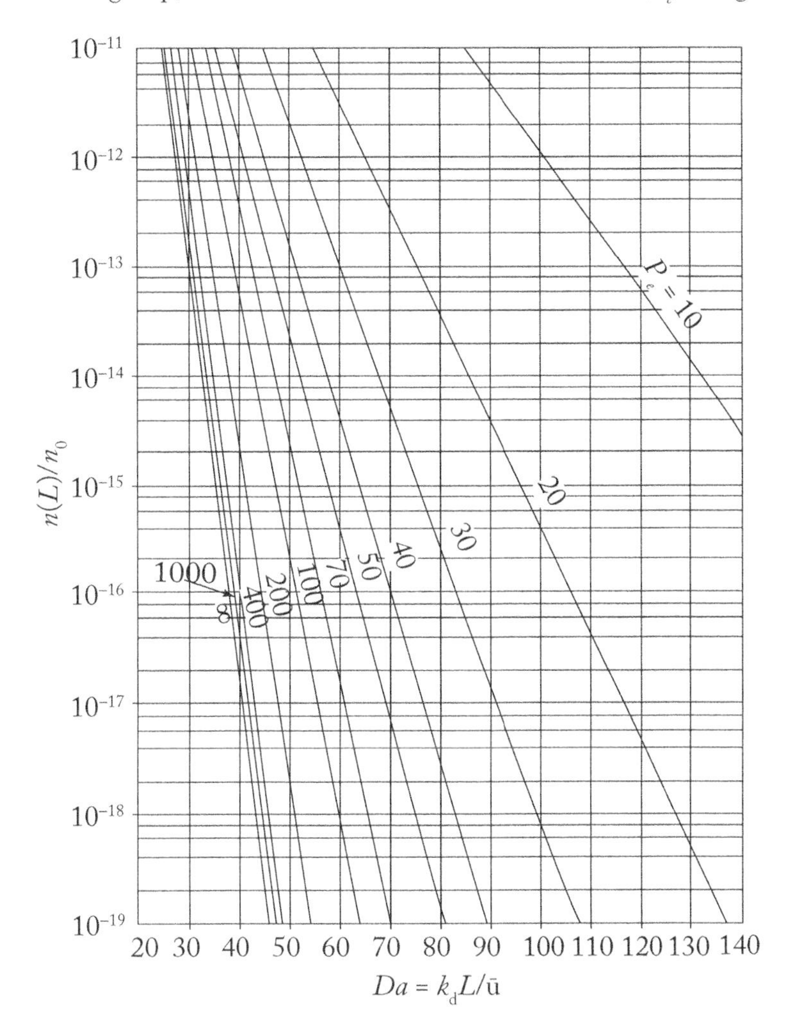

Fig. 8.5 Effect of axial dispersion on organism destruction in a continuous sterilizer

Problem 8.3 (Continuous sterilizations)

A continuous sterilizer with steam injection and a flash cooler will be employed to sterilize medium continuously at the rate of 2 m^3/hr. The time for heating and cooling is negligible with this type of sterilizer. The typical bacterial count of the medium is about $5 \times 10^{12} / m^3$, which needs to be reduced to such an extent that only one microorganism can survive during the three months of continuous operation. The heat resistant bacterial spores in the medium can be characterized by an Arrhenius Coefficient, k_{do} of $5.7 \times 10^{39}\ hr^{-1}$ and activation energy, E_d of $2.834 \times 105\ KJ / kmole$. The holding section of the sterilizer will be constructed with 20m long pipe with ID = 0.075m. Steam of 600kPa (gauge press) is available to bring the sterilizer to an operating temp of 125°C. The physical properties of the medium at 125°C are $C_P = 4.187\ KJ / kgK$, $\mu = 4\ kg / m\ hr$, $\rho = 1000 Kg/m^3$

(a) How much medium can be sterilized per hour if we assume ideal plug flow?

(b) How much medium can be sterilized per hr if the effect of axial dispersion is considered?

Solution: The Δ delfactor can be obtained as

$$\nabla = ln\frac{no}{n} = \frac{5 \times 10^{12} m^{-3} \times 2\ \frac{m^3}{hr} \times 24 \frac{hrs}{day} \times 30\ day}{1} = ln\left(7.2 \times 10^{15}\right) = 34.5$$

$$\text{Sterility level, } \frac{n}{no} = \left(7.2 \times 10^{15}\right) = 1.38 \times 10^{-14}$$

Raised temp of the medium = 125°C = 398°K

$$Kd = 5.7 \times 10^{39} exp\left[-\frac{2.834 \times 10^5}{(8.314)(398)}\right] = 363\ hr^{-1}$$

Again $\Delta = ln\frac{n_0}{n} = k_d t$

t_{hold} = 34.5/363 =0.095 hr

$$\text{Now } \bar{u} = \frac{2\ m^3 / hr}{0.7354(.075)^2 m^2} = 453\ m / hr$$

$$R_e = \frac{D\bar{u}\rho}{\mu} = \frac{.075(m) \times 453\ \frac{m}{hr} \times 1000 \frac{kg}{m^3}}{4\ kg / (m)(hr)} = 8.49 \times 10^3 \approx 10^4$$

From the plot of $\frac{D_L}{u\,dt}$ Vs Reynolds number

$$\frac{D_L}{\bar{u}\,dt} = 0.5 \text{ at } R_e \approx 10^4$$

$$\frac{D_L}{uL} = \frac{D_L}{\bar{u}\,dt} \cdot \frac{dt}{L} = (0.5)\left(\frac{.075}{20}\right) = 1.87 \times 10^{-3}$$

$$P_e = \frac{\bar{u}L}{D_L} = 535$$

From the chart, the sterility label, $\frac{n}{no} = 1.38 \times 10^{-14}$ at $P_e = 535$ gives

$$\frac{kdL}{u} = 33 \text{ Or, } L = \frac{33u}{kd} = \frac{(33)(453)}{363} = 41\ m$$

Corrected Peclet no.

$$P_e = \frac{uL}{D_L} = 1093$$

At $\frac{n}{no} = 1.38 \times 10^{-4}$ & $P_e = 1093$

$$\frac{kdL}{u} \approx 33$$

Or L= 33 * 453/363= 41m

So the required length at dispersion is 41m.

Now the medium sterilized per hour

$$= \frac{\pi}{4} d_t^2 u \rho$$

$$= 0.7854(.075)^2 m^2 (453) \frac{m}{hr} (1000) \frac{kg}{m^3}$$

$$= 2001 \frac{kg}{hr}$$

For part (a)

At plug flow condition and the desired sterility level

$$P_e = \infty,\ \frac{kdL}{u} = 32 \text{ (from the fig 8.5)}$$

$$L = \frac{32u}{kd} = \frac{(32)(453)}{363} = 40\ m$$

So the medium sterilized per hr is the same as in part (b).

From the plot as $P_e \rightarrow \infty$ the ideal plug flow is attained. The desired degree of medium sterility can be achieved with the sterilizer of short length (i.e. small value of Da)

8.4.3 Sterilization model based on residence time distribution

It is assumed that segregated reactor model may be applicable to the suspension of cells or spores subjected to thermal treatment. We assume that organisms are destroyed by first order kinetics in slugs of batch reactors having a definite age distribution $E(t)$ obtained from the residence time distribution from the reactor. The organism concentration, n, at exit can be expressed as

$$\bar{n} = \int_0^{\infty} n(t)E(t)dt \tag{8.25}$$

Where $n(t)$ is the concentration in each slug of batch reactor and can be given as

$$n(t) = no\ e^{-k_d t} \tag{8.26}$$

Substituting the equation (8.26) in the equation (8.25), we have

$$\frac{n}{no} = \int_0^{\infty} E(t)\ e^{-k_d t}\,dt \tag{8.27}$$

The right hand side can be evaluated if we know the RTD data, $E(t)$ vs t from tracer study, using some numerical technique and with known K_d value. Furthermore the equation (8.27) is identical to the Laplace transform of E, with Laplace transfer parameter s, replaced by k_d.

In summary it may be mentioned that high temperature and short residence time (HTST) feature in continuous sterilization has several advantages. Continuous treatment reactors provide more uniform, reproducible effluent than batch operation. This is very important in the food industry, where small change in treatment can alter the taste of the product. The sensitivity of sterilization affects the microorganism and medium components. Equipment size can be reduced by using continuous sterilization. There is scale up problem in batch fermentation as heating and cooling depends on surface to volume ratio of the fermenter. Some problems may be visualized for continuous sterilization. In the continuous method direct stream heating may lead to excess water to the medium which may affect the process requirement. Fouling may be a problem in heat-exchanger due to the presence of suspended solids. Foaming of fermentation media in continuous sterilization also causes serious problems.

8.5 Microbial Death Kinetics: Deterministic and Statistical Approach

For chemical species in a liquid medium the number of moles is of the order of 10^{23} moles per gm-mole. So the application of deterministic model is common. For microbial systems, cell concentration may not be very large. The microbial population may be as small as 10^5. Yet the system is satisfactorily described by the deterministic model.

The experimental data for thermal death of E.Coli have been obtained at different temperatures and their survival rate has been plotted against time.

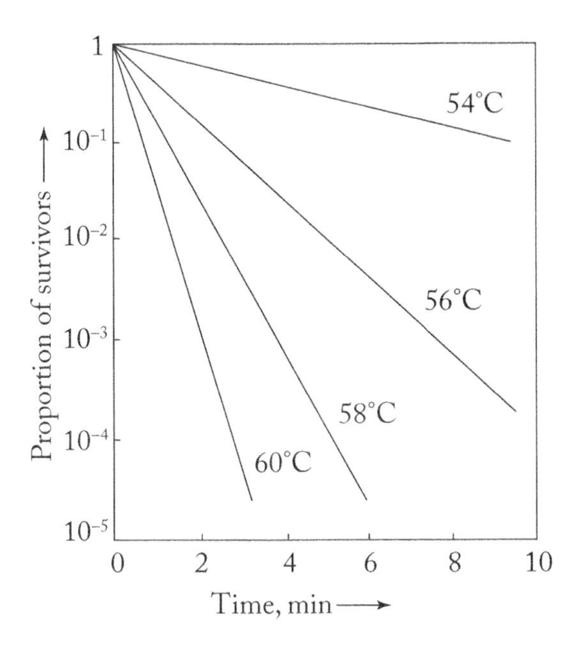

Fig. 8.6 Experimental data for E.Coli in buffer

The data fitted to a first order kinetic, model as

$$\frac{dn}{dt} = -k_d n \tag{8.28}$$

for constant k_d, integration yields,

$$n(t) = no\ e^{-k_d t} \tag{8.29}$$

k_d is a first order death constant which is function of temperature in the Arrhenius form,

$$k_d = k_{do} e^{-Ed/RT} \tag{8.30}$$

where E_d is the activation energy for the thermal death of microorganisms. E_d varies for many spores and vegetative cells in the range of 50 – 100 Kcals/mole.

In earlier time, a term was used by the name 'decimal reduction factor,' $DF = \frac{2.303}{kd}$, which indicates the time needed to reduce the viable population by a factor of 10.

In a deterministic model, the fraction of the viable n is given as

$$\text{Viable fraction} = \frac{n}{no} = e^{-k_d t} \tag{8.31}$$

The kinetic description by the above equation is reasonably good if the concentration of microorganisms is large in the statistical sense. Deviations from such prediction become more probable as the number involved diminishes drastically after sterilization.

8.5.1 Statistical model

A statistical approach to sterilization for a drastic change in microbial population from very large to a small number is appropriate. Theory indicates that the probability that at any time t, the remaining population contains N viable organisms, is expressed as

$$P_N(t) = \frac{No!}{(No-N)!N!} \cdot (e^{-k_d t})^N \times (1-e^{-k_d t})^{No-N} \tag{8.32}$$

where N_o is the initial viable number of organisms in the fluid being sterilized

The k_d appears here in equation (8.32) as the rate constant. In the statistical model, k_d may be interpreted as the reciprocal of the mean life span of the organisms. As the number of organisms fall to a low value, the assumption of deterministic approach characrized by k_d value begins to fail. In the stringent requirement of survival organisms, the statistical approach is very much desirable. Under such conditions, it is important to define the extinction probability i.e. the probability that all organisms are inactivated.

Setting $N = 0$ in the equation (8.32), we have EXTINCTION PROBABILITY, $P_0(t)$ is

$$P_O(t) = 1-(1-e^{-k_d t})^{No} \tag{8.33}$$

Now the probability of at least one organism's survival is

$$1-P_O(t) = (1-e^{-k_d t})^{No} \tag{8.34}$$

For usual situation, $N_o >> 1$

$$1-P_O(t) = 1-e^{-Nt} \tag{8.35}$$

where

$$N_t = No\ e^{-k_d t} \tag{8.36}$$

$1-P_O(t)$ is physically interpreted as the fraction of sterilization which is expected to fail to produce a contamination free product.

In the equation (8.33), No is the number of individual spores in reactor, not the concentration.

If no is the concentration of spores in number per m^3, for $1000 m^3$ tank, we have

$$1-P_O(t) = 1-[1-e^{-k_d t}]^{1000\ no} \tag{8.37}$$

Where $1-P_O(t)$ is the probability of an unsuccessful sterilization.

For example, let us assume, $k_d t = 15$ and N_0 = (10)10

So $$N_t = No\ e^{-k_d t} = 10^{10} e^{-15} = 3.1 \times 10^{-3}$$

$$1 - P_O(t) = 1 - e^{-Nt} = 1 - e^{-3.1\times10^{-3}} = 1 - 4.50 \times 10^{-5} \quad (8.38)$$

So the probability of extinction of the spores is 4.50×10^{-5}.

A sterilization chart can be constructed for equation 8.35 in fig.8.7

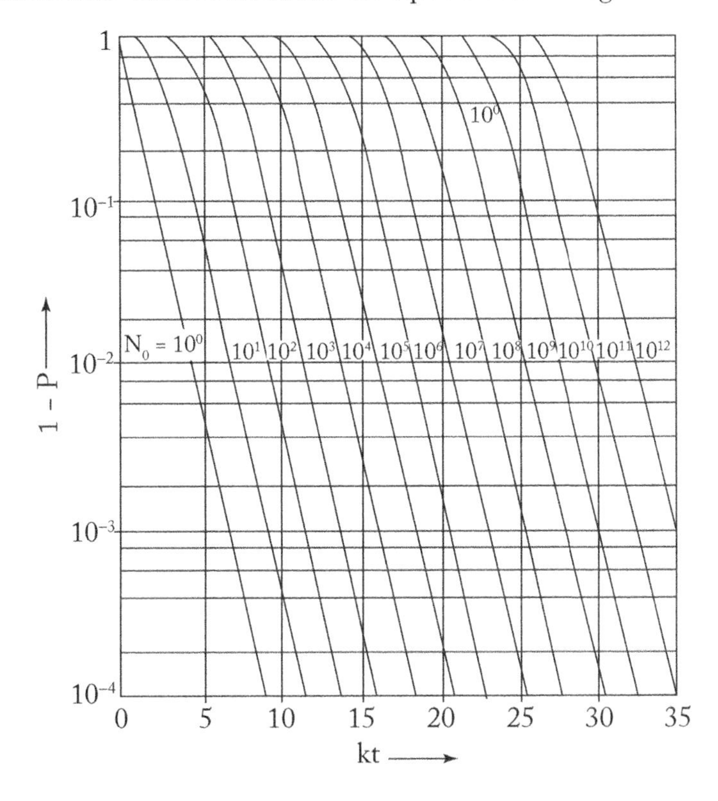

Fig. 8.7 Sterilization chart, 1 - P_0(t) vs kt

Problem 8.4

A medium containing vitamins is sterilized in two separate reactors of volume $0.01m^3$ and $10m^3$ separately at 121°C. The medium also contains the spores. The initial concentration of vitamin = 30 gm/m³ and the initial concentration of spores = 108 spores / m³.

The kinetic parameters of vitamin and spores are:

For inactivation of vitamins,

$kd_o = 10^4 min^\gamma$, $E_d = 4.15 \times 10^4 KJ / kmole$,

For the spores

$kd_o = 10^{36} min^\gamma$, $E_d = 2.70 \times 10^5\ KJ / kmole$,

(a) Compare the amount of active vitamin in the sterilized medium in $0.01m^3$ and $10m^3$ fermenters operating at 121°C. In both cases the probability of an unsuccessful sterilization is 0.001.

(b) Find out the times required for the two fermenters ($0.01m^3$ and $10m^3$) containing spores which are to be killed to the desired sterility level when the probability of unsuccessful sterilization of spores is 0.001.

Solution:

(a) For vitamins:

$$V_R = 0.01m^3, \; T = 121°C = 394K$$

$$k_d = 10^4 \, exp\left(-\frac{4.15 \times 10^4}{8.314 \times 398}\right) = 0.315 \; min^{-1}$$

$$No = \frac{30g}{m^3} \times .01 \; m^3 = 0.30 \; gms$$

Now utilizing the sterilizing chart, (fig 8.7) for $1 - P(t) = 0.001, \; No = 0.30 \; gm$

$$k_d t = 8.7, \; t = \frac{8.7}{0.315} = 27.6 \; min$$

$$N = No \; e^{-k_d t} = 0.30 \; e^{-8.7} = 5.0 \times 10^{-5}$$

$$n = \frac{5.0 \times 10^{-5}}{0.01 \; (m^3)} = 5 \times 10^{-7} \; g/m^3 \qquad \text{Ans.}$$

where V_R = $10m^3$

$$No = 30 \frac{gms}{m^3} \times 10 \; m^3 = 300 \; gms$$

Using the sterilization chart

$$1 - P(t) = 0.001, \; N_o = 300,$$

$k_d t$ value is obtained from the fig 8.7

$$k_d t = 10.4, \; t = \frac{10.4}{.315} = 33.0 \; mins$$

$$N = No \; e^{-k_d t} = 300 \; e^{-10.4} = 9.13 \times 10^{-3}$$

$$n = \frac{9.13 \times 10^{-3}}{10 \; m^3} = 9.13 \times 10^{-4} \frac{gm}{m^3} \qquad \text{Ans}$$

(b) For the spores

For reactor, V_R = $0.01m^3$

$$No = 10^7 \frac{spores}{m^3} \times 0.01m^3 = 10^5$$

K_d at 121°C (394°K)

$$= 10^{36} \exp\left(-\frac{2.7\times10^5}{8.314\times394}\right) = 1.6\,\text{min}^{-1}$$

From the sterilization chart,

$1 - P(t) = 0.0001$, $No = 10^5$, from the fig 8.7 $k_d t = 15.7$

$$\text{Or } t = \frac{15.7}{1.6} = 9.8\,minutes \text{ Ans.}$$

For reactor, $V_R = 10m^3$

$$No = 10^7 \frac{spores}{m^3} \times 10 = 10^8\,spores$$

From the sterilization chart

$1 - P(t) = 0.0001$, $No = 10^8$ from the fig 8.7, we get $k_d t = 25$

$$t = \frac{25}{1.6} = 15.6\,minutes \text{ Ans.}$$

Problem 8.5

For autoclave malfunctions, the temperature reaches only 119.5°C. The sterilization time at the maximum temperature was 20 mins. The jar contains $0.01m^3$ of complex medium containing 10^8 *spores/m*^{3}. At 121°C, $k_d = 1.0\ min^{-1}$ $E_d = 3.7 \times 10^5\ KJ/kmole$

What is the probability that the medium was sterile?

Solution:

At $T = 121$°C (394°K), $k_d = 1\ min^{-1}$

$E_d = 3.7 \times 10^5\ KJ/kmole$,

K_{do} can be calculated as

$$1 = k_{do} e^{-} \frac{3.7\times10^5}{(8.314)(394)}$$

Or $k_{do} = 1.13 \times 10^{49}\ min^{-1}$

At 119.5°C (392.5°K), k_d is calculated,

$$k_d = 1.13\times10^{49} exp\left(-\frac{3.7\times10^5}{8.314(392.5)}\right) = 0.647\,min^{-1}$$

$k_d t = 0.647 \times 20 = 12.94 \approx 13$

From the sterilized chart,

No = $10^8 \times 10^{-2} = 10^6$, $k_d t = 13$

$1 - P(t) \approx 0.8$ Ans.

So the probability that the medium was sterile is satisfactory.

8.6 Other Types of Death Kinetics

A number of death or deactivation kinetics are available for different deactivating compounds like phenol (50%), chlorine, methanol, formaldehyde etc.

(a) The first order deactivation rate is

$$\frac{dN}{dt} = -k_d N \tag{8.39}$$

Which is often referred as **Chick's Law** and has been used extensively to interpret the deactivation rate.

There are other deactivation rate equations which have been used for several cases. They may be presented as follows:

(b)
$$-\frac{dN}{dt} = kN + k'N(N_o - N) \tag{8.40}$$

Which is known as **logistic equation.**

Another equation is given as

(c)
$$-\frac{dN}{dt} = ktN \tag{8.41}$$

Where the log of the viable function, $ln\frac{N}{No}$ varies with t^2.

(d) Retardant

$$-\frac{dN}{dt} = \frac{kN}{1+\alpha t} \tag{8.42}$$

on integrating, we get

$$-ln\frac{N}{No} = \frac{k}{\alpha} ln(1+\alpha t) \tag{8.43}$$

All these models have limited applicability. Some specific deactivation process may be explained by one of these models.

Prob. 8.6 (Sterilization by Chlorine)

E.Coli is sterilized in a batch reactor using different chlorine concentrations. The following date have been reported

Cl concentration = 0.34 mg/L, $\frac{n}{no}$ = fraction of survival

t min	0.5	2	5	10	20
$\frac{n}{no}$, fraction of survival	0.52	0.11	0.007		
C_L = 0.07 mg/L					
n/no	0.80	0.56	0.30	0.005	0
C_L = 0.05 mg/L					
n/no	0.95	0.85	0.65	0.21	0.0031

The following kinetic models are given

1. $-\frac{dN}{dt} = kN$ 2. $-\frac{dN}{dt} = ktN$

Fit the best model to the above data sets

Solution: Integrating model (1)

We have

$$ln\frac{n}{no} = -k_d t \quad \text{(A)}$$

Integrating the model (2), we have

$$ln\frac{n}{no} = -\frac{kt^2}{2} \quad \text{(B)}$$

The following table is made to taste the models

C_L = 0.14 mg/L

t min	0.5	2	5	10	20
$\frac{n}{no}$	0.52	0.11	0.007		
$ln\frac{n}{no}$ C_L = 0.07 mg/L	-0.65	-2.20	-4.96		
$\frac{n}{no}$	0.80	0.56	0.30	0.005	

$ln\frac{n}{no}$ C_L = 0.05 mg/L	-0.223	-0.58	-1.24	-5.3	
$\frac{n}{no}$	0.95	0.85	0.65	0.21	0.031
$ln\frac{n}{no}$	-0.051	-0.16	-0.43	-1.56	-3.47
t^2	0.25	4	25	100	400

From the linear plot of $ln\frac{n}{no}$ vs t in fig 8.8a

1. For $C_L = 0.14 mg/L$ in 8.8 (a)

   $$-k = Slope = \frac{-5.0(-3.0)}{5-3} = \frac{-2}{2} = -1$$

   Or, k = 1 min^{-1}

2. For $C_L = 0.07 mg/L$

   $$-k = Slope = \frac{-2.6(-1.5)}{9-5.2} = \frac{-1.1}{3.8} = -0.29$$

   Or, k = 0.29 min^{-1}

3. For $C_L = 0.05 mg/L$

   $$-k = Slope = \frac{-1.0(-0.7)}{12-8.5} = \frac{-0.3}{3.5} = -0.085$$

   Or, k = 0.085 min^{-1}

4. From the 2nd model fig 8.8b

   $$-\frac{dN}{dt} = ktN$$

   The rate constant, k, obtain is

   $$-k = \frac{-0.98(-0.7)}{60-42.5} = -0.016$$

   k = 0.016 min^{-2}

   The third set of data are fitted best by the second model as shown in Fig. 8.8 (b).

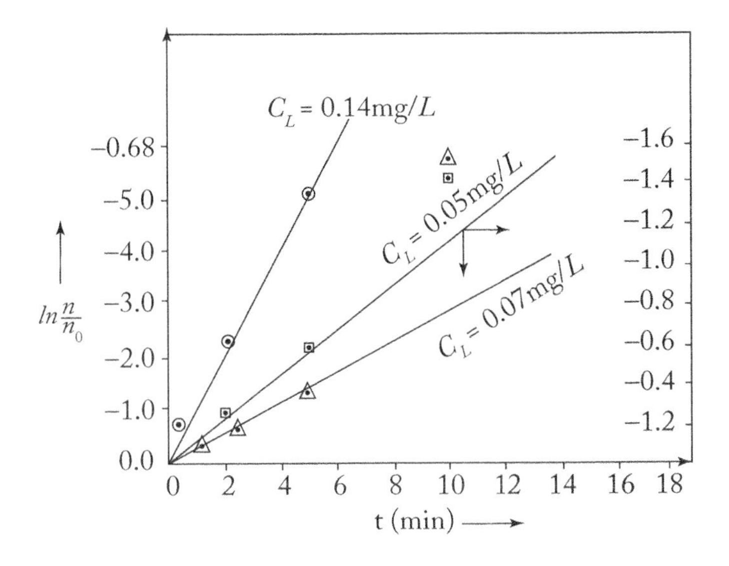

Fig. 8.8 (a) Plot of $ln\frac{n}{no}$ vs t

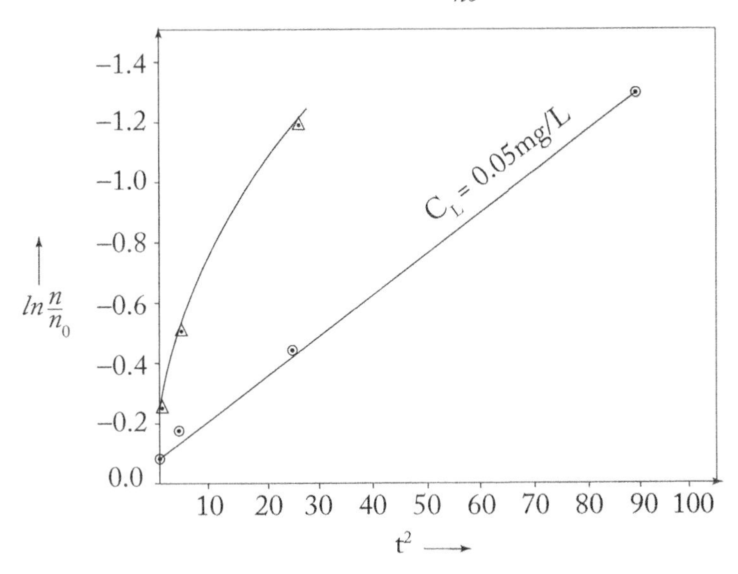

Fig. 8.8 (b) Plot of $ln\ n/n_0$ vs t^2

Problems (exercise)

8.1 A 10m^3 chemostat is operated with D = 0.2hr^{-1}. A continuous sterilizer with steam injection and flash cooling, sterilized the medium which is fed to the fermenter. Medium in the holding section of the sterilizer is kept at 125°C. The concentration of contaminants in the raw medium is 10^{11}/m^3 and the acceptable concentration risk is one organism in 2 months. The Arrhenius constant, k_{do} and the activation energy for thermal death, E_d are $k_{do} = 7.5 \times 10^{39} hr^{-1}$, $E_d = 290 \times 10^3\ KJ / kmole$ respectively. The sterilizer pipe diameter is $D_i = 0.10\ m$. At 180°C, Liquid density, ρl=1000Kg/m³, $\mu_L = 1.1 \times 10^{-3}$ Pas.

(a) What length of pipe is required if axial dispersion effects are considered
(b) Assuming perfect plug flow determine the length of the holding section
(c) If the sterilizer is constructed with length obtained at the part (b) and operated at 125°C, what will be the rate of removal of contamination?

8.2 A 10m³ fermenter is sterilized by batch sterilization and the heat up and cool-down periods are given by time-temperature data. The initial spore concentration is $10^6/m^3$. The kinetic data for sterilization are

$$k_{do} = 1 \times 10^{36}\ min^{-1},$$

$$E_d = 2.7 \times 10^5\ KJ / kmole$$

t min	0	10	30	50	60	65	70	100	140
T°C	30	40	70	121	121	106	98	64	30

(a) What is the probability of a successful sterilization?
Hint assume $k_d t = 18$, read the sterilization chart at $k_d t = 18$ and $no = 10^7$ and read the value of (1−p), the probability of unsuccessful sterilization.
(b) What is the fraction of spores deactivated in the heat-up and cool-down cycle?

8.3 In a batch fermenter containing medium of 20m³ at 25°C is to be sterilized by the direct injection of saturated steam. Initial contamination in medium is $3 \times 10^{12}/m^3$ which is to be reduced to contamination level of 1 in 1000. The saturated steam (345kPa pressure) is injected with a flow rate of 2000 kg/hr which will be stopped when the medium temperature reaches 121°C. During the holding time, the heat loss through the vessel is negligible. After a proper holding time, the fermenter will be cooled by passing 80 m³/hr of water at 20°C through the cooling coil in the fermentor until the medium reaches 30°C. The coil has a heat transfer area of 30m² and the overall heat transfer coefficient U for cooling is 2400 kJ/hrm²K.
The kinetics of the thermal deactivation of spores in the medium are: Arrhenius Coefficient, $k_{do} = 5.0 \times 10^{39} hr^{-1}$, activation energy, $E_d = 2.8 \times 10^5\ KJ / kmole$.
The heat capacity of the medium $C_{p_m} = 4.2 \times 10^5\ KJ / kg\ hr$ and the density, ρm = 1000Kg/m³.

(a) Calculate the total sterilization criterion, del factor hint: $\nabla_{total} = ln\frac{no}{n}$

Calculate ∇_{heat}

Hint: $T = T_o + \dfrac{Hmst}{Cp(M + mst)}$

Given $H = 2000\ KJ/kg$

When $T = 121 + 273 = 394\ K$

$T_o = 298K$, Calculate t_n

$$\nabla_{heat} = k_d \int_0^{t_n} exp\left[-\frac{E_d}{R}\left(T_o + \frac{Hmst}{Cp(M+mst)} \right)^{-1} dt \right]$$

Use numerical method to integrate

(b) Calculate ∇_{cool}

Hint: For the cooling process the change of temperature is given as

$$T = T_{co} + (T_o - T_{co})\left\{\left[1 - exp\left(\frac{UA}{m_o C_{P_o}}\right)\right] \times \frac{m_o t}{M}\right\}$$

Solving t_c for

$T = 30°C + 273 = 303\ K$

$$\nabla_{cool} = k_{do} \int_0^{t_c} exp\left[-\frac{E_d}{RT} \right] dt$$

∇_{cool} is obtained by numerical integration

(c) Calculate ∇_{hold} and holding time

$$\nabla_{hold} = \nabla_{total} - \nabla_{heat} - \nabla_{cool}$$

$$t_{hold} = \frac{\nabla_{hold}}{k_d(T = 394)}$$

8.4 A medium at the flow rate of $2m^3/hr$ is sterilized by heat exchange in a continuous sterilizer at the temperature of 121^0C. The liquid contains bacterial spores at a concentration of $10^{12}/m^3$ and the liquid volume is $10m^3$. The desired sterility level is 1 in 1000. The kinetics of deactivation of spores is given by the equation.

$$k_d = 10^{36} exp\left[-\frac{2.70 \times 10^5}{RT} \right] min^{-1}$$

It has been calculated that the dispersion effect in the flow is described by Peclet number whose magnitude is $P_e = 100$. If the diameter of the tube is 0.3m, calculate the length of the sterilizer required to achieve the above sterility level.

Hint: Calculate N/N_o and read the chart of N/N_o vs. Da (Damköhler no.) at different values of Peclet number and N_o. Read the value of Da or $\frac{k_d L}{u}$. Then calculate k_d and u to obtain L.

8.5 A medium of V=10m^3 contains spores, no. = $10^7/m^3$. The probability of an unsuccessful fermentation is $1 - P_o(t) = 10^{-3}$. If the sterilization is carried out at 121°C and the value of the deactivation constant at 121°C is $k_d = 1.0 min^{-1}$.
Calculate the residence time required to achieve the above probability of unsuccessful sterilization using a chart of $1 - P_o(t)$ vs $k_d\tau$ at different values of No.
Hint: Locate the value of $k_d\tau$ from the plot of $1 - P_o(t)$ vs. $k_d\tau$ at the value of No., initial number of spores.

References

Aiba, S. A.E Humphrey and N.F. Millis, Biochemical Engineering, 2nd ed Academic Press. New York, 1973.

Bader, F.G., Sterilization: Prevention of contamination in A.L. Demain & N.A. Solomon (eds), Manual of Industrial Microbiology, p. 345 – 362, Am. Society of Microbiology, Washington, D.C. 1986.

Bailey, J.E. and D.F. Ollis Biochemical Engineering Fundamentals, 2nd ed., Mc Graw Hill Book, 1986.

Chisty, Y., "Assure Bioreactor Sterility", Chem. Engg. Progr. 88, P. 80–85, 1992.

Cooney, C.L. "Media Sterilization in M. Moo-Young (Editor) Comprehensive Biotechnology, Vol. 2, P. 287–298, Pergamon Press, Oxford, 1985.

Deindoerfer, F.H. and A.E. Humphrey, "Analytical method for calculating Heat Sterilization time", Appl. Micro 7, P. 256–264, 1959a.

Deindoerfer, F.H. and A.E. Humphrey, "Principles in the Design of continuous sterilizers", App. Micro. 7, P. 264–270, 1959b.

Doran, P.M., Bioprocess Engineering Principles, Academic Press, P. 377–385, 1995.

Levenspiel, O., "Longitudinal Mixing of Fluids flowing in circular pipes", Ind. Eng. Chem 50, P. 343–346, 1958.

Richards, J.W., Introduction to Industrial Sterlization, Academic Press, London, 1968.

Shuler, K.L. and F. Kargi, Bioprocess Engineering: Basic Concepts, Chapter 9, Prentice Hall, N.J., 1992.

Quesnel, L.B., Sterilization and Sterility in Basic Biotechnology, edited by BuLock and B. Kristiansen, London, England, Academic Press, P. 197–215,

CHAPTER 9

Design and Analysis of Bioreactors

9.0 Introduction

Bioreactors are characterized by complex kinetics, degree of mixing, mass transfer, heat transfer and hydro dynamics. The actual design of such reactors is carried out by simplifications on heat transfer and mixing. In most of the packed bed reactors, the enzymes or microbial cells are immobilized on solid particles of definite sizes, which are placed in the reactor as packing. These packed systems involve biochemical reactions, interacting multiphase flow and intra particle diffusion.

Earlier two types of reactors have been defined such as ideal and non-ideal reactors. The ideal reactors are of two basic types viz plug flow and backmix systems. The non-ideal reactors have been modeled by dispersion model, tanks in series or combination of some ideal reactors.

The present chapter deals with plug flow and mixed flow reactor systems containing free and immobilized enzymes or cell systems, non-ideal reactors like plug flow with axial dispersion, tanks in series model. Other topics include residence time distribution, reactor dynamics, design of packed bed and fluidized bed reactor system with diffusion effect.

9.1 Ideal Plug Flow Tubular Reactor (PFTR)

When the flow in a long pipe is turbulent i.e. Reynolds number exceeds 2100, the flow is assumed to be plug, that is, no axial mixing is present. Assuming a differential length, dz in a tubular reactor, a mass balance can be made for a reactive component, C, for the steady state,

$$Auc\big|_{atZ} - Auc\Big|_{z+\Delta Z} - A\Delta Z r_c = 0 \tag{9.1}$$

Where A is the area of the reactor, u is the axial velocity, r_c is the rate of reaction. Rearranging and dividing by $A\Delta Z$, we have

$$uc\big|_{Z+\Delta Z} \bar{\ } uc\big|_{Z} = r_c \tag{9.2}$$

Taking the limit $\Delta Z \to 0$, we have

$$\frac{d}{dZ}(uc) = r_c \tag{9.3}$$

where r_c is the rate of reaction for the component, c.

For a constant velocity, u,

$$u\frac{dc}{dZ} = r_c \tag{9.4}$$

We consider the cell growth in a tubular reactor

where $$r_x = \frac{\mu_{max} SX}{K_s + s}$$

The equation (9.4) can be used for cell growth kinetics for plug flow reactor design,

$$u\frac{dX}{dZ} = \frac{\mu_{max} SX}{K_s + S} \tag{9.5}$$

In terms of substrate, S and yield coefficient, Yx/s, the plug flow design equation for a substrate is given as,

$$u\frac{dS}{dZ} = -\frac{1}{Y_{x/S}}\frac{\mu_{max} SX}{(S + K_s)} \tag{9.6}$$

This is equivalent to batch reactor, when $dt = \frac{dZ}{u}$

Equation (9.6) can be integrated by eliminating X by the following equation,

$$X = X_o + Y_{x/S}(S_0 - S) \tag{9.7}$$

Substituting equation (9.7) in equation (9.6) we have,

$$-u\frac{dS}{dZ} = \frac{\mu_{max}\left[X_0 + Y_{x/S}(S_0 - S)\right]S}{Y_{x/S}(S + K_s)} \tag{9.8}$$

Integrating equation (9.8),

$$-\int_{S_0}^{S} \frac{(S + K_s)Y_{x/S}dS}{\left[X_0 + Y_{x/S}(S_0 - S)\right]S} = \frac{\mu_{max}}{u}\int_0^L dZ \tag{9.9}$$

we get after integration and simplification,

$$X_0 + Y_{x/S}\left(S_0 + K_s\right) ln X_0 + Y_{x/S}\left(S_0 - S\right) - K_s Y_{x/S} ln\frac{S}{S_0} = \mu_{max}\left(X_0 + Y_{x/S} S_0\right)\frac{L}{u} \tag{9.10}$$

If the cell reaction is carried out in a packed bed with inert packing and a bed porosity of $\in$, the right hand side of the equation (9.10) will be given as

$$\frac{\left(1 - \in\right) L \mu_{max}\left(X_0 + Y_{x/S} S_0\right)}{\in u} \tag{9.11}$$

where $\in u$ is the effective velocity in the porous packed bed. The packed bed porosity is about 0.3.

The effluent concentration is S corresponding to residence time, $\tau = \frac{L}{u}$ as given by equation (9.10). In contrast to CSTR, sterile feed to PFTR shows zero biomass at the exit. This problem is solved by recycle, so that incoming stream is inoculated before entering the reactor.

9.2 Reactor Dynamics

For the steady state operation of a bioreactor, its dynamic characteristics must be evaluated. To find out those characteristics, a classical system of non-isothermal CSTR with first order kinetics has been considered. For a chemostat system, the unsteady state mass balance for the component, C can be given as

$$dC/dt = D(Co - C) + r_c \tag{9.12}$$

Where D (F/V) is a parameter called dilution rate and $1/D$ is equal to residence time, τ, r_c is the microbial kinetics for a component, C equivalent to cell mass.

9.2.1 Linearization of a set of non-linear differential equations

The dynamic reactor model consists of a set of first order differential equations. To evaluate the dynamic characteristics of a non-isothermal CSTR in the form of stability criteria, let us consider two first order linear differential equations of a non-isothermal CSTR with first order kinetics, obtained after linearization of the non-linear kinetic terms using Taylor series expansion in terms of deviation variable x (concentration) and y (temperature). They are defined as

$$x = X - Xs,\ y = T - T_s$$

where X, Xs are the concentrations at any time and at the steady state respectively. Similarly T and T_s are the temperatures at any instant, and at the steady state respectively.

These equations may be presented as

$$\tau\frac{dx}{dt} = a_{11}x + a_{12}y \tag{9.13}$$

$$\tau \frac{dy}{dt} = a_{21}x + a_{22}y \tag{9.14}$$

where a_{11}, a_{12}, a_{21} and a_{22} are constants containing process parameters like τ (residence time), $k(T_s)$ (rate constant at T_s), E (activation energy), T_s (steady state temperature, R (gas constant) in the Arrhenius equation.

9.2.2 Stability analysis in terms of eigen values

These two simultaneous first order differential equations in terms of deviation variables (x) & (y) can be expressed by a single second order differential equations with constant coefficients,

$$\frac{d^2x}{dt^2} - (a_{11} + a_{22})\frac{dx}{dt} + (a_{11}a_{22} - a_{12}a_{21})x = 0 \tag{9.15}$$

The general solution of equation (9.15) can be presented as

$$x(t) = C_1 e^{+\lambda_1 t} + C_2 e^{-\lambda_2 t} \tag{9.16}$$

The coefficients, C_1 and C_2 are evaluated from two initial boundary conditions. The quantities, $\pm\lambda$ are called eigenvalues. They are related to the coefficients by the following equations,

$$\lambda = \frac{1}{2}(a_{11} + a_{22}) \pm \frac{1}{2}\left[(a_{11} - a_{22})^2 + 4a_{12}a_{21}\right]^{1/2} \tag{9.17}$$

The stability of $x(t)$ is determined by whether the solution will come back to the steady state values of x, y (Xs, Ys) following a perturbation. If the solution is stable, the real part of λ should be negative.

In a more general way, the set of equations (9.13) and (9.14) can be presented in vector-matrix notation.

$$\dot{x} = Ax \tag{9.18}$$

where the matrix A can be given as

$$A = \begin{bmatrix} a_{11} & a_{12} \\ a_{21} & a_{22} \end{bmatrix} \tag{9.19}$$

Where $\dot{x}$ is a vector of concentrations with dimensions equal to the number of components.

All the solutions of the linearized equation(9.18) take the form:

$$x(t) = \sum_{i=1}^{m} \propto_i \beta_i e^{\lambda it} \tag{9.20}$$

The quantities, β_i and λ_i are the corresponding pairs of eigenvectors and eigenvalues of the matrix A respectively. αi are constants.

If $\lambda = \lambda_i$ satisfies the characteristic equation,

$$det(A - \lambda I) = 0 \tag{9.21}$$

where I is the identity matrix and β satisfies the equation

$$(A - \lambda_i I)\beta_i = 0 \tag{9.22}$$

where i = 1.... m

The $\propto_i$ constants of equation (9.22) are to be chosen to satisfy the following algebraic equation with the specific initial conditions.

$$\sum_{i=1}^{m} \propto_i \beta_i = x(0) \tag{9.23}$$

where $x(t)$ is a specified vector of initial deviation. The solution for x ($x(t)$) around xs leads to the local stability. It is determined in most cases by the eigenvalues of the matrix A. The steady state, xs is locally asymptotically stable if all the eigenvalues of A have negative real parts.

The determinant A of equation (9.22) can also be expanded to get an nth order algebraic equation,

$$\lambda^n + B_1\lambda^{n-1} + \cdots + B_{n-1}\lambda + B_0 = 0 \tag{9.24}$$

Now we can use the classical method of **Herwitz stability** criterion which give the negative real parts of the equation (9.24)

Coming back to the analysis of non-isothermal, linearized CSTR system, the solution of equation (9.15), $x(t)$ is given as function of time, t around the steady state value of $x(x_S)$ and is illustrated in fig. 9.1

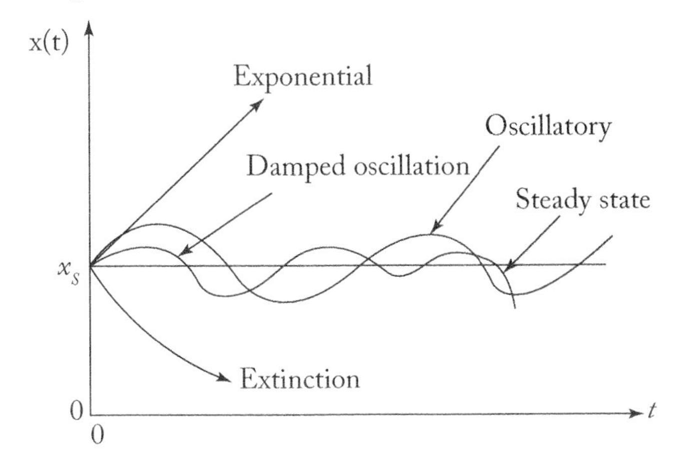

Fig. 9.1 Behavior of a CSTR for a small perturbation from the steady state, x_S

The solution will remain bounded in (x_S) if $\lambda < 0$, but not if $\lambda \geq 0$ If λ is complex, the system will exhibit either damped or growing oscillation depending on whether the real part of λ. is negative or positive.

The behavior of such system can be illustrated by the following example.

A linearized reaction system is given by two linear differential equations,

$$\dot{x}_1 = 4x_1 - 5x_2$$

$$\dot{x}_2 = x_1 - 2x_2$$

or in vector form

$$\dot{x} = Ax$$

$$A = \begin{vmatrix} 4 & -5 \\ 1 & -2 \end{vmatrix}$$

Where A is the matrix of the system. Show whether the system is stable.

The solution can be shown as follows:

We find out the characteristic equation from the determinant

$$det(\lambda I - A) = 0$$

Or

$$\begin{vmatrix} \lambda - 4 & 5 \\ -1 & \lambda + 2 \end{vmatrix} = 0$$

Or

$$\lambda^2 - 2\lambda - 3 = 0$$

Solving $\lambda_1 = -1, \lambda_2 = 3$

So the solution is

$$x(t) = C_1 e^{\lambda_1 t} + C_2 e^{\lambda_2 t}$$
$$= C_1 e^{-t} + C_2 e^{3t}$$

Since one eigenvalue is positive, $x(t)$ will not come to the steady state.

Routh criterion

Further if the characteristic equation is tested with Routh test, it will be found that the system is also not stable.

a_0	a_2	a_4
1	3	0
a_1	a_3	a_5
-2	0	0
b_1		
3		

$$b_1 = \frac{a_1 a_2 - a_0 a_3}{a_1} = \frac{-6-0}{-2} = 3.$$

Since all the elements in the first column of the Routh array (a_0, a_1, b_1, ...) are not positive, so the system is not stable.

9.2.3 Dynamics of monod chemostat model (Bailey & Ollis, 1985)

In the dynamic analysis of Monod Chemostat model, we consider the unsteady state mass balance equations of the model for the cell mass (X) and substrate (S) (assuming single substrate controlling). The equations are:

$$\frac{dX}{dt} = D(X_0 - X) + \frac{\mu_{max} SX}{K_s + S} \tag{9.25}$$

$$\frac{dS}{dt} = D(S_0 - S) - \frac{1}{Y_{x/s}} \cdot \frac{\mu_{max} SX}{K_s + S} \tag{9.26}$$

For sterile feed two steady states are possible, these are,

$$S_S = \frac{DK_s}{\mu_{max} - D} \tag{9.27}$$

$$X_S = Y_{x/s}\left(S_0 - \frac{DK_s}{\mu_{max} - D}\right) \tag{9.28}$$

and the other steady state is the washout solution when Xs = 0 and S_s = S_0 (initial substrate concentration).

Local stability can be evaluated by linearization of the equations (9.25 and 9.26) and solving the linearized equation in terms of perturbation variables using eigenvalues. The results of such analysis provided the following information.

For non-trivial steady state, the system is stable when

$$D < D_0 = \frac{\mu_{max} S_0}{K_s + S_0} \tag{9.29a}$$

The unstable condition prevails when

$$D > D_0 = \frac{\mu_{max} S_0}{K_s + S_0} \tag{9.29b}$$

For the Washout steady state, the system is stable when $D > D_0$ and becomes unstable when $D < D_0$, where D_0 is the dilution rate at $S = S_0$ in the Monod equation.

From the dynamic analysis of the above system, it has also been predicted that concentration cannot approach the steady state in a damped oscillation. Experimentally such oscillations have been observed indicating that the substrate and the cell model as given by Monod is not capable of predicting the dynamic features of such reactors. It has been suggested that structured models where more components are included are supposed to satisfy the dynamic requirements of the reactors.

9.3 Non-ideal Bioreactors in Terms of Residence Time Distribution

Non ideal reactors have been defined with respect to mixing. An ideal plug flow reactor is characterized by a flow without axial mixing. Similarly an ideal mixed reactor is a system with complete mixing. The residence time distribution in a vessel gives the extent of mixing with the vessel. The effluent stream in a vessel is a mixture of fluid elements which have resided in the vessel for different lengths of time. The distribution of these residence times in this exit stream is called Residence time distribution (RTD) which gives a valuable information about mixing and flow pattern inside the vessel.

9.3.1 Tracer experiments

In order to find out the mixing characteristics of the vessel, experiments are carried out by two input techniques viz. (1) Pulse input and (2) Step input. For pulse input experiment, an inert tracer at time, $t = 0$ is introduced quickly into the feed line, the system response is measured from the exit tracer concentration.

For step input, the response to the unit step tracer input is

$F(t)$

where
$$F(t) = \frac{C(t)}{Co} = \begin{cases} 0 t < 0 \\ 1 t \geq 0 \end{cases} \tag{9.30}$$

$F(t)$ is evaluated by diving the $C(t)$ function obtained in the above experiment by the tracer feed concentration, Co. The unit step function response of the vessel is called the F-function,

$$F(t) = \frac{C(t)}{Co} \tag{9.31}$$

Which is a response to the unit step input of tracer.

For pulse input of tracer, the response curve, also known as the residence time distribution function, $E(t)$ defined as

$E(t)$ = fraction of fluid in exit stream which remains in the vessel for the time between t and $t + dt$ (9.32)

on the basis of the above definition

$$\int_0^\infty E(t)\,dt = 1 \tag{9.33}$$

$$E(t) = \frac{C(t)}{\int_0^\infty C\,dt} \tag{9.34}$$

with known $E(t)$, the exit tracer concentration, $C(t)$ is the sum of fluid I and fluid II contributions.

Now fluid I contains tracer at concentration Co and fluid II is devoid of tracer. Mathematically

$$C(t) = Co\int_0^t E(t)\,dt + 0\int_t^\infty E(t)\,dt \tag{9.35}$$

$$\text{Or } \frac{c(t)}{C_0} = F(t) = \int_0^t E(t)\,dt \tag{9.36}$$

$$\text{Or } \frac{dF}{dt} = E(t) \tag{9.37}$$

which gives the relation between $F(t)$ and $E(t)$.

Now let us consider an ideal CSTR to which an inert tracer, C is introduced, tracer balance gives

$$dC/dt = F/V_R(Co - C) = 0 \tag{9.38}$$

For F experiment,

$$C(0) = 0$$

$$C = Co, \text{ at } t \geq 0 \tag{9.39}$$

The solution of the equation (9.44) with the given conditions may be shown as

$$F(t)=\frac{c(t)}{C_o}=1-\bar{e}^{t/\tau} \tag{9.40}$$

Where $\tau = V_R / F$, V_R = reactor volume and F = flow rate

Then,
$$E(t)=\frac{dF(t)}{dt}=\frac{1}{\tau}e^{-t/\tau} \tag{9.41}$$

9.3.2 Non-ideal reactor models

On the basis of residence time distribution, there are two important models viz. 1) Tanks in series model and 2) Axial dispersion model. Both of them are equivalent. The E and F functions can be used to assess the extent of deviation from an idealized reactor. The performance of a non-ideal reactor may be determined by any of the above two models. For batch reactors, RTD measurements have been used to determine gas-holdup and mixing behavior.

Another type of reactor model known as a combined model or a mixed model comprising inter-connected idealized reactor types.

9.3.2.1 Tanks-in-series model

In this model, the total reactor volume, V_R, is divided into N equal sized tanks, that is

$$V_1 = V_2 = \cdots = V_n = \frac{V_R}{N} \tag{9.42}$$

and these tanks are arranged in series.

A pulse input of tracer of concentration, *Co* is introduced into the first vessel where liquid flows at the rate of F. The residence time distribution for N number of such reactors has been deduced in Chapter 3 and may be given as,

$$E(t)=\frac{N^N}{(N-1)!}\left(N\frac{F}{VR}\right)^{N-1} t^{N-1} exp\left(-\frac{NF}{V_R}t\right) \tag{9.43}$$

The dimensionless variance of RTD, σ_0^2 from equation (9.49) is given as

$$\sigma_0^2 = 1/N \tag{9.44}$$

Where $N \rightarrow 1$, the system behaves as an ideal CSTR and if $N \rightarrow \infty$ the system predicted by RTD approaches ideal PFR. RTD for tanks-in-series model has been shown in fig. 9.2.

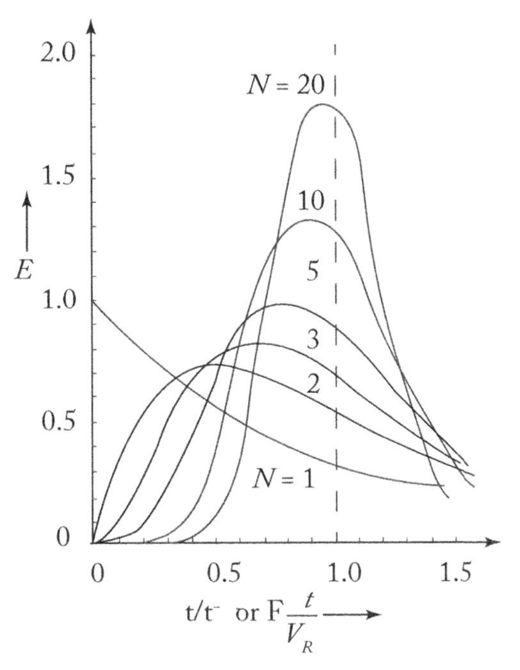

Fig. 9.2 Residence time distribution for Tanks-in-series model (Levenspiel, Chemical reaction engineering, Wiley, 1999)

A stirred tank train with recycle can be presented as below.

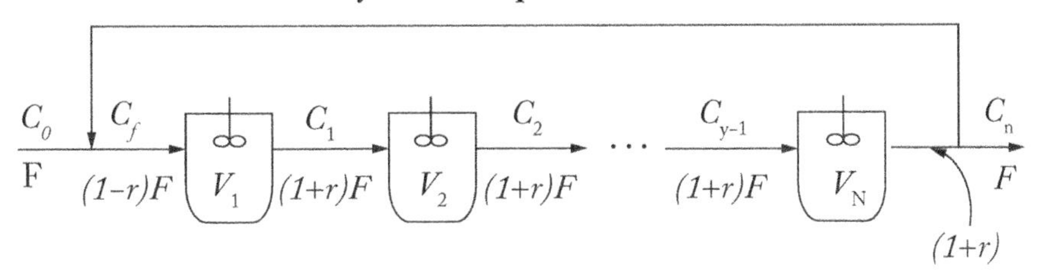

Fig. 9.3 Stirred tank train with recycle

For the above system, it can be deduced that the effect of staging and recycle on the washout dilution rate, $D_{washout}$ will be given by an expression in terms of N and other parameters,

$$\frac{\mu_{max}}{D_{washout}} = \frac{N\left(1+K_s/S_o\right)\left\{1-\left[r/\left(1+r\right)\right]^{1/N}\right\}}{1-\left(k_e/\mu_{max}\right)\left(1+K_s/S_O\right)} \tag{9.45}$$

where Monod model is used with maintenance= factor, k_e.

Where $$D = \frac{F}{V_R} = \frac{F}{NV_1} = \frac{D^*}{N} \tag{9.46}$$

D^* is the dilution rate of each stage and Monod model is used with maintenance coefficient

$$\mu = \mu_{max} S / (K_s + S) - ke$$

Setting $r = 0$ and $k_e = 0$, the equation (9.51) reduces to familiar expression,

$$D^* = \frac{\mu_{max} S_o}{K_S + S_O} \tag{9.47}$$

9.3.2.2 Dispersion model

This model is analogous to CSTR cascade model with backmixing. It is the modification of ideal PFTR by axial dispersion which is superimposed on the convective flow (uC) through the tube. Using an elemental length of dZ and making a shell balance with convective flow and dispersion (D_L) $\left(-AD_L dC / dZ\right)$ the following equation may be deduced as,

$$u\frac{dC}{dZ} = D_L \frac{d^2C}{dZ^2} \mp r_c \tag{9.48}$$

where r_C is the rate of reaction.

Or $$D_L \frac{d^2C}{dZ^2} - u\frac{dC}{dZ} \mp r_c = 0 \tag{9.49}$$

The above is a second order differential equation which can be solved with two boundary conditions

$$uc_o = \left(uc - D_L \frac{dC}{dZ} \right)_{Z=0} \tag{9.50}$$

Or $$\left.\frac{dC}{dZ}\right|_{Z=L} = 0 \tag{9.51}$$

when D_L is not large, the condition (9.50) may be replaced as

$$C /_{Z=0} = C_O \tag{9.52}$$

In this condition, the variance of RTD, σ_θ^2 is given for a closed vessel,

$$\sigma_\theta^2 = \frac{2}{P_e}\left[1 - \frac{1}{P_e}\left(1 - e^{-P_e}\right)\right] \tag{9.53}$$

where the axial peclet number, P_e is defined as

$$P_e = \frac{uL}{D_L} \tag{9.54}$$

The physical significance of Peclet number is given as the ratio of convective mass transfer $(\approx uc)$ to the mass transport by dispersion $(\approx D_L C / L)$

Now the values of P_e indicate the degree of mixing.

When P_e tends to zero, $\sigma_\theta^2 \approx 1.0$ which leads to ideal CSTR. Similarly when $P_e \rightarrow \infty$, the value of σ_θ^2 is zero, which corresponds to ideal plug flow.

Dispersion model can be applied to chemostat system with Monod kinetics containing maintenance parameter, k_e. The dilution rate at washout condition can be presented by the following equation (9.61)

$$\frac{\mu_{max}}{D_{washout}} = \frac{\frac{1}{4}P_e\left(1 + {K_S}/{S_o}\right)}{1 - (k_e / \mu_{max})(1 + K_S / S_o)} \tag{9.61}$$

As $P_e \rightarrow 0$ $D_{washout}$ decreases to zero. This agrees with the notion that an ideal PFTR with sterile feed will not support microbial population.

9.3.2.3 Axial dispersion, D_L

Axial dispersion, D_L, is different from molecular diffusion. It is a modeling parameter which represents the mixing effects in a vessel. We model three types of mixing behavior for three types of flow viz. laminar and turbulent flow and flow in packed beds.

For laminar flow, the dispersion coefficient, D_L is presented as

$$D_L = D + \frac{\bar{u}^2 dt^2}{192D} \tag{9.55}$$

where D is the diffusion coefficient, $\bar{u}$ is the average velocity in a pipe

$$\left(\bar{u} = \frac{1}{2}u_{max}\right) \tag{9.56}$$

For turbulent flow, $R_e > 2100$, macroscopic eddies of fluid provide important mechanisms, for mass, momentum and energy transport. The turbulent flow peclet number is usually of the order of 3. It falls with decreased Reynolds number. Intensity of mixing defined by $D_L / \bar{u}dt$ has been plotted as a function of $R_e \left(= dtu\rho/\mu\right)$ at different values of Schmidt (Sc) number in figure 9.4.

For packed bed, Peclet number based on particle diameter *dp* is defined as,

$$P_{ep} = udp / D_L \tag{9.57}$$

and a plot of $1 / P_{ep}$ vs particle Reynolds number $dp\ u\rho/\mu$ (has been shown in figure 9.5 at different values of Schmidt number (Sc). Theoretically, axial peclet number based on particle diameter, *dp* is $P_e = \dfrac{\bar{u}dp}{D_L} = 2.0$

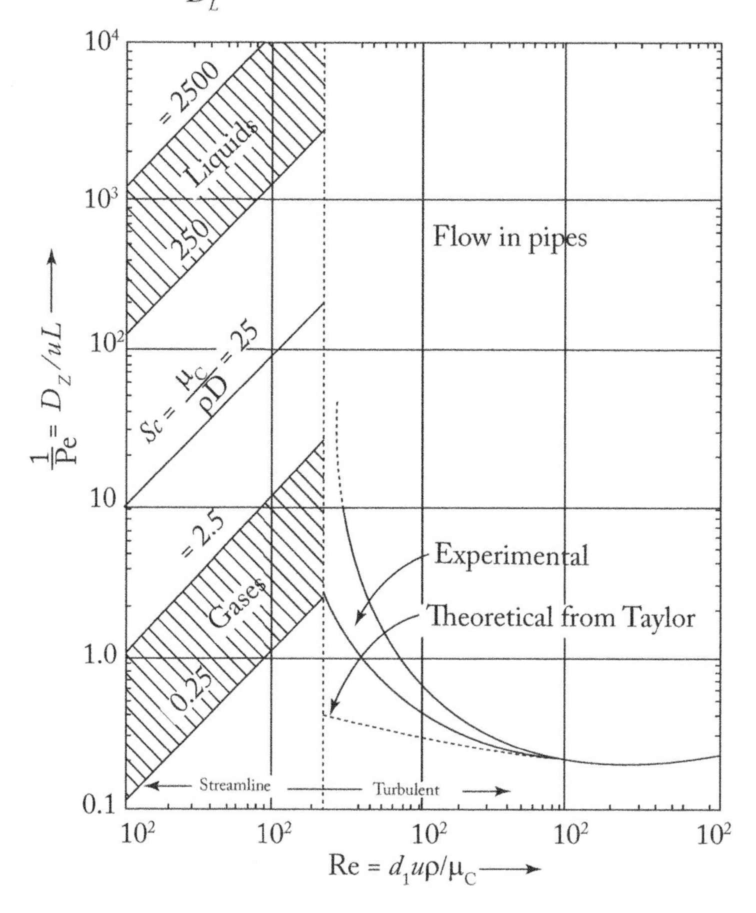

Fig. 9.4 Correlations for the axial Peclet number (P_e) in terms of Reynolds (R_e) and Schmidt (S_c) number for fluid flow in pipes (Levenspiel (1958b) p.310)

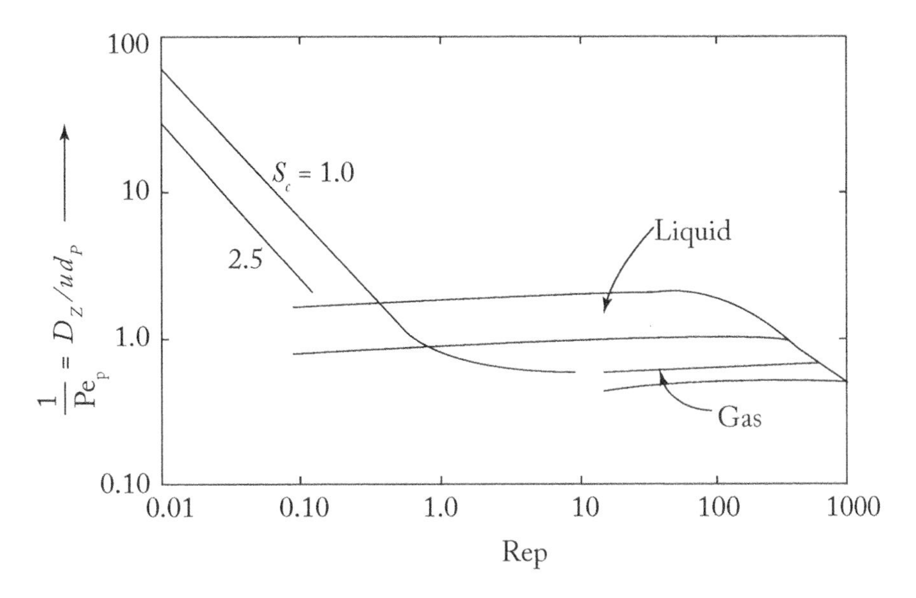

Fig. 9.5 Correlation for the axial Peclet number with Reynold number based on particle diameter, *dpuρ/μ* for flow in packed pipes (Levenspiel, 1999 p.311)

It is to be noted that the dimensionless groups are based upon particle diameter d_p

Regarding the choice of models, it is a general guideline that the dispersion model is preferable for small deviation from plug flow when $\sigma_\theta^2 \leq 0.05$

On the other hand the Tanks-in-series model is recommended when there is substantial backing as indicated by the values of $\sigma_\theta^2 \leq 0.2$

The dispersion model has been successfully used in continuous sterilization reactor as illustrated in a Chapter 8 and the tanks in series model satisfactorily explained the behavior of biological waste water treatment tank.

Problem 9.2

The RTD of a chemostat system with a recycle gives the variance,

$$\sigma^2 = 0.2$$

(a) It is required to find out the dilution rate for washout condition, D_{max}, from Tanks-in-series model. The data for the chemostat system are:
$\mu max = 0.8\ hr^{-1}$, $K_s = .2\ Kg/m^3$, r (the recycle ratio) = 0.5

(b) If the dispersion model is used, calculate the dilution rate for washout (D_{max})

Solution:

(a) For the tanks-in-series model

$$N = \frac{1}{\sigma_\theta^2} = \frac{1}{0.2} = 5$$

D_{max} for washout is given as

$$\frac{\mu_{max}}{D_{max}} = \frac{N\left(1+\frac{K_s}{S_o}\right)\left\{1+\left(\frac{r}{1+r}\right)^{1/N}\right\}}{1-\left(K_s/\mu_{max}\right)\left(1+\frac{K_S}{S_o}\right)}$$

Substituting the values into the above equation

$$\frac{\mu_{max}}{D_{max}} = \frac{5(1+0.2/10)\left\{\left(1-\frac{0.5}{1.5}\right)^{1/5}\right\}}{1-\left(0.2/0.8\right)\left(1+\frac{0.2}{10}\right)} = 6.3$$

$$\text{Or } D_{max} = \frac{\mu_{max}}{6.3} = \frac{0.8}{6.3} = 0.12\,hr^{-1}$$

(b) Dispersion model to calculate, D_{max} for washout

Calculation of Peclet number form σ_θ^2.

$$\sigma_\theta^2 = 0.2 = \frac{2}{P_e}\left[1-\frac{1}{P_e}\left(1-e^{-P_e}\right)\right]$$

Solving by trial and error

$P_e = 8$

Using Peclet No., D_{max} can be calculated from the following equation

$$\frac{\mu_{max}}{D_{max}} = \frac{\frac{1}{4}P_e\left(1+K_s/S_o\right)}{1-\left(k_e/\mu_{max}\right)\left(1+K_s/S_o\right)}$$

Substituting the values, we have

$$\frac{0.8}{D_{max}} = \frac{\frac{1}{4}(8)(1+0.2/10)}{1-(0.2/0.8)\left(1+\frac{0.2}{10}\right)} = 2.79$$

$$\text{Or } D_{max} = \frac{0.8}{2.79} = 0.28\,hr^{-1}$$

Since $\sigma_\theta^2 = 0.2$, the tanks-in- series model gives the better result

9.4 Some Reactor Design Models

9.4.1 Packed bed reactor

Immobilized cells on the support surface are fragile. So reactor with low hydrodynamic shear should be used. Examples are packed bed, fluidized bed or air lift reactors. Mechanically agitated fermenters may be used if the support matrix is strong. Schematic diagram of a packed column has been shown in the chapter 7 (fig 7.3a)

When the fluid circulation is high, the reactor is modeled as plug flow tubular reactor.

To develop the design of a packed bed reactor on the basis of plug flow, a material balance on the substrate(s) in an elemental reactor volume of *Adz*, is given as,

$$-F\frac{dS}{dZ} = N_s a A \tag{9.58}$$

and $N_s = r_m S/(K_s + S)$

Where, S is the bulk-liquid substrate concentration which decreases with height Z. N_s is the flux of substrate decomposition rate, F is the volumetric flow rate, a is the biofilm surface area per unit reactor volume, dZ is the differential height of the column, A is the cross section of the bed, r_m = max substrate consumption $\frac{\mu_{max} X}{r_{x/s}}$

Now introducing an effectiveness factor, η, for intra particle diffusion and substituting the design equation of a packed bed reactor is obtained from equation (9.58) as

$$-F\frac{dS}{dZ} = \eta\frac{r_m S}{K_s + S} L a A \tag{9.59}$$

Where r_m is the maximum rate of substrate utilization rate.

Integrating the above equation we have

$$K_s ln\frac{S_0}{S} + \left(S_o - S\right) = \frac{\eta r_m L a A H}{F} \tag{9.60}$$

where S_o is the inlet substrate concentration and S is the exit substrate concentration. L is the biofilm thickness, H is the required packed bed heights. If the porosity of the bed is considered, the right hand side is multiplied by $(1 - \in_b)$.

For low substrate concentration, the substrate consumption rate is first order and the design equation becomes

$$ln\frac{S_o}{S} = \frac{\eta r_m \left(1 - \epsilon_b\right) L a A H}{F K_s} \tag{9.61}$$

Problem 9.3

Lactose(dairy waste) is hydrolysed to glucose and galactose in a packed bed reactor of resins immobilized with lactose. The resin particles size is approximately 1 mm diameter. The packed bed volume is 1.0 m^3. The lactose concentration in the feed is 10 kg/m^3. 98% conversion of the substrate is required. The column is operated as plug flow for a total of 310 days per year. The other data are: K_M = 1.5 kg/m^3, $v_{max} = 40 Kg/(m^3)(hr)$

De (the effective diffusivity of lactose in resins = 1.2 x 10^{-10} m^2/s.

Calculate the following:

(a) The effectiveness factor of the systems, η

(b) What is the flow rate in the reactor?

(c) How many tons of glucose are produced per year?

Solution:

(a) Calculation of effectiveness factor, η for spherical particle

$$\eta = \frac{1}{\phi}\left[\frac{1}{\tanh 3\phi} - \frac{1}{3\phi}\right]$$

Where $\phi = \frac{R}{3}\sqrt{\frac{\mu_{max}}{k_m D_e}}$

$$R = \frac{1}{2} mm \times 10^{-3} \frac{m}{mm} = 0.5 \times 10^{-3} m$$

$$\phi = \frac{0.5 \times 10^{-3}}{3}\sqrt{\frac{40/3600}{1.5 \times 1.2 \times 10^{-10}}} = 1.3$$

Now $\eta = \frac{1}{\phi}\left[\frac{1}{\tanh 3\phi} - \frac{1}{3\phi}\right]$

$$= \frac{1}{1.3}\left[\frac{1}{\tanh 3.9} - \frac{1}{3.9}\right]$$

$$= \frac{1}{1.3}\left[1 - 0.2564\right] = 0.57$$

(b) $\frac{V}{F} = -\int_{S_0}^{S} \frac{ds}{\left(\frac{\upsilon_{max} S}{K_M + S}\right)},$

$$S_o = 10Kg/m^3,\ S = (.02)\times 10 = 0.2\frac{Kg}{m^3}$$

Integrating and rearranging and putting we get, $\tau = V/F$

$$\tau = \frac{K_M}{\eta v_{max}} ln\frac{S_o}{S} + \frac{S_o - S}{\eta v_{max}}$$

$$= \frac{1.5}{0.570\times 40} ln\frac{10}{0.2} + \frac{10-0.2}{.57\times 40} = 0.257 + 0.430 = 0.687\,hr$$

$$F = \frac{V}{\tau} = \frac{1m^3}{0.687\,hr} = 1.45\frac{m^3}{hr}$$

(c) $Lactose \quad + \quad H_2O \rightarrow glucose \quad + \quad galactose$
$\quad (342) \qquad (18) \qquad (180) \qquad (160)$

$$\text{Lactose decomposed} = \frac{(10-0.2)}{342} = 0.028\frac{k_{mol}}{m^3}$$

Glucose produced per yr.

$$= 0.028\frac{K_{mol}}{m^3}\times 180\frac{Kg}{K_{mole}}\times 1.45\frac{m^3}{hr}\times 24\frac{hrs}{day}\times\frac{310\,day}{yr}\times\frac{ton}{1000Kg} = 5.24\,tons$$

Problem 9.4

A packed bed of immobilized yeast cells entrapped in gel-beads is used to convert glucose to ethanol. The specific rate of ethanol production, q_P = 0.2 kg ethanol / (kgcells) (hr). The effectiveness factor for an average bead, η = 0.8. Beads contain 50 kg cells per m^3 of bed. The voidage of the column is $\in_b$ = 0.4. Growth of cells is negligible. The feed flow rate is F = 0.5m^3/hr. Glucose concentration in the feed is S_0 = 150 kg glucose/m^3. The diameter and length of the reactor column are 1m and 4m respectively. $Y_{P/S}$ = 0.4 kg ethanol/kg glucose

(a) Calculate the glucose conversion in the column
(b) Estimate the ethanol concentration (P) in the exit stream

Solution:

The design equation based on the plug flow

$$-Fds/dz = \eta q_P X/Yx/s\ (1 - \in_b)\ A$$

$$\text{or } -F\frac{dS}{dZ} = \eta\frac{q_P}{Y_{P/S}}\cdot X(1-\in_b)\text{A}$$

Integrating and rearranging

$$S = S_0 - \frac{\eta q_p \cdot q_x \left(1-\in_b\right) A.H.}{Y_{P/S}\, F} \tag{A}$$

Now $A = \frac{\pi}{4} D^2 = 0.7854 \times \left(1m\right)^2$

Putting the values in equation (*A*)

$$S = 150 - \frac{(0.8)(0.2)(50)(1-0.4)(0.7854)(4)}{(0.4)(0.5)}$$

$$= 150 - 75.3 = 74.61 \text{ Kg/m}^3$$

(a) Substrate conversion, δ

$$\delta = \frac{S_0 - S}{S_0} = \frac{150 - 74.61}{150} = 0.5$$

(b) Ethanol concentration, P

$$P = Y_{P/S}\left(S_o - S\right) = 0.4\left(150 - 75.39\right) = 29.84 Kg / m^3$$

Problem 9.5

In order to remove urea from the blood of patients with renal failures, a prototype fixed bed reactor is set up with urease immobilized with 2mm gelatin beads. Buffered urea solution is recycled rapidly through the bed so that the system is well mixed. The urease reaction is

$$\left(NH_4\right)_2 CO + 3H_2O \rightarrow 2NH_4^+ + HCO_3^-$$

K_M for the immobilized urease is 0.54 kg/m^3 and the total amount of urease is 10^{-7} kg and the turnover number is 11000 kg NH_4^+ (kg enzyme^{-1} sec^{-1}). The volume of the beads in the reactor is 0.25 x 10^{-3} m^3. The effective diffusivity of urea in the gel is D_e = 7 x 10^{-10} m^2/sec. External mass transfer effect is negligible. The feed stream contains 0.42 kg/m^3 urea and the effluent urea concentration is 0.2 kg/m^3. There is no enzyme deactivation. What volume of urea solution can be treated in 30 minutes.

Solution:

Evaluation of v_{max} from turnover number:
Urease loading in the column

$$= \frac{10^{-7}\, Kg}{0.25 \times 10^{-3}\, m^3} = 4 \times 10^{-4} \text{ Kg urease/ m}^3 \text{ of resin}$$

$$v_{max} = 11000 Kg\, urea \times 0.25 \times 10^{-3}$$

$$= 2.2 \frac{kg\; urease}{(m)^3\; of\; resin(s)}$$

Calculation of effectiveness factor, η:

ϕ Thiele parameter

$$\phi = \frac{R}{3}\sqrt{\frac{V_{max}}{K_M D_e}}$$

$$\text{or } \phi = \frac{1\times 10^{-3} m}{3}\sqrt{\frac{2.2}{.54\times 7\times 10^{-10}}} = 25.4$$

Since ϕ is large, the effectiveness factor, η

$$\eta \approx \frac{1}{\phi} = \frac{1}{25.4} \approx 0.04$$

Calculation of Flow rate:

We know the plug flow design equation for immobilized enzymes applicable to packed bed reactor.

$$\eta v_{max} \frac{V}{F} = K_M\, ln \frac{So}{S} + (S_0 - S)$$

$$\text{Or } \frac{V}{F} = \frac{K_M}{\eta v_{max}} ln \frac{So}{S} + \frac{(S_o - S)}{\eta v_{max}} = \frac{0.54}{(0.84)(2.2)} ln \frac{0.42}{0.02} + \frac{(0.42 - .02)}{(.04)(2.2)} = 23.22 \text{ Seconds}$$

$$F = \frac{V(volume\; of\; packing)}{23.22\, Sec} = \frac{0.25\times 10^{-3} m^3}{23.22} = 1.07\times 10^{-5} \frac{m^3}{Sec}$$

So the volume of urea solution treated in 30 min.

$$= 1.07\times 10^{-5} \frac{m^3}{Sec} \times 60 \times 30 = 1.926\times 10^{-3} m^3 / (1.926L)$$

9.4.2 Fluidized bed bioreactor

In the fluidized bed column reactor (shown in fig. 9.6), liquid flows upwards through a vertical cylindrical column with a top of larger diameter. Biocatalysts particles with immobilized cells or enzymes are suspended in the fluid of the column by the drag forces exerted by the rising liquid. Entrained particles are released at the top of the tower by the reduced liquid drag at the expanded cross-section of the column and the particles fall back into the bed. The biocatalysts are retained in the tower when the medium substrate flows through the column. Continuous fluidized bioreactors are more complicated than the CSTR or PFTR. There are concentration gradients of immobilized cells from the bottom to the top of the tower. The bottom concentration may reach 35% and the cells concentration at the top may be 5 to 10%, so there is rapid fermentation at the bottom and fermentation slows down to the top.

A simple model for such fluidized bed reactors may be developed with the following assumption: (1) the immobilized particles are uniform in size (2) the liquid moves upward through the column (3) substrate utilization rate is first order with respect to biomass concentration (4) the catalyst particle Reynolds number based on the terminal velocity is small, (5) the fluid velocity does not change along the height, z.

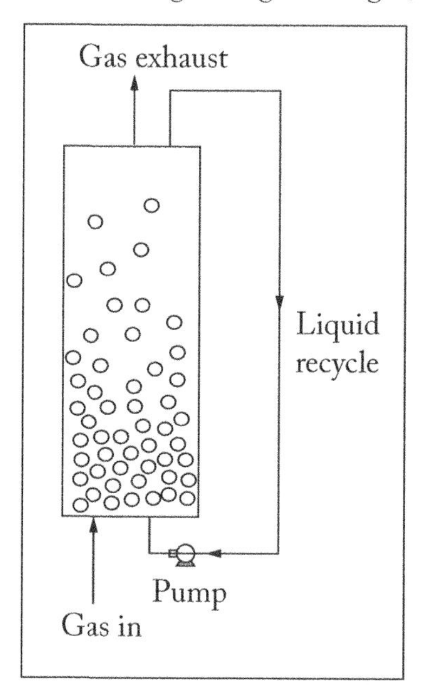

Fig. 9.6 Fluidized bed reactor

With these assumptions, the substrate mass balance in a differential element of length, dZ is,

$$u\frac{dS}{dZ} = -kX \tag{9.62}$$

For stokies law, the concentration of biomass (X) can be related to the liquid velocity in a fluidized bed by the expression,

$$X = \rho_o\left[1-\left(\frac{u}{u_t}\right)^{1/4.65}\right] \tag{9.63}$$

where ρ_o is the microbial cell density on a dry wt. basis, u_t is the terminal velocity of a spherical particle for stokes regime and is given as

$$u_t = \frac{g D_P^2 (\rho_L - \rho_g)}{18\mu_L} \tag{9.64}$$

Now combining the equation (9.100) & (9.101) and integrating, we obtain,

$$+\int_{S_o}^{S} ds = -\frac{k\rho_o}{u}\left[1 - \frac{u}{u_t}\right]^{1/4.65} \int_0^L dZ \tag{9.65}$$

Or
$$S = S_o - k\rho_o\left[1 - \left(\frac{u}{u_t}\right)^{1/4.65}\right]\frac{L}{u} \tag{9.66}$$

Instabilities in the flow pattern within the bed may cause significant backmixing adversely affecting, the reactor's performance. In biological fluid bed reactors, relatively low linear velocities are required because the immobilized particles are small and the density difference between the fluid and catalyst particles is also low.

Insertion of static mixing elements in the fluidized bed bioreactor enhances the bed expansion characteristics and reduce undesired fluid backmixing.

Fluidized bed bioreactors have certain merits over packed bed reactors. It provides effective aeration, no build-up of CO_2 gas in the reactor, good liquid-solid contacting, good heat and mass transfer.

However, fabrication and maintenance costs are higher than those of packed bed reactor.

Problem 9.7

Ethanol is produced by fermentation of glucose, using yeasts in a fluidized bed reactor at 30°C and pH = 5.0. Kinetic parameters given:

$$\mu_{max} = 0.5hr^{-1}, Y_{X/S} = 0.02kgX / KgS$$

$$K_S = 0.025kg / m^3, Y_{P/S} = 0.44kgP / KgS$$

So (initial substrate concentration) = 50 kg/m^3, dp = 1 mm

The following design equations are given

Plug flow model: $u\dfrac{dS}{dZ} = -kX$

Suspended bio mass (X) is related to the liquid flow velocity, u

$$X = \rho_o \left[1 - \left(\frac{u}{u_t}\right)^{\frac{1}{4.65}}\right] \frac{L}{u}$$

and the substrate concentration, S, is given as

$$S = S_o - k\rho_o \left[1 - \left(\frac{u}{u_t}\right)^{\frac{1}{4.65}}\right] \frac{L}{u}$$

ρ_o = microbial density (1100 kg/m^3)

(a) Calculate the residence time of the bed for 90% conversion

(b) Calculate the length of the fluidized bed

(c) Calculate the cell mass concentration (x)

Solution:

(a) Determination of k

We know

$$-u\frac{dS}{dZ} = \frac{1}{Y_{X/S}} u\frac{dX}{dZ} = \frac{1}{Y_{X/S}} \frac{\mu_{max} SX}{K_s + S}$$

For $S >> K_s$

$$-u\frac{dS}{dZ} = \frac{\mu_{max} SX}{Y_{X/S}} \cdot X = kX$$

So $k = \dfrac{\mu_{max}}{Y_{X/S}} = \dfrac{0.50}{0.02} = 2.5hr^{-1}$

We know

$$S = S_o - k\rho_o \left[1 - \left(\frac{u}{u_t}\right)^{1/4.65}\right] \frac{L}{u}$$

Dividing by S_o

$$\frac{S}{S_o} = 1 - \delta = 1 - \frac{k\rho_o}{S_o}\left[1 - \left(\frac{u}{u_t}\right)^{0.215}\right]\tau$$

Assume, $u/u_t = 0.9$

$$1-0.9 = 1-\frac{(25)(1100)}{50}\left[1-(0.9)^{0.215}\right]\tau$$

$$0.1 = 1-12.31\tau$$

$$\text{Or } \tau = 0.073\,hr$$

(b) $L = u\tau = (0.9u_t)\tau$

$$u_t = \frac{gD_P^2(\rho_P-\rho_L)}{18\mu}$$

$$= \frac{9.8\left(\dfrac{m}{S2}\right)(1\times10^{-3})m^2(1100-1000)\dfrac{Kg}{mS}}{18\times10^{-3}\,Kg\dfrac{m}{S}}$$

$$= \frac{0.054m}{s} = 194.4\,m/hr$$

$$L = 0.9(194.4)(.073)m = 12.77\,m$$

(c) $X = Y_{X/S}(S_o - S)$

$$= 0.02(50-5) = 0.9Kg/m^3$$

Problem 9.8

A fluidized bed biofilm reactor contains spherical plastic particles where the cells are attached, forming a biofilm of average thickness, L = 0.5 mm. The bed is used to remove carbon compounds from a waste water stream. The feed flow rate is $F = 2 \times 10^{-3}m^3/hr$ and inlet concentration S_o = 2 kg/m^3. The column diameter, D_t = 0.1m, the kinetic constant of the microbial population, $r_m = \dfrac{50kgs}{(m^3)(hr)}$ and K_s = 25 kg/m^3. The specific surface area of the biofilm, = 250 m^2/m^3. Assume first order kinetics and average effectiveness factor, η = 0.7 throughout the column.

Determine the required height of the column when the effluent concentration is S = 0.1 kg/m^3.

Solution:

The desired equation is based on the plug flow model.

$$-F\frac{dS}{dZ}=\eta\cdot\frac{r_m S}{K_S}La\,A$$

Where A is the cross-section of the tower.

Integrating

$$-\int_{S_o}^{S}\frac{dS}{S}=\frac{\eta r_m La\,A}{F\,K_S}\int_{0}^{H}dZ$$

$$\text{Or } ln\frac{S_o}{S}=\frac{\eta r_m La\,A\,H}{F\,K_s} \qquad \text{(A)}$$

$$\text{Then } H=\frac{F\,K_s\,ln\frac{So}{S}}{\eta\,r_m\,La\,A}$$

Data given are:

$F=2\times10^{-3}m^3/hr, K_S=25\,Kg/m^3$

$r_m=\frac{50KgS}{m^3(hr)}, S_o=\frac{2Kg}{m^3}, S=0.1kg/m^3$

$\eta=0.7, L=0.5\times10^{-3}m, a=250\,m^2/m^3$

$A=0.785(0.1)^2=7.85\times10^{-3}m^2$

Putting the values of all parameters in equation (A)

$$H=\frac{(2\times10^{-3})(25)ln2/0.1}{(0.7)(50)(0.5\times10^{-3})(250)(7.854\times10^{-3})}=\frac{149.78}{34.37}=4.35\,m$$

Problem 9.9

Glucose is converted to ethanol in a fluidized-bed immobilized-cell reactor containing *Z. mobilis* cells immobilized in *K.* Carrageenan gel beads. The dimensions of the bed are 10 cm diameter and 2 m height. Since the feed is introduced from the bottom due to CO_2 evolution, the substrate and the cell concentration decreases with the height of the column. The average cell concentration at the bottom of the column is X_o = 45 kg/m^3 and the average cell concentration decrease with column height according to the following equation.

$$X = X_o (1 - 0.00Z)$$

where Z is the column height in m. The specific rate of substrate consumption is $q_s = \frac{2KgS}{(Kg)(hr)}$ The feed rate is 0.05m³/hr with initial substrate concentration, S_o = 160 kg/m³.

Calculate the following:

(a) Substrate (glucose) concentration at the exit

(b) Ethanol concentration at the exit and the ethanol productivity in kg/(m³) (hr), where $Y_{P/S} = 0.4 \frac{KgP}{KgS}$

(a) The design equation of the fluidized bed is given as

$$-FdS/dZ = q_S XA = q_S AX_0 (1-.00Z)$$

Now integrating,

$$-\int_{S_0}^{S} dS = \frac{q_S AX_0}{F} \int_{0}^{Z} (1-0.00Z)\,dZ$$

$$\text{Or } S = S_0 - \left\{ \frac{q_s A X_0 \left(Z - \frac{.005Z^2}{2} \right)}{F} \right\}$$

$$\text{Now, } A = \frac{\pi}{4} D_t^2 = 0.7854(.1)^2 = 7.854 \times 10^{-3} m^2$$

Z = 2 m, S_O = 160 Kg/m³

q_S = 2 KgS/(Kg Cell) (hr)

X_0 = 45 Kg/m³, F = 0.05 m³/hr

Substituting the values in the equation (α),

$$S = 160 - \frac{(2)(7.854 \times 10^{-3})(45)(2 - .005 \times 2)}{0.05}$$

$$= 160 - 28 = 132\, Kg / m^3$$

(b) P (ethanol) = $Y_{P/S}(S_o - S) = 0.48(160 - 132) = 13.44\,Kg/m^3$

Volume of the reactor

$$V = 0.7854(0.1)^2(2) = 0.0157\,m^3$$

$$D = F/V = \frac{0.05}{0.0157} = 3.18\,hr^{-1}$$

$$\text{Productivity} = PD = 13.44 \times 3.18 = 42.8\frac{Kg}{m^3 hr}$$

9.4.3 Trickle bed reactors

Trickle-bed reactors involve three phases contained in a packed bed reactor and the solid phase is heterogeneous catalyst of immobilized cells, one or more reactants are present in the liquid phase and the gas phase is usually oxygen or air. The performance of such reactors depends upon the physical state of gas-liquid flow through the fixed bed and the associated mass transfer. A schematic diagram of Trickle bed reactor is shown in Fig. 9.8.

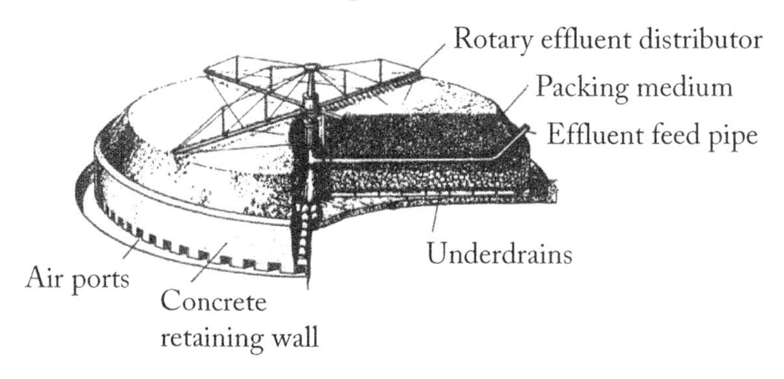

Fig. 9.7 Trickling bed filter (Bailey and Ollis, 1986, p.941)

Important features of such reactor are the surface area of the packing, wetting of the catalyst surface in the flowing liquid phase, gas-liquid flow pattern, mass transfer of sparingly soluble reactant gas from the gas to the liquid phase, mass transfer of both reactants (substrate and dissolved oxygen) to the catalyst surface and also into intra-particle catalyst sites for a porous catalyst.

One such reactor known as trickle filter has been successfully used for waste-water treatment. In this system a rotating distributor feeds the liquid in the form of spray over a circular bed of gravels with immobilized microbial cells. The liquid trickles down along the bed in a laminar flow, when air rises through the bed by natural convection due to heat generated by the microbial reaction. An identical operating design has been employed for the manufacturing of vinegar by biological oxidation of ethanol in a rectangular column packed with wood chips.

Trickling biological filter is a nice example of Trickle bed reactors. Dissolved oxygen limitation is possible for high density of cells, unfavorable hydro dynamic conditions and diffusion within the film may be predominant. But the oxygen transfer problem is nearly eliminated by using liquid with saturated oxygen in pre-aerated feed stream.

The design aspect is simply based on substrate mass balance on a differential height of a trickling biological filter dZ. It can be given as

$$-F\frac{dS}{dZ} = \eta\frac{r_m S}{K_S + S} L\,a\,A$$

The above equation has been used in the packed bed reactor design given by the equation (9.59) and has been integrated to obtain the design equation given by (9.60).

Problem 9.10

A packed bed reactor containing immobilized bacteria on small ceramic particles is used to treat waste water in an upflow mode containing substrate, S_o = 2 kg/m^3 at the flow rate, F = 1 m^3/hr. The effluent desired substrate concentration, S = 0.03 kg/m^3. The rate equation is,

$$r_s = \frac{kXS}{K_S + S}$$

Calculate the required height of the column. The following data are given:

$$k = 0.5hr^{-1}, X = 10Kg/m^3, K_S = 0.2Kg/m^3$$

$$L = 0.2\times 10^{-3}m, a = 100m^2/m^3$$

$$A = 4m^2, \eta = 0.8$$

Solution:

The design equation is

$$-F\frac{dS}{dZ} = \frac{\eta(kX)S\times a\times L\times A}{K_S + S}$$

Integrating

$$\int_{S_o}^{S} -\frac{(K_S - S)ds}{S} = \frac{\eta(kX)a\,L\times A}{F}\int_0^H dZ$$

$$\text{or } K_S ln\frac{S_o}{S} + S_o - S = \frac{\eta(kX)aLAH}{F}$$

$$= \frac{\eta(kX)aLAH}{F} \ldots\ldots(\alpha)$$

Now putting the values in equation (α),

$$0.2ln\frac{2}{.03} + 2 - .03$$

$$= \frac{0.8(0.5 \times 10)(100)(0.2 \times 10^{-3}) \times 4}{1} \text{H}$$

Or 2.81 = 0.32 H

Or H = 8.78 m.

9.4.4 Air lift fermenter with external loop

Schematically the air lift fermenter can be shown as given below:

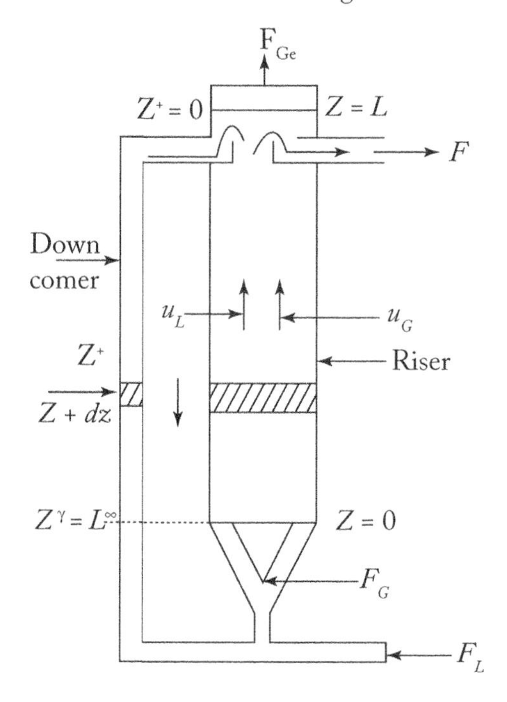

Fig. 9.8 Schematic diagram of an air-lift fermenter with external downcomer

The right hand side tower is called riser where upward two-phase flow of gas and liquid occurs. Gas is separated at the top of the column and the liquid stream is recycled through the loop on the left, called Downcomer and flows down the column and reaches the bottom of the column where gas is sparged to the system. At the bottom, the descending liquid again encounters the gas stream and is carried up to the top through the riser. Air lift system can handle more viscous fluids than bubble column and coalescene is minimum. The liquid and gas ascend the tower with different velocities, U_L and U_g respectively for liquid and gas. The volume fractions occupied by liquid and gas are $\in_L$ and $\in_G$ respectively.

Axial dispersion is superimposed on the convective axial transport in plug flow for liquid phase.

Substrate and oxygen utilization takes places due to microbial reactions in the fluid phase. Steady state approximations are imposed. The dissolved oxygen concentration, C_L can be presented as

$$D_L \frac{d^2C_L}{dZ^2} - u_L \frac{dC_L}{dZ} - \frac{r_{max} S}{K_S} \cdot Y_{02/S} + k_{La}\left(C_L^* - C_L\right) = 0 \tag{9.67}$$

where first order kinetics is assumed.

The maximum substrate utilization rate, $r_{max} = \frac{1}{Y_{X/S}}(\mu_{max} X)$

Similarly for the substrate (S) and cell mass (X) balance may be made on the elemental axial length, dZ.

Extensive experimental studies and parameter evaluation for the model have been described by K. Schugert (1982).

9.5 CSTR for Immobilized Cells or Enzymes

Though immobilized cells or enzymes are used in packed bed or fluidized bed, but CSTR can also be used for the immobilized system if the agitation in the tank is low preferably at 20 rpm, so that the shear effect on the immobilized particles is minimum.

Reactor design for those systems can be developed as follows:

Substrate balance can be presented for the CSTR

$$F\left(S_o - S\right) = \eta\left(\frac{r_{max} S}{K_S + S}\right) V_P \tag{9.68}$$

where F is the volumetric flow rate, η is the effectiveness factor for the immobilized particles, r_{max} is the maximum substrate utilization rate, K_S is the saturation constant, S_o is the initial substrate concentration. S is the exit concentration of the substrate, V_P is the volume of the immobilized particles. For low values of S, we assume the reaction to be first order with respect to substrate.

Rearranging the equation (9.68), we get,

$$S = \frac{S_0}{\left(1 + \eta k' \frac{V_P}{F}\right)} \tag{9.69}$$

Where $k' = r_{max} / k_s$

For chemical reaction in a CSTR, the liquid volume is used as reactor volume. In this case, volume of immobilized beads is used, since the biochemical reaction takes place inside the immobilized beads where cells or enzymes are uniformly distributed.

Problem 9.11

A CSTR containing yeast immobilized in Ca-alginate beads of diameter, $D_p = 0.5$ cm in a liquid volume of 0.05m^3. The volume of the beads suspended in the liquid is $V_P = 0.005$m^3. The reactor is gently agitated by a paddle stirrer at the speed of 20 rpm. The flow rate of feed and initial substrate concentration are $F = 2 \times 10^{-3}$m^3/hr and $S_o = 5$ kg/m^3 respectively. The kinetic parameters are:

$r_{max} = 100$ kgs / (m^3) (hr)

$K_s = 15$ kg/m^3, De = 1.0×10^{-10} m^2/sec.

(3.6×10^{-7} m^2/hr)

(a) Calculate the effectiveness factor, η for the bed, $D_p = 0.5$ cm
(b) Calculate the exit concentration of glucose in the CSTR
(c) If a packed bed reactor is used with the same flow what is the volume of the reach for the same outlet concentration of substrate?

Solution:

(a) Determination of effectiveness factor

$\phi = r_0/3\ (k'/D_e)^{1/2}$, $k' = r_{max} / K_S$

$$\phi = \frac{0.25 \times 10^{-2} m}{3} \sqrt{\frac{6.67 hr^{-1}}{3.6 \times 10^{-7} m^2 / hr}} = 3.58$$

$$\eta = \frac{1}{\phi_3}\left[\frac{1}{tan\, h\phi_s} - \frac{1}{3ds}\right] = \frac{1}{3.58}\left[\frac{1}{tan\, h\, 3.58} - \frac{1}{10.74}\right] = 0.25$$

(b) For the CSTR design equation

$k' = \text{r}_{\text{max}} / \text{K}_{\text{s}} = 100 / 15 = 6.67$

$$S = \frac{S_o}{\left(1 + \eta k' \frac{V_P}{F}\right)} = \frac{5}{1 + (0.25)(6.67)\left(\frac{.005}{.002}\right)} = 0.96 Kg / m^3$$

(c) For a packed bed reactor

$$-F\frac{dS}{dZ} = -\eta \frac{r_{max}}{K_s} S\, A$$

Integrating and rearranging

$ln\ S_0/S = r_{max}\ A\ z\ /\ FKs$

$$\text{Or } ln\frac{5}{0.96} = \eta\frac{r_{max}V}{F\,kg}, \text{ where Az = V}$$

$$= \frac{0.25 \times 100V}{(0.002)\times 15}$$

Or $V = 1.98 \times 10^{-3}\ \text{m}^3 = 1.98\ \text{L}$

9.6 Summary

This chapter deals with non-ideal plug flow reactors and the non-ideality has been explained using residence time distribution in the reactor vessel. Two reactor models such as Tanks-in-series or dispersion model have been utilized to predict the performance of non-deal reactors.

Packed bed reactors with immobilized cells on supports as packing have been designed on the basis of plug flow behavior and the actual rate has been evaluated by multiplying the intrinsic rate by effectiveness factor which takes into account the effect of intra particle diffusion. Effectiveness factor, η, has been shown to be a unique function of a lumped parameter known as Thiele modulus, φ and saturation parameter, β.

Some aspects of reactor dynamics have been focused with linearized design equations of a CSTR system which may be solved in terms of eigenvalues and these eigenvalues determine the local stability around the steady state value.

A few types of commercial bioreactors with immobilized cell supports, such as packed bed, fluidized bed, trickle bed and airlift fermenter have been described and mathematical expressions have been presented for their design.

Problems (Exercise)

9.1 Hydrolysis of benzoyl organic ethyl ether (BAEF) was carried using papain immobilized in porous iron oxide particles of 170 – 250 μm, in a fluidized bed reactor.

For immobilized papain the kinetic parameters are:

$K_M = 0.012\ K\ moles\ /\ m^3,\ S_o = 10\ kg/m^3$

$$v_{max} = 0.05\frac{k_{mol}}{(mm)m^3 of\ suppor}$$

$De = 1.5 \times 10^{-10} m^2\ /\ S$

(a) Calculate Thiele parameter ϕ And the effectiveness factor

$$\text{Hint: } \phi = \frac{R}{3}\sqrt{\frac{u_{max}}{KMDC}}$$

$$\text{Or } \eta = \frac{1}{\phi}\left[\frac{1}{tan\,h3\phi} - \frac{1}{3\phi}\right]$$

(b) For 90% conversion of the substrate, calculate the wt, of the support immobilized with papain

$$\text{Hint: } F\frac{dS}{dW} = -\eta\frac{u_{max}S}{KM + S}$$

9.2 Yeast cells entrapped in Ca- alginate beads are used to convert glucose to ethanol in a packed bed reactor with a feed rate of 0.12 m^3/hr with initial substrate concentration S_o = 5 kg/m^3. The particle size of Ca-alginate bead is D_p = 0.5 cm. The rate equation is $r_s = r_m S / (K_S + S)$

where r_m (the maximum substrate utilization rate)

= 100 kg S / (hr) (m^3 of beads)

K_s = 10 kg/m^3, D_e = 1 x $10^{-10}m^2/s$

The cross section area of the bed = 0.01m^2, Assume first order reaction as

$S_o << K_M$

(a) Calculate the effectiveness factor for Ca-alginate beads

$$\phi = \frac{R}{3}\sqrt{\frac{r_m}{K_M\,De}}$$

(b) Calculate the required height of the packed bed for 80% conversion of glucose solution.

$$\text{Hint: } -F\frac{dS}{dZ} = \eta\frac{r_m}{K_s}S\,A$$

9.3 In a packed bed system of Ca-alginate beads of 0.5 cm diameter, glucose is converted to ethanol. It is suspected that there may be external diffusion effect for the following data:

r_{max} (maximum substrate utilization) = 10 kg/m^3 hr

k_L = 0.05 cm / sec, S_o = 5 kg / m^3

predict whether there is external diffusion effect in the system.

9.4 A waste water stream is treated in a reactor containing immobilized cells in porous particles. For particle size, $D_p = 4$ mm, the following data were obtained at different initial substrate concentration:

S_o kg/m^3	0.1	0.25	0.5	1.0	2.0
r kg / m^3 hr	0.085	0.20	0.36	0.63	1.0

Determine r_{max} and K_s for the microbial system

Hint: $r = \dfrac{r_{max} S}{K_S + S}$

9.5 A packed bed column containing Ca-alginate beads of diameter, $D_p = 0.5$ cm immobilized with yeast cells, is used to convert glucose to alcohol. The nutrient flow rate is 0.25 m^3/hr. The desired residence time is 2 hrs, the rate constants are

$r_m = 100\ K_s\ S / (\text{m}^3)\ (\text{hr})$

$K_S = 10$ kg / m^3, $S_o = 5$ kg / m^3

The rate expression

$$r_s = \frac{r_m S}{K_s + S}$$

Assume first order kinetics for low substrate concentration. The extent of dispersion in the bed is given by Peclet No., $P_e = 10$.

(a) Calculate the exit concentration of the substrate using dispersion model, Neglect intra-pellet diffusion.

(b) What will be the length of the bed if the cross section of the bed is 0.01m^2.

Hint for (a)

k (first order rate constant) $= r_m / k_s$

Design equation is

$$\frac{S(L)}{S_o} = \frac{4 exp(0.5P_e)}{(1+\gamma)^2\ exp(0.5Y\ Pe) - (1-\gamma)^2\ exp(-0.5YP_e)}$$

Where $\gamma = \sqrt{1 + \dfrac{4Da}{P_e}}$

$Da = kt = kL/u$

9.6 A strain of Escherichia Coli has been genetically engineered to produce human protein. A batch culture is started by inoculating 0.012 kg cells into a 0.1 m^3 bubble column fermenter containing 10 kg/m^3 glucose. Data given are

$$\mu_{max} = 0.9 hr^{-1}, Y_{X/S} = 0.6$$

(a) Estimate the time required to reach stationary phase
(b) What will be the final cell density if the fermenter is stopped after 70% of the substrate is consumed?

9.7 A batch culture gave the following data:

t, hrs	0	5	10	15	20	25
x, kg/m^3	0.5	2.1	4.8	7.7	9.6	10.4
S, kg/m^3	100	85	58	30	12	5
P	0.0	7.5	20.0	34.0	43.0	47.5

The growth model is given by logistic equation,

$$\frac{dX}{dt} = \mu X\left(1 - \frac{X}{X_m}\right)$$

Given $X_m = 10.8 Kg/m^3$ where X_∞ is the maximum cell concentration

(a) Determine μ from the data using logistic equation
(b) Determine yield coefficients,

$Y_{P/S}$ and $Y_{X/S}$

Hint for (a):

$$\frac{1}{X}\frac{\Delta X}{\Delta t} = \left(1 - \frac{\overline{X}}{X_\infty}\right)$$

$$\text{And } Log\left(\frac{1}{\overline{X}}\frac{\Delta X}{\Delta t}\right) = log\,\mu + log\left(1 - \frac{\overline{X}}{X_\infty}\right)$$

Use the logistic equation

$$\frac{dX}{dt} = \mu X\left(1 - \frac{X}{X_{max}}\right)$$

With the following data

$\mu = 0.8\, hr^{-1}, X_{max} = 10.6\, kg/m^3$

X_o (in feed) = 0

Solve X graphically for

$D = 0.75\ hr^{-1}$, and $0.25\ hr^{-1}$

Hint:

$$D\left(X_o - X\right) = r_x = \frac{dX}{dt}$$

Prepare a graph r_x vs X. Locate a slope of D from the inter section of the rate curve and then locate X.

9.9 Methylomonas methanolica (biomass) was grown in methanol in presence of oxygen in a chemostat at 30°C, pH of 6.0. For the system the following kinetic parameters are given:

$\mu_{max} = 0.83\ hr^{-1}$, $Y_{S/X}$ (methanol yield coefficient) = 0.48 g/g, $Y_{O_2/S} = 0.53\ go_2/g$ substrate, $Y_{X/S} = 0.57$ g x/g substrate, O_2 quotient = 0.90 mol of O_2 / mol CH_3oH

Respiratory quotient, RQ = 0.52 mole CO_2 per mole of O_2, ke (maintenance co-efficient) = 0.35 gm of CH_3OH/gm of cell mass. $K_s = 0.002$ g/L. The specific yield factors correspond to a dilution rate, $D = 0.50\ hr^{-1}$.

$k_e = k_d / Y_{X/S}$

Or $k_d = 0.35 \times 0.57 = 0.2\ hr^{-1}$

(a) Write down equations for CSTR for cell-growth of cell mass, the oxygen consumption and CO_2 production Vs. dilution rate

(b) At the given $D = 0.60\ hr^{-1}$, Calculate S, X, oxygen consumption rate, CO_2 production rate

for $S_o = 10$ gm/L, $C_L^* = 8$ ppm (8 mg/L)

hint for (a)

$$\mu = \frac{\mu_{max} S}{K_s + S} - kd = D$$

$$S = \frac{K_s\left(D + kd\right)}{\mu_{max} - D - kd}$$

$$X = Y_{X/S}^{M}\left(S_o - S\right)$$

Oxygen consumption,

$$C_L^* - C_L = Y_{O_2/S}\left(\frac{g_{o_2}}{g_S}\right)\left[S_o - S\right]\frac{1}{32}\ \text{moles of oxygen}$$

CO_2 production rate

$$= RQ\left[C_L^* - C_L\right]$$

9.10 A simple model for a diauxic growth on two substrates: glucose (G) and a second carbohydrate (S) is given as

$$r_x = \mu(g,S)X$$

$$= \left[\frac{\mu_G g}{K_G + g} + \frac{\mu_3 S}{K_S + S}\cdot\frac{K_R}{K_R + g}\right]X$$

(a) For typical condition including glucose repression it is expected $\mu_S \leq \mu$ and $K_S \sim K_G \gg K_R$ ketch the expected variation of g, s and lnx vs. time in a batch fermentation

(b) For diauxic fermentation aimed at complete utilization of cell substrate, would you use a plug flow reactor? Tanks in series, or a single chemostat? Why?

(c) Since $K_R << K_G$, K_S, the growth equation is simplified depending on whether $g > K_g$, or $K_R > G = 0$. Develop a plug flow model and integrate it analytically to give x(z), g(z), and s (z)

(d) For $\mu_G = 1\ hr^{-1} = 1.1\ \mu_s$, $K_G = K_s = 10mm$, $K_R = 0.1$ mm, $X_o = 0.1$ g/L, $Y_G = Y_S = 0.5$ g/g, go = $S_o = 0.1$ mm.

Plot the results of part C for u_o (flow velocity) / ......reactor with 1 = 1 hr^{-1}, 3 hr^{-1}, 5 hr^{-1}
Hint for part (a) for batch fermentation.

$$\frac{dlnX}{dt} = \mu(g,s)$$

Since $g > kg$, $s > kg$, $\dfrac{K_R}{K_R + g} \approx 1$

$$\mu(g,s) = \mu_g + \mu_s$$

$$ln\frac{X}{X_O} = (\mu_g + \mu_s)t$$

For variation of g,

$$\frac{dg}{dt} = -\frac{1}{Y_{X/g}}\mu(S,g)X$$

$$= -\frac{1}{Y_{X/g}}\mu(S,g)\left[X_0 + Y_{X/g}(S_0 - S)\right] = s(t)$$

(e) For diauxic fermentation aimed at complete utilization of substrate, PFR is preferable since glucose is first converted and then the second carbohydrate. CSTRs in series or a single CSTR is not suitable for complete conversion of the substrates.

(f) Since $K_R << K_G$, K_g, the growth equation will be simplified for plug flow model

$$u\frac{dX}{dZ} = \mu(g,s) = (\mu_G + \mu_S) = X(Z)$$

$$u\frac{dg}{dZ} = \frac{1}{Y_{X/g}}\mu(g,s)\left[X_0 + Y_{X/g}(g_0 - g)\right] = g(z)$$

$$u\frac{dS}{dZ} = -\frac{1}{Y_{X/S}}\mu(g,s)\left[X_0 + Y_{X/g}(S_0 - S)\right] = S(z)$$

9.11 When a continuous culture is fed with substrate of concentration, S_o = 1.5 g/L, dilution rate at washout is 0.3 h^{-1}. This changes to 0.1 h^{-1} if the same organisms is used but the feed concentration is 3.2 g/L. Calculate the effluent concentration when in each case the fermenter is operated at its maximum productivity. A microbial reaction is carried out in a chemostat with a growth model of nutrient inhibition. Use substrate inhibited model

Given the kinetic parameters: μ_{max} = 1.17 hr^{-1}, K_s = 2.5 kg/m^3, K_I = 25.6 kg/m^3.

Using a sterile feed, the dilution rate used is D = 0.5 hr^{-1}

(a) Under these conditions, how many steady states of substrate concentration are possible?
(b) Draw the variation of S with D,
(c) Show that there are two steady states.

9.12 A linearized reactor system is given in matrix – vector form:

$$\dot{x} = Ax$$

$$\text{Where } A = \begin{bmatrix} 3 & -2 \\ 4 & -1 \end{bmatrix}$$

From the Characteristic equation, find out the eigenvalues and from their sign, state whether the system is stable or not.

9.12 When a pilot scale fermenter is run with feed flow rate of 60 l/hr, the effluent substrate from the fermenter contains 10 mg/L. The same fermenter is connected to a settler/thickener which can concentrate the effluent biomass by a factor of 3.0 and a recycle stream with enhanced biomass is fed to the inlet of the reactor and the flow rate of the recycle stream is 40 l/hr and the fresh feed rate is at the same time increases to 100 l/hr. Calculate the concentration of the final clarified liquid effluent from the system. The system follows Monod Kinetics with the kinetics with the kinetic parameters, μ_m = 0.15 hr^{-1}, K_S = 100 mg/L.

9.13 A CSTR is operated at a series of dilution rate at constant sterile feed concentration, pH, aeration rate and temperature. The data given below were obtained when the limiting substrate concentration was 1400 mg/L and the working volume of the fermenter was 10L.

Estimate the kinetics parameters K_S, μ_m, k_d used in the modified Monod model

$$\mu = \frac{\mu_m S}{K_S + S} - kd$$

And also $Y_{x/s}$ (yield coefficient)

F, L/hr	0.79	1.03	1.31	1.78	2.4
S, mg/L	37	49	64.5	93.5	139
X, mg/L	485	490	480	482	472

Hint:

$$D(S_0 - S) = \frac{\mu_{m\,SX}}{Y(K_S + S)}$$

Or $$\frac{\mu_{mSX}}{Y(K_S + S)} = YD(S_0 - S)/X$$

$$D + R_d = \frac{\mu_{m\,S}}{K_S + S} = YD(S_0 - S)/X$$

Or $$\frac{(S_0 - S)}{X} = \frac{kd}{Y} \cdot \frac{1}{D} + \frac{1}{Y}$$

References

Aiba, S., E. Humphrey and N.F. Millis, Biochemical Engineering, 2nd ed., Academic Press, N.Y. 1973 Amundson, N.R. and R. Aris, "An Analysis of Chemical Reactor Stability and Control", Chem. Eng., Sci. 7, 121, 1958

Atkinson, B. Biological Reactors, Pion Limited, London, 1974

Bailey, J.E. and D.F. Ollis, Biochemical Engineering Fundamentals, 2nd ed. Mc Graw Hill Book Co., 1986

Blanch, H.W. and D.S. Clark, Biochemical Engineering, Marcel Dekker, Inc, New York, 1996

Charles, M. "Technical Aspects of the Rheological Properities of Microbial Cultures" in Biochemical Engineering Vol. 5 edited by T.K. Ghosh, A Fiechter, B. Blakebrough, Springer Verlag New York, 1978

Chibata, I., T., Tosu and T. Sato, "Methods of Cell Immobilization" in Manual of Industrial Microbiology edited by A.L. Demain and N.A. Solomon, Amercan Society of Microbiology Washington, D.C., p. 217–229, 1980

Erickson, L.E. and G. Stephanopoulos, "Biological Reactors", in Chemical Reaction Engineering, edited by J.J. Carberry and A. Verma, Marcel Dekker, Ini, New York, 1985

Fyeld, M., O.A. Ashorsen and K.J. Astrom, "Reaction invariants and their importance in the analysis of eigen vectors, state observability and controllability of the continuous stirred Tank Reactor", Chem. Eng. Sci., 29, 1918, 1974

Gbewonyo, K. and D.I.C. Wang, "Confining Mycelial growth in porous Microbeads, A novel technique to alter the morphology of the Non-Newtonian mycelia cultures", Biotechnol. Bioeng, 25, 967, 1983

Jefferson, C.P. and J.M. Smith, "Stationary and non-stationary models of Bacterial Kinetics in well mixed flow reactors", Chem. Eng. Sci., 28, 629, 1973

Levenspiel, O. Chemical reaction Engg, 1999, Wiley.

Moo Young, M. "Bioreactor, Immobilized Enzymes and Cells: Fundamental Applications, Elsevior Science publishing Inc., New York, 1988

Nagata, S. Mixing Principles and Applications, Wiley, New York, 1975

Roels, J.A. "Mathematical Models and the Design of Bioreactors", J. Chem. Tech. Biotechnol. 32, 59, 1982

Shinar, R., "Residence time and contact time Distributions in Chemical Reactor Design" in Chemical Reaction and Reactor Engineering, edited by J.J. Carberry and A. Verma, Marcel Dekker, Inc., New York, 1985

Schugerl, K. "Characterization and performance of single multistage tower reactors with outer loop for cell mass production", p. 93 in Advances in Biochemical Engineering 22A, edited by A. Fiechter, Springer Verlag, New York, 1982

Schugerl, K., J. Tliicka, U. Oels, "Bubble column Bioreactors (Tower Bioreactors) without mechanical agitation" in Advances in Biochemical Engineering, Vol. 7, edited by T.K. Ghosh, A Fiechter and N. Blakebrough, Springer Verlag, New York, 1977

Vardar, F. "Problems of mass and momentum Transfer in large fermenters", Process Biochem. 38, 21, 1983

Venkat Subramanian, K. (editor), "Immobilized Microbial Cells", ACS symposium series 106, American Chem. Soc., Washington, D.C. 1979

Villemaux, J., "Mixing in Chemical Reactors" p. 135 in Chemical Reaction Engineering, Plenary Lectures, edited by J. Wei, C. Georgakis, American Chem. Soc., Washington, D.C. 1983

Walter, C.F., "Kinetics and Biological and Biochemical Control mechanisms", p. 335 in Biochemical regulatory mechanisms in E.Coli cells, John Wiley & Sons, New York, 1972

Wang, D.I.C. and R.C.J. Fewkes, "Effect of operating and geometric parameters on the behavior of non-Newtonian, Mycelial Antibiotic fermentations", Develo Ind., Microbial, 18, 39, 1977

Young, T.B., D.F. Bruley and H.R. Bungay, III, "A dynamic mathematical model of the chemostat", Biotechnol. Bioeng. 12, 747, 1970

CHAPTER 10

Reactors for Animal Cells, Recombinant Proteins and Plant Cells

10.0 Characteristic Features of Animal Cells

Animal cells act as important catalysts for many bioprocesses viz production of therapeutic proteins like interleuken, lymphokines, virus vaccines, enzymes, hormones, growth factors, insecticides. With recombinant DNA techniques, significant levels of production of the above compounds are achieved, compared to very low productivity by using simple animal cells.

Animal cells have sizes in the range of 10–30 μm of spheres or ellipsoids. They do not have cell wall, but cells are surrounded by a thin, fragile membrane, made of proteins, lipids and carbohydrates. The animal cells are very shear sensitive.

The metabolism of nutrients by animal cells is quite different from that of other cells. The growth medium of animal cells contain glucose, glutamine, essential and non-essential amino-acids, serum (horse or calf), mineral salts etc. In the metabolic path, glucose is converted to pyruvate by glycolysis, and also to biomass production

Pyruvate is converted partly to CO_2 and H_2O by the TCA cycle, partly to lactic acid and fatty acids. Glucose is also consumed as carbon and energy source, as in glutamine. Part of the glutamine is de-aminated to ammonia and the remaining glutamate is converted to other amino acids. The release of lactate and ammonia as waste products of metabolism leads to major problems in high cell density culture systems. Both lactate and ammonia in high concentration are toxic to the cells and inhibit the growth of cells.

10.1 Growth in Animal Cell Culture

There are two types of animal cell cultures. The primary culture cells are grown in small T-flasks with medium containing serum and other nutrients and small antibiotics. Tissues extracted from specific organs of animals such as lung and kidney under aseptic conditions are transferred to the growth medium. These cells grow as monolayers on support surfaces (e.g., glass surface of flasks). The cells growing on support surfaces are called anchorage dependent cells. However, some cells also grow in suspension culture and are called non-anchorage dependent cells.

The cells from the primary cell-culture are further treated to get cells in a culture known as secondary culture. Cells are removed from the surface of the flasks using a solution of EDTA, trypsin, collagenase or pronase. Serum is added to the culture bottle. The serum containing suspension is centrifuged, washed with buffered isotonic saline solution and used to inoculate the secondary culture.

Most differentiated mammalian cell lines (e.g. human fibroblasts such as WI-38 and MRC-5) that are licensed for human vaccin production are called Mortal. These cell lines show a phenomenon called SENESCENSE where cells will divide for a limited number of generations (about 30 generation for MRC-5 cells).

The other type of cells that can propagate indefinitely are called continuous immortal or transformed cells (e.g. cancer cells). Mortal or non-transformed cells form only a mono-layer on the surface and the transferred cells do not sense the presence of other cells and go on dividing forming multilayers.

Besides mammalian cell lines, other cells of insects, fish crustacia are having evolving technology. The bacilovirus that affects insect cells is an ideal vector for genetic engineering. Insect cells or fish cells are not transformed.

Another cell culture is the culture of hybridomas. They are obtained by fusing lymphocytes (blood cells that make antibodies) with myeloma (cancer) cells. After fusion with myeloma cells, hybridomas become immortal, can reproduce indefinitely and produce antibodies. Using hybridoma cells, highly specific monoclonal antibodies (Mab) can be produced against antigens.

A typical growth medium for mammalian cells contain serum (5–20%), inorganic salts, nitrogen sources, carbon and energy sources, vitamins, trace elements, growth factors. Serum (fresh bovine serum or calf serum or horse serum) has the following functions such as 1) stimulate cell growth, 2) enhance cell attachment by certain proteins, 3) provide transport proteins carrying hormones, lipid etc.

Serum is very expensive and also cause further complications in downstream processing like foaming. Serum is heat-sensitive and must be sterilized by filtrations. Contaminations with viruses, prions and mycoplasma are serious problems. Serum composition is variable. A batch of serum will remain effective for only one year. To overcome these difficulties, serum- free medium has been developed, containing basal salts, vitamins, growth factors and hormones. It reduces the cost, eliminates the problems of product purification.

10.2 Environmental Conditions and Kinetics

Mammalian cells grow at 37°C and pH 7.3, doubling time, td, varies from 12 to 20 hrs. 5% CO_2 enriched air is used to buffer the medium pH nearly at 7.3. A carbonate buffer ($H_2CO_3^{2-}$ / $H_2CO_3^{-}$) is used to control pH at 7.3. Bicarbonate is consumed by the cells. So, CO_2^- enriched air is provided to balance carbonate equilibrium. Other buffers such as HEPES may also be used to maintain the desired pH level.

The kinetics of the growth of mammalian cells follows Monod model like other microorganisms. The stationary phase is comparatively short and the concentration of viable cells decreases sharply after the stationary phase, due to accumulation of toxic metabolite such as lactate and ammonium. Cell concentration reaches a peak value within 3–5 days. Product formation such as monoclonal antibodies from hybridoma cells may continue under non-growth conditions. Most of the products from mammalian cell cultures are mixed growth associated and the kinetics are given by Luedeking-Piret equation. The specific product formation, q_p is given as,

$$\frac{1}{X}\frac{dp}{dt} = q_p = \alpha\mu + \beta \tag{10.1}$$

Where α is the growth associated factor , β is the non-growth associated factor, μ is the specific growth rate of cells. The magnitude of growth rate depends on the cell type, medium composition and other environmental conditions viz. dissolved oxygen (DO), pH, CO_2 level, ionic strength etc.

Regarding oxygen requirement of cells, animal cells require 0.06 to 0.2×10^{-12} mole O_2 / (hr)(cell) which is five times less than oxygen requirement of many plant cells and much lower than microbial cells. Typical cell concentration in suspension culture media is 0.1 to 1 Kg/m^3(5×10^5 to 5×10^6 cells/ml for suspension culture). For a cell culture density of 10^6 cells/ml, the O_2 requirement is of the order of 0.1 to 0.6 mmole O_2/(L)(hr).

Animal cells are very shear-sensitive. Chemicals like pluronic F-68 can be added to culture to provide shear protection. Animal cells are produced in spinner flasks of 0.5 to 10 litre in the laboratory. These flasks are provided with magnetically driven stirrers operating at 10 to 60 rpm. Aeration is done by surface-aeration using 5% CO_2 enriched filtered air. Spinner flasks are placed on a magnetic stirrer plate in a CO_2 incubator.

Shear effects may cause apoptosis or programmed cell death.

10.3 Bioreactors for Animal Cell Culture

Some of the important features of animal cells must be taken into consideration for designing their reactors.

The growth rate of mammalian cells is very low as indicated by the doubling time, td which varies in the ragne of 10 to 50 hrs. Some animal cells are anchorage dependent, requiring surfaces of glass, natural polymers like collagen, dextran etc.

Product concentration is very low (μg/ml). Toxic metabolites such as ammonium and lactate inhibit the cell growth. Animal cells are shear sensitive. These aspects lead to the following considerations for reactor design:

1. The reactors should be gently aerated and agitated. If mechanical agitators are used, the agitation speed should be 20 rpm or less. Bubble columns and air lift fermenters operate at low aeration rate, and protect the cells from shear damage and thus these reactors are suitable for animal cell culture.

2. The environmental conditions like temperature, pH, D.O, redox potential, and a supply of 5% CO_2 enriched air should be strictly maintained at optimum values.
3. Large surface of the support material is necessary for anchorage dependent cells.
4. The removal of toxic products of metabolism such as lactic acid and ammonium is essential and the higher yield of high value products like Mab, vaccines and lymphokines, is to be maintained.

Reactors with a high surface- volume ratio are the micro carriers systems, hollow -fiber reactors, ceramic matrix systems, micro-porous beads are normally used for anchorage dependent cells.

For suspension culture, modified stirred reactors, bubble columns and airlift reactors have been successfully employed.

Membrane bioreactors, Micro encapsulated systems have been developed for simultaneous cell cultivation, product concentration and toxic product removal.

10.3.1 Roller bottles

Roller bottles have been used for anchorage dependent cells. There are important components of the system which can be used for scale up purposes. In the roller bottles liquid covers about 25% of the surface area and they are rotated along the axis at 1 to 5 rpm. Cells adhere to the walls of the bottle and are exposed to liquid 25% of the time and to the gas (5% CO_2 and 95% air) 75% of the time. When the rotating bottle is exposed to the liquid, nutrients are transported into the cells and in the gas phase aeration takes place. However Roller bottles cannot be used for large scale production.

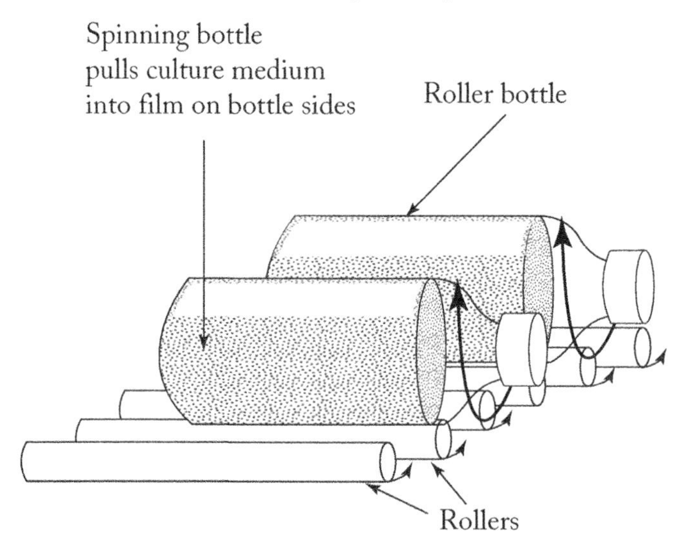

Fig. 10.1 Roller bottle for animal cell culture

10.3.2 Microcarriers

Micro carriers are spherical beads with large surface area. Anchorage dependent mammalian cells may grow on the surface of the beads which provide surface area per unit volume of reactor about 70,000 cm^2/L and allow high cost cell concentration in the medium (10^7 cells/ml). The micro carrier beads immobilized with cells are placed in a gently agitated tank reactor or in a fluidized bed reactor, or air lift fermenter, or bubble columns. Agitation of large scale vessels with micro carrier beads is difficult for balancing aeration and mixing against the shear sensitivity of cells.

Macroporous microcarriers: They provide large surface area and also protect the cells from the shear effect. But diffusion limitations and heterogeneous growth conditions may cause problems in the reactor design.

Problem 10.1

Mammalian cells are immobilized in spherical micro carrier beads (8mm diameter at a concentration of 0.018 kg cells mass/m^3 of bead volume. 100 such beads are immersed in a CSTR of 1 litre capacity which is stirred at 20 rpm. Kinetics of the system can be approximated as a first order with specific rate constant, $k_1 = 3.11 \times 10^5$ sec^{-1} per kg cell mass. The effectiveness factor, is $\eta = 0.7$. If the substrate is fed to the reactor with initial substrate concentration, $S_o = 3.21 \times 10^{-3}$ kg/m^3,

(a) what is the feed rate, F for 80% conversion of the substrate?
(b) If $\eta = 0.7$, what are the values of thiele parameter, φ and the effective diffusivity, De?

$$\text{Given: } \eta = \frac{1}{\phi}\left[\frac{1}{tanh3\phi} - \frac{1}{3\phi}\right]$$

$$\phi = \frac{R}{3}\sqrt{\frac{k}{De}}$$

Solution

(a) Radius of each bead = 4×10^{-3}m

Volume of 100 bead

$$= 100 \times \frac{4}{3}\pi R^3 = 100 \times \frac{4}{3} \times 3.14 \times \left(4 \times 10^{-3}\right)^3$$

$$= 2.68 \times 10^{-5} m^3$$

Total amount of cell mass present

$$= 2.68 \times 10^{-5}\ m^3 \times 0.018\ \frac{kg}{m^3}$$

$$= 4.9 \times 10^{-7} kg$$

k_1 (fist order rate constant)

$=3.11 \text{ x } 10^5 \dfrac{sec^{-1}}{kg} \text{ x } 4.9 \text{ x } 10^{-7}\text{kg}$

$= 0.152 \text{ sec}^{-1}$

Now the design equation of CSTR, $V/F = (S_o - S)/ k S$

After 80% conversion,

$$S = 0.2\left(3.2\times10^{-3}\right) = 0.64\times10^{-3}\frac{kg}{m^3}$$

$$F = \frac{\eta V k_1 S}{(S_o - S)} = \frac{0.7\left(2.68\times10^{-5}\right)m^3\left(.152\right)S^{-1}\times0.64\times10^{-3}\frac{kg}{m^3}}{3.2\times10^{-3} - 0.64\times10^{-3}}$$

$$= 7.1\times10^{-7}\, m^3 / sec = 2.55\times10^{-3}\, m^3 / hr$$

$$= 2.56\, L / hr$$

(b) The effectiveness factor for a spherical bead, η

$$\eta = \frac{1}{\phi}\left[\frac{1}{tanh 3\phi} - \frac{1}{3\phi}\right] \qquad \text{(A)}$$

Now $\eta = 0.7$, the value of φ, the thiele parameter is to be obtained by iteration.

1. Assume $\phi = 1.5, 3\phi = 4.5$

$$RHS\ of\ Eqn(A) = \frac{1}{1.5}\left[\frac{1}{tanh 4.5} - \frac{1}{4.5}\right]$$

$$\frac{1}{tanh 4.5} = \frac{e^{4.5} + e^{-4.5}}{e^{4.5} - e^{-4.5}} = \frac{90.011}{89.989} \approx 1.0$$

$$\eta = \frac{1}{1.5}\left[1 - 0.2222\right] = 0.518$$

2. Assume $\phi = 0.9, 3\phi = 2.7$

$$RHS\ of\ Eqn(A) = \frac{1}{0.9}\left[\frac{1}{tanh 2.7} - \frac{1}{2.7}\right]$$

$$\frac{1}{tanh 2.7} = \frac{e^{2.7} + e^{-2.7}}{e^{2.7} - e^{-2.7}} = 1.009$$

$$\eta = \frac{1}{0.9}\left[1.009 - 0.37.3\right] = 0.7$$

So $\phi = 0.9$

Now $\phi = \dfrac{R}{3}\sqrt{\dfrac{k_1}{De}}$

Squaring

$$\phi^2 = \frac{R^2}{9} \cdot \frac{R_1}{De}$$

$$De = \frac{R^2}{9} \frac{k_1}{\phi^2}$$

$$= \frac{\left(4 \times 10^{-3}\right)^2 (.152)}{(9)(0.9)^2}$$

$$= 0.55 \times 10^{-6} m^2 / sec$$

10.3.3 Modified CSTR

General CSTRs are modified to reduce shear rates on the cells in suspension by removing the baffles and reducing the rotation speed of the stirrer (10–40 rpm), and using the different types of stirrers viz sail type, axial flow hydrofoil agitators. Since cells are significantly smaller than the turbulent eddies created by stirring, cell lysis is reduced. Cells at the gas-liquid interface are susceptible to cell breakage. The breakage of air bubbles may also damage the cells accumulating at the interface of gas bubbles and medium. Shear protective agents like serum or pluronic F-68 are used to reduce shear effects by preventing cells from accumulating at the gas-liquid interface. 5% CO_2 enriched air can be introduced to the reactor by membrane or silicone rubber tubing. Some cell lines like chinese hamster ovary (CHO) cells, widely used for protein production are comparatively resistant to shear damage. In general, shear damage of cells causes changes in physiology and possible induction of apoptosis or programmed cell death. Moreover, cell damage may also alter apparent growth rate, reduces productivity and product quality.

Problem 10.2

A CSTR of $2m^3$ with $L/D = 2$ equipped with a paddle stirrer is used to cultivate animal cells with agitation speed of the stirrer of 15 rpm. It is desired to maintain a cell population of 1×10^{13} cells/m^3. When the oxygen consumption rate is 0.1×10^{-12} g.moles O_2 / (hr) (cell). The 5% CO_2 enriched air is introduced from the top through a tube extended to the bottom and tube tip is perforated so that small bubbles are produced.

(a) Calculate the volumetric mass transfer coefficient, k_{La} in hr^{-1}

(b) Calculate the volumetric flow rate of air to attain the above k_{La} value.

Data given:

Saturated conc. of dissolved oxygen, $C_L^* = 8$ ppm,

$$\rho_{H_2O} = 1000 kg/m^3,\ \mu_{H_2O} = 8.9\times10^{-4}\frac{kg}{(m)(s)}$$

(a) We know, oxygen uptake rate (OUR)

= oxygen transfer rate (OTR),

In mathematical form

$$q_{O_2}\cdot X = k_{L_a}\left(C_L^* - C_L\right) \text{...........(A)}$$

Now $C_L^* = 8$ ppm = 8 gm/m^3

$$C_L = 0.1C_L^*$$

Substituting the values in equation (A)

$$\left(0.1\times10^{-12}\right)\frac{g\cdot mole}{(hr)(un)}\times\left(1\times10^{13}\right)\frac{cell}{m^3} = k_{L_a}\frac{1}{hr}(0.25-.025)\frac{g\cdot mole}{m^3}$$

$$\text{So } k_{L_a} = \frac{\left(0.1\times10^{-12}\right)\left(1\times10^{13}\right)}{0.225} = 4.4hr^{-1} = 1.22\times10^{-3}\,Sec^{-1}$$

(b) For stirred vessel with coalescing bubble, we know

$$k_L a = 2.6\times10^{-2}\left(P/V\right)^{0.4}\left(u_{gs}\right)^{0.5}$$

Calculation of dimension of the reactor.

$$V = 2m^3 = 0.7854D^2(2D)$$

Solving, D = 1.083m.

Impeller diameter, $D_I = 0.3D = 0.325$m

Impeller based Reynold number

$$R_{eI} = \frac{\rho N D_I^2}{\mu} = \frac{1000(0.25)(0.325)^2}{8.9\times10^{-4}} = 2.96\times10^4$$

So the flow in the CSTR is turbulent and the power number, N_p is constant.

$$N_P = 5 = \frac{P}{\rho N_i^3 D_I^5} \text{ (from the plot of } N_P \text{ vs } R_{eI} \text{ given in chapter 7)}$$

$$\text{Or } P = 5\rho N_i^3 D_I^5 = 5(1000)(0.25)^3(0.325)^5 = 0.28KW = 280.0Watt$$

$$k_{L_a} = 0.026(P/V)^{0.4}(u_{gs})^{0.5}$$

$$1.22\times10^{-3}(sec^{-1}) = 0.026\left(\frac{280}{2}\right)^{0.4}(u_{gs})^{0.5}$$

$$\frac{1.22\times10^{-3}\times7.2}{0.026} = ugs^{0.5}$$

$$ugs = 0.114\,m/s$$

A,Cross section of the reactor($\pi/4D2$)

$$= 0.7854(1.083)^2 = 0.92m^2$$

F = Volumetric flow rate (uA)

$$= 0.92 \times .114 = 0.1049\ m^3/s$$

10.3.4 Hollow fiber bioreactor (HFBR)

Hollow- fiber bioreactors provide high surface volume ratio and high cell concentrations. A HFBR consists of a number of hollow tubes inside a shell. Cells are immobilized on the external surface of hollow fiber tubes or spongy solid matrix, and nutrients pass through the tubes. The micro environmental conditions inside the reactor where cells are immobilized on fiber surface are controlled carefully. HFBRS are constructed with fibers of known molecular wt. cut-off to control the flux of components with different Molecular weight products into the effluent stream.

A schematic diagram of a typical HFBR is shown in Fig. 10.2

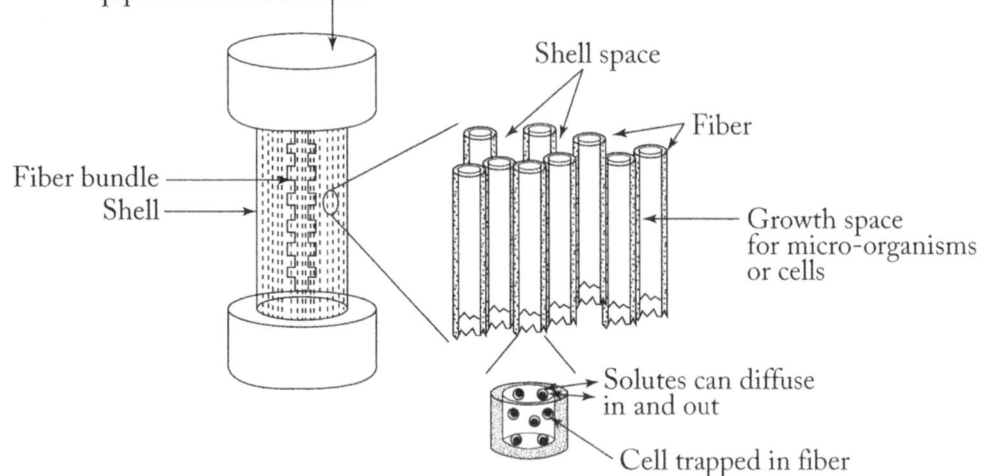

Fig. 10.2 Hollow Fiber reactor

The production of some toxic products such as lactate and ammonia causes problems, but they are removed by passing through the membrane as they are low molecular weight compounds. The metabolic product, a protein molecule is retained by the membrane.

Hollow fiber reactors have been used for the production of monoclonal antibodies from hybridoma cells. Antibody concentrations in the order of 5 to 50 mg/ml have been obtained in HFBR, since the protein, a high mol.wt. compound is retained by the membrane, while medium flows through membrane tube.

Hollow fibers reactors are well suited for perfusion operation with continuous feed. Mab production from hydridomas can be continued for long period (100 days) as stationary phase cells release the product. HFBRS are not suitable for large scale production, as there is strong mass transfer limitation along with control problems.

To overcome these difficulties, several modified types such as axial flow hollow fiber reactors, radial flow or cross flow hollow fiber reactors have been developed.

Problem 10.3

Interleuken is to be produced in a HFBR containing 500 lumens. Each lumen has the length of 1 meter having i.d. of 1 mm and O.D. of 1.4 mm. Animal cells are immobilized on the sprongy matrix in the shell side. The microbial reaction is first order with respect to substrate. The rate constant is given as $k' = \frac{r_{max}}{K_s} = 0.002\, sec^{-1}$

The initial substrate concentration is S_o = 10kg/m^3.

(a) For 95% conversion of the substrate, calculate flow rate of the substrate
(b) If *Yp/s* =4mgp/kg Substrate, Calculate *P* & Productivity (*DP*)

Solution: The working design equation of HFBR is given as

$$\frac{S}{S_o} = exp\left[-2.03\left(\frac{r_{max}}{K_S}\right)\left(\frac{V_R}{F}\right)\right]$$

Solution:

$V_R / F = \tau$ (The residence time)

$$\frac{S}{S_o} = exp\left[-2.03(0.002)\tau\right]$$

Or $\frac{0.5}{10} = exp\left[-2.03(.002)\tau\right]$

Solving $\tau = 747\, sec = 12.45 min$

Reactor Volume,

$V_R = n\pi r^2 L = 500(3.142)\left(0.5\times 10^{-3}\right)^2 \times 1m^3$

$$= 3.92 \times 10^{-4} m^3$$

$$\tau = \frac{V_R}{F}$$

$$\text{So, } F = \frac{V_R}{\tau} = \frac{3.92 \times 10^{-4}}{12.45} \frac{m^3}{min}$$

$$= 0.31 \times 10^{-4} m^3 / min$$

$$= 1.86 \times 10^{-3} m^3 / hr = 1.86 L / hr$$

(b) $P = Y_{P/S}(S_o - S)$

$$= 4(10 - 0.5) = 38 \frac{mgP}{m^3}$$

$$D = F/V = \frac{1.86 \times 10^{-3} m^3 / hr}{3.92 \times 10^{-4} m^3} = 4.7 hr^{-1}$$

DP = 4.7 X 38 =178.6 $mg\ P/(m^3)(hr)$

10.3.5 Fixed bed or fluidized bed reactor

Fixed bed or fluidized bed reactors may be used for animal cell culture. The bed contains the mammalian cells immobilized in gel beads of agar or alginate or collagen or polyacrylamide. In these systems the shear effects are eliminated. High cell densities ensure increased volumetric productivity. However, the control of environmental conditions inside the bead particles and the removal of toxic metabolites inside the beads are difficult.

A tubular ceramic matrix has been employed for the cultivation of hybridoma cells using immobilized cells. High cell and Mab concentrations have been achieved in these reactors. The quantification of cell concentration and the control of the micro environmental conditions inside the heterogeneous cells are difficult to achieve. Scale up of a ceramic matrix reactor is difficult if long tubes are used. The heterogeneous nature of the bed prevented the modelling of the system as done in the case of packed bed system.

Problem 10.4

Hybridoma cells immobilized on surfaces of sephadex beads are used in a packed column for production of monoclonal antibodies (mab). Hybridoma cells concentration is X = 5 kg/m^3 in the bed. The flow rate of the synthetic medium and the glucose concentration are F = 2 x 10^{-3} m^3/hr and S_o = 40 kg/m^3 respectively. The rate constant of Mab formation is $k = \frac{1K_g X}{m^3 day}$

There is no diffusion limitations and glucose is the limiting substrate. The bed diameter is D_t = 0.2 m. The growth of hybridoma is negligible and the process follows first order kinetics.

(a) Calculate the height and the volume of the packed bed reactor for 95% glucose conversion.

(b) If $Y_{P/S}$ is $\frac{4\,gMab}{kg\ glucose}$, determine the effluent Mab concentration and the productivity of the reactor

Solution:

We have the design equation for a packed bed reactor with first order kinetics.

$$-F\frac{dS}{dZ} = kSA\ldots\ldots\ldots\ldots\ldots\ldots\ldots(\propto)$$

$$\text{Now } k = 1\frac{KgX}{m^3 day} = 1\frac{kgX}{(m^3)(24)hr}\times\frac{1\ m^3}{5\ Kg}$$

$$= \frac{1}{120} = 8.33\times 10^{-3}\,hr^{-1}$$

Integrating the equation (α) we have

$$F\,ln\frac{S_O}{S} = kA\,H\ldots\ldots\ldots\ldots\ldots\ldots\ldots(\beta)$$

For 95% conversion, S = 0.05 x 40 = 2 Kg/m^3

$$S_O = 40\frac{kg}{m^3}$$

Putting the values of known parameters in equation (β)

$$2\times 10^{-3}\left(\frac{m^3}{h}\right)ln\frac{40}{2} = \left(8.33\times 10^{-3}\right)hr^{-1}\times(.0314)m^2 H/m$$

$$A = 0.7854(0.2)^2 = 0.0314m^2$$

$$\text{Or } H = \frac{\left(2\times 10^{-3}\right)(2.995)}{\left(8.33\times 10^{-3}\right)(.0314)} = 22.9m$$

$$V = (.0314)(22.9) = 0.72m^3$$

$$P = Y_{P/S}\left(S_O - S\right) = 4\frac{g\ Mab}{Kg\ glucose}(40-2)\frac{Kg\ glucose}{m^3}$$

$$= 152g\ Mab/m^3$$

D (the dilution rate) $= F/V \; hr^{-1}$

$$= \frac{2 \times 10^{-3} m^3 / hr}{0.72 m^3} = 2.78 \times 10^{-3} hr^{-1}$$

PD (the productivity)

$$= 152 \left(\frac{gMab}{m^3} \right) \times 2.78 \times 10^{-3} \left(\frac{1}{hr} \right) = 0.422 \frac{gMab}{(m)^3 (hr)}$$

10.3.6 Microencapsulation

Hybridoma cells have been encapsulated within spherical membranes of polylysine alginate for production of Mabs. A capsule size may be 300–500 μm and the capsule membrane has the molecular wt – cut off, 60 to 70 Kda. Microcapsules act as small membrane bioreactors in which very high cell concentration viz. 10^9 cells/ml can be attained. Using the appropriate capsule membrane with a desired MW cut off, toxic products such as lactate and ammonium can be eliminated from the inside of the capsule in the culture media. High molecular wt. products like Mab or lymphokines are concentrated inside capsules. Cells are protected from the shear effect and direct aeration is possible through permeable membrane. However, diffusion limitation of substrate and dissolved oxygen inside the large capsule is to be taken into consideration in rate process. The capsule size may be reduced to 200 μm to remove the mass transfer limitation. By varying the concentration of the product and the average Mol. Wt. of poly l-lysine, the pore size of the capsule membrane may be controlled in the range of 30–80 Kda.

Another associated problem is the monitoring and control of micro environmental conditions like pH, D.O, and Temperature. In practice, capsules containing microbial cells are placed in a fluid bed or air-lift- fermenter, where shear effects are minimum.

10.4 Reactors for Recombinant Proteins

Recombinant Proteins are produced with the help of recombinant DNA technology. The latter involves the manipulation of genetic material of individual cells. By inserting foreign genes into fast growing micro-organism, foreign gene products (recombinant proteins) are produced with higher rates and good product yield, which is not possible with any other cellular systems. This technology is popularly known as GENETIC ENGINEERING, as it comprises the manipulation of genetic materials. This section deals with the basic principles involved in recombinant DNA technology, growth kinetics of the recombinant cells, problems of reactor design in cultivating the genetically engineering cells.

10.4.1 Recombinant DNA technology

A brief procedure for cloning a segment of DNA is outlined below.

The steps are carried out in the test tube in vitro and are presented in figure 10.3 (Bailey and ollis, 1986).

1. A foreign DNA is obtained by cutting a larger piece of DNA into fragments by restriction enzymes
2. The foreign DNA is joined with plasmid vector.
3. The recombinant plasmid is introduced into host cell.........................
   ....
   ....

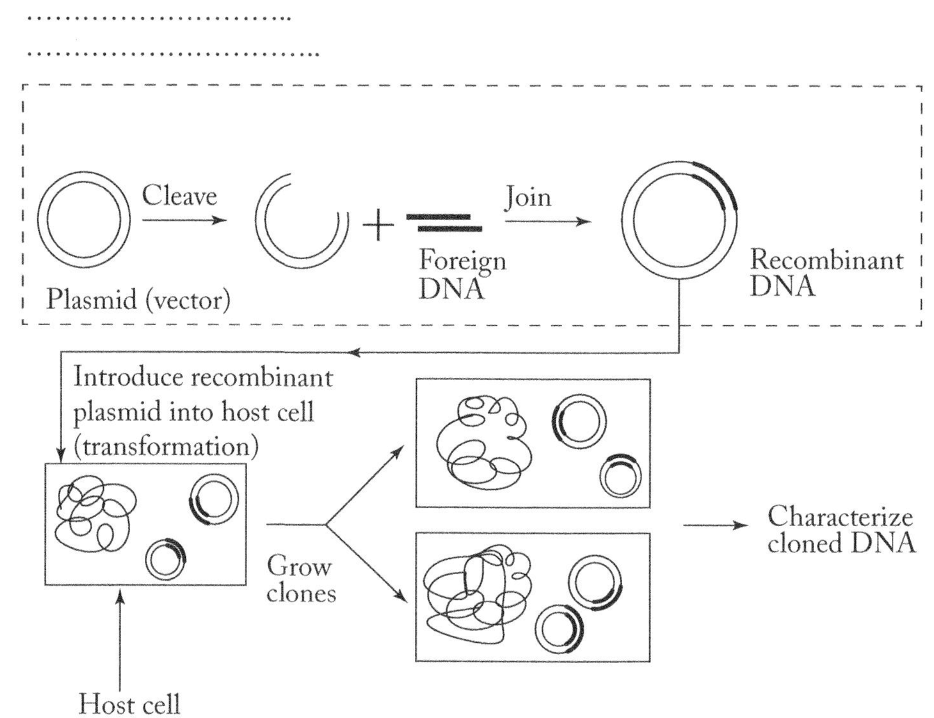

Fig. 10.3 Major steps in cloning of foreign segment (Bailey and Ollis, 1986, p.341)

The foreign DNA is joined in the test tube (in vitro) to a vector which will carry the foreign DNA as a passenger into a bacterial cell. The vector in this case is a bacterial plasmid which has been utilized to facilitate the cloning process. The recombinant DNA molecule formed by the foreign DNA plasmid component is introduced into the bacterial cell (E.Coli) by transformation.

Cloning a DNA fragment gives the enough of the DNA sequence for detailed analysis and for use as a reagent in subsequent biochemical and genetic manipulations. Several restriction enzymes have been discovered, which are used to cut, alter and join DNA molecules in the test tube. If the joint is ligated after base pairing, the fragments are joined permanently by DNA ligase (also called polynucleotide ligase). The latter binds the recombinant molecules by creating phosphodiester band between 5-PO_4 end of one polynucleotide and 3'-OH end of another. The recombinant plasmids are introduced into bacteria by co-cultivation of plasmids and bacteria.

10.4.2 Genetic instability

The formation of large amounts of target protein is always detrimental to the host cells. They lose the capacity to make the target protein and grow most rapidly and can displace the original ones which is most productive. This phenomenon is known as **genetic instability**. This genetic instability can occur due to 1) segregation loss, 2) structural instability, and 3) the variation of growth rate ratio (α) of plasmid-free to plasmid containing unaltered cells ($\alpha = \mu^- / \mu^+$).

- **Segregation loss** occurs when a cell divides such that one of the daughter cells received no plasmids. There are two types of plasmids – low copy number plasmids and high copy number plasmids. The latter is distributed randomly among daughter cells, following a binomial distribution. For high copy numbers, almost all the daughter cells receive some plasmids and the probability of forming plasmid-free cells is low. But a large reactor may contain so many cells that some plasmid-free cells will also be present.
- **Plasmid structural instability** occurs when some cells retain plasmids, but alter the plasmid to reduce its harmful effects in the cell. Cells containing structurally altered plasmids can normally grow much more rapidly than cells with original plasmids, without producing the target protein. This is known as **structural instability**.

Most cell Mutations make the host cells less useful as production systems. These mutations often change cellular regulation, reducing target protein synthesis. Promoters like lactose or a chemical analog of lactose e.g. IPTG are used in plasmid construction to control the synthesis of a plasmid encoded protein. These host-cell mutations increase the growth of the mutant, dominating in the culture. In this case, the mutant cells containing unaltered plasmids will produce very little target protein.

Growth-rate dominated Instability

Let us consider the two rates – the growth rate of plasmid-containing cells and that of plasmid-free cells. In actual situation, the plasmid free cells grow so rapidly that they outgrow the original host-vector system, producing little gene product (i.e. protein).

The growth rate ratio (α) is maintained favourably by the choice of medium. The use of selective pressure such as addition of antibiotic to the culture helps to kill the plasmid-free cells. The inducible promoter can be added at the end of a batch growth cycle when only one or two more cell-doublings may occur. Before induction, the formation of target protein by metabolic path is very poor and the growth rate ratio (α) of the altered to the original host-vector system is nearly one or less than one (with pressure of selective pressure).

This two-phase fermentation is usually carried out as a modified batch system or in a multistage chemostat. In the first stage, the production of viable plasmid-containing cells is optimised and the product formation takes place in the second stage in the presence of an inducer.

The problem of genetic instability becomes a very significant in commercial reactors compared to laboratory units. Moreover, the use of antibiotics as a selective agent is not desirable in a large scale reactor due to cost and regulatory constraints on product quality. The fraction of cells containing plasmids in the total population has been shown as a function of the number of generations of growth. As the size of the fermenter increases the number of generations of growth increases and the fraction of cells carrying plasmid and the product yield decrease.

10.4.3 Fermentation kinetics of the recombinant cells (in a batch reactor)

It is assumed that the probability of plasmid-containing cells (n^+) to produce plasmid-free cells (n^-) is "p" after one division. The N plasmid carrying cells will produce $N(1-p)$ plasmid carrying cells and NP plasmid-free cells after one division. The total number of n^+ cells will be $N + (1-p)$ or $N(2-p)$.

During the exponential growth period, the growth rate of plasmid carrying cells is,

$$\frac{dn^+}{dt} = (1-p)\mu^+ n^+ \tag{10.1}$$

where μ^+ is the specific growth rate of the plasmid-carrying cells, n^+ is the number of plasmid carrying cells per unit volume. Now the growth rate of plasmid-free cells (n^-) is given as,

$$\text{dn-/dt} = p\mu\text{+n+} + \mu\text{-n-} \tag{10.2}$$

Assuming μ^+ and p are constants, the equation (10.1) is integrated with the initial boundary conditions

At $t = 0$, $n_0^+ = N + 0$ and $n_0^- = 0$

We get

$$n^+ = n_0^+ exp\,(1-p)\,\mu^+ t \tag{10.3}$$

where n_0^+ is the initial concentration of the plasmid-carrying cells.

For plasmid-free cells, n^-, the equation (10.3) is substituted in equation (10.2) and we have,

$$\frac{dn^-}{dt} - \mu^- n^- = p\mu^+ n_0^+ exp\left[(1-p)\mu^+ t\right] \tag{10.4}$$

The above linear first order differential equation is solved with $n^+ = n_0^+$ at $t = 0$ and we get,

$$n^- = \frac{p\mu^+ n_0^+}{(1-p)\mu^+ - \mu^-}\left\{exp\left[(1-p)\mu^+ t\right] - exp\left(\mu^- t\right)\right\} + n_0^+ exp\left(\mu^- t\right) \quad (10.5)$$

Now the fraction of plasmid carrying cells in the total population, f^+, can be expressed as

$$f^+ = \frac{n^+}{n^+ + n^-} \quad (10.6)$$

Substituting equation (10.3) and (10.5) in equation (10.6), and after simplification we get,

$$f^+ = \frac{exp\left[(1-p)\mu^+ t\right]}{exp\left[(1-p)\mu^+ t\right] + \dfrac{P\mu^+ n_0^+}{(1-p)\mu^+ - \mu^-}}\left\{exp(1-p)\mu^+ - exp\left(\mu^- t\right)\right\} \quad (10.7)$$

Now it can be shown that f^+ varies with respect to the number of generations (n). During the exponential growth period, the number of generation of plasmid-carrying cells (n) can be calculated from the following equation,

$$n = \mu^+ t / \ln 2 \quad (10.8)$$

Combining the equation (10.7) & (10.8) we get,

$$f_n^+ = \frac{1 - \propto - p}{1 - \propto - p\left[2^n\left(\propto + p - 1\right)\right]} \quad (10.9)$$

where α is the growth rate ratio, defined as

$$\propto = \frac{\mu^-}{\mu^+} \quad (10.10)$$

The equation (10.9) can be used to predict the change of f^+ with respect to the number of generation for the series of batch fermentations assuming exponential growth of cells during each batch cultivation.

The figures (10.4) illustrates the effects of p and α on f^+ which decreases with the increase of p and α.

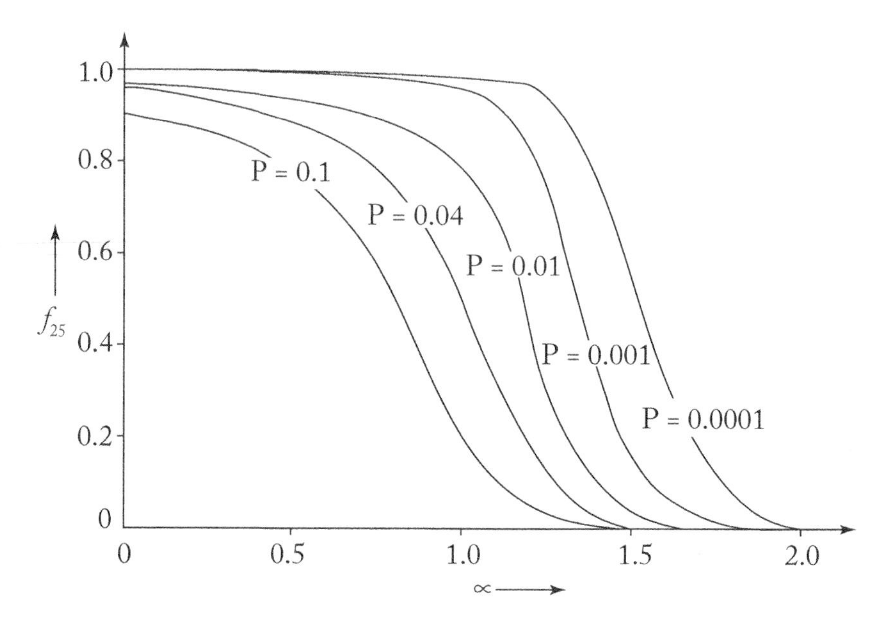

Fig. 10.4 The fraction of plasmid carrying cells after 25 generations

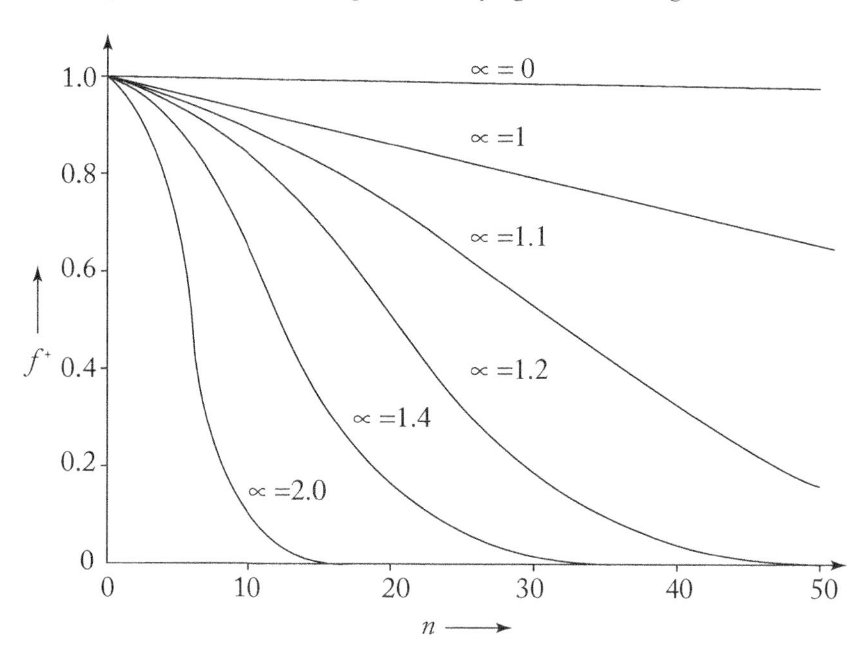

Fig. 10.5 The variation of f^+ with n for different values of α at constant p = 0.01

If $p \leq 0.01$ and $\alpha < 1$, f_{25} is close to 1, indicating that the plasmid-carrying cells are very stable. Further as α approaches 2, f_{25} becomes zero. The figure (10.5) indicates the variation of f^+ as a function of n, and when α is 1.4, all of the plasmid-carrying cells lost plasmids after 33 generations.

10.4.4 Stirred-tank fermenter for recombinant cells

Let us consider the stability of recombinant cells in the continuous stirred tank reactor (chemostat). The material balance for the plasmid carrying cells around a CSTR is

$$\frac{dn^+}{dt} = -Dn^+ + (1-p)\mu^+ n^+ \tag{10.11}$$

Similarly for the plasmid-free cells, the materials balance in the CSTR is

$$\frac{dn^-}{dt} = -Dn^- + p\mu^+ n^+ + \mu^- n^- \qquad 10.12)$$

The addition of equation (10.11) and (10.12) give the equation of total cell concentration as

$$\left(\mu^+ n^+ + \mu^- n^-\right) - D\left(n^+ + n^-\right) = \frac{d\left(n^+ + n^-\right)}{dt} \tag{10.13}$$

If the CSTR is operated so that the total concentration of cells is constant with time, we get,

$$\mu^+\left(n^+ + \propto n^-\right) = D\left(n^+ + n^-\right) \tag{10.14}$$

If $\propto = 1$, equation (10.14) reduces to $\mu^+ = D$ (10.15)

Now integrating the equations (10.11) & (10.12) we get

$$n^+ = n_0^+ exp(-pDt) \tag{10.16}$$

$$\text{and } n^- = n_0^- + n_0^+\left[1 - exp(-pDt)\right] \tag{10.17}$$

Therefore from the equation (10.16), it is evident that the concentration of plasmid-carrying cells will be reduced. The equation (10.17) shows that the plasmid-free cells will increase with time t.

If $\alpha \neq 1$, μ^+ is no longer constant for a constant dilution rate (D) during the steady state operation of CSTR, rather μ^+ depends on n^+ and n^- and α changes according to equation (10.14).

Now let us see how f^+ will decrease with time. Substitution the equation (10.14) into equation (10.11) and dividing by $(n^+ + n^-)$, we get,

$$\frac{df^+}{dt} = -Df + \frac{(1-p)Df^+}{\propto + (1-p)f^+} \tag{10.18}$$

The numerical solution of equation (10.18) gives the variation of f^+ with time. The initial value of f^+ for the solution of equation(10.18) can be obtained from the equation (10.9). The figure (10.6) indicates the variation of the fraction of plasmid-carrying cells (f^+) with time. It shows that f^+ decreases with time and with the increase of dilution rate, D.

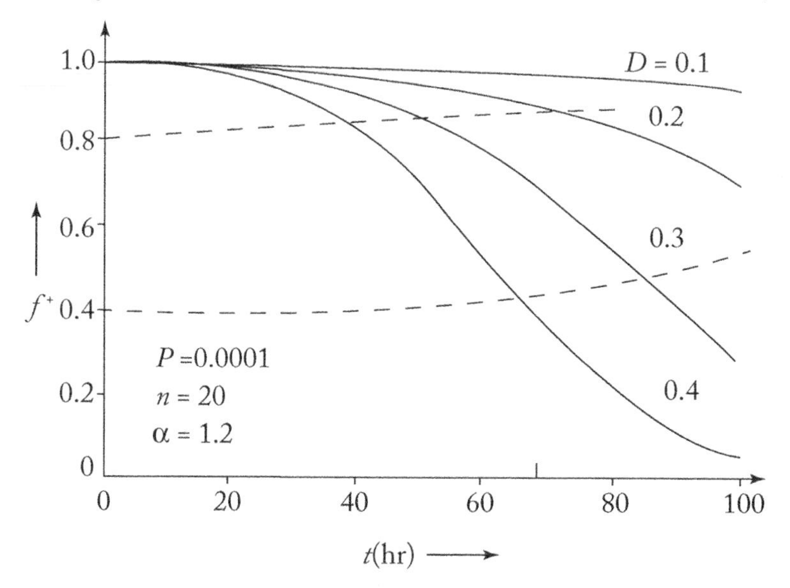

Fig. 10.6 The effect of dilution rate (D) on f^+ during the steady state operation of CSTR

10.4.5 Dynamics of plasmid containing cells

The dynamics of plasmid containing cells, will be considered when the plasmid-free host is auxotropic for metabolite M. The metabolite (M) is formed from the plasmid containing cells in the medium of a chemostat. The mass balances for n^+, n^- are already given and those of substrate (S) and metabolite (M) are expressed as below.

$$\frac{dn^+}{dt} = -Dn^+ + (1-p)\mu^+ n^+ \tag{10.19}$$

$$\frac{dn^-}{dt} = -Dn^- + p\mu^+ n^+ + \mu^- n^- \tag{10.20}$$

$$\frac{dS}{dt} = D(S_o - S) - \frac{\mu_+ n_+}{Y_{S_+}} - \frac{\mu_- n_-}{Y_{S_-}} \tag{10.21}$$

$$\frac{dM}{dt} = \delta\mu_+ n^+ - \frac{\mu_- n^-}{Y_M} - DM \tag{10.22}$$

$$\mu_{+} = \mu_{+max} \frac{S}{K_{S_+} + S} \tag{10.23}$$

$$\mu^{-} = \mu_{-max} \frac{S}{K_{S_-} + S} \cdot \frac{M}{K_M + M} \tag{10.24}$$

where M is the metabolic product, δ is the stoichiometric coefficient relating the production of M to the growth of n^+. The above set of equations are to be solved numerically. The solution becomes simpler if the M was supplied in the medium such that residual level of $M >> K_M$.

Problem 10.5

Estimate the number of generations of growth required for genetically modified microorganisms from 1 ml culture of 20 m^3 production scale fermenter. The inoculum size used in each stage of the scale-up is 5 percent except the first inoculum step.

Solution:

For the cells to grow 5% to 100% of saturated cells, they have to multiply 100/5 = 20 times, which require the cells to double n times as $2^n = 20$, solving n = 4.3 ≈ 4.0. Typical inoculum steps with about 20 times volume increase can be as follows:

1. 1 ml – 20 ml (Erlenmeyer flask)
2. 1 litre – Large flask
3. 20 litre – (bench scale fermenter)
4. 1000 litre – (Pilot scale fermenter)
5. 20,000 litre (20m^3) – Production scale

So the total number of generations needed for 20 m^3 fermenter = 5 x 4 = 20 Ans.

Problem 10.6

A 10 m^3 batch fermenter with 5×10^{16} cells/m^3 is the desired scale-up operation. Inoculum for the large tank is prepared through a series of seed tanks and flasks. Starting with a single pure colony growing on agar and a colony of 10^6 plasmids-carrying cells is picked and placed in 1 ml of culture in a test tube.

(a) Calculate how many generations will be required to achieve the required cell density in the large fermenter.
(b) What fraction of the total population will be plasmid-free cells.

Given: $\mu^+ = 1.0\ hr^{-1}$, $\mu^- = 1.2\ hr^{-1}$

P (probability) = 0.0005

Solution:

(a) $$\frac{N}{N^o} = \frac{\left(10\times10^6\right)ml\cdot5\times10^{16}\frac{cells}{ml}}{\left(10\times10^6\right)ml\times10^{16}\frac{cells}{ml}} = 5\times10^4$$

We know, $\frac{N}{N^o} = e^{\mu^+ t}$

Or $ln\frac{N}{N^o} = \mu^+ t$

$$t = \frac{ln\left(5\times10^4\right)}{1.0} = 10.81 hrs.$$

So the number of generations, n_g, required is

$$ng = \frac{\mu^+ t}{ln2} = \frac{1\times10.81}{0.693} = 15.6$$

(b) $$\propto = \frac{\mu^-}{\mu^+} = \frac{1.2}{1} = 1.2$$

The fraction of plasmid containing cells over total population is

$$f^+ = \frac{1-\propto-P}{1-\propto-P.2^{ng(\propto+P-1)}}$$

$$= \frac{1-1.2-0.0005}{1-1.2-0.0005.2^{15.6(1.2+.005-1)}}$$

$$= -\frac{0.2005}{-0.20108} = 0.99$$

f^- = fraction of plasmid-free cell,

$$f^- = 1-f^+ = 1-0.99 = 0.01$$

Problem 10.7

An inoculum with 95% plasmid containing cells and 5% plasmid-free cells is present in a 2L reactor with a total population of 2×10^{10} cells/ml. This inoculum is used for a 1 m^3 reactor and achieve a final population of 4×10^{10} cells/ml. Given $\mu^+ = 0.69\ hr^{-1}$, $\mu^- = 1.0\ hr^{-1}$, P = 0.0002.

Estimate the fraction of plasmid containing cells.

Solution:

$$\propto = \frac{\mu}{\mu^+} = \frac{1}{0.69} = 1.45$$

$$\frac{N}{N^o} = \frac{\left(1\times10^6\right)ml \times \left(4\times10^{10}\frac{cell}{ml}\right)}{\left(2\times10^3\right)ml \times 2\times10^{10}\frac{cell}{ml}.95} = 1.05\times10^3 \text{ , } ln\ NO/N = \mu + t$$

$$\text{Or } t = \frac{6.956}{0.69} = 10 hrs.$$

$$\text{Now } n_G = \frac{\mu^+ t}{ln 2} = \frac{0.69\times10}{0.693} = 9.9$$

$$f^+ = \frac{1-\propto -P}{1-\propto -P\left[2^{n_{G(\propto+P-1)}}\right]}$$

$$= \frac{1-1.45-0.0002}{1-1.45-0.0002\left[2^{9.9(1.2+.0002-1)}\right]}$$

$$= \frac{-0.4502}{1-1.45-.00078} = 0.998$$

Problem 10.8

For recombinant micro organisms (E.Coli) containing plasmids, the probability for producing a plasmid-free cell (P) is 0.001 and the ratio of specific growth rates,

$$\alpha = \mu^- / \mu^+ = 1.2.$$

Calculate the fraction of the plasmid carrying cells when the cells are fully grown in a 64 litre fermenter. The inoculation was carried out by several steps from 1 ml stock culture. The inoculation size in each step is 2.5% percent.

Solution:

For the cells to growth from 2.5% to 100% of saturated cells, they have to multiply 100/2.5 = 40 times double n times,

$$2^n = 40 \text{ or } n = 5.3 \approx 5.$$

A typical inoculation step will be 40 times volume increase and can be as follows:

1. Incolum for 1 ml to 40 ml
2. Large flask (1L)
3. Bench scale fermenter (40L)
4. Productions scale from (80l) / 64 l

So the total number of generation needed for 64l fermenter will be (5 x 4) = 20
Now the function of plasmid containing cells, f^+ is,

$$f^+ = \frac{1-\propto-P}{1-\propto-P\left[2^{ng(\propto+P-1)}\right]}$$

$$= \frac{1-1.2-0.001}{1-1.-0.001\left[.2^{20(1.2+.001-1)}\right]}$$

$$= \frac{-0.2010}{-0.2162} = 0.93$$

Problem 10.9

The recombinant microorganism (E.Coli) are cultivated in a continuous fermenter of 64L with a flow rate of 15L/hr. The total cell concentrations is constant. If the fraction of plasmid containing cells (f^+) in CSTR is 0.8 initially, how long it will operate until the fraction, f^+ drops to 0.40.

Given: p (the probability) = 0.001, α = 1.2

Solution:

We have the working equation

$$\frac{df^+}{dt} = -Df^+ + \frac{(1-p)f^+}{\propto+(1-\propto)f^+} \quad \text{(A)}$$

Now D (dilution rate) = $F/V = \frac{15}{64}$ = 0.23 hr^{-1}

p (probability) = 0.0001

$\alpha = \mu^- / \mu^+ = 1.2.$

Putting the values of the parameters in the equation (A), we have

$$\frac{df^+}{dt} = -0.23f^+ + \frac{(1-0.001)f^+}{1.2+(1-.001)f^+}$$

$$\text{Or } \frac{df}{dt} = -0.230f + \frac{0.999f^+}{1.2+0.999f}$$

$$= \frac{-0.23f\left(1.2+0.999f\right)+0.999f}{1.2+0.999f}$$

$$\text{Or } \frac{dt}{df} = \left(\frac{1.2+0.999f}{0.723f-0.23f^2}\right)$$

$$\int_{t_1}^{t_2} dt = \int_{0.8}^{f=0.4} \left(\frac{1.2+0.999f}{0.723f-0.23f^2}\right) dt$$

From the plot of f vs t at different values of D

We have t_1=70 hrs

At D = 0.23

P = 0.0001

$\propto$ = 1.2

When f decreases from f = 0.8 to 0.4

The corresponding time, t_2

t_2 = 100 hrs.

So after (100-70) or 30 hrs, the fraction of plasmid containing cells reaches to 0.4.

Problem 10.10

A reactor of 1L containing 1 x 10^{10} cell/ml of 100% plasmid containing cells is scaled up to 20,000L containing 5 x 10^{10} cells/ml at which point the over production of target protein is induced. Cells are harvested 6 hours after induction. The value of P = 0.0005. Before induction, μ^+ = 0.95 hr^{-1} and μ^- = 1.0 hr^{-1}. After induction, μ^+ =0.15 hr^{-1}, (a) what is the fraction of plasmid containing cells at induction? (b) What is the fraction of plasmid containing cells at harvest?

Solution:

$$\frac{N}{No} = \frac{\left(20{,}000\times 10^3\right)ml\times 5\times 10^{10}\,\frac{cell}{ml}}{\left(1\times 10^3\right)ml\times 1\times 10^{10}\,\frac{cell}{ml}}$$

$$= \frac{1\times 10^{18}}{10^{13}} = 10^5$$

$$ln\left(10^5\right) = 095t$$

Or t = 12 hrs.,now $ng = \mu + t/\ln 2$ or

$$ng = \frac{0.95 \times 12}{0.693} = 16.5$$

$$\propto = \frac{\mu^-}{\mu^+} = \frac{1}{095} = 1.05$$

$$P = 0.0005,\ n_g = 16.5$$

$$f^+ = \frac{1 - \propto - P}{1 - \propto - P\left[2^{ng(\propto + p - 1)}\right]}$$

$$= \frac{1 - 1.05 - 0.0005}{1 - 1.05 - (.0005)\left[2^{16.5(1.05 + 0.0005 - 1)}\right]}$$

$$= \frac{-0.0505}{1 - 1.05 - .00089} = 0.99$$

(b) t = 6hrs, $\propto = \frac{\mu^-}{\mu^+} = \frac{1.0}{0.5} = 6.6$

$$n_g = \frac{0.15 \times 6}{0.693} = 1.3$$

$$f^+ = \frac{1 - 6.6 - .0005}{1 - 6.6 - .0005\left[2^{1.3(6.6 + .0005 - 1)}\right]}$$

$$= \frac{-5.6005}{-5.6776} = 0.98$$

Problem 10.11

A strain of recombinant microorganisms is maintained as a stock culture. After three consecutive sub-cultures, the fraction of plasmid containing cells was found to be 0.74. In each sub culturing, 3 percent inoculum is used. The ratio of the sp. growth rates is $\alpha = \mu^-/\mu^+ = 1.38$.

Calculate the probability (P) per generation of producing a cell without plasmid.

Solution:

For the cells to grow from 3 percent to 100 percent of saturated cells, the cells should multiply

100/3 = 33 times. Which require the cells to double n times or $2^n = 33$ or $n = 5.04$

Now the total number of generations after 3 subcultures = 3 x 5 = 15 (n_g)
We have n_g = 15, α = 0.35 we know,

$$f^+ = \frac{1-\propto -P}{1-\propto -P\left[2^{ng(\propto+P-1)}\right]}$$

Now assuming $P = 0$ in the value of P in the power of 2.

$$f^+ = \frac{1-\propto -P}{1-\propto -P\left[2^{ng(\propto-0-1)}\right]}$$

Substituting the values of the parameters

$$0.74 = \frac{1-1.38-P}{1-1.38-P\left[2^{15(1.38-1)}\right]}$$

Solving P = 0.0026

Now putting the value of P where it is assumed zero

$$0.74 = \frac{1-1.38-P}{1-1.38-P\left[2^{15(1.38+.0026-1)}\right]} \quad 0.74(0.38+53.36) = (0.38+0.00283)$$

Solving for P

$$P = 0.0026$$

10.5 Plant Cells and Recombinant Plant Cells

10.5.1 Plant cell culture

Many thousands of chemicals are obtained from plants. A small number of characterized plants have produced many important compounds. In the West, 25% of pharmaceuticals have been prepared by extraction of whole plants. The rest is obtained as medicines derived from medicinal plants. Besides, plant products are available as dyes, food flours, food flavours, insecticides, herbicides etc.

The plant-derived products are chemically complex and are normally non-protein. Plant tissue culture can achieve several useful functions. Enzymes and complicated metabolic pathways in plants can be used for biotransformation. Digoxin, an important heart stimulant drug, has been derived by 1,2 β-hydroxylation of digitoxin, using cells of foxglove (Digitalis). Another important anticancer agent, paclitaxel (Taxol) has been produced by plant tissue culture. In the beginning paclitaxel was produced by extraction of the bark of the pacific yew tree (Taxusbrevifolia). But the growth of the trees is very slow e.g. 100 years old tree can supply enough paclitaxel to treat one patient. By a semisynthetic method, a precursor compound is extracted from branches of more common yews and the precursor is chemically converted to paclitaxel. But commercial production of paclitaxell is now possible by suspension culture in a large stirred tank vessel of 30 m^3.

The plant tissue culture has a number of important advantages:

1. Control of supply of product
2. Cultivation under controlled and optimized conditions
3. Strain improvement is analogous to microbial systems.
4. Novel compounds, not present in nature can be synthesized.

The production of dye, shinkonin, is an example illustrating the above advantages.

Bioreactors for plant cell tissue culture are not only used for chemical production, but also transgenic cell cultures have been employed for production of proteins such as vaccines. The production of artificial seeds is economically important with respect to labour intensive micro propagation of plants.

10.5.2 Plant cell culture methodology

The primary difference between plant cells and microbial cells is the ability of the plant cells to undergo differentiation organization after extended culture in the undifferentiated state. The capacity of plants to regenerate whole plants from undifferentiated cells under appropriate environmental conditions is called TOTIPOTENCY. This helps plant micro propagation and secondary metabolite formation. Callus and suspension cultures have been developed from hundreds of different plants. A callus can be formed from any portion of the plant containing dividing cells. The excised plant material is placed on solid medium containing nutrients and hormones that help rapid cell differentiation. The callus formed is a differentiated tissue containing mixture cell types. Suspension culture is maintained in the dark. Typical media contain carbon energy source such as sucrose, inorganic nutrients, vitamins and hormones such as auxins, cytokinins, gobbrellins, Ethylene acts as a plant hormone and is produced by the culture itself. The suspension culture contains friable callus. Plant cells are large with diameter in the range of 10 to 100 μm; have slow growth of doubling time from 20 to 100 hrs. Typical respiration rates are 0.5 mmole O_2 / (hr) (g dry weight of plant cell).

Plant cell culture can have high cell density, upto 70% of the total reactor volume. Plant cells secrete food compounds inside their vacuoles. Most of them are secondary metabolites of commercial importance; but may be cytotoxic if not removed from cytoplasm. In a whole plant, there are number of defensive mechanisms (elicitors) used by the plants to save themselves from the attack of pathogenic fungi, bacteria and viruses. Defence, mechanisms are activated if the breakdown products of fungal or plant cell walls are present. The exposure of suspension cultures to elicitors lead to rapid increases in the accumulation of secondary metabolites. The most common elicitor is methyl jasmonate, fed to large bioreactors, regulate the expression of a wide variety of plants. Genetic engineering of plant cells is also important for biosynthetic pathways.

10.5.3 Bioreactors for plant cells culture

10.5.3.1 Reactors for suspension culture

Plant cells are large compared to animal cells. But Cells may be damaged in a turbulent shear field where the eddy size approaches the cell size. Plant cells can withstand far more shear than animal cells. Modified stirred tanks may be used for plant cell suspension culture. The reactor size may be very large such as 75 m^3. Plant cell cultures can have high cell densities and high viscosity. Air-lift reactors for low or moderate density (<20 kg dry wt/m^3) or CSTRS with paddle type of helical ribbon impellers are used to make a compromise between the need of a good mixing and shear sensitivity of plant cells.

Mixing depends on a combination of sparging and mechanical agitation. Over sparging is to be avoided. Elevated CO_2 levels can enhance productivity. Volatile hormone, ethylene, produced by the plant cells, affects productivity. The quick removal of ethylene is desirable to sustain productivity. The optimization of the gas composition, sparging rate, mechanical agitation are essential requirements. The maintenance of aseptic conditions in the plant cell culture for 2-4 week required for completion of the fermentation, is difficult to maintain. Another limitation is the genetic instability. After 4 to 6 months, the suspension and callus culture show decreased productivity, probably due to inadequate supply of nutrients or genetic instability. Improved productivity has been obtained with two phase culture. The first phase makes a medium optimized for growth while the second phase uses a different medium optimized for product fermentation. The two phase commercial process has been used for the production of shinkonin (a dye) in Japan.

10.5.3.2 Reactors for immobilized plant cells

Immobilized plant cells reactors operate as a two-phase system. At first cells are grown and they are immobilized. The immobilized cells can be used in a continuous process under optimized conditions.

Plant cell are preferably attached to or placed within a porous matrix. They have been entrapped in gels or between membranes. The cell-to-cell contact due to immobilization or the contact of the cell surface with the surrounding gel phase may alter the cell physiology. In some cases, immobilization enhances the production rates by more than one order. In a fine suspension of cells, small aggregates are formed without mass transfer limitation. If the cells are concentrated and entrapped, they form a pseudo tissue of size ranging from 1 to 10 mm having high diffusion effects. Immobilization may be coupled with other strategies to enhance product formation.

10.5.3.3 Bioreactors for organized tissues

In many cases, suspension or immobilized cultures will not release satisfactory amounts of products. Organ cultures from some plants may give good results with respect to yield, product spectrum, enhanced product secretion, genetic stability, self-immobilization. Disadvantages are: 1) slow rates of growth, 2) difficult to maintain micro environmental conditions, 3) reactor design difficulties. However, the above limitations have been overcome by recent advances.

Large scale culture of roots has the problems of forming root mats. These mats prevent internal mass transfer, entrap gas and float, creating significant problems for uniform environment. Inspite of these problems, large scale units of root culture have been established in Japan for ginseng roots.

An advantage of using organ culture over whole plants is the use of precursors and elicitors, which enhance the formation of flavour compounds.

Shoot cultures require light, while roots are grown in the dark. The role of light is to make cellular regulation. Exposure of light of certain wave length is essential for inducing synthesis of some enzymes which are crucial for secondary metabolites.

Though suspension culture is not possible for shoot, but recently it has been shown that tobacco shoots will grow in submerged state, in a standard shake flaks in three-day doubling time if the liquid phase is well mixed.

However, very few commercial reactors for plant cells cultures have been developed, though recently are reported some large scale systems constructed in Japan and South Korea (1.2 to 30 m^3) and Germany (75 m^3).

10.5.3.4 Genetic engineering of plant cells

The main problem of plant tissue cultivation techniques is the very slow growth of plant cells and low yield of products.

Recent development in recombinant DNA technologies has the great promise for overcoming the above problem. The genetically modified plant cells have immensely increased the production of secondary metabolites. The genes responsible for economically important proteins have been identified and inserted into microorganisms, producing the proteins as gene products.

Now the technology of introducing foreign gene into plant cells has been developed for plant cell culture. The production of foreign proteins from the genetically modified tobacco cells have been established.

Recently Hiet et al (1989) demonstrated the great potential of foreign proteins from plants. Immunoglobulins have been produced with an assembly of functional antibodies in genetically modified tobacco. They transformed tobacco leaf segments by using complementary DNA from a mouse hybridoma messenger RNA and regenerated the segments to mature plants. A foreign gene should be altered for properly expressing in plant cells, because neither bacterial nor animal genes can be expressed in plants without modification.

10.5.3.5 Kinetics of plant cell growth

Little information is available about the kinetic parameters of plant cells growth, though Monod model has been used as the kinetic model. Wilson and co-workers (1980) studied plant cell cultivation in chemostats and reported the yield coefficients and doubling times for two different plant cells in media with different limiting components. Their results have been shown in the following table:

Table: 10.1 Doubling times and yield coefficients for culture plant cells

Cell Culture	Limiting nutrients	Yield Coefficient 10^6 cells / μmole	Doubling time (h)
Galium	Phosphate, PO_4^-	3.95	40
Acer	Phosphate, PO_4^-	3.47	183
Acer	Nitrate, NO_3^-	0.263	109
Acer	Glucose	0.039	109

Extremely long doubling times create a problem in the economic exploitation of plant cells as biocatalysts.

The substrate saturation constants of Monod model, K_s for acer (sycamore) plant cells in a culture on the limiting substrates NO_3^-, glucose and PO_4^- are reported as 0.13, 0.5 and 0.032mM respectively.

In the process development by Mitsu Petrochemical industries, the dye and the pharmaceutical, Shinkonin was obtained by culturing Lithospermumcrythrorhizen over a period of more than 3 weeks in successive operations. After growing cells in an initial aerated tank, they are transferred to a tank of 200L where, a medium denoted as M-9 (9 days) is added, which stimulates production of shinkonin. Subsequent transfer to third tank of 750 L followed by a reaction for 14 days results in the accumulation of the desired product in the cell. An independent estimate gives the productivity of a single batch of 5 kg product.

Problem 10.12

Codeine is to be produced in a packed bed column of gel-immobilized cells of opium popy from Codeinone (the substrate). The rate constant of codeinone consumption is first order with a rate constant, $k = 2.5 \times 10^{-8}$ m³/(kg dry cell) (s). Gel particles of 4 mm diameter with 25% volume loading of cells & 95% of water. The diffusivity of codeinon into gel is $De = 0.2 \times 10^{-9}$ m²/sec. The density of gel is $.\rho_p = 1100$ kg/m³. The feed rate is $F = 2 \times 10^{-3}$ m³/hr and initial substrate concentration is $S_o = 5$ kg/m³.

(a) Calculate the effectiveness factor, η of the system.

(b) For 90% conversion of the substrate, what is the length of the column if the cross-section of the column is $A = 0.008$m².

Solution:

(a) $$\phi = \frac{R}{3}\sqrt{\frac{k\rho}{De}}$$

$$= \frac{2 \times 10^{-3}(m)}{3}\sqrt{\frac{2.5 \times 10^{-8}\dfrac{m^3}{(kg)(S)} \cdot 1100\dfrac{kg}{m^3}}{0.2 \times 10^{-9}\, m^2/_S}} = 0.247$$

$$\eta = \frac{1}{\phi}\left[\frac{1}{tanh3\phi} - \frac{1}{0.74}\right]$$

$$3\phi = 3 \times .247 = 0.74$$

$$\eta = \frac{1}{.247}\left[\frac{1}{tanh0.74} - \frac{1}{0.74}\right] = 0.93$$

(b) The packed bed design equation is

$F\dfrac{dS}{dZ} = -\eta k \rho_p SA$ for first order integrating

$$ln\frac{So}{S} = \frac{\eta \rho_p AH}{F}$$

Putting the values of

$$F = 2 \times 10^{-3}\frac{m^3}{hr} = 5.55 \times 10^{-7}\frac{m^3}{S}$$

$S = 0.1 \times 5 = 0.5\dfrac{Kg}{m^3}$, we have,

Or, $ln\dfrac{5}{.5} = \dfrac{0.93(2.5 \times 10^{-8})(H00)(0.008)H}{5.55 \times 10^{-7}}$

Solving for H

Or $H = 6.25$m.

Problem 10.13

In a bubble column, air is sparged into the culture of catharanthisroseus at the rate of 0.24 m^3/hr. The rate of oxygen consumption by the plant cells is q_{O_2} = 0.2 × 10^{-3}mol / (kg dry wt. of cells) (hr) and the critical oxygen concentration (C_L) is 10% of the saturation (8ppm).

What value of k_{La} is to be maintained, so that plant cell of X = 20 kg/m^3 will be produced in the medium?

Solution:

$$C_L^* = 8 \times 10^{-3}\frac{KgO_2}{m^3} = 0.25 \times 10^{-3}\frac{kmol}{m^3}$$

We have the mass transfer equation for cells

$$k_{La}\left(C_L^* - C_L\right) = q_{o_2}X$$

$$k_{La} = (0.9) \times 0.25 \times 10^{-3} = 0.2 \times 10^{-3} \times 20$$

$$\text{Or, } k_{La} = \frac{0.2 \times 10^{-3} \times 20}{0.9 \times 0.25 \times 10^{-3}} = 17.8 hr^{-1}$$

Problem 10.14

C. roseus cells are immobilized in Ca-alginate beads and used in a packed bed with a flow rate of $F = 2 \times 10^{-3}$ m³/hr. Glucose is the rate limiting substrate. The product is indole alkaloids. The following data are available

$D_p = 1$ mm, $S_o = 20$ kg/m³, $x = 5$ kg/m³, $k' = 5$ m³ / (kg dry wt. of cell), $Y_{P/S} = 0.15$ kg/kg

$D_e = 1.0 \times 10^{-10}$ m²/s, $k_s = 0.5$ kg/m³

Diameter of the reactor = 0.2 m

(a) Calculate the effectiveness factor, η

(b) If 90% conversion of the substrate is desired, calculate the following?

 (i) The height of the bed,

 (ii) Volume of the bed

 (iii) Hydraulic residence time,

 (iv) If $Y_{p/s} = 0.15$ kg/ks, Calculate P, DP

 (v) If the reactor is to be used as a fluidized bed, what should be the approximate fluid velocity or flow rate.

Solution:

$$\text{(a) } \phi = \frac{r_0}{3}\sqrt{\frac{k}{De}} = \frac{0.5 \times 10^{-3}}{3}\sqrt{\frac{2.89 \times 10^{-4}}{1.0 \times 10^{-9}}} = 0.073$$

$$r_o = 0.5 mm = 0.5 \times 10^{-3} m$$

$$k = k'X = \frac{5 \times 5}{24 \times 3600} = 2.89 \times 10^{-4} Sec^{-1}$$

For $\phi = 0.023$, $\eta \approx 1.0$

(b) Packed bed design equation

$$F\frac{dS}{dZ} = -\frac{kS}{(K_s + S)} \cdot A$$

Integrating we have

$$K_s ln\frac{So}{S} + (S_o - S) = \frac{kAH}{F}(\propto)$$

$$A = 0.7854(0.2)^2 = 0.0314m^2$$

$$F = 2\times10^{-3} m^3 / hr = 5.55\times10^{-7} m^3 / Sec$$

Putting the values in equation (α), we get,

$$0.5ln\frac{20}{2} + (20-2) = \frac{2.89\times10^{-4}\times.314H}{5.5\times10^{-7}}$$

$$19.15 = 16H \text{ Or, } H = 1.2\text{m}$$

(ii) $V = 0.7854(.02)^2(1.2) = 0.037m^3$ residence time, $\tau = \frac{V}{F} = \frac{.037}{2\times10^{-3}} = 18.5hrs.$

$\frac{F}{V} = D$ (dilution rate) = 0.05 hr^{-1}

(iii) $P = Y_{P/S}(S_o - S) = .015(20-2) = 0.27\frac{kg}{m^3}$

$$DP = 0.05(0.27) = 0.0135\frac{kg}{(m^3)(hr)}$$

(iv) For fluidization $\frac{u}{u_t} = 0.8$

$$u_t = \frac{gD_p^2(\rho_p - \rho)}{18\mu} = 9.8(1\times10^{-3})^2(1100-1000)/18\times10^{-3}$$

$$= .054 m/sec$$

$$u = 0.8\times.054 = .043\frac{m}{sec}$$

Problems (Exercise)

10.1 Each spherical microcapsule of diameter of 0.5 mm contains 1.5 × 103 cells. Such 200 cells are suspended in a gently stirred CSTR for animal cell culture. The oxygen consumption rate is q_{o2} = 0.2 × 10^{-5}kmole / (hr) (cell). Air is sparged into the reactor. The critical oxygen concentration is 10% of the saturation (8 ppm). Calculate k_{La} to maintain the required D.O. in the reactor.

10.2 Hybridoma cells immobilized on the surface of cephadex beads are placed in a packed column of volume, V = 0.05 m^3 to produce monoclonal antibodies (Mab). Hybdridoma cells are distributed in the packing and its concentration is X = 10 kg/m^3 of the bed. The controlling nutrient is glucose and the initial feed concentration is S_o = 50 kg/m^3 in a packed bed reactor. The rate constant of Mab fermentation is

$k = 0.8$ kg x / (m^3) (day). There is no diffusion limitation

(a) Determine the flow rate of the feed for 90% conversion of the substrate

(b) If $Y_{p/s} = 5 \times 10^{-3}$ kg Mab / kg of glucose, evaluate the effluent concentration of Mab (p) and its productivity (DP).

10.3 An industrial batch fermenter of 20 m^3 with 4×10^{16} cells/m^3 is the desired scale up operation. Inoculum for the large tank is prepared through a series of seed tanks and flasks, beginning with a single pure colony growing on a agar slant. It is assumed that 10^7 plasmid containing cells are placed in a test tube with 10 ml medium.

(a) Calculate how many generations will be required to achieve the cell density in 20 m^3 fermenter?

(b) What fraction of total population will be plasmid-free cells, if

$\mu^+ = 0.8$ hr^{-1}, $\mu^- = 1.0$ hr^{-1}& P = 0.001

10.4 (a) Estimate number of generations of cells during a scale up from shake flask through seed fermenters into production fermenter. Cells are maintained at constant growth rate of plasmid-containing cell, $\mu^+ = 0.7$ hr^{-1} during scale up. The total time for the scale up is 30 hrs. It is assumed that the shake flask contains 100% plasmid containing cell. If the growth rate for plasmid free cells is $\mu^- = 1.0$ hr^{-1}& $P = 0.002$, (b) Calculate the fraction of plasmid containing cells in the scaled up reactor.

10.5 The k_{La} of a bubble column of diameter of 0.2 m has been found to be 15 hr^{-1}. If the rate of oxygen consumption by some plant cells is 0.2×10^{-3} kmole O_2 / (kg dry cell) x (hr) and the critical oxygen concentration (C_L) is about 10% of the saturation

($C_L^* = 0.25 \times 10^{-3}$kmole/$m^3$)

(a) Calculate the maximum plant cell mass (X) in the reactor

(b) If $k_{La} = (0.32)(Ug_s)^{0.7}$, where Ug in m/hr. Calculate the flow rate of gas to achieve $k_{La} = 15hr^{-1}$

10.6 The gel immobilized cells of opium popy is packed in a column to make codein (P) from codeinone (S). The rate of codeinone consumption is first order, with a rate constant, $k' = 4.0 \times 10^{-8}$ m^3 / (kg cell) (hr). The cell mass distributed in the bed, $X = 10$ kg/m^3.

(a) If the flow rate is 2×10^{-3} m^3/hr and initial substrate concentration, S_o = 15 kg/m^3, what is the volume of the reactor for 90% conversion of the substrate.

(b) If $Y_{p/s} = 0.015$ kg codein / kg substrate, what is the exit cone of codein (P) & the productivity (DP)?

References

Animal Cell Culture

Acton, R.T. and D. Lynn (editors) Cell culture and its applications Academic Press, New York, 1977.

Acton, R.T. and J.D. Lynn, "Descriptions and operations of a large Mammalian cell suspension culture facility", in Advances in Biochemical Engineering, Vol. 7 edited by T.K. Ghosh, A. Fiechter, and N. Blakebrough, Springer Verlag, New York, 1977.

Bailey, J.E. and D.F. Ollis, Biochemical Engineering Fundamentals, 2nd ed., McGraw Bill Book Co., 1986.

Chalmers, J.J. Animal Cell Culture: Effects of Agitation and aeration on cell adaptation in Encyclopaedia of Cell Technology, edited by R.Spier, J.B. Griffiths and A.H. Seragg, Wiley, New York, 2000.

Feder, J. and W.R. Tolbert, "The Large scale cultivation of Mammalian Cells", Scientific American, 248, 36, 1983.

Fiechter, A., "Batch and continuous culture of Microbial Plant and animal cells", P. 453 in Biotechnology Vol. 1, edited by J.J. Rehm and G. Reed, VerlagChemie, Weinheim, 1982.

Glacken, M.W., R.J. Fleischaker and A.J. Sinskey, "Large scale production of Mammalian Cells and their products: Engineering principles and Barriers to scale-up", Annals N.Y. Acad. Sci. 413, 355, 1983

HO, C.S., and D.I.C. Wang, Animal Cell Bioreactors, Butterworth – Heinemann Press, Stoneham, MA, 1991

Johson, I.S. and G.B. Beder, "Metabolities from Animal and Plant cell Culture", Adv. Microbiol 15, 215, 1973

Ku, K., M.J. Kuo, J. Delente, B.S. Wilde and J. Feder "Development of a hollow fiber system for large scale culture of Mammalian Cells", Biotechnol.Bioeng., 23, 79, 1981.

Lim, F. and R.D. Miss, "Microencapsulation of living cells and Tissues", J. Pharm, Sci. 70, 351, 1980.

Scattergood, E.M., A.J. Schiabach, W.J. McAleer and M.R. Hilleman, "Scale up of chick cell growth on microcarriers in Fermenters for vaccine production", Annals N.Y. Acad. Sci 413, 332, 1983.

Schuler, M.L. and F. Kargi, Bioprocess Engineering: Basic concepts, Pearson Education, 2004.

Tolbert, W.R., R.A. Schoenfeld, C. Lewis, and J. Feder, "Large scale Mammalian cell culture : Design and use of an Economical Batch suspension system", Biotechnol. Bioeng.34, 1671, 1982.

White, R.J., F. Klem., J.A. Chen. And R.M. Stroshane, "Large scale production of Human Interferons" in Annual Reports on Fermentation processes, Vol. 4, edited by G.T. Tsao, Academic Press, 1980.

Genetically Engineered Cells

Aiba, S. and T. Imanaka, "A perspective on the application of Genetic Engineering: Stability of Recombinant Plasmids", Ann. N.Y. Acad. Sci, 364, 1, 1981.

Deretic, V., O. Francetie and V. Glisin, "Instability of plasmid carrying Penicilin acyalse gene from E. Coli: Conditions inducing insertion activations", FEMS Microbiology Letters, 24, 173–177, 1984.

Cooper, N.S., M.E. Brown and C. A. Caulcott, "A mathematical method for Analyzing Plasmid Stability in Microorganisms", J. Gen, Microbiol., 133, 1871, 1987.

Georgion, G. and J.J. Wu, "Design of large scale contaminant facilitates for Recombinant DNA fermentation", Trends in Biotechnol, 4, 60, 1986.

Georgiou, G., "Optimizing the production of Recombinant proteins in Microorganisms" AIChE J., 34, 1233, 1988.

Kadam, K.L., K.L. Wokweber, J.C. Grosch and Y.C. Jao, "Investigations of Plasmid instability in Amylase: Continuous Culture", Biotechnol. Bioeng. 29, 859–872, 1987.

Schuler, K.L. and F. Kargi, Bioprocess Engineering: Basic Concepts; Chap. 14, Pearson Education, 2004.

Stephanopoulos, G., J. Nielsen, A. Aristidou, Metabolic Engineering: Principles and Methodologies, Academic Press, 1998.

Plant Cell Culture

Doran, M., "Design of Mixing systems for plant cell suspensions in stirred reactors", Biotechnol. Prog.15, 319–355, 1999.

Flores, H.E., Plant roots as Chemical Factories", Chem&Ind (May 18) 374–377, 1992.

Fujita, J.Y. Hara, C. Sugan, T. Morimoto, " Production of Shinkonin Derivatives by Cell Suspension Cultures of Lithospermumerythrorhizon, II. A new method for the production of shinkonin derivatives", Plant Cell reports I, 61–63, 1981.

Hsiao, T.Y., F.T. Bacani, E.B. Carvaho and W.R. Curtis, "Development of a low capital investment Reactor system: Application for plant cell suspension culture", Biotechnol. Prog. 15: 114–122, 1999.

Kargi, F. "Plant Cell Bioreactors, Present states and Future trends", Biotechnol. Prog., 3; 1, 1987

Ketchum, R.E., D.K. Gibson, R.B. Crcteau, and M.L. Shuler, "The kinetics of Taxoid accumulation in cell suspension culture of Taxus following elucitation with methyl jasmonate", Biotechnol. Bioeng 62(1), 97–105, 1999.

Lindsey, K., M.M. Yeeman, G.M. Black and F. Mavituna, " A novel method for the immobilization and culture of plant cells", FEBs Letters, I 55, 143–149, 1983

Mac Loughlin, P.F., D.M. Malone, J.T. Murtagh and P.M. Kieran, "The effect of turbulent jet flow on plant cell suspension cultures", Biotechnol. Bioeng. 58, 595–604, 1998.

Mavituna, F. and J.M. Park, "Size distribution of Plant Cell aggregates in Batch Culture", Chem. Eng. J. 35, B9 - B14, 1987.

Roberts, S.C. and K.L. Shuler, "Large scale plant cell culture: Current opinion, Biotechnol. 812, 154–159, 1997.

Shuler, M.L., "The production of secondary Metabolities from Plant Tissue Culture – Problems and Prospects", Ann. N.Y. Acad. Sci. 369, 65–79, 1981.

Wilson, G., "Continuous culture of plant cells using chemostat principle" in Advances in Biochem. Engg. Vol. 16, edited by A. Fiechter, Springer Verlag, New York, 1–25, 1980.

Zenk, K., H. Ecstagic, H. Arens, J. Stockigt, E.W. Weiler and B. Deus, "Formation of the Indole alkaloids, Serpentine and Ajmalicine in cells suspension cultures of Catharanthusnosens" in Plant Tissue Culture and its Biotechnological Applications, edited by W. Barz, E. Reinhardt, K.H. Zenk, Springer Verlag, New York, P 27–43, 1977.

CHAPTER 11

Membrane Bioreactors

11.0 Introduction

Membrane bioreactors provide a means of retaining enzymes or cells and carry out biochemical reactions, allowing the product and substrate to pass out. In a typical shell-and-tube arrangement, the bio-catalyst is retained on the shell side immobilized on the solid matrix and the reacting substrate flows through the tubes (lumens). The reaction process involves diffusion of substrate through the membrane and reaction in presence of biocatalyst. The product and unreacted substrate diffuse through the membrane back to the flowing stream in the lumen. With modern asymmetric membranes, diffusion paths can be reduced dramatically by immobilizing the catalyst in the pores close to the inner membrane.

Membranes are permeable to both substrates and low molecular weight products. These membranes can be used as a physical barrier to separate a continuous phase for a batch catalyst system. In some cases the enzymes or whole cells may be either suspended in the substrate phase or attached to the membrane.

11.1 Classification and Applications

Membrane bioreactors may be classified under the following categories:

(a) Tubular membrane reactor with enzyme or cells immobilized on inner membrane wall e.g. gel immobilization of enzymes on the inside of membrane surface.

(b) Hollow-fiber membrane reactors where microbial cell-reactions, enzymatic reaction or animal cell products-formation take place.

(c) Ultrafiltration reactor, where continuous enzymatic hydrolysis of cellulose is carried out. The system may be applied to protein hydrolysis also.

(d) Multilayer enzyme filters, where enzymes are immobilized on nylon fabric filters. Hollow fiber biochemical reactors (HFBR) have many applications such as (a) production of L-alanine from fumarate using 2 hollow-fiber membranes, (b) continuous saccharification and fermentation with cell retention.

(e) Continuous ethanol production from whey using a hollow-fiber membrane reactor

11.2 Tubular Membrane Reactor

A tubular membrane reactor contains the biocatalyst in the form of enzymes or cells immobilized on the inner wall. It may be a simple plastic with enzymes immobilized on the wall and the substrate which is flowing through the tube, diffuses to the wall and reacts with the biocatalyst and the products come back to the flowing stream of the tube.

The analysis of such a system is worked out by considering a radially well mixed segments of reactor, consisting of ΔV_t in volume and Δz in length as shown below. The reactor is divided into n number of circular zones of ΔV_t and Δz.

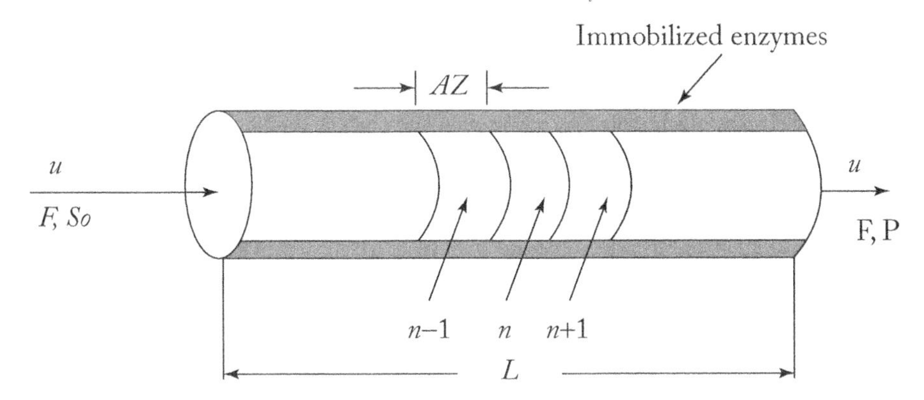

Fig. 11.1 Tubular Membrane reactor

The component balance for substrate, S in the volume element ΔV_t in the nth zone is given as

$$\Delta V_t \frac{dS_n}{dt} = F(S_{n-1} - S_n) - r_s \Delta A_w \tag{11.1}$$

Where F is the flow rate and ΔA_w is the elemental area on the reactor wall, r_s is the specific surface enzymatic reaction.

Here the external and internal mass transfer to the wall are neglected. Perfect mixing in the radial direction is assumed.

Now ΔA_w and ΔV_t are given as

$\Delta A_w = \pi d_t \Delta Z$ and $\Delta V_t = \pi dt^2/4 * \Delta Z = A_t \Delta Z$

So the equation (11.1) may be written as

$$\frac{dS_n}{dt} = \frac{F(S_{n-1} - S_n)}{A_t \Delta Z} - \frac{r_s dt}{A_t} \tag{11.2}$$

At the steady state, $d\dfrac{Sn}{dt} = 0$,

$$0 = \frac{4F}{\pi dt}\frac{(S_{n-1} - S_n)}{\Delta Z} - 4r_s \tag{11.3}$$

Taking the limit $\Delta Z \rightarrow 0$

$$\frac{dS}{dZ} = \frac{\pi d_t}{F} r_s \tag{11.4}$$

The above equation is similar to that of plug flow reactor with homogeneous kinetics. If r_s is used for enzymatic reaction rate in kmoles/(m^2) (sec) of Michaetis-Menten type,

$$r_s = v_{max}\, S/(K_M + S) \tag{11.5}$$

Substituting equation (11.5) in equation (11.4) and separating the variables and integrating we get,

$$-\int_{S_0}^{S} \frac{(K_M + S)\,dS}{S} = \frac{v_{\max}\, \pi d_t}{F} \int_0^H dZ \tag{11.6}$$

Or $K_M \ln S_0/S + (S_0 - S) = v_{max}\, \pi d_t\, H/F$ (11.7)

If the cells are immobilized, the rate is given by the equation using Monod model.

$$r_s = (\mu_{max}\, X/Yx/s)\; S/\,(K_s + S) = r_{max}\; S/\,(K_s + S) \tag{11.8}$$

where r_{max} = maximum substrate utilization rate = $\mu_{max}\, X\,/\,Y_{x/s}$ (11.8(a))

Substituting the equation (11.8) into equation (11.4), and integrating as above, we get

$$K_s \ln S_0/S + (S_0 - S) = r_{max}\, \pi d_t\, H/F \tag{11.9}$$

where K_s is the saturation constant of Monod model, H = length of the reactor, F = flow rate of substrate.

Problem 11.1

An enzymatic decomposition of urea is carried out in a tubular membrane reactor with ureas-enzyme immobilized on the inner wall of the reactor. The kinetics of the reaction is given by Michaelis-Menten equation with the following kinetics and other parameters.

$$v_{\max} = 1.54 \times 10^{-2} \frac{kg\ urea}{(m^2\ inner\ surface)(hr)}$$

$K_M = 0.2\frac{kg}{m^3}$, $F(flow\,rate) = 1 \times \frac{10^{-3}\,m^3}{hr}$, S_o (initial feed concentration) = 2 Kg/m^3,

ID = 3 cm.

What length of reactor is necessary for 90% decomposition of urea?

Solution:

The design equation for the system is given in differential form:

$$-\,d_s/dz = \pi\ dt\ rs/\ F \tag{A}$$

Where $r_s = v_{max}\ S/\ (K_M + S)$

Substituting r_s in equation (*A*), $-\ ds/dz = \pi dt\ v_{max}\ S\,/(K_M + S)\ F$

Integrating, we have

$K_M lnS_0/S + (S_0 - S) = \pi d_t\ v_{max}\ H/F$

Or

Now putting the values in the above equation, we have

$$\frac{0.2\ln 2}{0.2} + (2 - 0.2) = \frac{3.14(3\times10^{-2})(1.54\times10^{-2})H}{1\times10^{-3}}$$

Or 2.26 = 1.45 *H*,

Or *H* = 1.55 m

11.3 Shell-and-Tube Membrane Reactor

Enzymes or whole cells can be trapped behind the membranes which allow the passage of the substrate and product to a flowing stream in the tube side. In this reactor, the substrate enters the tube and diffuses radially through the porous membrane wall to the liquid on the shell side where the enzyme or whole cells are located. After reaction the product passes through the porous tube wall to the flowing liquid. A schematic of the reactor is shown in Fig. 11.2

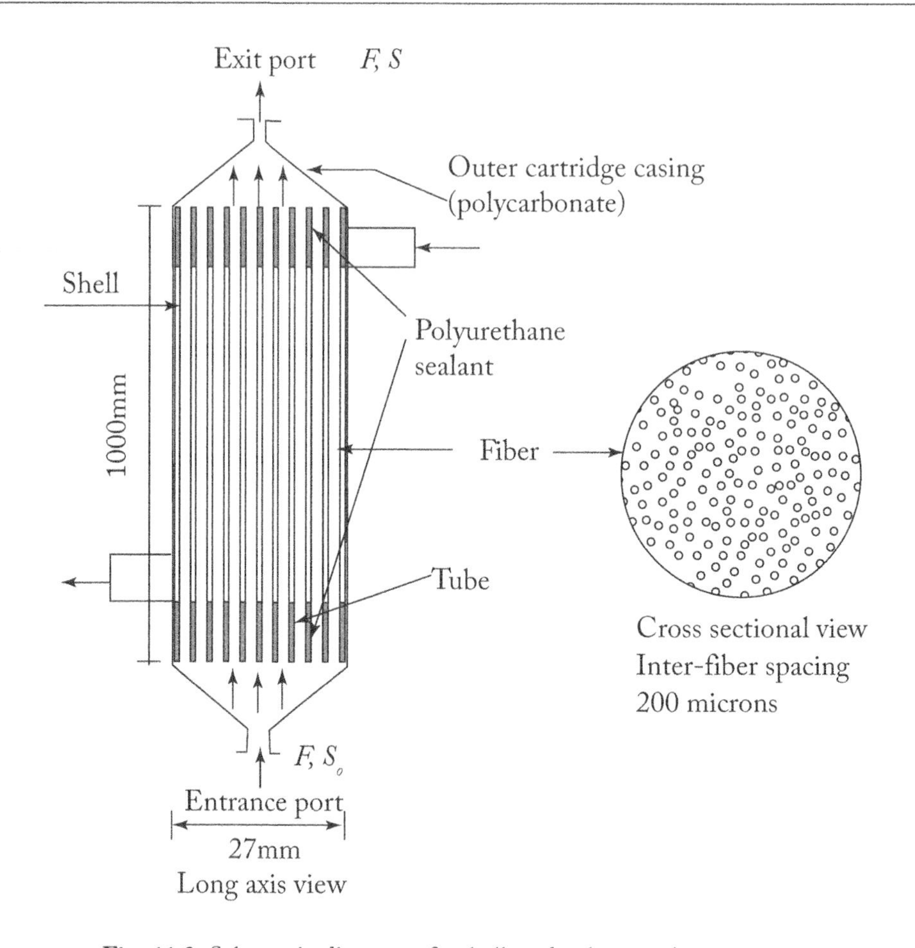

Fig. 11.2 Schematic diagram of a shell-and-tube membrane reactor

Substrate balance on the tube side

On the tube side, substrate balance may be given on the tubular segment volume, ΔV_t (= $A_t \Delta Z$) as,

$$\Delta V_t \frac{dS_n}{dt} = F(S_{n-1} - S_n) - J_{sn} \Delta^{Am} \qquad (11.10)$$

S_n, S_{n-1} are the concentration of the substrate on n and n-1 segments, J_{sn} is the mass flux of substrate on the nth segment, ΔAm (= $\pi d_t \Delta Z$) is the elemental membrane area. The mass balance is based on the convective mass flow and diffusive mass flow in the unsteady state.

Now, using the overall mass transfer coefficient, k_m, J_{sn} can be presented as

$$J_{sn} = k_m (S_n - S_a) \qquad (11.11)$$

where k_m is the mass transfer coefficient. The subscript, n, refers to the segment and a to the annulus.

Like the tubular membrane reactor, it is assumed that volume elements are radially well mixed. The mass balance on the annulus is given as

$$\Delta V_a \frac{dS_n}{dt} = J_{Sn} \Delta A_m - r_{sa} \Delta V_a \tag{11.12}$$

where $J_{sn} \Delta A_m$ is the amount of substrate transferred to the membrane and $r_{sa} \Delta V_a$ is the amount of substrate consumed by enzymatic reaction in the annulus.

Now,

$$\Delta A_m = \pi dt \, \Delta Z \tag{11.13}$$

$$\Delta V_a = \frac{\pi}{4}\left(d_0^2 - d_t^2\right)\Delta Z \tag{11.14}$$

d_0 = diameter of the annulus

d_t = diameters of the membrane tube

Now combining equations (11.10) and (11.11) and assuming steady state, we get,

$$\frac{dS}{dt} = \frac{\pi dt}{F} K_m (S - S_a) \tag{11.15}$$

Assuming the annulus substrate concentration, S_a to be constant, the equation (11.15) is integrated as,

$$-\int_{S_0}^{S} \frac{dS}{S - S_a} = \frac{\pi \, dt \, K_m}{F} \int_0^Z dZ \tag{11.16}$$

Or,

$$\ln\left(\frac{S_0}{S - S_a}\right) = \frac{\pi dt K_m}{F} Z \tag{11.17}$$

Rearranging, we have

$$S(z) = S_a + S_0 \, exp\,(-\,\pi dt K_m \, Z/F) \tag{11.18}$$

Mass balance in the Annulus:

The steady state mass balance around the annulus assuming no axial gradient (i.e. S_a = Constant)

$$O = r_{sa} V_a + \sum_{n=1}^{N} J_{Sn} \Delta Am \tag{11.19}$$

The summation indicates adding the flux from each element.

The equation (11.19) can be expressed in integral form after substituting $J_{sn} = k_m [S(z) - S_a]$, we have

$$O = r_{sa} V_a + \int_0^Z k_m \{S(Z) - S_a\} \pi dt\, dZ \tag{11.20}$$

Where V_a is the volume of the annulus, and given as $V_a = \frac{\pi}{4}\left(d_0^2 - d_t^2\right) Z$.

For simple kinetics, the above equation can be integrated and solved for S_a.

When the kinetics depend on S only as in Michaelis-Menter equation, the numerical integration of the balances for all n segments must be undertaken. If the liquid flows on the shell side necessary to maintain pH, the balance for this region should contain a flow term.

11.4 Hollow Fiber Biochemical Reactor (Kim and Cooney, 1976) (HFBR)

A hollow fiber biochemical reactor has the biocatalyst (enzymes or cells) separated from the reactant by a semipermeable membrane which permits passage of substrate and products (low mol.wt), but restricts microorganisms or protein molecules(product). A schematic of HFBR is shown in fig. 11.3

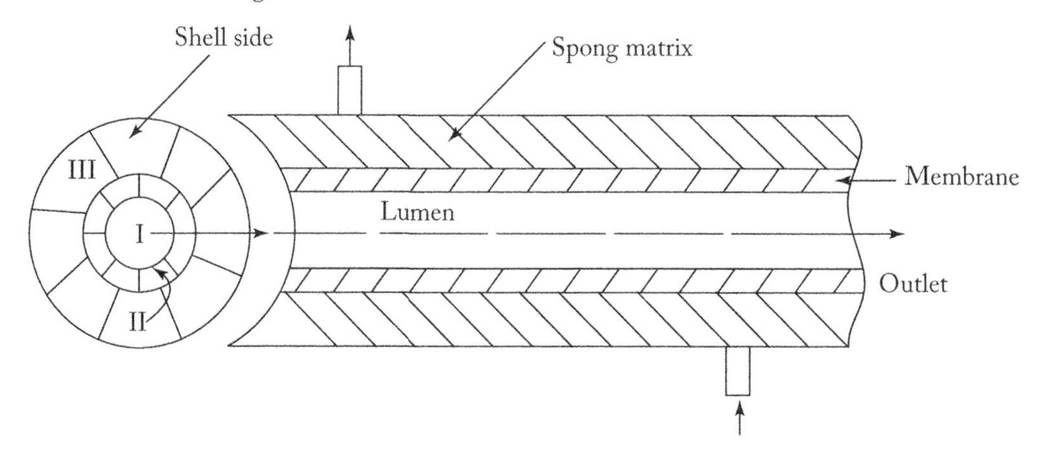

Fig. 11.3 Schematic Membrane of a single hollow fiber with shell side

Entrapment of catalyst, usually in the annular region can be accomplished by using a carrier liquid with cells to the spongy matrix as counter current to the feed stream in the lumen. A porous matrix for enzyme immobilization may be of several geometries such as planar, cylindrical (commonly used) spherical (microencapsulation).

It is important to consider the mode of reactor operation and the choice of the type of solute, solvent and the membrane, since enzyme or cells retention characteristics depend on the diffusion ratio, convective flux, fluid shear stresses, axial flow rate, enzyme characteristics such as shape, size and deformability.

Size of the apparent membrane pores and the corresponding enzyme retentions, may be controlled by the interactions of membrane flux, flow, solvent and solute. For example, certain solvents may cause membrane swelling that alters the apparent pore size or enzyme shape and elasticity. In addition, radial volume flux may modify the membrane structure leading to solvent leakage at higher flux values.

The membrane is assumed to function as an ideal semi-permeable membrane where concentration polarization and radial volume fluxes have been neglected.

11.4.1 Mathematical modeling

A representative hollow fiber of which hundreds may be packed into one tube has been shown schematically in the figure 11.2 and 11.3. A laminar feed stream in the lumen (region 1) is a continuous source of substrate molecules that diffuse through the thin membrane (region II) into the stagnant annular region (spongy matrix, region III) in which microbial catalysts are distributed.

The governing equation for the substrate transport and uptake in the hollow fiber reactors may be given as

$$\frac{\delta c}{\delta t} + v_z \frac{\delta c}{\delta Z} + v_r \frac{\delta c}{\delta r} = \frac{D_r}{r} \frac{\delta}{\delta r}\left(r \frac{\delta c}{\delta r} \right) + D_z \frac{\delta^2 c}{\delta Z^2} - kc^n \qquad (11.21)$$

In the present analysis it is assumed that $v_z = 0$ and $D_z \frac{\delta^2 c}{\delta Z2} = 0$ & $n = 1$ and Kim and Cooney (1976) considered the steady state condition, $\left(\frac{\delta c}{\delta t} = 0 \right)$.

Webster and Shuler (1981) solved the reduced form of the above equation without the convection term for the annular region. The sink term, k is the first order rate constant for enzymatic reaction of the substrate and it is the linearized form of Michaelis-Meinten equation, where $k = v_{max} / K_m$. An analytical solution of the simplified equation with parameter sensitivity analysis has been presented by Kleinstrener and Powigha (1984).

A comprehensive computer simulation model of the transient two dimensional transport phenomena of a HFBR was worked out by Kleinstrener and Agarwal (1986).

Kim and Cooney (1976) considered a HFBR containing a bundle of parallel hollow fibers (lumens) inside a cylindrical shell. Substrate solution is fed to tube or lumen side while the enzymes or cells are placed on the shell side.

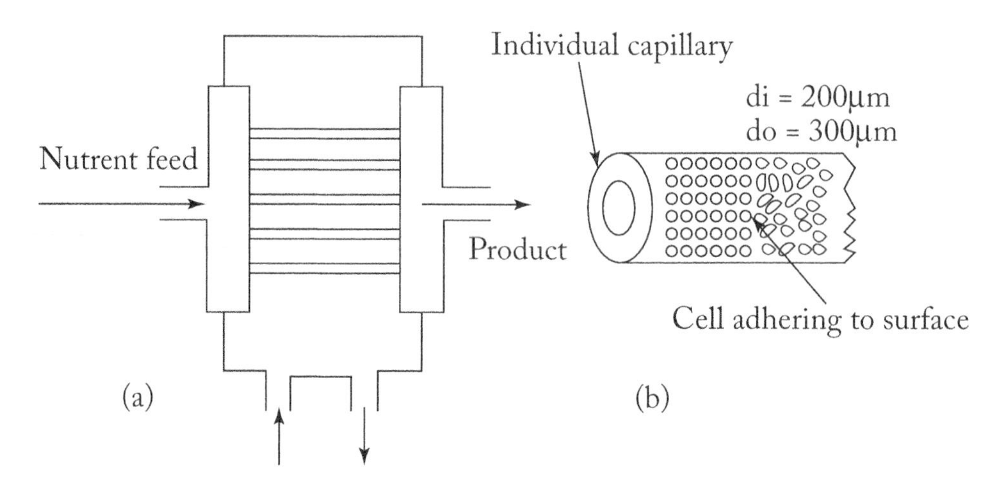

Fig. 11.4 (a) Schematic diagram of a hollow fiber module

(b) Structure of a single fiber

During operation, substrate diffuses through the fiber walls, reacts under the catalytic influence of enzyme or cells and the resultant product diffuses back into the lumen interior. The product is carried out of the reactor by the flowing tube side stream. In the system the enzymes or cells are effectively confined to the quiescent shell side, since its molecular size does not permit it to pass through the pores in the fiber wall.

Recently asymmetric microporous hollow fibers have been developed for ultrafiltration and dialysis, which have been found to be suitable for enzymatic reaction. These fibers consists of a thin (0.5 μm thick) skin having pores of the size range from 10Å to 200 Å plus a 75 μm thick annular open celled spongy layer. The spongy layer has a pore size of 5 – 10 μm having 80 – 90% void space which is suitable for containing the enzyme solution phase.

It is presumed that neither the skin nor the spongy matrix has much diffusional resistance. Thus an enzyme reaction carried out in a hollow fiber device will eventually be limited either by diffusion on the tube side or reaction kinetics on the spongy matrix. In the present analysis of HFBR a first order enzyme kinetics has been assumed.

It will be shown that the relevant (partial differential equations) equations are reduced to ordinary differential equation and explicit and simple expressions for concentration field in all regions are readily obtained.

Furthermore, it has been deduced that the bulk concentration vs axial distance can be determined as a functions of a single parameter. The method is also applicable to other problems such as analysis of a type of artificial kidney systems.

11.4.2 Model equations and solutions

Designating the fiber lumen, membrane skin and spongy matrix as regions I, II, III respectively, the mass balance equations characterizing these regions are given below:

Fiber lumen (regime I)

$$D_\tau \frac{1}{r}\frac{\delta}{\delta r}\left(r\frac{\delta C_1}{\delta r}\right) = v_0\left(1-\frac{r^2}{\alpha^2}\right)\frac{\delta C_2}{\delta Z} \tag{11.22}$$

Membrane (skin) region (II):

$$D_2 \frac{1}{r}\frac{\delta}{\delta r}\left(r\frac{\delta C_2}{\delta r}\right) = 0 \tag{11.23}$$

Spongy matrix (region III):

$$D_3 \frac{1}{r}\frac{\delta}{\delta r}\left(r\frac{\delta C_3}{\delta r}\right) = \frac{v_{\max}}{K_m} C_3 \tag{11.24}$$

Where D_1, D_2, D_3 are diffusivities in three regions. Assuming first order Michaelis-Mentein kinetics, the rate is,

$$v = \frac{v_{\max} C_3}{K_M + C_3} \approx \frac{v_{\max}}{K_M} C_3 \tag{11.25}$$

$$\text{Where } C_3 << K_M,\ v = kC_3,\ k = \frac{v_{\max}}{K_M} \tag{11.25a}$$

Lewis and Middleman (1974) found that the first order kinetics pertained for urease in an asymmetric hollow fiber system in which the substrate (urea) concentration was 2 kg/m^3.

The integration of membrane equation (11.23) gives

$$\frac{\delta C_2}{\delta r} = \frac{f(Z)}{r} \tag{11.26}$$

Where $f(Z)$ is some function of axial distance, Z.

A second integration of equation (11.26) gives

$$C_2 = f(Z) \ln r + g(Z) \tag{11.27}$$

where $g(Z)$ is another function of z.

Now using the boundary conditions

$$\left.\begin{aligned} C_2 &= C_2a(Z)\ at\ r = a \\ C_2 &= C_2b(Z)\ at\ r = b \end{aligned}\right\} \tag{11.28}$$

Where C_2a and C_2b are the concentrations of the diffusing component at inner surface of the lumen (a) and that at the outer surface of the lumen (b) respectively and (b-a) is the membrane thickness.

Substituting the values of $f(z)$ and $g(z)$ in equation (11.27), we get,

$$C_2 = \left[\frac{C_{2a}(Z) - C_{2b}(z)}{\ln \frac{a}{b}} \ln \frac{r}{a} + C_{2a}(Z) \right] \tag{11.29}$$

differentiating with respect to r and evaluating the derivative at $r = a$, we get

$$\left. \frac{dC_2}{dr} \right| r = a = \frac{C_{2a}(Z) - C_{2b}(Z)}{a \ln \frac{a}{b}} \tag{11.30}$$

Defining a partition coefficient, k as

$$\mathrm{K} = \frac{\text{Concentration in the membranne}}{\text{Concentration in the adacent field}} \tag{11.31}$$

We can consider diffusional mass balance in the membrane and the flux can be given as

$$-D_2 \left. \frac{dC_2}{dr} \right| r = a = \frac{-D_2 K_a}{a \ln \frac{a}{b}} \left[C_1 - \left(\frac{Kb}{Ka} \right) C_3 \right] \tag{11.32}$$

Where K_a and K_b are the partition coefficients at a and b. C_1 is evaluated at $r = a$ and C_3 is evaluated at $r = b$.

Now the continuity of the flux of the diffusing component requires

$$-D_2 \left(\frac{\delta C_2}{\delta r} \right) = -D_1 \left(\frac{\delta C_1}{\delta r} \right) \tag{11.33}$$

Using the equation (11.32 and 11.33) we get,

$$-D_1 \left. \frac{dC_1}{dr} \right| r = a = h(C_1 - k' C_3) \tag{11.34}$$

Where

$$h = -\frac{KaD_2}{a \ln \frac{a}{b}} \tag{11.35}$$

$$k' = \frac{K_b}{K_a}$$

For first order kinetics, explicit expressions for the concentration fields on both the enzyme (C_3) and lumen sides (C_1) can be obtained by numerical integrations of appropriate equations by numerical method.

The results indicate that the lumen side profiles of bulk concentration vs. axial distance can be expressed as a function of a single parameter, an "overall mass transfer resistance, designated as σ.

Limiting forms of the equations and system's behaviour can be predicted from very low and high values of Thiele modulus, λ^2, which is defined as

$$\lambda^2 = a^2 \frac{v_{max}}{K_M D_s} \quad (11.36)$$

where a is the inner lumen radius, the first order kinetic rate constant of an enzymatic reaction $\left(k = \frac{v_{max}}{K_M}\right)$ and D_s is the diffusion coefficient of substrate.

The dimensionless equations for lumen and shell sides can be presented using the following dimensionless variables.

$C = \frac{C_1}{C_0}$, $X = \frac{r}{a}$ for substrate in the lumen side.

For the shell side (enzymatic)

$$Z = z/aP_e, \quad P_e = av_0/D_1, \quad (11.37)$$

where C_0 is the inlet substrate concentration, v_0 is the fluid velocity, P_e = Peclet number, D_1 = diffusity in the lumen side.

Now the lumen side (region 1) equation (11.22) and spongy matrix (region III) side equation (11.24) can be written in terms of the above dimensionless variables:

$$\frac{1}{X}\frac{\delta}{\delta X}\left(X\frac{\delta C_1}{\delta X}\right) = (1-X)\frac{\delta C_1}{\delta Z} \quad (11.38)$$

$$\frac{1}{X}\frac{\delta}{\delta X}\left(X\frac{\delta C_3}{\delta X}\right) = \lambda^2 C_3 \quad (11.39)$$

The appropriate boundary conditions are:

Boundary conditions:

$$C_1 = 1 \text{ at } Z = 0$$

$$\frac{dC_1}{dX} = 0 \text{ at } X = 0$$

$$\left(\frac{\delta C_1}{\delta X}\right) = \alpha\left(\frac{\delta C_3}{\delta X}\right) \text{ at } X = 1$$

$$-\left(\frac{\delta C_2}{\delta X}\right) = h'(C_1 - k'C_3) \; at \; X = 1$$

$$\left(\frac{\delta C_2}{\delta X}\right) = 0 \; at \; X = \beta \tag{11.40}$$

Where $\quad h' = \frac{ha}{D_1}, \propto = \frac{D_3 b}{D_1 a}, \; \beta = \frac{d}{b}$

where d is the diameter of the shell.

Using Laplace transformation and other manipulation, the solutions were given by Kim and Cooney (1976) for the Lumen side (region I), C_1 and shell side (region III), C_3

$$C_1(X, Z) = \sum_{n=2}^{\infty} A_{1n} \exp\left(-\Lambda_1^2, nZ\right) \times \exp\left(-\Lambda_1^2, n\frac{X}{2}\right) M\left(\frac{2-\Lambda_{1,n}}{4} \cdot 1 \cdot \Lambda_1^2, nX\right) \tag{11.41}$$

where A_{1n} are eigen constants, $\lambda^2_{1,n}$ are the eigen values for the lumen side and M is the Kummer function. All these parameters were evaluated.

The shell side concentration field has been shown as

$$C_3(X, Z) = \sum_{n=2}^{\infty} A_{3n} \; exp\left(-\Lambda_{3n}^2 \; Z\right) \times \left\{ I_0(\lambda \; X) + \frac{I_1(\lambda \; \beta)}{K_1(\lambda \; \beta)} \cdot K_0(\lambda \; X) \right\} \tag{11.42}$$

Where A_{3n} = eigen constants λ^2_{3n} = eigen values for the shell side.

I_0, I_1 = modified Bessel function of the first kind of zero order and that of first order respectively. K_0 and K_1 = modified Bessel functions of the second kind of zero order and that of first order respectively.

11.4.2.1 Overall mass transfer coefficient σ

A characteristic parameter, σ "the overall mass transfer coefficient" has been derived and can be expressed as

$$\sigma = \frac{I_0(\lambda)K_1(\lambda\beta) + I_1(\lambda\beta)K_0(\lambda)}{\lambda[I_1(\lambda\beta)K_1(\lambda) - \mathrm{I}_1(\lambda)K_1(X\beta)]} \tag{11.43}$$

The magnitude of σ will depend directly on λ where

$$\lambda = b\left(\frac{V_{max}}{K_M D_1}\right)^{\frac{1}{2}} \tag{11.44}$$

It can be shown that since $K_1(\lambda\beta) \rightarrow 0$ as $\beta \rightarrow \infty$ then $\sigma \rightarrow K_0(\lambda) / \lambda K_1(\lambda)$ as $\beta \rightarrow \infty$. It is evident that $\beta \rightarrow 1$ (zero spongy matrix thickness), $\sigma \rightarrow \infty$, since no reaction can occur as the enzyme side is non-existent,

Where $\beta = \frac{d}{b}$, $\propto = \frac{D_3 b}{D_1 a}$.

Kinetic control regime:

When $\lambda \to 0$, the system becomes kinetically controlled where no radial concentration gradients exist.

A simple mass balance yields,

$$\frac{v_0}{2}\frac{dC_1}{dZ} = \frac{v_{max}}{K_M}(\beta^2 - 1)\left(\frac{b}{a}\right)^2 C_1 \tag{11.45}$$

$$\text{Or } \frac{dC_1}{dZ} = k^2(\beta^2 - 1)\left(\frac{b}{a}\right)^2 C_1 \tag{11.46}$$

Where
$$k^2 = \frac{2v_{max}a^2}{K_M D_1} \tag{11.47}$$

$\beta^2 = \frac{d^2}{b^2}$ and k^2 is a type of Thiele modulus.

The equation (11.46) is integrated between the limits C_0 (z=0) and C_1 (z=z) to give,

$$\frac{C_1}{C_0} = exp\left[-k^2(\beta^2 - 1)\left(\frac{b}{a}\right)^2 Z\right] \tag{11.48}$$

The above equation is the design equation of Kim and Coony (1976)...

11.4.3 Model of hollow fiber biochemical reactor by waterland et al (1974)

Let us consider a steady state convection down a porous tube with radial diffusion and first order biochemical reaction. The steady state mass balance based on radial diffusion and first order biochemical reaction is given as,

$$D_m \frac{1}{r}\frac{\delta}{\delta r}\left(r\frac{\delta C_m}{\delta r}\right) = \frac{u_{max}}{K_M} C_m \tag{11.49}$$

where C_m is the concentration of the substrate in the matrix and D_m is the diffusivity in the matrix. For slow kinetics, Thiele modulus is small, implying $\lambda^2 < 1$, the flowing substrate concentration in the lumen has a radially uniform concentration profile, so that at the fiber wall

at $r = a$,
$$\frac{v_0 a}{4}\frac{\delta C}{\delta Z} = D_m \frac{\delta C_m}{\delta r} \tag{11.50}$$

where v_0 is the superficial velocity in the lumen.

Other boundary conditions are

$r = 0,\ \frac{dC}{dr} = 0$ (no flux at outer fiber boundary)

$Z = 0,\ C = C_O$ (initial condition)

$$r = 0,\ \frac{dC}{dr} = 0 \text{ (symmetry along lumen axis)} \tag{11.51}$$

The analytical solution of equation (11.49) has been presented as

$$\frac{C_e}{C_0} = \exp(-4B\lambda\xi) \tag{11.52}$$

where C_e = exit concentration and the Thiele modulus, λ^2 is expressed as

$$\lambda^2 = a^2 \frac{U_{max}}{K_M D_S} \tag{11.53}$$

and the dimension less distance, ξ is

$$\xi^2 = \frac{ZD_S}{a^2 v_0} \tag{11.54}$$

and

$$v_0 = \frac{2Q}{a^2 \pi N} \tag{11.55}$$

Where Q is the volumetric flow rate, N = number of lumens, a = inner radius of the lumen and B is the constant given by

$$B = -\frac{K_1(\lambda\phi)I_1(\lambda) + I_1(\lambda\phi)K_1(\lambda)}{K_1(\lambda\phi)I_0(\lambda) + I(\lambda\phi)K_0(\lambda)} \tag{11.56}$$

where K and I are the Bessel functions of first order and zero order, with $\phi = r_o/a$, r_o is the outside radius of the hollow fiber and a is the inner radius of the fiber,

N = number of fibers (lumens)

When $\phi = 1.75$, $0 < \lambda < 0.5$, that is the system is kinetically controlled and B becomes $B = 1.031\lambda$.

Substituting the values of B, λ, ξ, v_0 in the equation (11.46), we get

$$\ln\frac{C_e}{C_0} = -4(1.031\lambda)\xi \tag{11.57}$$

Now substituting the value of λ & ξ and simplifying we get,

$$\ln\frac{C_e}{C_0} = -2.06\left(\frac{V_{max}}{K_M}\right)\left(\frac{V_{reactor}}{Q}\right)$$

or $C_e = C_0 \exp(-2.06k'\tau)$ (11.58)

where $(V_{max}/K_M) = k'$ (first order rate constant), $\tau = V_{reactor} / Q$ (residence time).

The above analysis is very much simplified. Much more sophisticated models have been presented for specific systems such as ultrafiltration HFBR (Piret and Cooney, 1990).

11.4.4 Hollow fiber biochemical reactor for animal cell culture

HFBRs provide a high growth surface-area-volume ratio, leading to high cell concentrations (10^6 cells/ml) in animal cell culture. Cells may be immobilized on the external surface of hollow fibers and the nutrients flow through the tubes. Reactors are constructed using fibers of known Mol. Wt. cut-off to control the different molecular weight products into the effluent stream. Toxic products formed in animal cell culture such as lactate and ammonia are low mol. wt. compounds and pass through the membrane to the flowing stream in the tube.

HFBRS have been used for the production of monoclonal antibodies (Mabs) from hybridoma cells immobilized on the shell side, and antibodies formed on the shell side have the concentrations of 5 to 50 mg/ml and are retained on the cell surface, which are washed out from the shell side by using a suitable solvent or buffer solution. The reactors have severe mass transfer and other control problems. So HFBRS are not suitable for large scale production of animal cell products..

These difficulties have been overcome by the change in the reactor design. New designs are axial-flow hollow fiber reactors, radial flow or cross-flow hollow fiber unit. HFBRs have been designed to operate with perfusion system.

With continuous feed, Mab production from hybridomas is sustained for extended period such as 100 days, because during stationary phase cells will synthesize products.

Problem 11.2

A hollow fiber biological reactor (HFBR) of length 1 m containing 200 number of fiber (a, lumen inside radius = 0.5 mm r_o, lumen's outside radius = 0.7 mm) is to be used to carry out an enzymatic reaction with enzymes immobilized on spongy matrix on the shell side. The substrate concentration in the feed is such that the first order Michaelis-Menten kinetics is satisfied. The following data are given: $\frac{v_{max}}{K_M} = 0.001 \text{ sec}^{-1}$. The total flow rate through the bundle, $F = 10^{-3}$ m^3/h, D_s (the substrate diffusivity) = 1.9×10^{-9} m^2/s, S_o = 10 kg/m^3. Calculate the following:

(a) Thiele modulus, $\lambda^2 = a^2 \frac{v_{max}}{K_M D_S}$
(b) Concentration of the substrate conversion at the exit of the HFBR
(c) How do you operate the HFBR by changing the variables to obtain higher conversion?
(d) What is the shell diameter of the system?
(e) Calculate the substrate conversion using the Kim and Cooney's model:

$$\frac{C}{C_0} = exp\left[-k^2(\beta^2 - 1)\left(\frac{b}{a}\right)^2 Z\right]$$

$$k^2 = \frac{2v_{max}a^2}{K_M D_S}, \quad \beta = \frac{d}{a}$$

Solution:

(a) $\lambda^2 =, a^2 . \dfrac{v_{max}}{K_M D_S} = \dfrac{(0.5\times10^{-3})^2 m^2 (.001)}{(1.9\times10^{-9})} = 0.1315$

(b) HFBR design Eqn.

$$\frac{C(L)}{C_0} = exp\left[-(2.06)\frac{v_{max}}{K_M}, \frac{V_R}{F}\right]$$

$$\frac{v_{max}}{K_M} = 0.001 \text{ sec}^{-1}$$

$$\begin{aligned} V_R &= n\frac{\pi}{4}D^2L \\ &= (200)(.07854)(1\times10^{-3})^2 m^2 \\ &= 1.57\times10^{-4} m^3 \end{aligned}$$

$$\begin{aligned} F(\textit{flow rate}) &= 1\frac{L}{hr} = \frac{1\times10^{-3}}{3600}\frac{m^3}{\text{sec}} \\ &= 2.77\times10^{-7}\frac{m^3}{\text{sec}} \end{aligned}$$

$$\begin{aligned} \frac{V_R}{F} &= \tau \ (\textit{residence time}) \\ &= \frac{1.57\times10^{-4} m^3}{2.77\times10^{-7}\dfrac{m^3}{\text{sec}}} = 566 \text{ sec.} \end{aligned}$$

$$C_0 = 10\frac{kg}{m^3}$$

$$C(L) = 10\ exp[-2.06(.001)(566)]$$

$$= 3.116\frac{kg}{m^3}$$

$\delta(\textit{substrate conversion})$

$$= \frac{C_e - C(L)}{C_0} = \frac{10 - 3.116}{10} = 0.688$$

(c) In the reactor the only variable is the residence time which also depends on reactor volume and flow rate. Reactor volume depends on number of tubes.
If n is increased to 500 from 200 & keeping flow rate constant

$$\tau = \frac{500 \times .7854(1 \times 10^{-3}) \times 1}{2.77 \times 10^{-7}} = 1417 \text{ sec.}$$

$$\frac{C}{C_0} = exp[-2.03(.001)(1417)] = 0.056$$

$$1 - \delta = 0.056, \delta = 0.944$$

So the conversion increased from 0.688 to 0.944

If the number of lumens is kept constant and flow rate is increased, the conversion will decrease, $F = 2\frac{l}{hr} = \frac{2 \times 10^{-3}}{3600} = 5.55 \times 10^{-7} \frac{m^3}{\text{sec}}$

$$\tau = \frac{1.57 \times 10^{-4}}{5.55 \times 10^{-7}} = 282 \text{ sec.}$$

$$\frac{C}{C_0} = 1 - \delta = \exp[-2.03(.001)(282)] = 0.564$$

Or $\delta = 0.436$ (substrate conversion)

(d) Shell diameter:

Based on the outside diameter of the lumen. Outside radius of the lumen = 0.7mm

Volume of 200 lumens based on outside diameter:

$$= n\pi r_0^2 L$$
$$= 200 \times 3.142(00.7 \times 10^{-3})^2 \times 1$$
$$= 3.079 \times 10^{-4} m^3$$

Taking 30% extra, reactor. Volume,

$$V = 1.3 \times 3.079 \times 10^{-4} m^3$$
$$= 4.0 \times 10^{-4} m^3$$

$$V = \frac{\pi}{4} D_0^2 L$$

Where D_o shell diameter

$$D_0 = \left(\frac{4.0 \times 10^{-4}}{0.7854 \times 1}\right)^{1/2} = 2.25 \times 10^{-2} m$$
$$= 22.5 \; mm = 2.25 \; cm$$

(e) HFBR design equation of Kim & Cooney:

$$\frac{C}{C_0} = exp\left[-k^2(\beta^2 - 1)\left(\frac{b}{a}\right)^2 Z\right]$$

$$k^2 = \frac{2a^2 \cdot v_{max}}{K_M} \cdot \frac{1}{D_S}$$

$$= 2(0.5 \times 10^{-3})^2(.001)\frac{1}{1.9 \times 10^{-9}} = 0.263$$

$$\beta^2 - 1 = \left(\frac{0.7}{0.5}\right)^2 - 1 = 0.96$$

$$\left(\frac{b}{a}\right)^2 = \left(\frac{0.7}{0.5}\right)^2 = 0.96$$

$$\frac{C}{C_0} = \exp[-0.263(0.96)(1.96) \times 1]$$

$$Or\ 1 - \delta = 0.61$$

$$Or\ \delta = 1 - 0.61 = 0.39$$

Where δ is the substrate conversion and is equal to $(C_0\text{-}C)/C_0$

Problem 11.3

A HFBR with each fiber of 1 metre length containing 400 number of fibers (inside radius of lumen, a = 0.5 mm, $O.D$ is 0.6 mm) has been used for the production of monoclonal antibodies (Mabs) from hybridoma cells immobilized on the spongy matrix on the shell size. The flow rate of the synthetic medium to the tube side is $F = 2 \times 10^{-3}$ m^3/hr with initial substrate concentration of $S_o = 20$ kg/m^3. The rate constant of the Mabs formation is $k = 1 \times 10^{-3}$ sec^{-1} and the glucose is the rate limiting reactant.

(a) Calculate the Thiele modulus, λ, given $D_s = 2 \times 10^{-9}$ m^2/s
(b) Determine the conversion of glucose at the exit of the HFBR.
(c) If $Y_{p/s}$ = 4 g Mab / *Kg* glucose converted, determine the Mab concentration (*P*) and the productivity of Mab (*DP*)

Solution:

(a) $$\lambda^2 = a^2 \frac{k}{D_S} = \frac{(0.5 \times 10^{-3})^2 10^{-3}}{2 \times 10^{-9}} = 0.125$$

$$\lambda = 0.353$$

(b) Since $\lambda < 0.5$, the system is chemical reaction controlling, the following design equation is applicable

$$\frac{S}{S_0} = \exp[-2.06k'\tau]$$

where k' is the first order rate constant, τ is the residence time.

$$\tau = \frac{V}{F}$$

$$V = n\pi a^2 L$$

$$= 400 \times (3.142) \times (.5 \times 10^{-3})^2 \times 1 \text{ m}^3$$

$$= 3.142 \times 10^{-4} \text{ m}^3$$

$$\tau = \frac{3.142 \times 10^{-4}}{5.55 \times 10^{-7}} = 5.66 \times 10^2 \text{ sec} = 566 \text{ sec} = 9.43 \text{ mm}$$

$$\frac{S}{S_0} = 1 - \delta = \exp[-2.06(1 \times 10^{-3})566] = 0.31$$

or $\delta = 0.69$ (substrate conversion)

where substrate conversion δ, is defined as $(S_0 - S)/S_0$

(c) $P = Y_{P/S}(S_0 - S)$

Or $P = 4\ (20 - 6.2)$

$= 55.2$ g/m^3 Mab

$$PD = 55.2 \times 0.106 = 5.85 \frac{g\ Mab}{(m^3)(\min)}$$

$$\text{Where } D = \frac{F}{V} = \frac{1}{9.43} = 0.106 \text{ min}$$

And D is the dilution rate

Problem 11.4

An HFBR is used to convert glucose to ethanol by yeast cells immobilized in calcium alginate in the shell side. The glucose solution is fed to the fiber tube at the initial concentration of $S_o = 5$ kg/m^3. The nutrient flow rate (F) is 2×10^{-3} m^3/hr. the rate constant for the conversion are:

$r_{max} = 100$ mgs / cm^3 hr & $K_s = 10$ mgs/cm3. The fibre radius is $a = 0.5$ mm and length is 1 metre. It is desired to convert glucose to ethanol by 80 percent.

(a) Calculate the number of fibers required for the above conversion. The substrate diffusivity, $D_s = 10^{-9}$ m^2/s
(b) Calculate the Thiele modulus, $\lambda^2 = a^2 k' / D_s$. k' is the first order rate constant.
(c) Calculate the number of fibers required for 80% conversion.

Solution:

(a) Thiele modulus, λ

$$\lambda^2 = a^2 \frac{\gamma_{max}}{K_M D_S}$$
$$= (0.5\times10^{-3})^2\left(\frac{100}{10\times3600}\right)\frac{1}{10^{-9}}$$
$$= 0.675$$
$$\lambda = 0.82$$

(b) For first order kinetics & chemical reaction control, the design equation of HFBR is:

$$\frac{S}{S_0} = \exp[-2.06k'\tau]$$

$$\text{Now } k' = \frac{r_{max}}{K_M} = \frac{100\frac{mgS}{cm^2 hr}}{10\frac{mg}{cm^3}} = 10\ hr^{-1} = 2.7\times10^{-3}\sec^{-1}$$

$$S_0 = \frac{5kg}{m^3}$$

$$\tau = \frac{V}{F},\ F = \frac{2\times10^{-3}}{3600} = 5.55\times10^{-7}\frac{m^3}{S},$$

we know, $\frac{S}{S_0} = 1-\delta = \exp[-2.06k\tau]$ where δ = substrate conversion.

$$1-\delta = 1-0.8-\exp[-2.06(2.7\times10^{-3})\tau]$$

Solving for τ, residence time

$$\tau = 289\ \sec$$
$$V = \tau F = (289)\sec\times(5.57\times10^{-7})m^3/\sec$$
$$= 1.6\times10^{-4}m^3$$

$$V = n\frac{\pi}{4}D^2L$$
$$= n\ .7854(1\times10^{-3})^2\times1(m)$$
$$= n\ 7.854\times10^{-7}m^3$$

$$\text{So } n = \frac{1.6\times10^{-4}}{7.854\times10^{-7}} = 203.7 \approx 204$$

Where n is the number of fibers or lumens.

11.4.5 Merits and demerits of HFRBs

These reactors have little steric hindrance and low deactivation of enzymes. For high surface-area to volume ratio, high volumetric productivity is obtained. As the flow is in the laminar range, there is no possibility of washout. Using membranes of suitable Mol. Wt. cut-off bio chemical reactions and physical separation of catalyst from the product may take place in the same unit. There is also the possibility of using multi culture conversions.

Reactors with HFBR configuration have been employed in therapeutic application. Bundles of hollow fibers or a single large fiber in a cylindrical module are used to separate blood flowing within the lumen and mammalian pancreatic islets as assistance to diabetic patients. They are also used as blood detoxifiers.

However, there are also a number of problems in the operation of HFBR. One problem is the fouling of the membrane due to deposition on the pores of the membrane. Certain solvents may cause membrane swelling that alters the apparent pore size of enzyme shape and elasticity. Moreover, radial volume flux may modify the membrane structure leading to solute leakage at higher flux values. Besides there may be severe diffusion limitation for substrate and products. The fouling phenomenon has been analysed by Kleinstrener and Chin (1984). In many cases, membrane can no longer be treated as simple mechanical sieve-separators based on different molecular weights.

11.4.6 Artificial kidney

11.4.6.1 Hemodialysis

In hemodialysis, membranes are used for making artificial kidneys for people suffering from renal failure. It removes toxins from patients' blood such as urea, creatin, uric acid etc that are usually eliminated by kidney in the form of liquid urine. The membrane function as a dialyzer similar to that of nephron. Nephron consists of glomerules and tubules. The filtration of plasma through the glomerule leads to urine formation. The glomerular ultra-filtrate is absorbed through the tubule with 99% water and important electrolytes through diffusion and active transport.

In hemodialysis, the patient is connected to extra corporeal circuit by a veno-arterial shunt. Arterial blood goes through the dialyzer with a flow rate of 200 ml/min while the buffer solution goes through the dialyzer in opposite direction on the other side of the membrane shown in Fig. 11.5.

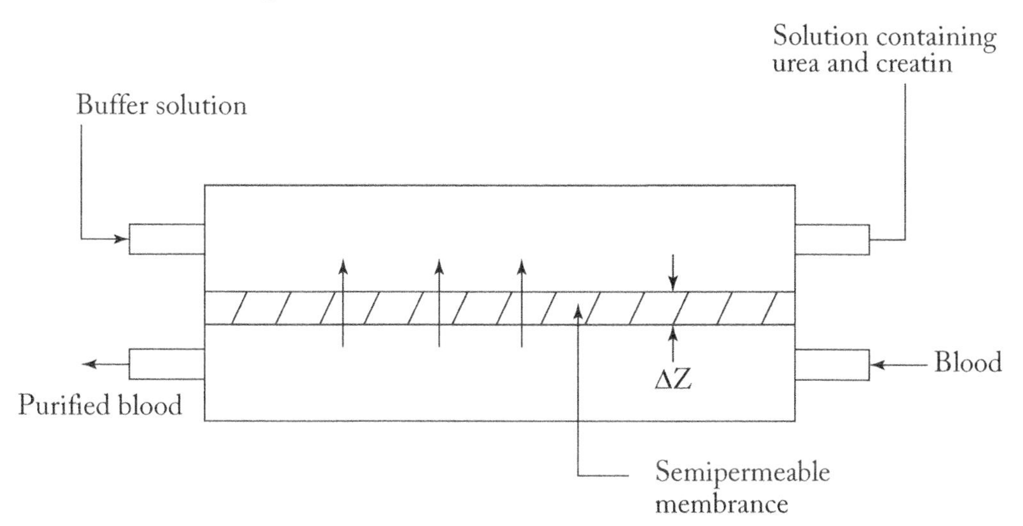

Fig. 11.5 Schematic diagram of Hemodialyzer with a flow rate of 500 ml/min

Separation of metabolic toxins from the patients' blood is effected by differences in diffusion rates across the membrane. The solute flux can be given as

$$J_i = D_i \frac{\Delta C_i}{\Delta Z}$$

neglecting pressure gradient, electric driving force and assuming ideal solution.

D_i = Diffusivity of solute through the membrane

ΔC_i = Concentration gradient across the membrane

ΔZ = Thickness of the membrane

Chemical species with molecular masses between 5000 and 12,000 Daltons which are eliminated by natural kidney are not removed by haemodialysis treatment. New developments in dialysis membrane are based on high flux dialyzer and preparing highly permeable membranes such as polymethyl methacrylate (BK-F) (which is comparable with the conventional PMMA and cellulose acetate) has displayed satisfactory dialytic removal of solutes including β-micro globulin, which causes dialysis failure.

11.4.6.2 Hemofiltration

New techniques based on hemofiltration, have been developed for removing toxins from patient's blood. Membranes used in the process are closer to those of renal glomerule membranes having a pore size range of 40 to 50 KDa. In hemofiltration, excess water can be removed in contrast to the diffusion process, effective in conventional dialysis for removing small solutes. In the hemofiltration, uremic middle sized molecules are effectively eliminated compared to dialysis. It has been realized that the small solutes are not effectively removed by the hemofiltration membrane having higher molecular weight-cutoff. This defect is eliminated by a new dialysis method combining both high rate of ultrafiltration and efficient diffusion as used in dialysis. Hemofiltration is sometimes used in combination with hemodialysis, when it is termed hemo-dia filtration. Blood is pumped through the blood compartment of a high flux dialyzer and a high rate of ultrafiltration is used, so that there is a high rate of movement of water and solutes from blood to dialysate that must be replaced by a substitution fluid that is infused directly into the blood line. However, dialysis solution is also run through the dialysate compartment of the dialyzer. The combination is theoretically useful because it results in good removal of both large and small molecular weight solutes.

11.4.6.3 Mathematical model for dialysis

This porous membranes are used in dialysis for selectively removing low molecular wt. solutes from a solution to diffuse into a region of lower concentration. There is no pressure difference across the membrane and flux of each solute is proportional to the concentration gradient across the membrane. Solutes of high molecular wt. are mostly retained in the feed solution because their diffusivity is low in the small pores of the membrane. Concentration gradients for a typical dialysis experiment can be shown in the Fig. 11.6

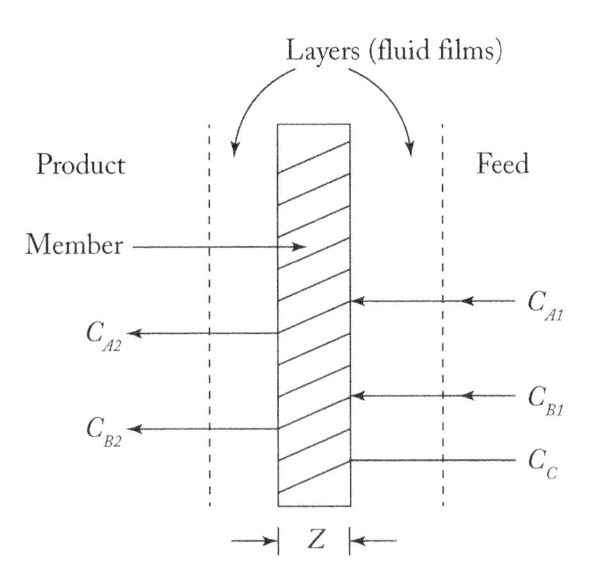

Fig. 11.6 Concentration gradient in dialysis membrane

The feed is assumed to contain a solute, A of low mol. wt and solute B of high molecular wt and C, a colloid. On both sides of the membrane there are concentration boundary layers and they contribute significantly to the overall resistance if the membrane is thinner than the boundary layers. The concentrations C_A and C_B are the concentrations of A & B in the pore fluid on both sides of the membrane. In the pore fluid in the product stream $C_C = 0$, since the colloid products are larger than the pore size. The flux of solutes J_A, J_B can be given on the basis of three resistance in series as shown in the Fig. 11.8,

$$J_A = K_A\,(C_{A1} - C_{A2}) \tag{11.59}$$

$$J_B = K_B\,(C_{B1} - C_{B2}) \tag{11.60}$$

where C_{A1}, C_{B1} are the concentrations of A & B in the feed that C_{A2} & C_{B2} are those in the product stream, and the overall mass transfer coefficients, K_A & K_B are:

$$\frac{1}{K_A} = \frac{1}{k_{1A}} + \frac{1}{k_{mA}} + \frac{1}{k_{2A}} \tag{11.61}$$

$$\frac{1}{K_B} = \frac{1}{k_{1B}} + \frac{1}{k_{mB}} + \frac{1}{k_{2B}} \tag{11.62}$$

Where k_1' s are liquid film mass transfer coefficients and k_m is the mass transfer coefficient in the membrane.

The coefficients, k_1 & k_2 for the feed and product depend on the flow rates, physical properties and membrane geometry and, they can be predicted by the following correlations.

For laminar flow:

$$Sh = 1.62\ G_Z^{1/3} \tag{11.63}$$

and

$$G_Z = 0.7854\ R_e S_c \frac{D}{L} \tag{11.64}$$

Where Sh (Sherwood number) = $k_L\ d_t/D$ and

$$G_Z(\text{Graetz number}) = \frac{\pi D}{4L} \cdot P_e$$

For turbulent flow:

$$Sh = 0.023\ R_e^{0.81}\ S_c^{0.44} \tag{11.65}$$

Where, $R_e = d_t\ u\rho / \mu$, $S_c = \mu / \rho D$

The membrane coefficient, K_M depends on effective diffusivity, D_e and membrane thickness, Z

$$K_M = \frac{De}{Z} \tag{11.66}$$

$$De = \frac{DK_A \in}{\tau} \tag{11.67}$$

Where € is the porosity and τ is the tortuosity factor.

where DK_A is the Knudsen diffusivity of component A through the membrane pore, $\bar{a}$ and is calculated from the relation

$$D_K = 9.70 \times 10^3 \bar{a} \left(\frac{T}{MA} \right)^{\frac{1}{2}} \tag{11.68}$$

where $\bar{a}$ is the average pore radius in cm. M_A is the molecular weight of the solute.

11.4.6.3.1 Design of artificial kidney

The principle of dialysis has been used in the design of artificial kidney to remove waste products like urea, creatin etc. from the blood of the persons with kidney disease. In the device hollow fiber cellulosic membranes have been used, and blood passes through the fibers while buffer solution is circulated on the shell side. Urea and other small molecules diffuse through the membrane to the external buffer solution, while protein and cells are retained in the blood. The dialyzing solution has added salts and glucose to prevent loss of these materials from the blood. A hollow fiber module for artificial kidney has been shown in Fig. 11.7

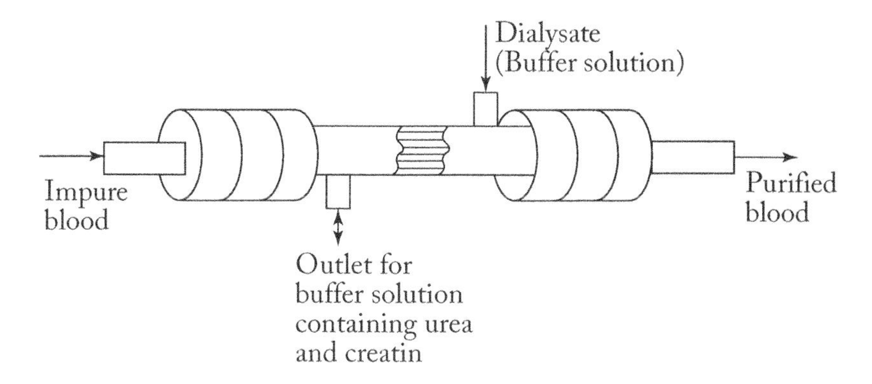

Fig. 11.7 Schematic diagram of artificial kidney

Modern dialysers typically consists of a cylindrical rigid casing enclosing hollow fibers extruded from a polymer or a copolymer. The combined area of the hollow fibers is typically 1–2 square meters. It is necessary to optimize the blood and dialysate flows moving counter currently in order to achieve efficient transfer of wastes from blood to the dialysate stream. Cellophane is the common membrane material for the dialyzer.

A simple mass transfer model can be presented as follows:

Based on the concentration of a toxic material, C, the mass balance for the system can be given as for a differential length, dZ of a dialyser,

$$u\frac{dc}{dZ} = -k_{LA}(C - C_1)$$

where C_1 is the average concentration of the toxic material over the entire length of the dialyser, Z.

The appropriate boundary conditions are

at $Z = 0$, $C = C_o$, $C_1 = 0$

at $Z = Z$, $C = C_e - C_1$

Integrating and rearranging, we get

$C_e - C_1 = C_o \exp(-k_{La} Z / U)$

where C_e is the exit concentration in the lumen k_{La} is the volumetric mass transfer coefficient, U is the linear velocity of blood and Z is the length of the lumen.

Problem 11.5

An artificial kidney consists of hollow fiber membrane of shell-and-tube type. Arterial blood passes through the tube-side with a flow rate of 200 ml/min and the buffer solution flows through the shell side of the dialyzer at the rate of 500 ml/min. the blood contains urea (A, Mol. wt. 60), glucose (B, Mol. wt. 180) and Myoglobin (C, Mol. wt. 17,000).

The membrane has a porosity of 0.45 (tortuosity factor, $\tau = 2$), a mean pore size of 0.05 μm (radius), and a thickness of 30 μm. The feed solution contains 1.0% of A, 1.0% of B and 1.0% of C. Predict the initial fluxes of A, B & C through the membrane. Neglect the boundary layer resistances.

Solution:

In order to calculate the individual overall mass transfer coefficients, K_A, K_B, K_C we first calculate the diffusivities of three components, D_{KA}, D_{KB} & D_{KC} through the membrane pore,

$$\bar{a} = .05\,\mu m = .05 \times 10^{-6} \times 10^{2}\,\text{m}$$

$$T = 25°C = 298$$

We know

$$D_K = 9.7 \times 10^3\, \bar{a} \sqrt{\frac{T}{M_L}}\, \frac{cm^2}{sec}$$

Where $\bar{a}$ pore radius in cm

$$DK_A = 9.7 \times 10^3 (5 \times 10^{-6}) \sqrt{\frac{298}{60}}$$

$$= 0.108 \frac{cm^2}{sec}$$

$$DK_B = 9.7 \times 10^3 (5 \times 10^{-6}) \sqrt{\frac{298}{180}} = 6.24 \times \frac{10^{-2} cm}{\text{sec}}$$

$$K_M = \frac{DK \in}{\tau Z}, \text{ Where } Z = \text{membrane thickness}$$

$$Z = 30\ \mu m = 30 \times 10^{-6} \times 10^2 = 30 \times 10^{-4}\,\text{cm}$$

$$\in = 0.45,\ \tau = 2$$

$$KM_A = \frac{0.107 \times 0.45}{2 \times 3 \times 10^{-3}} = 8.025 \frac{cm}{\text{sec}}$$

$$KM_B = \frac{6.24 \times 10^{-2} \times 0.45}{2 \times 3 \times 10^{-3}} = 4.68 \frac{cm}{\text{sec}}$$

$$KM_C = \frac{6.42 \times 10^{-3} \times 0.45}{2 \times 3 \times 10^{-3}} = \frac{0.48'5 cm}{\text{sec}}$$

The fluxes of three components

$$J_A = K_{MA}(C_{A_1} - C_{A_2})$$

$$J_B = K_{MB}(C_{B_1} - C_{B_2})$$

$$J_C = K_{MC}(C_{C_1} - C_{C_2})$$

Assuming C_{A2}, C_{B2}, C_{C2} =0

$$J_A = 8.025 \times .01 = .08025 \frac{gm}{cm^2 \sec}$$

$$J_B = 4.68 \times .01 = .0468 \frac{gm}{cm^2 \sec}$$

$$J_C = 0.4815 \times .01 = 0.0004815 \frac{gm}{cm^2 \sec}$$

11.5 Bioartificial Lever Device

The first bioartificial liver was developed by Dr. Kenneth Matsumura in 2001. Live cells obtained from an animal were used instead of developing a piece of equipment for each function of the lever. The structure and function of the first device resembles that of today's 'BALS'. Animal lever cells are suspended in a solution and a patients' blood are separated by a semipermeable membrane that allow toxins and blood products to pass but restricts an immunological component. The purpose of the Bal-type device is to serve as a supportive device either allowing the lever to regenerate properly upon acute lever failure or to bridge the individual's lever function unit and a transplant.

Hollow fiber System

One type of ball is similar to kidney dialysis systems that employ a hollow fiber cartridge. A hollow fiber system uses porcin (pig) hepatocytes. Such cells are relatively easy to obtain in large quantities and maintain a satisfactory level of differentiated cells activity for detoxification.

The liver performs many metabolic functions such as metabolism of carbohydrate, fat and vitamin, production of plasma proteins, conjugation of bile acids, and detoxification. Of all these activities of lever, detoxification is very critical. A failing lever may recover if the metabolic and detoxification demands are reduced, or an artificial lever may provide service for self-repair of a liver. The other alternative is the lever transplant.

Problems (Exercise)

11.1 Lactic acid is to be produced from glucose using streptococcus lactis in a hollow fiber reactor containing a number of lumens. Each lumen is of diameter D = 0.5 mm and L = 1 metre. Streptococcus Lactis is immobilized in a spongy matrix on the shell side. Feed is introduced to the reactor at the rate of $F = 2 \times 10^{-3}$ m^3/hr with initial glucose concentration of S_o = 5 kg/m^3. The kinetic parameters are:

r_{max} (maximum substrate consumption rate) = 2.5 kg/(m^3) (hr)

K_s = 0.2 kg/m^3, $Y_{P/S}$ = 0.2 kg P / Kg S,

(a) For 90% conversion of the substrate, what is the number of lumens required?

Hint: $S / S_0 = exp\,[- 2.06\,(r_{max} / k_s)\,V_R / F]$

(b) What is the product concentration (P) & the productivity (DP), where D is the dilution rate.

(c) Since it is an aerobic process, buffer solution saturated with oxygen is passed into the shell side. If one kg of cells immobilized per m^3 of packing and the oxygen consumption rate is 2×10^{-3} kmole O_2 / (kg dry wt. cell) (hr), What k_{La} value should be maintained in the shell side?

The saturated concentration of oxygen in the buffer solution is C_L^* and $C_L^* = 8$ ppm (8 mg O_2/1) and the critical oxygen concentration is 10% of the saturation.

11.2 Indole alkaloid (a plant metabolite) is produced in a hollow fiber biological reactor with plant cells of catharanthus roseus immobilized in the gel matrix in the shell side of the reactor. The HFBR unit consists of 250 lumens and each lumen is of 0.5 mm diameter having length of 1 metre.

The following kinetic information are available:

The rate constant,

$k' = 10$ m^3 / (kg cell) (day)

$X = 10$ kg/m^3 spongy matrix

$D_e = 1.5 \times 10^{-10}$ m^2/s

$S_o = 5$ kg/m^3, a (lumen radius) = 0.25×10^{-3} m

(a) Calculate the Thiele modulus, λ^2

$$\text{where } \lambda^2 = \frac{a}{3}\sqrt{\frac{k' X}{D_e}}$$

(b) Calculate the flow rate of the feed for 95% conversion of the feed in m^3/hr

(c) If Yp/s =.02 kg P/kgS, Calculate product concentration (P) and the productivity (DP)

11.3 A tubular membrane bioreactor has been used to transform lactose to lactic acid by Lactobacilus rhamosus which are immobilized on the inner wall of the reactor. The kinetics of the system are:

$\mu_{max} = 0.25$ hr^{-1}, Ks = 1.0 kg/m^3,

$Y_{x/s} = 0.4$ k_g X / k_g S, $X = 2$ k_g/m^2 of the inner surface

$S_o = 2$ k_g / m^3, $F = 1 \times 10^{-3}$ m^3/hr

$D_t = 3 \times 10^{-3}$ m (tube diameter)

(a) For 95% conversion of lactose, calculate the length of the reactor. The rate is first order with respect to substrate

(b) If $Yp/s = 0.3$, Calculate the exit concentration of the product (P) & the productivity (DP).

References

Cheryan, Munir, Ultrafiltration and Microfiltration Handbook, Techonic Pub. Co. 1998.

Cooney, D.O., Biomedical Engineering Principles, Marcel Dekker, New York, 1980.

Drioli, E. and L. Giorno, Biocatalytic Membrane Reactor, Taylor and Francis, 2004.

HO, W.S.H., and K.R. Sirkar, Membrane Handbook, Van Nostrand, Reinholde, 1992.

Kim, S.S. and D.O. Cooney, "An improved theoretical model for hollow fiber enzyme reactors", Chem. Eng. Sci., 31, 289–294, 1976.

Kleinstreuer, C. and S.S., Agarwal, "Analysis and simulation of hollow fiber dynamics", Biotechnol. Bioeng. 28, 1233–1246, 1986.

Kleinstreuer, C. and T. Poweigha, "Hollow fiber Reactors", in Advances in Biochem. Biotechnol. Vol. 30, edited by A. Fiechter, Springer Verlag, Berlin, 1984.

Ku, K. and M.J. Kuo, J. Delente and F. Feder, "Development of Hollow Fiber system for large scale culture of Mammalian cells", Biotechnol. Bioeng. 23, 79, 1981.

Piret, J.M. and C.L. Cooney, Mammalian Cells and protein distributions in hollow fiber Bioreactors", Biotechnol. Bioeng. 36, 902–910, 1990.

Porter, M. (editor), Handbook of Industrial Membrane Technology, 401–481, Noyes Publications, Park Ride, N.J., 1989.

Prenosil, J.E., I.J. Dunn and E. Hernzle, "Biocatalytic Reaction Engineering in Chap 10, Biotechnology edited by H.J. Rehm and G. Reed, VCH, Vol. 78, 1987.

Waterland, L.R., A.S. Michaelis and C.R. Robertson, "A theoretical model for enzymatic catalysis using asymmetric hollow fiber membranes" AIChE J. 20, 50–59, 1974

Webster, I.A. and M.L. Shuler, Biotechnol. Bioeng. 23, 447, 1981.

CHAPTER 12

Biological Reactors for Waste Water Treatment

12.0 Introduction

Waste materials derived from different sources contain various compounds like hydrocarbons, carbohydrates lipids and aromatics, cellulosics. Industrial wastes from various industries have characteristics components from organic compounds to inorganic materials.

Domestic wastes include ground garbage, laundry waste, excrement, and their flows vary with time in a period of 24 hours. Agricultural wastes include waste plants (straw), carbon rich cellulosic materials, poultry manure with high nitrogen content.

12.1 Waste Water Treatment Methods

Treatment methods for these waste materials depend upon their characteristic ingredients. Three major treatment methods are employed:

1. Physical treatment involves screening, flocculation, sedimentation, filtration, flotation in order to remove insoluble components.
2. Chemical treatment include chemical oxidation by oxygen and chlorine, chemical precipitation by $CaCl_2$, $FeCl_2$, $Ca(OH)_2$, $Al_2(SO_4)_3$
3. Biological treatment consists of aerobic and anaerobic treatment of waste water containing soluble carbon, nitrogen & phosphorous compounds by a mixed culture of microorganisms.

12.2 Characteristics of Waste Water

Major carbon compounds in industrial waste water contain carbohydrates, lipid-oils, hydrocarbons, proteins. Other compounds such as phenols, surfactants, herbicides, pesticides, assymetric compounds are present in relatively small concentration (< 1 kg/m^3) and are difficult to degrade by biological methods.

The carbon content of a waste water sample can be expressed by several ways such as biological oxygen demand (BOD), chemical oxygen demand (COD) and total organic carbon (TOC).

BOD is defined as the amount of oxygen consumed by a sample sewage incubated for a specific length of time. BOD_5 is that value of BOD obtained from 5 day-incubation.

The amount of dissolved oxygen in an incubation which is continued until carbonaceous biological oxidation ceases is called ultimate BOD (BOD_u).

The chemical oxygen demand (COD) is equal to the number of milligrams of oxygen by which a litre of waste water sample can be chemically oxidised. For the same sample of waste water, BOD value is smaller than COD. COD has the advantage of being measured in about 2 hours by a conventional method or in a few minutes by using sophisticated analytical instruments.

In BOD, the biochemical reaction involved can be shown as below, based on the organic compound like glucose:

$$C_6H_{12}O_6 + 6O_2 \rightarrow 6\ CO_2 + 6\ H_2O \tag{12.1}$$

Based on the stoichiometry of the reaction 1.07 gm of oxygen is required for the oxidation of 1 gm of glucose. A typical chemical oxidation reaction for COD is given as

$$C_6H_{12}O_6 + Cr_2O_7^{2-} + 2\ H2 \rightarrow 2\ Cr^{3+} + 6\ CO_2 + 6\ H_2O \tag{12.2}$$

The TOC (total organic carbon) of waste water sample is determined by a TOC analyser. After proper dilution, samples are injected into a high temperature (900°C) furnace and all organic carbon compounds are oxidized to CO_2, which is measured by infrared analyser. Waste water samples should be acidified to remove inorganic carbonate compounds. The total carbon content of waste water is determined before and after acidification, and the difference is inorganic carbon content.

The concentration of biomass in waste water is measured as mixed liquor volatile suspended solids (MLVSS). Waste water is filtered and the collected solids are dried and weighed to give mixed liquor suspended solids (MLSS).

The above material is then volatilized by burning in air at 600°C. The weight of the remaining incombustible inorganic material is the fixed solids. The difference between the original mass before combustion and the fixed solids is the volatile portion. The volatile portion or MLVSS is assumed to be primarily the mass of microbes.

A complete waste water treatment comprises three major steps:

1. Primary treatment which removes the coarse solids and suspended matter by screening, sedimentation, filtration, pH adjustment etc.
2. Secondary treatment is the main biological method, involving biological oxidation, anaerobic treatment of soluble & insoluble organics.
3. Tertiary treatment involves the removal of remaining inorganic compounds such as phosphate, sulphates, ammonium and other refractory organic compounds by physical separation methods like carbon adsorption, deep bed filtration, membrane separation like reverse osmosis, electro dialysis etc.

12.3 Biological Waste Treatment Reactors

Biological waste treatment uses a suitable mixed culture of microorganism collected from specific sources. The process may be aerobic or anaerobic. The major reactor types are (1) activated sludge, (2) trickling filter, (3) rotating biological contactor (4) oxidation pond, (5) anaerobic filter (6) aerated lagoon

Activated sludge process employs a well agitated and aerated continuous flow-reactor tank provided with a settling tank. It may be operated as PFR or CSTR. A long narrow tank with single feed approaches PFR and a circular basin with recycle sludge is modelled as CSTR.

12.3.1 The activated sludge process (CSTR with sludge recycle)

The reactor employed in the activated sludge process is a continuous flow aerated basin connected to a sedimentation tank for clarifying the exit liquid. A portion of the sludge collected in the sedimentation tank is recycled to the reactor to maintain a continuous sludge inoculation. The recycling enhances the mean sludge residence time and the microorganisms present adapt to the available nutrients. The long residence time of the sludge in the aerobic reactor favours the adsorbed organic compounds to be oxidized.

The mechanism of substrate removal in the unit depends on the nature and morphology of the community of mixed microorganisms. A common bacterium in the activated sludge population is Zoogloea ramigera which has the most important property of synthesizing and secreting a polysaccharide gel. Due to this gel, the microorganisms tend to agglomerate into flocs which are called 'activated sludge'. The activated sludge has the high affinity for suspended solids along with colloidal materials. The initial step in removing suspended solids from waste water is attachment to the flock, followed by the oxidation of biodegradable components of the adsorbed particulates by the floc organism.

Contact stabilization is a variation of the conventional process in which the recycle settled sludge is subjected to an additional aeration before being mixed with the waste influent to the aeration tank. In this system organics are removed entirely by the physical attachment. The system is shown in Figure 12.1

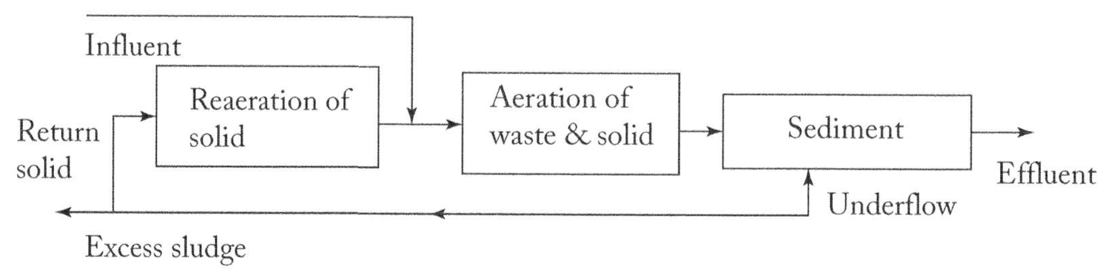

Fig. 12.1 Schematic of contact stabilization

Other variation of the activated sludge process is the 'step feed process' which involve mixing and aeration.

In the step-feed process the influent stream is split and introduced at different points of aerated basin. By distributing the feed along the reactor length, the system is made to behave more like a well-mixed tank reactor. The schematic flow sheet of step feed process is shown in Fig 12.2

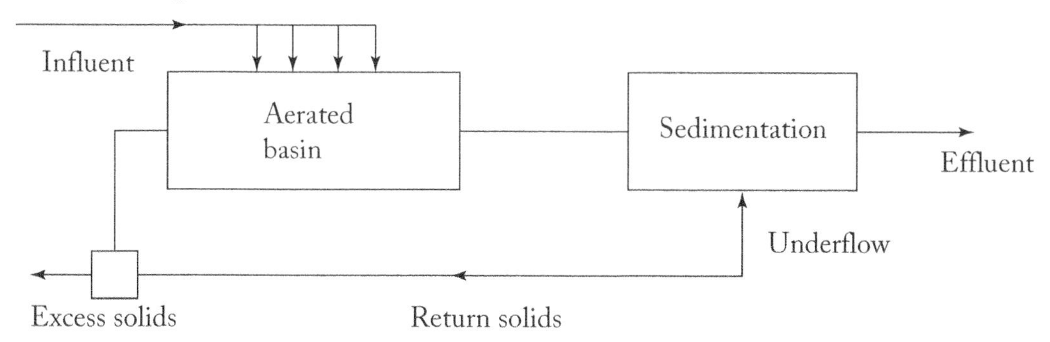

Fig. 12.2 Schematic flow sheet for step feed process

The conventional activated sludge aeration basin is a long narrow channel which behaves roughly as a tubular reactor with some dispersion. A better approximation to a backmix reactor is achieved by using a circular tank which is vigorously aerated to provide mass transfer and mixing, minimizing the gradients of dissolved oxygen and nutrient concentration in the reactor.

However aeration system may vary widely. Air may be bubbled into the vessel through diffusers on the bottom or sides of the reactor. Another alternative is that the surface of the basin may be brushed with rotating blades to great turbulence and promoted gas absorption. Another method uses a simple cone which draws liquid from the near the bottom of the basin and sprays it on the tank liquid surface. All these methods provide good aeration and agitation to provide oxygen to microorganism and to suspend and mix the sludge. Other particulates strip out volatile metabolic product such as CO_2.

The activated sludge system may have many uncontrolled disturbances such as waste water flow and composition. Such disturbances can lead to system failure. One type of disturbance is called 'Shock loading', indicating a sudden pulse of a high concentration of a toxic material. Another disturbance is 'sludge bulking'. When the process bulks, the effluent will not meet the necessary standards. A good sludge should settle rapidly. The microbial constituents of poor sludge reveals the presence of filamentous bacteria and flagellate protozoa. Healthy sludge does not contain a significant population of filament of any organism. The protozoa serve a valuable function in the overall process by preying on free,. unfloculated bacteria and thereby clarifying the effluent.

12.3.2 Reactor design and modeling of activated sludge process

The schematic diagram of an activated sludge unit is presented below in Fig. 12.3 containing aerated basin with a sedimentation tank, fresh feed inlet, recycle sludge. We want to develop the volume of activated sludge-reactor, using kinetics of microbial growth, substrate utilization, and material balance for biomass and substrate, with the objective of BOD removal of certain strength. The sludge reactor is operated with optimum parameters.

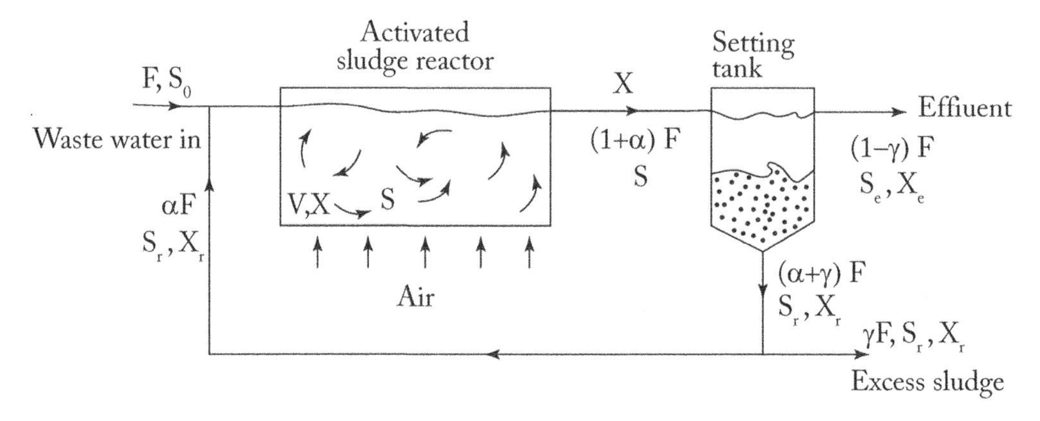

Fig. 12.3 Schematic diagram of an activated sludge unit

Though the tank contains a mixed culture of microorganisms the true kinetics of substrate removal in the tank are complex. The kinetics of the system used are based on pure culture kinetics and is given by Monod equation with endogenous respiration rate, as given below:

$$\mu = \frac{\mu_{max} S}{K_s + S} - k_d \tag{12.3}$$

Steady state mass balance for biomass (X) and rate limiting substrate (S) in an activated sludge reactor with a recycle ratio, α are given as:

Biomass (X):

$$(1+\propto)FX = \propto FX_r + \left(\frac{\mu_{max} S}{(K_s + S)} - k_d\right)XV \tag{12.4}$$

Substrate (S):

$$FS_o + \propto FS_r - \frac{1}{Y_x X/S}\left(\frac{\mu_m S}{K_s + S}\right)XV = (1+\propto)FS \tag{12.5}$$

Assuming no substrate utilization and cell growth in the settling tank (short residence time). Material balance around the settling tank gives,

Biomass: (X)

$$(1+\propto)FX=(1-\gamma)FX_e+(\alpha+\gamma)FX_r \tag{12.6}$$

Substrate (S):

$$(1+\propto)FS=(1-\gamma)F\,S_e+(\propto+\gamma)FS_r \tag{12.7}$$

where α = the ratio of sludge recycle rate to feed flow rate

γ = the ratio of excess sludge flow to feed flow rate.

Now we assume that the substrate is not separated in the settling tank leading to $S = S_e = S_r$, equation (12.7) is eliminated.

Now rearranging the equation, (12.6), we get,

$$(1+\propto)FX-\propto FX_r=(1-\gamma)FX_e+\gamma FX_r \tag{12.8}$$

Substituting equations (12.8) and (12.3) in equation (12.4), we obtain

$$\mu VX=(1-\gamma)FX_e+\gamma FX_r \tag{12.9}$$

Defining $\mu = I/\theta_c$, where θ_c is the 'cells or solid residence time', we express θc as

$$\theta_c=\frac{1}{\mu}=\frac{VX}{F(1-\gamma)X_e+\gamma FX_r} \tag{12.10}$$

The equation (12.10) is used to evaluate solid (cell) residence time (θ_c) in the sludge tank. 'Hydraulic or liquid residence time', θ_H, can be expressed using equation (12.9) as follows:

$$\theta_H=\frac{V}{F}=\frac{\theta_c(1-\gamma)X_e+\gamma X_r}{X} \tag{12.11}$$

Eliminating θ_c, equation (12.11) reduces to

$$\theta_H=\frac{(1-\gamma)X_e+\gamma X_r}{\mu X} \tag{12.12}$$

Further substituting
$S_r = S$ in equation (12.5), we get

$$F(S_o-S)=\frac{1}{Y_{X/S}}\mu XV \tag{12.13}$$

$$\text{Or, } V = \frac{Y_{X/S} F(S_o - S)}{\mu X} = \frac{Y_{X/S} \theta_c F(S_o - S)}{X(1 + kd\theta_c)} \tag{12.14}$$

Where
$$\mu = \frac{(1 + kd\theta_c)}{\theta_c} \tag{12.15}$$

The reactor volume V, can be given in terms of sludge recycle ratio, α, by substituting equation (12.8) into equation (12.11) and simplifying, reactor volume,V is given as

$$V = F\theta_c \left(1 + \propto - \propto \frac{X_r}{X} \right) \tag{12.16}$$

The kinetic parameters of the active microorganisms in the activated sludge reactor are μ_{max}, K_s, $Y_{X/S}$, which can be used to calculate the design parameters like V,, α, X_r, X and S.

Typical values of hydraulic residence, θ_H are in the range of 4 to 12 hrs and the sludge age in the range of 3-10 days.

Sludge settling characteristics must be determined experimentally to find X_r, the sludge concentration in the recycle stream.

Problem 12.1

The waste water feed with BOD value of S_o = 0.03kg/m3

μ_{max} = 1.5 day^{-1}, k_s = 0.4 Kg/m^3

$Y_{X/S}$ = 0.5 k_g dry mass / kg BOD, k_d = 0.07 day^{-1}

Calculate the following:

(a) Solids (cell) residence time (θ_c)
(b) Required reactor volume (V)
(c) Biomass concentration in recycle (X_r)
(d) Hydraulic residence time, θ_H
(e) Daily oxygen requirement given: O_2 required = 50 m^3 / kg COD removed

Solution:

(a) $$\frac{1}{\theta_c} = \mu = \frac{\mu_{max} S}{K_S + S} - kd$$

$$= \frac{1.5(0.03)}{0.4 + .03} - 0.07 = 0.0346\, day^{-1}$$

$$\theta_c = \frac{1}{.0346\, day^{-1}} = 28.9\, days$$

(b) Reactor volume, V

$$V = F\theta_C \left(1 + \propto - \propto \frac{X_r}{X} \right)$$

$$= 2 \times 10^4 \frac{m^3}{day} (28.9) day \left(1 + .5 - \frac{.5X_r}{5} \right)$$

Now Xr/X must be less than 3

Taking $\frac{Xr}{X} = 2.5$

$$V = 2 \times 10^4 \times 28.9(1.5 - 0.5 \times 2.5)$$

$$= 1.45 \times 10^5 m_3$$

(c) Biomass concentration in the recycle ration, Xr

$$X_r = 2.5 \times 5 = 12.5 \frac{Kg}{m^3}$$

(d) Hydraulic residual time,

$$\theta_H = \frac{V}{F} = \frac{1.45 \times 10^5 m^3}{2 \times 10^4 \frac{m^3}{day}} = 7.25\, days$$

(e) O_2 required:

$$\text{BOD removed} = F(C_o - C) = 2 \times 10^4 \frac{m^3}{day} (0.3 - .03) \frac{Kg}{m^3}$$

$$= 0.54 \times 10^4 Kg / day$$

Daily O_2 required

$$= \frac{50m^3}{Kg\ BOD\ remained} \times 0.54 \times 10^4 \frac{Kg}{day}$$

$$= 2.7 \times 10^5 m^3 / day$$

Problem 12.2

An activated sludge plant yields the following data with the specific growth rate, as

$$\mu = \frac{\mu_{max} S}{k_s + S} - kd$$

and the following data are available:

$$F = \frac{0.5m^3}{hr}, \propto = 0.5, \gamma = 0.1$$

$$X_e = 0, V = 1.5m^3, \mu_{max} = 1.0hr^{-1}$$

$$K_s = 0.01\frac{Kg}{m^3}, kd = 0.05hr^{-1},$$

$$S_o = \frac{1Kg}{m^3}, Y_{\frac{X}{S}} = 0.5\, kg\, dw.cells / kg \text{ substrate}$$

Calculate the following:

(a) Substrate concentration (S) in the reactor at the steady state
(b) Cell concentration (X) in the reactor
(c) Calculate X_r and S_r in the recycle stream

Solution: We know

(a) $$V = F\theta_c\left(1 + \propto - \propto\frac{X_r}{X}\right)$$

Putting the values in the above equation

$$1.5 = 0.5\theta_c\left(1 + 0.5 - 0.5\frac{X_r}{X}\right)$$

$X_{r/X}$ must be less than 3

Let us take $X_{r/X} = 2.5$, solving for

$$\theta_c = \frac{1.5}{(0.5)(.25)} = 12\, hrs.$$

$$\frac{1}{\theta_c} = \mu = \frac{\mu_{max}S}{K_S + S} - kd, \text{ putting the values of the parameters}$$

$$\frac{1}{12} = \frac{1 \times S}{0.01 + S} - 0.05$$

$$0.083 + .05 = \frac{S}{0.01 + S}$$

Solving for S, $S = 0.015\, kg / m^3$

(b) We know

$$F(S_o - S) = \frac{1}{Y_{X/S}} \mu X V$$

$$\text{Or } 0.5(1 - .015) = \frac{1}{0.5}(0.083) \times (1.5)$$

$$\text{Or } X = 1.978\, Kg / m^3$$

(c) $\frac{X_r}{X} = 2.5$

$$\text{Or } X_r = 2.5 \times 1.978 = 4.945 \frac{Kg}{m^3}$$

To find Sr, substrate balance around the settling tank

$$(1+\propto)FS = (1-\gamma)FS_e + (\propto + \gamma)F S_r$$

$$\text{Or } (1+\propto)S = (1-\gamma)^S + (\propto + \gamma)S_r$$

$$(1+0.5)S = (1-0.1)S_e + (0.5+.1)S_r$$

$$\text{Or, } 1.5(.015) = 0.9S_e + 0.6S_r$$

Assuming $S_e/S_r = 2$

$0.0225 = 24\ S_r$

$$\text{Or } S_r = 0.0093 \frac{Kg}{m^3}$$

Problem 12.3

An activated sludge basin is to be designed to reduce the amount of BOD from 1 kg/m^3 to 0.02 kg/m^3 at the exit. The sedimentation unit concentrates biomass by a factor of 3. The kinetic parameters are:

$\mu_{max} = 0.2$ hr^{-1}, $K_s = 0.08$ kg/m^3

$k_d = 0.01$ hr^{-1}, $Y_{X/S}{}^M = 0.5$ kg MLVSS/ kg BOD. The feed rate of waste water is 10 m^3/hr and the volume of the basin = 50 m^3.

(a) What is the value of solid residence time, θ_c?

(b) What value of the recycle ratio (α) must be used.

Solution:

(a) $\frac{1}{\theta_c} = \mu = \frac{\mu_{max} S}{K_s + S} - kd$

$= \frac{(0.2)(0.02)}{0.08 + .02} - 0.01$

Solving, $\frac{1}{\theta_c} = 0.03hr^{-1}, \theta_c = 33.3hrs.$

(b) $V = F\theta_c \left[1 + \alpha - \propto \frac{X_r}{X}\right]$

Putting the values of

$V = 50m^3, F = 10\frac{m^3}{hr}$, we get

$\theta_c = 33.3hrs., \frac{X_r}{X} = 3$

$50 = (1c)(33.3)\left[1 + \propto -3 \propto\right]$

Or, $0.150 = 1 - 2\propto$

Or $\propto = 0.425$

Problem 12.4

A well agitated waste treatment vessel of 1 m^3 is used for BOD removal with a feed rate of 0.1 m^3/hr. The separator concentrates by a factor of 2. The recycle ratio, α is 0.7. The kinetic parameters are:

$\mu_{max} = 0.5\ hr^{-1}$, $K_s = 0.02\ kg/m^3$

$Y_{X/S}^{M} = 0.5\ kg / kg$, $k_d = 0.05\ hr^{-1}$

What is the exit substrate concentration?

Solution:

$V = F\theta_\epsilon \left[1 + \propto - \propto \frac{X_r}{X}\right]$, Putting the values of parameters,

$1 = 0.1\theta_c\left[1 + 0.7 - 0.7(2)\right]$

or $\theta_c = \frac{1}{(0.1)(0.3)} = 33.33hr1$

$$\frac{1}{\theta_c} = 0.03 = \frac{\mu_{max} S}{K_s + S} - k_d$$

$$\text{Or, } .03 = \frac{0.5S}{0.2 + S} - .05$$

$$\text{Or, } 0.08 = \frac{0.5S}{0.2 + S}$$

Solving for S, $S = 0.038\, Kg / m^3$

Problem 12.5

An activated sludge reactor is designed to reduce the input 5 day BOD of $S_o = 0.25$ kg/m^3 to the outlet BOD of $S = 0.015$ kg/m^3. The following data are given:

Active biomass concentration,

$X = 3$ kg/m^3, Recycle ratio, $\alpha = 0.46$

$F = 4.5 \times 10^6$ gal/day $= \dfrac{4.5 \times 10^6\, gl / day}{264\, gl / m^3}$

$= 1.7 \times 10^4$ m^3/day

The kinetic parameters are:

$\mu_{max} = 3.7$ day^{-1}, $K_s = 0.022$ kg/m^3

$Y_{X/S} = 0.67$, $k_d = 0.07$ hr^{-1}

(a) Calculate
   (a) Sludge age, θ_c
   (b) Reactor volume needed, V
   (c) Concentration of active biomass in recirculation line X_r
   (d) Daily aeration rate assuming 7.5% oxygen utilization

(b) Evaluate X_r (optimum) if all operating costs are divided between the digester and the settler and the ratio of such costs (1 m^3 digester /1m^3 settler) is γ ($1.0 \le Y \le 10.0$) (Neglect construction costs).

Solution:

(a) 1) Sludge age, θ_c

$$\frac{1}{\theta_c} = \mu = \frac{\mu_{max} S}{K_S + S} - kd$$

$$= \frac{3.7(.015)}{0.022+.015} - 0.07$$

$$= 1.43\,day$$

$$\theta_c = 0.7\,day$$

2) Reactor volume, V

$$V = F\theta_c\left(1 + \propto - \propto \frac{X_r}{X}\right)$$

$$= 1.7\times10^4\left(\frac{m^3}{day}\right) * 0.7\,day * \left[\left(1 + .46 - 0.46\frac{X_r}{X}\right)\right]$$

The ratio of X_r/X should be less than 3.5 at which V becomes zero. Taking $X_r/X = 3$,

$$V = \left(1.7\times10^4\right)(0.7)(1.46 - 1.38)m^3$$

$$= 9.5\times10^2 m^3$$

3) $X_r = 3X = 3\times3 = 9\frac{kg}{m^3}$

4) O_2 req. Per day

$$= F\left(S_0 - S\right)\times 0.075\frac{kg}{day}$$

O_2 (volume) required per kg $= \frac{22.4}{32}\frac{m^3}{kg} STP$

$$= 1.7\times10^4\left(0.25 - .015\right)\times 0.07\times\frac{22.4}{32}$$

$$= 5.83 m^3 / day$$

where O_2 volume per kg $= \frac{22.4}{32} m^3 / kg$ at STP

(b) Assuming different values of X_r/X, the minimum reactor volume is determined. By iteration, the value of reactor volume was found to be 0.18 $\times10^2$ or 18 m^3 at $X_r/X = 3.18$.

12.3.3 Structured model for activated sludge process

The reactor model discussed above is based on unstructured model of cell growth (Monod Model) which is not suitable for control strategies or for simulation studies of plant's dynamics and control systems. Everyday most plants encounter large disturbances which require control adjustment during operation.

Andrew's Structured Model

The structured model for activated sludge process developed by Andrew (1974a, b, 1975) has significantly improved the operation of the plant compared to the previous design based on the unstructured kinetic model. According to Andrew's model, the total biomass (X_T) is assumed to be composed of three components which are derived from substrate according to the following scheme:

$$\text{Substrate} \xrightarrow{\text{mass transfer}} X_1 \xrightarrow{\text{monod}} X_2 \xrightarrow{\text{first order}} X_3$$

Where X_1 is stored mass, X_2 is active mass and X_3 is inert mass. The corresponding three rate equations are:

$$\frac{dX_1}{dt} = km\left(X_T \frac{f_s S}{S + K_{S1}} - X_1\right) \tag{12.17}$$

$$\frac{dX_2}{dt} = \frac{\mu_2 X_1}{K_{S2} + X_1} X_2 Y_2 \tag{12.18}$$

$$\frac{dX_3}{dt} = k_1 X_2 Y_c' \tag{12.19}$$

where k_m = mass transfer coefficient

f_s = maximum fraction of MLSS (storage product)

K_{s1} & K_{s2} = saturation constants

Y_2 = ratio of X_2 / X_1, Y_i = yield coefficient (mass of i formed per mass of X_2 consumed.) k_1 = first order rate constant, μ_2 = maximum specific growth rate for conversion of X_1 to X_2.

Busby and Andrews (1975) made the simulation study of the system using appropriate values of parameters (k_m, f_s, K_{s1}, K_{s2}, μ_2, Y_i, k_1). The simulation was carried out for a period of 18 hours with $S_o = 0.75\ k_s/m^3$.

The important features of the simulation has been found to make great improvement in the substrate consumption which occurs with decreasing stored mass. Such increased efficiency is achieved by the contact stabilization process in which the stored mass is largely converted to other mass forms in the aeration tank, which is concentrated in the clarifier.

12.3.4 Bioreactor for nitrification for tertiary treatment of waste water

In the activated sludge process nitrogen containing wastes are oxidized biologically to ammonia. This effluent ammonia is then oxidized to nitrite by a microorganism and the nitrite is further oxidized to nitrate by another microorganism, thus yielding an effluent of sufficiently low oxygen demand. The microbial reactions are brought out by two microbial species such as Nitrosomonas and Nitrobacter:

$$NH_3 + CO_2 + O_2 \xrightarrow{Nitrosomonas} cells + NO_2^- \tag{12.20}$$

$$NO_2^- + CO_2 + O_2 \xrightarrow{Nitrobactr} cells + NO_3^- \tag{12.21}$$

Kinetic parameters for these two reactions which follow Monod Kinetics are given by Lawrence and McCarty (1970) as given below:

Table 12.1 Kinetic parameters for Nitrification reactions by Nitrosomonos & Nitrobacter

Microorganism	Y_N (g cells / g_N)	μ_{max}, day^{-1}	K_s, mg/L
Nitrosomonas	0.05	0.33	1.0
Nitrobacter	0.02	0.14	2.1

If the cell residence time in the activated sludge reactor, θ_c is too short, a second aerated basin may be used for complete nitrification

12.3.4.1 Design equations for simultaneous BOD and NO_3 removal

BOD removal on the basis of CSTR model can be given as

$$\frac{F(S_o - S)}{V_R} = \frac{\mu X}{Y_{obs}} = \frac{X}{Y_{obs}\theta_c} \tag{12.22}$$

$$\theta_C = \frac{1}{\mu} = \frac{K_m + S}{\mu_s S} \tag{12.23}$$

where μ_s is the maximum substrate utilization rate in hr^{-1}.

The rate of biomass formation is,

$$\frac{X}{\theta_C} = \frac{Y_{obs}\mu_s S X}{K_s + S} \tag{12.24}$$

The minimum cell age, θ_c (min), may be calculated on the basis of BOD at the exit (S)

$$\theta_{min} = \frac{(K_m + S)}{Y_{obs}\,\mu_S\, S} \tag{12.25}$$

Given, $Y_{obs} = \dfrac{0.4}{1 + 0.06\theta_{C(min)}}$ (an optimum value) (12.26)

Combining the equation (12.25) and (12.26) and solving for θ_c (min), we have,

$$\theta_{c(min)} = \frac{K_m + S}{0.4\mu_s S - 0.06(K_m + S)} \tag{12.27}$$

Actual θ_c must be greater than θ_c min

For nitrification ($NH_3 \rightarrow NO_2^-$) with Nitrosomonas cells, sludge residence time, θ_{C_1}, can be given as

$$\frac{1}{\theta_{c_1}} = \frac{(\mu_{max} NH_3) S_{NH_3}}{K_{NH_3} + S_{NH_3}} \tag{12.28}$$

For conversion of nitrite to nitrate using nitrobactor, θ_{C_1} can be used as

$$\frac{1}{\theta_{c_1}} = \frac{(\mu_{max} NO_2) S_{NO_2}}{K_{NO_2} + S_{NO_2}} \tag{12.29}$$

Solving for SNO_2 from the above equation (12.29), we get

$$S_{NO_2} = \frac{K_{NO_2}}{\theta_{c_1}(\mu_{max} NO_2) - 1} \tag{12.30}$$

Using the relevant kinetic parameters as given earlier, it has been shown that BOD removal is faster than nitrification. So if we use θ_1 for nitrification for desired SNO_2, BOD removal will be complete by that time (θ_{C_1}).

Problem 12.7

A CSTR is to be designed for both BOD removal and nitrification. The nitrification consists of two steps as shown below.

$$NH_3 + CO_2 + O_2 \xrightarrow{Nitrosomonas} cells + NO_2^-$$

$$NO_2^- + CO_2 + O_2 \xrightarrow{Nitrosobaeter} cell3 + NO_3^-$$

Specific growth rates for each of the above species follow a simple model and their parameters are given below:

Microorganism	**Y_N (g cells / g_N)**	**μ_{max}, day^{-1}**	**K_s, mg/L**
Nitrosomonas	0.05	0.33	1.0
Nitrobacter	0.02	0.14	2.1

BOD removal rate also follows Monod Model and the parameters are:

$\mu_s = 2.0$ day^{-1}, $k_m = 5.0$ mg/L

$Y_{obs} = 0.4 / (1 + 0.06\ \theta_c)$

Initial BOD = 50 ppm

Organic N = 60 ppm

Outlet BOD = 15 ppm

Outlet N = 5 ppm

(a) Calculate the minimum cell age (θ_c) needed to effect outlet BOD of 15 ppm
(b) Calculate the nitrification biomass cell age (θ_1) for effluent Nitrogen (NH_3) is 0.5 ppm and the concentration of NO_2 at θ_{C_1}.

Solution:

(a) We know

$$\theta_{c(minimum)} = \frac{K_m + S}{0.4\mu_S S - 0.06(K_S + S)}$$

$$\text{or } \theta_{c(minimum)} = \frac{(5+15)}{0.4(2)(15) - 0.06(5+15)} = 1.85\, day$$

(b) The material balance for nitrosomonas

$$\frac{1}{\theta_1} = \frac{\mu_{max(NH_3)} S_{NH_3}}{K_{NH_3} + S_{NH_3}} = \frac{0.33(0.5)}{1+0.5} = 0.11\, day^{-1}$$

Or $\theta_1 = 9\, day$

For nitrite conversion:

$$\frac{1}{\theta_{C_1}} = \frac{(\mu_{max}\ NO_2) S_{NO_2}}{K_{NO_2} + S_{NO_2}}$$

$$\text{Solving for } S_{NO_2} = \frac{K_{NO_2}}{\theta_{c_1} \mu_{maxNO_2} - 1} = \frac{2.1}{9(0.14) - 1} = 8.07\, mg / L$$

12.3.5 Trickling biological filter (TBR)

The trickling (or percolating) biological filter is an alternative to the activated sludge process. It is a packed bed of inert support particles like sand or plastic with a film or slime of mixed microorganisms, loosely packed with a porosity of 0.5. The bed is shallow with high diameter to height ratio (D/H = 3.0) to eliminate clogging problems and axial dispersion.

The waste liquid is fed from the top of the bed continuously through fixed nozzles or intermittently through a rotating distributor as shown in the figure 12.4. The liquid flow rate should be low to trickle down the slime-covered packing to supply oxygen to the aerobic organisms in the outer film of the packing.

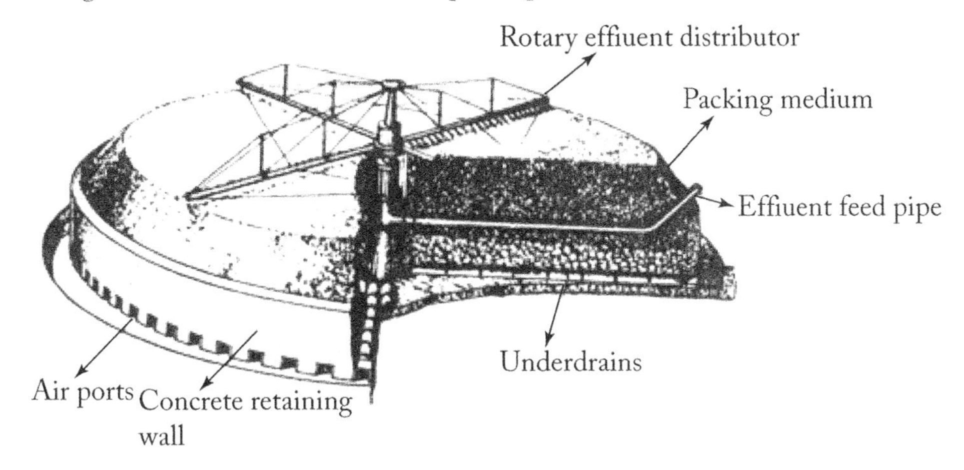

Fig. 12.4 A tricking biological filter (Bailey and Ollis, 1986, p.941)

Air enters the bed from the bottom and moves upward due to natural convection. The driving force for air circulation is the temperature difference caused by the heat release due to biological oxidation. The reaction medium conditions like temperature, pH, dissolved oxygen, nutrient profile vary throughout the length of the bed. There may be strong dissolved oxygen transfer limitation which is minimized by preaeration of the feed waste water stream to saturate with oxygen and high liquid circulation rate.

As the liquid trickles down the column on the surface of the microbial film, the organic compounds diffused through the film and are consumed by the organisms.

The design of the filter is based on the plug flow behaviour. The substrate consumption rate, N_s is given as

$$N_S = -D_e \frac{dS}{dZ} = \eta \frac{r_m S}{K_S + S} L \tag{12.31}$$

where D_e is the effective diffusivity of substrate, η is the effectiveness factor, L is the thickness of biofilm (mm), r_m is the maximum substrate utilization rate given as

$$r_m = \mu_{max} X / Y_{X/S} \text{ in kg S/ (m}^3\text{)(hr)}$$

S is the substrate concentration (kg/m^3)

The design equation based on the plug flow in differential form,

$$-F \frac{dS}{dZ} = \eta \frac{r_m S}{K_S + S} L a A \tag{12.32}$$

where F is the volumetric flow rate (m^3/hr), a is the biofilm surface area per unit volume (m^2/m^3), A is the cross sectional area of the bed (m^2), Z is the bed height. Since the substrate concentration is low, the rate is approximated to first order with respect to S.

$$r_p = \frac{r_m S}{K_S + S} \approx \frac{r_m}{K_S} S$$

The equation (12.32) is then integrated with first order kinetics to give

$$\frac{S}{S_u} = (1 - \delta) = exp\left[\frac{\eta r_m L a A H}{F K_S}\right] \qquad (12.33)$$

where δ is the fractional conversion of substrate, defined as

$$\delta = \frac{S_O - S}{S_O} \qquad (12.34)$$

H is the bed height, F is the flow rate, K_s is the saturation constant.

η is the effectiveness factor which may be calculated approximately by the equation

$$\eta = \frac{tan\, h\phi_L}{\phi_L} \qquad (12.35)$$

where φ_L is thiele parameter defined as

$$\phi_L = L\sqrt{\frac{r_m}{K_S D_e}} \qquad (12.36)$$

$$D_e\left(effective\ diffusity\right) = \frac{D_S \epsilon}{\tau} \qquad (12.37)$$

D_S is the substrate diffusing

$\in$ = porosity, τ = tortuosity factory

Trickling filters have certain advantages over activated sludge process. Trickling filters are more stable against shock loads than activated sludge reactors. Operating costs of trickling filters are low and give better effluent clarity compared to activated sludge process. The demerits of TBR are high capital cost, large space requirement, less effluent BOD level. Another major problem is the heterogeneous nature of the bed with respect to temperature, pH and dissolved oxygen, aeration control..

Other important design parameters of TBR are hydraulic residence times (0.5 to 4 hrs), liquid film thickness (0.01 mm), biofilm thickness (0.25 mm), r_m values (0.2 to 0.5 gm S / (m^3)(s)

Problem 12.8

A trickling biological filter is used to reduce the BOD of the waste feed stream, $S_o = 0.5$ kg/m^3 to the effluent BOD, S = 0.01 kg/m^3. The following kinetic parameters for the biocatalyst used in the filter are:

$$r_m = 0.020 \text{ kg S/(m}^3\text{)(hr). F} = 10 \text{ m}^3\text{/hr}$$

$K_s = 0.2$ kg S/m^3. The biofilm thickness, $L = 1.0 \times 10^{-3}$ m. The cross-section area of the filter is A = 2m^2, the biofilm surface area per unit volume of the bed, $a = 5 \times 10^4$m^2/m^3.

It is assumed that the dissolved oxygen is the rate limiting substrate and the diffusivity of oxygen is $Do_2 = 2 \times 10^{-9}$ m_2/s, the bed porosity is 0.4. Assume first order biochemical kinetics. Calculate the required length of the bed.

$$D_e = \frac{D_S \in}{\sigma} = 2\times10^{-9}\frac{\times.4}{2} = 0.4\times10^{-9}m^2 / S$$

Where σ (the tortuosity factor) = 2

Or $D_e = 1.44\times10^{-6}m^2 / hr$

$$\phi = L\sqrt{\frac{r_m}{K_S D_e}} = 1\times10^{-3}\sqrt{\frac{0.02}{2\times1.44\times10^{-6}}} = 0.26$$

$$\tan h\phi = \tan h\, 0.26$$

$$\eta = \frac{\tan h\phi}{\phi} = \frac{0.25}{0.26} = 0.96$$

The design equation of tricking filler first order kinetics

$$l_n\frac{S}{S_O} = \eta\frac{r_{max} La AH}{FK_S}$$

$$\text{Or } l_n\frac{0.01}{0.05} = -\frac{(0.96)(0.02)(1\times10^{-3})\times(5\times10)\times2\times H}{(10)(0.2)}$$

Or -3.91 == - 0.96 H

Or H = 4.07m

12.3.6 Bioreactors for anaerobic digestion

Microbial anaerobic digestion is normally applied to degrade solid waste and excess sludge produced in the aerobic waste water treatment, resulting the production of methane, an important biofuel.

Anaerobic digestion is characterized by slow kinetics compared to aerobic process; having residence time in the range of 30 – 60 days. The biochemical reactions involved in the process are very complex. The process involves several steps as presented below:

1. Solublization of insoluble organic compounds
   The solid waste may contain cellulosic materials, starches (protein waste) and other insoluble organic compounds which are solubilized by enzymes like cellulases, amylases, glucoamylases, lipases, proteases etc. in the first step of anaerobic treatment.

2. Formation of volatile acid
   In the second step, facultative anaerobic bacteria such as enteric bacteria and clostridial species and their mixture known as acid forming bacteria, act upon the solubilized organic compounds, producing organic acids like acetic, butyric and propionic along with some short chain fatty acids and also some small amount of alcohols.
   The optimal operating conditions are: Temperature = 35°C, pH = 4-6 for this stage of acid formation

3. Formation of methane
   In the third step, strictly anaerobic 'methanogenic bacteria' degrade volatile acids and alcohols to methane, CO_2 and H_2S.

The methanogenic bacteria contain methane bacterium (non-spore forming rods), methano-bacillus (spore forming), methano coccus, methane-sarcina, which act best at temperature range of 35–40°C and pH of 7–7.8.

Experimental studies indicate that bacterial synthesis of short chain fatty acids and volatile acids from soluble organic compounds, takes place at a relatively rapid rate. SO the methane formation from volatile acids is the rate limiting step.

If a single digester is used for both acid formers and methane bacteria, the operating conditions are maintained at pH = 7.0 and temperature of 35°C. An external heat exchanger is used to maintain the temperature in the mesophilic range which maximizes the rate of sludge digestion. The solids residence required for anaerobic sludge digestion at mesophilic temperatures is 10 to 30 days in a well agitated tank.

The composition of a typical product gas from anaerobic digestion is 70–75% CH_4, 20 to 25% CO_2 and 5% H_2S, and other gases (NO_2, H_2, CO). Part of methane produced is used to heat the digester to fermentation temperature. The calorific value of the digester gas is 24 to 28 million J/m^3 and the production rate of methane is 0.75 to 1.2 m^3 (standard) /kg of organic material decomposed. The foul and corrosive H_2S gas is to be removed before any use of the product gas.

The sludge produced in the anaerobic digester is easily dewatered by rotating drum or vacuum filter. The sludge is then dried and may be used as a fertilizer or dumped or incinerated.

However, the process encounters many operational problems leading to complete failure of the unit. The causes of failures are (a) hydraulic (b) organic and (c) toxic overloading.

In the beginning, the dilution rate (D) exceeds the growth rate (μ) of digester microorganisms, leading to washout of cells. (hydraulic over-loading).

High organic substrate concentrations increase the concentration of volatile acids, causing inhibition of methane bacteria, lowering the pH of the medium (organic over loading).

If the substances toxic to the methane bacteria enter the digester with the feed in large concentration, leads to the failure of the overall process (toxic overloading).

12.3.6.1 Model of graef and andrews (1976) for anaerobic digestion

To overcome these operational problems, Graef and Andrews (1974) developed an elaborate mathematical model for anaerobic digestion to study the dynamic characteristics of the units and developed a suitable control strategy.Transient mass balances for several components in the liquid and gas phases have been considered.

For the biological phase, the mass balance equations have been presented as

Liquid phase:

$$\frac{dX}{dt} = \frac{F}{V}\left(X_0 - X\right) + \mu X - k_T\left[toX\right] \tag{12.38}$$

$$\frac{dS}{dt} = \frac{F}{V}\left(S_0 - S\right) - \frac{\mu X}{Y_{X/C}} \tag{12.39}$$

$$\mu = \mu_{max} / \left[1 + \frac{K_s}{(hs)} + \frac{(hs)}{K_i}\right] \tag{12.40}$$

Gas phase:

$$R_B = Y_{CO_{2/X}}\,\mu X \tag{12.41}$$

$$\phi_{CH_4} = \frac{V}{\rho_g} Y_{CH_{4/X}}\,\mu X \tag{12.42}$$

where (hs) is the non-ionized volatile acids, R_B is the rate of CO_2 formation, CH_4 is the rate of methane formation, k_T is the death rate constant for toxic material. Graef and Andrews (1976) assumed that all volatile acids can be represented as acetic acid and the composition of methane bacteria was approximated by the empirical formula as $C_6H_7NO_2$. They also presented the following stoichiometry for the gasification reactions:

$$CH_3COOH + 0.032NH_3 \rightarrow 0.032C_5H_7NO_2 + 0.92CO_2 + 0.92CH_4 + 0.96H_2O \tag{12.43}$$

The limiting substrate for the above reaction is assumed to be non-iodized volatile acids (HS).

They have also presented the mass balance equations of different species in the liquid and gas phases.

The overall simulation results predict several suggestions for controller design, leading to the development of design and operational strategies for improved performance.

12.3.6.2 Two-tank anaerobic digester model

Poland and Ghosh (1971) considered the interaction between acid formers and methane formers in two separate tanks. The detailed stoichiometry has been expressed as follows:

Acid formation:

$$4C_3H_7O_2N_S^{+8H_2O} \rightarrow 4CH_3COOH + 4CO_2 + 4NH_3 + 8H^+ + 8e^- \quad (12.44)$$

Methane formation:

$$8H^+ + 8e^- + 3CH_3COOH + CO_2 \rightarrow 4CH_4 + 3CO_2 + 2H_2O \quad (12.45)$$

If the two tanks are kept in series, the environmental control for methane formers is easier.

The growth kinetics for two stages may be given by Monod model.

In the first stage of acid formation is given as

$$\mu_1 = \frac{\mu_{max_1} S_1}{K_{S_1} + S1} \quad (12.46)$$

where μ_1 is the specific growth rate of acid forming bacteria (X_1), μ_{max1} is the maximum specific growth rate, K_{S1} is the saturation constant, S_1 is the substrate concentration in the first stage or tank.

Growth rate for methane former (X2)

$$\mu_2\left(acid\ former\right) = \frac{\mu_{2max} S_2}{K_{S_2} + S_2 + \frac{S_2}{K_I}} \quad (12.47)$$

where S_2 = concentration of acetic acid, K_I is the substrate (acetic acid) inhibition constant, K_{S2} is the saturation constant

In the first CSTR, the mass balance for S_1 (organic chemical) is given as

$$ds_1/dt = D(S_1^0 - S_1) - \mu_1 x_1/(\gamma x_1/S_1) \quad (12.48)$$

For the second tank,

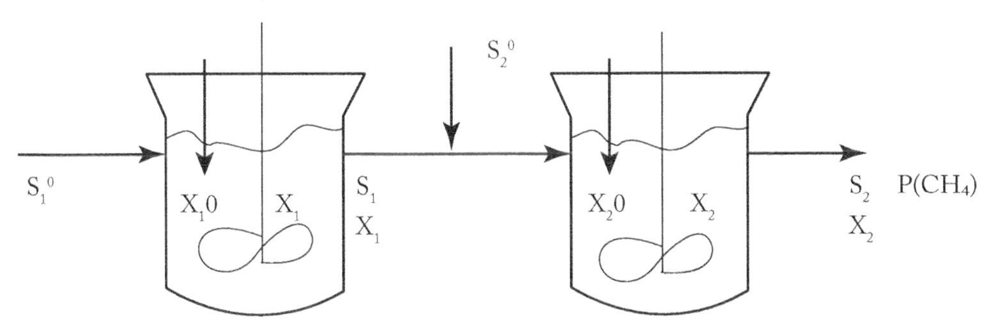

Fig. 12.5 Two Tank Anaerobic digesters in series

$$\frac{dS_2}{dt} = D\left(S_{20} - S_2\right) - \frac{1}{Y_{\frac{X_2}{S_2}}}\mu_2 X_2 \tag{12.49}$$

From the first tank, the product (P) is the acetic acid which is fed to the second reactor as initial substrate (S_2^O)

$$P_1\left(acetic\ acid\right) = S_2^0 = \frac{1}{Y_{\frac{org}{acetic}}}\left(S_1^0 - S_1\right) \tag{12.50}$$

$$X_1\left(con.of\ acidogenetic\ bacteria\right) = \left(S_{10}^c - S_1\right)\cdot Y_{X/acetic\ acid} \tag{12.51}$$

In the second tank, the product P_2 $(CH_4) = Y_G$ $(S_2^O - S_2)$ (12.52)

where Y_G = yield coefficient of methane to acetic acid & $(S_2^O - S_2)$ is the acetic acid consumed in the second reactor to form methane.

$$X_2\left(methanogenic\ bacteria\right) = Y_{X_2/Ad+}\left(S_2^o - S_2\right) \tag{12.53}$$

These equations predict the complete performance of two digesters with biomass (X_1 & X_2) and desired product, methane.

Problem 12.9

Two-Tank anaerobic digesters are used to treat the solubilized organics of concentration, S_1^O = 1 Kg/m^3. The first tank contains acid producing bacteria (X_1) which convert soluble organics to mainly acetic acid (S_2) which is fed to the second tank containing methanogenic bacteria producing useful gas methane (CH_4).

The reactions in the two tanks have been assumed to proceed as follows:

Acid formers (1st tank)

$$4C_3H_7O_2NS + 8H_2O \rightarrow 4CH_4COOH + 4CO_2 + 6NH_3 + 4H_2S + 8H^+ + 8e^-$$

Methane formers (2nd tank)

$$8H^+ + 8e^- + 3CH_3COOH + CO_2 \rightarrow 4CH_4 + 2CO_2 + 2H_2O$$

Kinetic parameters

- 1st Tank, Acid former (X1)

  $\mu_{max_1} = 1.25 hr^{-1}$, $K_{S_1} = 22.5 \times 10^{-3} \frac{kg}{m^3}$, $Y_{\frac{org}{Actt}} = 0.5 kg / kg$ (yield coefficient of organic to acetic acid)

- 2nd Tank, methane formers (X_2)

  $\mu_{max_2} = 0.14 hr^{-1}$, $K_{S_2} = 0.6 kg / m^3$, $Y_{\frac{org}{Actt}} = 0.05 kg / kg$

(a) If the dilution rate is D = 0.2 hr^{-1} Calculate the exit cell mass concentration (X_1) and substrate concentration (S_1) in the first tank and those (X_2) & (S_2) in the second tank.

for (i) $V_1 = V_2$ (ii) $V_2 = 5V_1$

(b) If the flow rate is F = 1 m^3/hr, calculate the yield of CH4 per hr from the second tank

Solution:

S_1^O = initial concentration of soluble organics, 1 kg/hr

S_1 = concentration of soluble organics at the exit of the 1st tank

$$P\left(acetic\,acid\right) = \frac{S_1^0 - S_1}{Y_{\frac{org}{AcA}}} = \frac{\left(1.0 - S_1\right)}{0.5}$$

Assuming sterile feed in the 1st tank,

$$D_1 = \mu_1 = 0.2 = \frac{\mu_{max} S_1}{K_S\left(1 + \frac{P}{K_I}\right) + S_1} \tag{A}$$

Using the product inhibition model.

Substituting the values of μ_{max}, K_s, K_I in equation (A), we get

$$0.2 = \frac{1.25 S_1}{2.25 \times 10^{-2}\left[1 + \frac{(1 - S_1)}{0.5}\right] + S_1}$$

Solving for S1

$S_1 = 0.0107$ kg/m^3

$$X_1 = Y_{X1/S1}\left(S_1^0 - S_1\right) = 0.2(1 - .0107)\frac{kg}{m^3} = 0.198\frac{kg}{m^3}$$

Second reactor:

Assuming sterile feed with respect to methanogenic bacteria.

$$D_2 = \mu_2 = \frac{\mu_{max_2} S_2}{K_{S_2} + S_2}$$

Putting the values of $D_2 = D_1$

$$\mu_{max_2} = 0.14\, hr^{-1}, \quad K_{S_2} = 0.6\, kg / m^3$$

$$0.2 = \frac{0.14 S_2}{0.6 + S_2}$$

Solving for S_2, S_2 = -2 kg/m^3

So the dilution rate, D_2 is to be changed

$$D_2 = \frac{F}{V_2} = \frac{F}{5V_1} = \frac{0.2}{5} = 0.04\, hr^{-1}$$

$$0.04 = \frac{0.14 S_2}{.6 + S_2}$$

Solving for S_2, S_2 = 0.24 kg/m^3

Initial concentration of acetic acid, S_2^0 can be obtained as

$$S_2^O = \frac{1}{Y_{\frac{org}{AcA}}}\left(S_1^0 - S_1\right) = \frac{1}{0.5}(1 - .0107) = 1.978\, kg / m^3$$

Biomass in the 2nd tank, X2 is

$$X_2 = Y_{\frac{X_2}{AcA}}\left(S_2^0 - S_2\right) = 0.05(1.978 - 0.24) = 0.087\, kg / m^3$$

(b) Calculation of 2nd reactor volume:

$$V_2 = \frac{F}{D_2} = \frac{1m^3/hr}{0.04\,hr^{-1}} = 25m^3$$

From stoichiometry

3 kmoles of acetic acid gives 4 k moles of methane

(3 × 60) kg of acetic acid gives (4x16) kg of CH_4

1 kg of acetic acid gives 0.35 kg CH_4 per 1 kg acetic acid

$$\text{So } CH_4 \text{ formed} = (1.978 - 0.24)\frac{kg}{m^3} \times \frac{0.35\,kg\,cl^{11}}{kg.accte} \times \frac{1m^3}{hr} = 0.608\frac{kg}{hr}$$

Volume of CH_4 produced / hr

$$= \frac{0.608}{16}\frac{(k.moles)}{hr} \times 22.4\frac{m^3}{k.mole} = 0.85\,m^3/hr \text{ at } 0°C \text{ \& 1 atm pressure.}$$

12.4 Bioscrubbers

A bioscrubber may be described as a reactor system involving mass transfer and microbial reactions. It operates in the mode of gas-liquid scrubber used for the removal of H_2S or CO_2 in air by utilizing a stream of alkali solution.

In a bioscrubber, an effluent air stream containing refractory organic compounds like benzene, toluene and xylene is brought in contact with a stream of an appropriate microbial liquid suspension, effecting the entrapment of the organic compounds and subsequent microbial degradation. The scrubbing microbial suspension may be recycled with necessary fresh cell loading and the pH and temperature are adjusted according to the desired values.

Benzene, toluene and xylene are carcinogenic. These are emitted from coating factories, paints manufacturing and petrochemical refining.

A process study on the removal of these compounds was reported by Lee et al (2002) using a bacterium (stenotrophomonas) maltophilia T3-C) grown in a mineral salt medium.

The mechanism of removing those compounds involves the transport of solutes from the bulk liquid to the surface of the microbial particles where microbial reaction takes place and organic compounds are degraded. The biomass concentration in the liquid phase is assumed to be constant throughout the column and the concentration of organic compounds, present in ppm level, decrease along with length of the scrubber.

The bioscrubber, a vertical column with L/D = 10 is used to scrub the organic compounds. The microbial suspension liquid is fed from the top and polluted air stream containing the carcinogenic organic compounds is introduced from the bottom.

12.4.1 Model based on mass transfer and microbial reaction

The model assumes that the dissolved solution (C_{Li}) is transferred from the bulk liquid to the microbial cell surface and then consumed by biochemical reaction. Another mass transfer from the gas interphase to the bulk liquid is neglected.

The flow of the dissolved solute is assumed to be plug flow and the system operates at the steady state. Since the liquid flow is from the top and the degradation rate increases as the concentration of the substrate increases downward, the rate equation for any component i can be given as

$$u\frac{dC_{L_i}}{dZ} = k_{La}\left(C_{L_i} - C_{L_i}^S\right) \tag{12.54}$$

$$= \left[\frac{\mu_{max}\,\overline{X}}{Y_{X/C}\,K_S}\right] C_{L_i}^S = k^{'} C_{L_i}^5 \tag{12.55}$$

$C_{L_i}^S$ is the concentration at the surface, CLi is the concentration in the bulk of the liquid for the component, i.

It is assumed that biomass concentration in the liquid phase (X) is constant.

Solving for surface concentration, $C_{L_i}^S$ from the two equations, and then eliminating $C_{L_i}^S$ from the equations (12.54) & (12.55) we have,

$$-u\frac{dC_{L_i}}{dZ} = k_o C_{L_i} \tag{12.56}$$

where k_o, the overall rate constant is given as

$$\frac{1}{k_o} = \frac{1}{k_i} + \frac{1}{k_{La}} \tag{12.57}$$

Integrating the equation (12.56) with the boundary conditions:

At $Z = 0$, $C_{Li} = C^*_{Li}$ (saturated concentration of component i)

At $Z = L$, $C_{Li} = C_{Li}$

$$\frac{C_{L_i}}{C_{L_i}^*} = exp\left[-\frac{koL}{u}\right] = exp\left(-ko\tau\right) \tag{12.58}$$

Where $C_{L_i}^* = \dfrac{p_i}{H_i}$, $\tau = L / u$ (12.59)

where p_i is the partial pressure of the component, i, H_i is the Henry's constant of the component, i

The above derivation is based on various simplification. We have not considered the velocities of the two streams. The velocity(u) used in the model is that of the liquid, containing cell suspension.

12.4.2 Biofilters for treatment of waste gases

Biofilters are packed bed system containing micro-organisms can be used for the treatment of waste gases.

Food and chemical factories often produce lot of organic volatiles such as acetone or hydrocarbons. These compounds can be adsorbed efficiently by a large filter mesh of micro-organisms, which are to be kept moist and supplied with nitrogen (say urea) and trace elements. The filter is self-sustaining with micro-organisms, getting carbon from the gases and oxygen from air. This is often cheaper than conventional chemical scrubbers.

Problem 12.10

A bio-scrubber is used to scrub an air stream containing 1% benzene & 1% Toluene (by volume) by a liquid stream of microbial cell suspension (stenotropho monas maltophilia) containing cells of 10 gm/L, fed from the top of the scrubber at the rate of 0.5 m^3/hr. Air enters from the bottom of the scrubber volume is 0.5 $m^{3/hr}$. Calculate the exit concentration of benzene and toluene. Data given:

For Benzene

$\mu_{max} = 0.05\ hr^{-1}$, $Y_{X/S} = 1.0$ gm/gm

$K_s = 0.15$ gm/L, $k_{La} = 20$ hr-1, $C_L^* = 7$ ppm (7 mg/L), $\bar{X} = 10$ gm/L

For Toluene

$\mu_{max} = 0.02\ hr^{-1}$, $Y_{X/S} = 0.8$ gm/gm

$K_s = 0.15$ gm/L, $k_{La} = 20$ hr-1, $C_L^* = 6$ ppm (6 mg/L), $\bar{X} = 10$ gm/L

Solution:

(a) Benzene removal

$$k_1' = \frac{\mu_{max}\bar{X}}{Y_{X/s}K_S} = \frac{(0.05)(10)}{(1)(0.1)} = 5\,hr^{-1}$$

$$k_{La} = 20\,hr^{-1} \text{ (given)}$$

$$\frac{1}{k_o} = \frac{1}{k_i} + \frac{1}{k_{La}} = \frac{1}{5} + \frac{1}{20} = 0.25\,hr$$

$$k_o = 4\,hr^{-1}$$

$$\frac{C_L}{C_L^*} = exp(-Ko\tau) = exp(-\Delta \times 1) = 0.0183$$

$$C_L = 0.0183 C_L^* = (.0183)(7 mg/L) = 0.128 mg/L$$

For Toluene:

$$k_1' = \frac{\mu_{max} \frac{\bar{X}}{\gamma_{X/S}}}{K_S} = \frac{(0.02)(10)}{(0.8)(0.15)} = 1.67\, hr^{-1}$$

$$k_{La} = 20\, hr^{-1} \text{ (given)}$$

$$\frac{1}{k_o} = \frac{1}{k_i} + \frac{1}{k_{La}} = \frac{1}{1.67} + \frac{1}{20} = 0.649\, hr$$

$$k_o = 1.54\, hr^{-1}$$

$$\frac{C_L}{C_L^*} = exp(-k_o \tau) = exp(-1.54 \times 1) = 0.21$$

$$C_L = (0.21)(6 mg/L) = 1.26 mg/L$$

So the benzene removal is slightly higher than toluene.

12.5 Rotating Biological Contactor (RBC)

Rotating Biological contactor (RBC) is used in the treatment of waste water following primary treatment, and shown in fig. 12.6.

The process involves allowing the waste water to come in contact with a biological medium in order to remove the pollutants.

It consists of closely spaced parallel discs mounted on a rotating shaft which is supported first above the surface of the waste water. Micro-organisms grow on the surface of the discs where biological degradation of waste-water-pollutants takes place. The degree of waste water treatment is related to the media surface area and the quality and volume of the involving waste water. The rotating packs of discs are contained in a tank or a trough and rotate at 2-5 rpm.

Commonly used plastics for discs are polyethylene, PVC or expanded polystyrene. The rotating shaft is aligned with the flow of waste water so that the discs rotate at night angles to the flow. Several packs are usually combined to make up a treatment train. About 40% of the disc area is immersed in waste water.

Microbial growth is attached to the surface of the disc forming a slime layer. The discs contact the waste water with atmospheric air for oxidation as they rotate. The rotation helps the sludge off excess sludge. The system is staged. The culture of the later stages is acclimatized to the slowly degraded materials.

The circular disc consists of plastic sheets ranging from 2–4 meters in diameter and thickness may be up to 10 mm. Several modules may be arranged in parallel or series and meet the flow and treatment requirements. The discs are submerged in waste water to about 40% of their diameters. Approx 95% of the surface area is thus alternatively submerged in waste water and then exposed to the atmosphere above the liquid. Carbonaceous substrate is removed in the initial stage of the RBC.

Carbon conversion may be completed in the first stage of a series of modules with nitrification being completed after the 5th stage. Most design of RBC systems will include a minimum of 4 or 5 modules in series to obtained nitrification of waste water.

Aeration is provided by the rotating action which exposes the media to the air after contacting them with waste water.

12.5.1 Modeling of a rotating biological contactor

It is assumed that the rotating contactor contains n number circular discs of radius R of width, B and they rotate in trough containing waste water at the rate of N rpm. Microbial growth takes place on the surfaces of the discs,

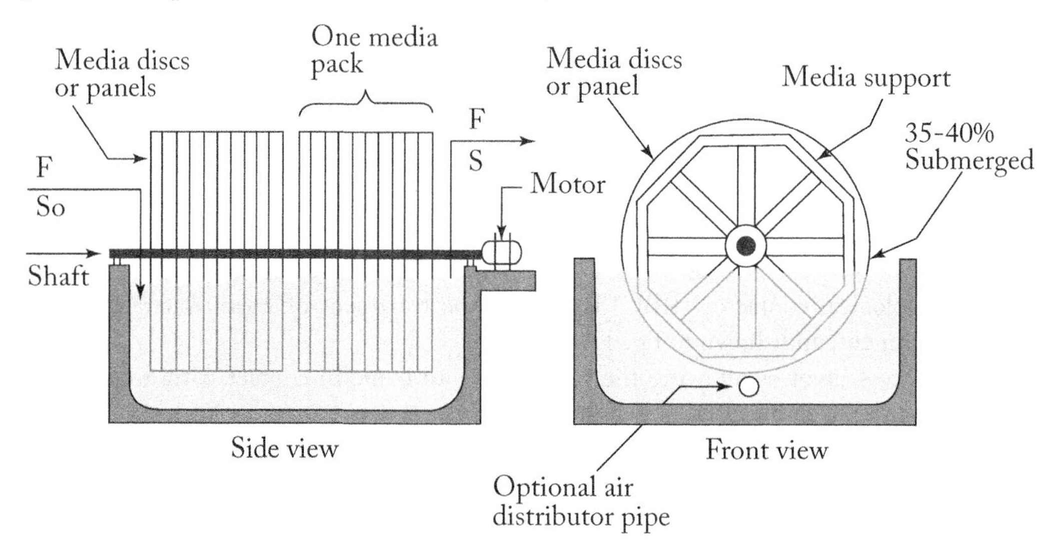

Fig. 12.6 Rotating biological contactor

where biological degradation of waste water pollutants occurs. To model the contactor for a continuous process, the system is assumed to be a mixed flow system. The rate of substrate consumption is given by a first order kinetics,

$$r_s = k\,S \tag{12.60}$$

where r_s is the rate of substrate consumption, k is the first order rate constant

Mass balance of the substrate:

$$F\,(So - S) = k\,S\,Vn \tag{12.61}$$

where F is the flow rate, S_o is the initial substra concentration, S is the substrate concentration at the exit, V_n is the slime volume deposited on n, number of circular discs.

Rearranging the above equation, we get

$$\frac{S}{S_o} = \frac{1}{\left(1 + k\frac{V_n}{F}\right)} \tag{12.62}$$

The first order rate constant, k may be evaluated by the relation,

$$k = \frac{\mu_{max}\overline{X}}{k_S Y_{x/s}} \tag{12.63}$$

where μmax is the maximum specific growth rate, X is the cell mass deposited on the discs assumed constant, k_s is the saturation constant, $Y_{X/S}$ is the yield coefficient.

Problem 12.11

A waste water stream of BOD (500 ppm) is fed to a system of rotating biological contactor containing 20 circular discs of 2 meter diameter with width of 1 cm. the discs covered with microbial films are rotated at the speed of 10 rpm. The flow rate is 1 m^3/hr. The first order rate constant, $k = 2 \times 10^{-4}$ sec^{-1}. There is no external or intra-film diffusion.

Calculate effluent concentration of the waste stream. Given the ratio of cell mass volume to n, number of circular discs volume = $\alpha = 0.20$ (arbitrary).

Solution:

$k = 2 \times 10^{-4}\ sec^{-1} = 0.72\ hr^{-1}$, $F = 1.0\ m^3/hr$

$V'_n = n(\pi R^2 b)$, R = radius of the disc.

$= (20)(3.142)(2^2)(0.01)\ m^3$

$= 2.5136\ m^3$

Where b = width of the column

$$S = S_o \frac{1}{\left(1 + k\frac{0.2 \times V'_n}{F}\right)} = 500\frac{1}{\left(1 + 0.72 \times \frac{0.2 \times 2.5136}{1}\right)} = 367\ ppm$$

12.6 Aerated Lagoons or Basins (Oxidation Bond)

There are two types of aerated lagoons:

1. Suspension mixed lagoons – the system has less energy provided by the aeration equipment to keep the sludge in suspension
2. Facultative lagoons

Here also aeration equipment provides sufficient energy to keep the sludge in suspension, but the solids settle to the lagoon's bottom. The biodegradable solids then degrade as in anaerobic lagoon.

Suspended mixed lagoon sludge will have a residence time(or sludge age) of 1 to 5 days. The objective of the lagoon is to act as a biologically assisted floculator which converts the soluble biodegradable organics in the influent to a biomass which is able to settle as a sludge. The effluent is then put in a second pond where the sludge can settle. The effluent can then be removed by a pump from the top with a low COD while the sludge accumulates on the floor and undergoes anaerobic stabilization.

Methods of aerating lagoons are of several types such as 1) motor driven submerged or floating jet aerators, 2) motor driven floating surface aerators, 3) motor driven fixed-in-place surface aerators, 4) injection of compressed air through submerged diffusers

12.6.1 Modeling of aerated lagoon

Consider a lagoon covering a land area A. Microorganisms for a sludge layer of thickness L at the bottom of the lagoon. This sludge remains essentially undisturbed by the movement of the liquid. As shown in the figure 12.7, the distance from the bottom of the lagoon is measured by a coordinate, Z.

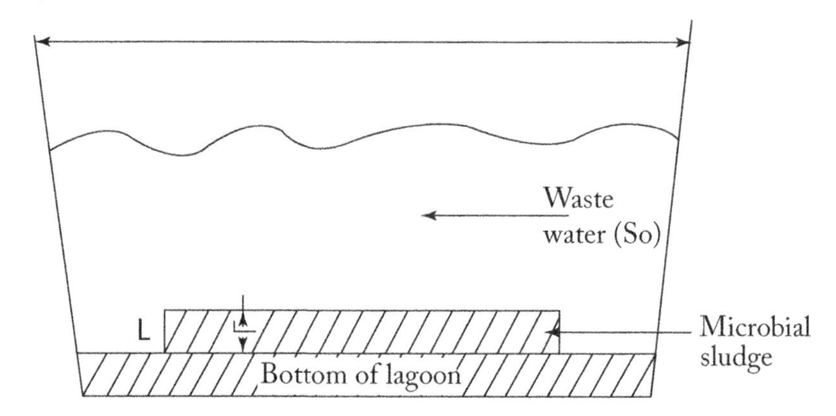

Fig. 12.7 Lagoon for waste water treatment

Assume that the micro-organisms are uniformly distributed in the sludge.

Waste water is fed into the lagoon so that bulk concentration of digestible substrate remains constant S_0. Cells consume substrate diffusing into the sludge layer and establish a concentration gradient across thickness L.

A shell mass balance on the substrate is made on a thin slice of sludge of thickness, Δz perpendicular to the direction of diffusion, assuming diffusion and biochemical reaction and we get

$$D_e \frac{d^2S}{dZ^2} = rs = k_1 S \tag{12.64}$$

where De is the effective diffusivity, $r_s = k_1 s$ and r_s is the rate of microbial reaction per unit volume of sludge and k_1 is the first order rate constant (sec-1), S is the concentration of substrate in the sludge.

The above equation is solved with the boundary conditions

at $Z = 0,\ S = S_O$

$$\text{at } Z = L,\ \frac{dS}{dZ} = 0 \tag{12.65}$$

And the solution is

$$\frac{S}{S_o} = \frac{\cos h\left(\frac{Z}{L}\phi_L\right)}{\cos h\left(\phi_L\right)} \tag{12.66}$$

$$\text{Where } \phi_L = L\left(k_1 / D_e\right)^{1/2} \tag{12.67}$$

and the effectiveness factor, η is given as

$$\eta = \frac{\tan h\phi_L}{\phi_L} \tag{12.68}$$

Where ϕ_L is the Thiele parameter.

The approximate values of the kinetic parameters are:

$k_1 = 1.5 \times 10^{-4}\ sec^{-1}$, $D = 6.0 \times 10^{-10}\ m^2/s$, So = 50 ppm.

12.7 Summary

This chapter deals with conventional reactors for waste water treatment like activated sludge tank reactor, single stage and double stage anaerobic digesters, trickling biological filter.

Other reactors, not commonly used are rotating biological contactor, bio scrubber, bio-filter, aerated lagoons etc.

Reactor designs for most of the systems have been presented on the basis of microbial kinetics, mass transfer and hydrodynamic features. Simulation of some reactor systems has also been presented.

Problems (Exercise)

Problem 12.1

An activated sludge tank reactor is to be designed to reduce an influent BOD of 500 ppm to effluent BOD of 20 ppm. The following data are given

Active solid concentration (X_a) = 3 Kg/m^3, Recycle ratio, $\alpha = 0.5$,

$F = 1.5 \times 104\ m^3/day$

$\mu_{max} = 1.3$ day^{-1}, $K_s = 0.3$ kg/m^3

$k_d = 0.05$ day^{-1}, $Y_{X/S} = 0.5$ kgMLss / Kg BOD

Calculate the following

(a) Sludge age, θ_c, (b) reactor volume (V), (c) concentration of active solids in the recirculation line (X_r), (d) daily aeration rate assuming 7.5 percent oxygen utilization, (e) the optimum ratio of X_r / X_t (α) where the X_r is active solid concentration.

Problem 12.2

A waste water stream of initial BOD of 500 ppm is to be reduced to BOD to 30 ppm in a trickling Biological filter. The following kinetic parameters for the biocatalyst (mixed micro-organism immobilized on the porous solid).

r_m (the maximum substrate consumption rate) = 0.03 kg s/(m^3) (hr), $K_s = 0.2$ kg s /m^3

The biofilm thickness, L = 1 mm = 1×10^{-3} m. The cross-section of filter = 4 m^2, H = 6m. The biofilm surface area per unit volume of the bed, a = 5×10^4 m^2/m^3.

Assume first order kinetics with respect to carbon compounds(s).

Calculate the flow rate of the feed (F) to achieve the BOD removal from 500 ppm to 30 ppm.

Problem 12.3

In an activated sludge process substrates were carbonaceous material (Xs), organic-N and nitrite-N which were converted by, respectively corresponding to carbon, organic-N and Nitrite-N biomass (X_C, X_O, X_N). Plant operating conditions were specified as follows: F = 2.0×10^4 m^3/day, $\alpha = 0.15$, $\beta = 0.0015$, $S_{Ca} = 16.8$ mg/L (5 day BOD), $S_{Co} = 1811$ mg/L, $S_{Ca} = 250$ mg/L, $S_{No} = 0.5$ mg/L.

Data for Monod plus endogenous metabolism kinetics:

μ_{max} (day^{-1}), K_e(day^{-1}), K(ms/L), Y (g/g) = Carbonaceous (5.0, 0.005, 100, 0.5)

Nitrosomonas [0.33, 0.05, 1.0, 0.05]

Nitrobacter [0.80, 0.05, 2.1, 0.02]

(a) Calculate the sludge age for the carbonaceous biomass

(b) Calculate sludge age (θc_1) for the nitrification and calculate the exit level concentration for Nitrite N (S_{NO2})

(c) Calculate VX_A (the active biomass) and reactor volume, V for carbonaceous biomass

Hint for

(a) $$\frac{1}{\theta_c} = \frac{\mu_{max} S_{ca}}{K_S + S_{ca}} - k_c$$

(b) $\frac{1}{\theta_1} = \frac{(\mu_{max} NH_3) S_{au}}{K_{NH_3} + S_{au}} - ke$ & $\frac{1}{\theta_1} = \frac{\mu_{max} NO S_{NO_{2a}}}{K_{NO_2} + S_{NO_{2a}}} - ke$

Solve for SNO_{2a}

(c) $VX_a \left(active\ biomas \right) = \frac{Y F\theta_c (S_o - S_a)}{1 + ke\theta_c}$

$$V = F\theta_c \left(1 + \propto - \propto \frac{X_r}{X_a} \right)$$

Problem 12.4

A nitrification unit is used for the removal of NH_3 and BOD using Nitrosomonias bacteria. Initial BOD, So = 400 mg/L, organic N = 60 mg/L

For BOD removal:

$\mu_m = 2.0$ day^{-1}, $K_m = 5.0$ mg/L

$K_e = 0.05$ day^{-1}, Y_{obs} $(Y_{S/X}) = 0.4 / (1.0 - 0.6\ \theta_c)$

(a) If the cell age, θ_c, is determined from ammonia effluent requirement and its values is $\theta_c = 9$ days, what is the effluent BOD?
(b) Also calculate exit concentration of NH_3 if $\theta_c = 9$ days. Given μm, $NH_3 = 0.33$ day^{-1}, $K_{NH_3} = 1$ mg/L, $k_e = 0.05$ day^{-1}, $Y_{N/S} = 0.05$

Problem 12.5

The hydraulic residence time of an activated sludge waste water treatment plant is 6h. The fresh feed has a BOD of 300 mg/L and the settler produces a recycle stream containing 6000 mg/L. Using a sludge age of 6 days, Calculate the recycle ratio and the final effluent BOD. $Y_{x/S} = 0.55$ and the specific growth rate, μ is given by

$$\mu = \frac{\mu_m S}{K_s + S} - k_d$$

$$\mu_m = 0.5\,h^{-1}, K_s = 90\,mg\left(BoD\right)l^{-1}$$

and endogenous metabolic coefficient, $k_d = 0.01$ h^{-1}

Problem 12.6

A waste water feed containing 500 BOD (500 mg/L) is fed to an anaerobic digester of 1 m^3 volume at the feed rate of 100 L/hr and BOD is to be reduced to BOD 50 (50 mg/L).

Calculate the methane formation per hr.

Data given:

$$\theta_{CH4} = \frac{V}{\rho_g} Y_{CH4/X} \mu x$$

Where V = volume of the digestor

Density of methane gas, $\rho_g = 0.64\, g / L$

$$Y_{CH4/X} = 0.37 \frac{gm CH_4}{gm\ cell}$$

$$Y_{X/S} = 0.2 \frac{gm\ cell}{gm\ substrat}$$

$$\mu X = DY_{X/S}(S_1 - S)$$

Problem 12.7

A waste water stream is fed to a rotating biological contactor containing 10 disecs of 2 m diameter with 10 cm width. The inlet concentration of the organics is 500 BOD and the exit BOD desired is 50 BOD.

The total cell mass deposited is 1.5 kg. The kinetic parameters are :
$\mu_{max} = 0.05$ hr^{-1}, $K_S = 0.1$ kg/m^3, $Y_{X/s} = 0.4$, film thickness, L = 3 mm.
Calculate the flow rate required for the system.

Problem 12.8

Activated sludge process is to be modelled by structured model. The biomass, x_T is divided into three components – x_1 (stored mass), x_2 (active mass) and x_3 (inert mass). Their kinetic rates are given by the following equations:

$$\frac{dx_1}{dt} = k_L \left(X_T \frac{f_s S}{K_S + S} - X_1 \right)$$

$$\frac{dx_2}{dt} = \frac{\mu_{m_2} X_1}{K_2 + X_1} \cdot x_2 \cdot Y_{X_2}$$

$$\frac{dx_3}{dt} = kdX_2 Y_{X_3}$$

Parameter values:

$k_L = 3$h^{-1}, $f_1 = 0.45$, $K_s = 150$ mg/L

$\mu_2 = 0.06$ h^{-1}, $K_2 = 80$ mg/L, $Y_{X2} = 0.66$

$k_d = 0.03h^{-1}$, $Y_{XS} = 0.35$

$x_T = x_1 + x_2 + x_3$

$S = S_o = Y_{X1/S}(X_{To} - X_1)$

$S_o = 500$ mg/L, $X_1^0 = 300$ mg/L

At $t = 0$, $S = S_o$, $X_{TO} = 300$ mg/L, $X_2 = 0$, $X_3 = 0$

Determine the variation of S, X_1, X_2, X_3 vs. t (t = 0 – 10 hrs)

References

Andrews, J.F. "Dynamic methods and control strategies for Waste water treatment Processes", Water Res. 8, 261–289, 1974

Bailey, J.E. and D.F. Ollis, Biochemical Engineering Fundamentals, 2nd ed., Mc Graw Hill Book Co., 1983

Blanch, H.W. and I.J. Dunn, Modeling and Simulation in Biochemical Engg in Advances in Biochemical Engg, 3, 159–162, 1974

Busby, J.B. and J.F. Andrews " Structured Kinetic Model for Activated Sludge Process". Water Pollution Control. Asso. 47, 1067, 1975

Chiu, S.Y., L.E. Erickson, L.T. Fan and F.C. Kao, Kinetic model identification in Monod population, using continuous culture data, Biotechnol. Bioeng. 14, 217–231, 1972

Graef, S.P. and J.F. Andrews, "Mathematical modelling of anaerobic digestion" CEP Symposium series, 136, 70, 101–127, 1974b

Lawrence, A.W. and P.L. Mc Carty, "Unified basis for Biological Treatment: Design and Operation", J. Sanitary Eng. Proc. ASCE 96, 757, 1970

Petersen, R.B. and M. Denn, "Computer aided Design and Control of an activated sludge process", The Chemical Eng. J., 27, 1313, 1983

Pohland, F.G. and S. Ghosh, "Developments in Anaerobic treatment Processes", Biotechnol. Bioeng. Sympo. 2, P.85, 1971

Schroeder, E.R., "Water and Waste water Treatment" Mc Graw Hill Book Co. 1977

Schuler, M.L. and F. Kargi, Bioprocess Eng: Basic concepts, Pearson pub., 2002

Sundstrom, D.W., and H.E. Klei, Wastewater Treatment Prentice Hall, Engle wood Cliffs, NJ, 1979

Tehobanoglous, G., Waste Water Engineering Treatment, Disposal and Reuse (Metcalf & Eddy, Inc) 3rd ed., Mc Graw Hill Book Co., New York, 1991.

CHAPTER 13

Miscellaneous Bioreactors

13.0 General Classification of Bioreactors

A general classification of Bioreactors may be presented as given below using alphabetical order:

A

- Algae Bioreactor
- Algae Fuel
- Algae PARC
- Anaerobic membrane bioreactor
- Auxostat

B

- Bioelectrochemical Reactor
- Biostat

C

- Chemostat

F

- Fed Batch culture

H

- Hollow fiber reactor

M

- Membrane bioreactor
- Moss bioreactor

O

- Oscillatory baffled reactor

P

- Photo bioreactor

S

- Single-use Bioreactor

T

- Turbibostat

U

- Ultraviolet sterilizing bioreactor

13.1 Algae Bioreactor

An algae bioreactor or photobioreactor is used for cultivating algae with a purpose to fix CO_2 or produce biomass.

Specifically algae bioreactors can be used to produce fuels such as biodiesel and biomethane, to generate animal feed or to reduce pollutants such as NOx and CO_2 in the flue gas of power plants.

This type of bioreactor is based on the photosynthetic reaction which is performed by the chlorophyll containing algae itself, using dissolved carbon dioxide and sunlight energy. The carbon dioxide is dispersed into the reactor fluid to make it accessible to algae.

The bioreactor has to be made out of transparent material.

The algae are photo autotrophy organism which perform oxygenic photosynthesis:

$$6CO_2 + 6H_2O \xrightarrow[algae]{Sunlight} C_6H_{12}O_6 + 6O_2, AH = 2870KJ / mole$$

13.2 Auxostat

It is a continuous culture device which during operation, uses feedback from a measurement taken on the growth chamber to control the medium flow rate, maintaining the concentration of biomass at a constant value.

Most typical auxostats are pH auxostat. Other auxostats may measure oxygen tension, ethanol concentration and sugar concentration.

Continuous culture systems in which unlike chemostats the growth rate and dilution rate may vary.

Auxostats are useful for studying cultures growing at rates close to their maximum value and they can be used to select fast growing culture.

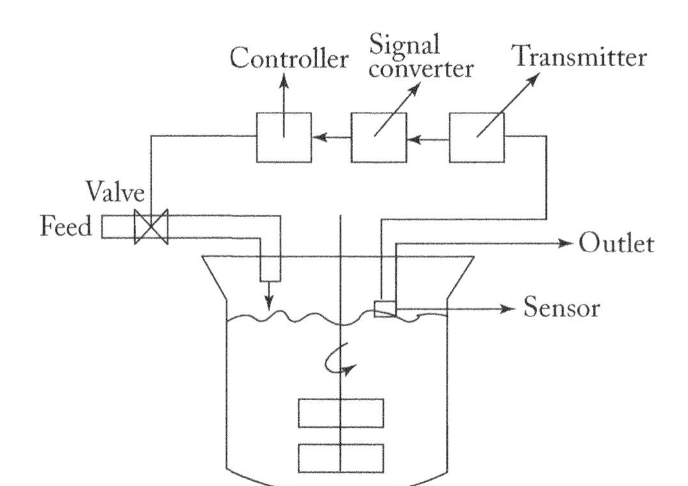

Fig. 13.1 Auxostat

The pH auxostat is a tool for studying microbial dynamics in a continuous culture.

The medium should have a nutrient concentration half of that feed medium to reduce any detrimental effect on the transition phase between the batch and the continuous processes. The transition phase is a period of adaptation for the organism similar to lag phase that is seen in the batch culture. Feeding begins when the organisms enters the exponential phase. The dilution rate should not exceed μ_{max}. Once the organisms has adopted to the new growing condition, a steady state is reached.

An auxostat can be adjusted to keep up automatically with bacterial growth, and so maximize the growth rate. At such high growth rates, bacteria that can grow fast are selected over ones that grow slowly. Thus, natural selection acts on the population, selecting fast growing variants of the bacteria.

In practice, all large, continuous industrial fermentation systems are auxostats, rather than chemostats, since they have many feedback controls, which allow the operator to adjust what materials the fermenter receives as it proceeds.

13.3 Photobioreactors

There are several types of photobioreactors such as a) Plate Photobioreactor, b) Tubular Photobioreactor, c) Bubble column photo bioreactor.

A photobioreactor is a reactor system which uses some type of light source viz. natural sunlight or some artificial illumination.

These reactors are usually employed to grow some small phototrophic organisms such as cyanobacter, algae or moss plants. These organisms use light through photosynthesis as their energy source and do not require sugars or lipids as energy source.

Risk of contamination with other organisms like bacteria or fungi is lower in photo reactors compared to bioreactors for heterotroph organisms. Tubular photobioreactor has been designed for algae culture. Principles of fluid mechanics, gas-liquid mass transfer (k_{La})

and irradiance – controlled algae growth are integrated into a method for designing tubular photobioreactors in which culture is circulated by an airlift pump. A 0.2 m^3 photobioreactor has been used in continuous outdoor culture of the micro algae (Phaeodactylum tricornutum). The reactor performance was assessed under various conditions of irradiance, dilution rates, and liquid velocities through the tubular solar collector. A biomass productivity of 32 gm m^{-2} d^{-1} (1.9 gL^{-1} d^{-1}) could be obtained at a dilution rate of 0.04h^{-1}. Photoinhibition was observed during hours of peak irradiance and the photosynthetic activity of the cells were recovered a few hours later. Linear liquid velocities of 0.50 and 0.35 ms^{-1} in the solar collector gave similar biomass productions, but the culture collapsed at lower velocities. The effect of dissolved oxygen concentration was quantified in indoor conditions. D.O. levels higher or lower than air saturation values reduced productivity.

Bubble columns or airlift fermenters with transparent reactor bodies have been designed with appropriate k_{La} and holdup.

Biodiesel from microalgae seems to be most important renewable biofuel and has the potential to displace petroleum – derived transport fuels. This source of fuel is much more effective than those of palm oil. Similarly bioethanol from sugarcane is no match for microalgae biodiesel.

13.4 Turbidostat

It is a continuous microbial culture device similar to a chemostat or an auxostat, which has feed-back control between the turbidity of the culture vessel and the dilution rate.

A turbidostat dynamically adjusts the flow rate to make the turbidity constant. At equilibrium, operation of both the chemostat and turbidostat are identical. It is only when classical chemostat assumptions are violated (i.e out of equilibrium) or the cells are mutating that a tubidostat is functionally different.

One case may be when the cells are growing at the maximum growth rate, which case it is difficult to set a chemostat to the appropriate constant dilution rate. Most turbidostats use a spectrophotometer or tubidometer to measure the optical density for control purpose.

The theoretical relationship between the growth in a chemostat and growth in a turbidostat is somewhat complex. A chemostat has fixed volume, constant flow rate and thus has a fixed dilution rate.

13.5 Biostat

A culture vessel in which physical, physiochemical and physiological conditions as well as cell-concentrations are kept constant is known as Biostat.

13.6 Ultraviolet Sterilizing Bioreactor

Ultra violet germicidal irradiation (UVGI) is a disinfection method that uses short wave length ultraviolet (UV) light to kill or inactivate micro organisms by destroying nucleic acids and disrupting their DNA unable to perform their cellular function. UVGIs are used in various applications such as food, air and water pollution. It has been used to sterilize the drinking water and waste water, as the holding facilities are enclosed and can be circulated to ensure a higher exposure to the UV. In recent years UVGI has found renewed applications in air purifier.

A low pressure mercury vapour discharge tube floods the inside of a biosafety cabinet with short UV light (260–27 nm); when not in use, sterilizing micro biological contaminants from infrared surface.

A UVGI system is designed to expose environments such as water tanks, sealed rooms, forced air system to germicidal exposure from germicidal lamps which emit UV electro magnetic radiation at the current wave length.

UV dose cannot be measured directly but can be inferred based on the known or estimated inputs to the process:

- Flow rate (Contact time)
- Transmittance (light reaching the target)
- Turbidity
- Lamp age

UV dose:

$$\mu ws / cm^2 = \text{uv intensity } \mu w/cm^2 \times (\text{exposure time}) (\text{seconds})$$

Microbes die according to the first order rate law; the rate constant, k is proportional to the local UV intensity, I. UV illumination may be assumed to radially symmetric and UV intensity varies radially according to Beer's Law,

$I(r) = I(R)e^{-\alpha(R-r)}$, where R is the inner glass tube radius r is any radius. For laminar flow, a tube length L, the ratio of viable organism (N/No) averaged over the entire exit section may be evaluated, α is a constant.

13.7 Anaerobic Membrane Bioreactor (AMBR)

Membranes provide exceptional removal of suspended solids and complete biomass retention that improve the biological treatment process. But their commercial application to anaerobic treatment has been limited. AMBRs have been tested with synthetic food processing, removal of industrial solid, municipal waste water treatment. COD removal ranges from 56 to 99% and the reported design membrane fluxes are from 10 to 40 L/m^{2}h.

AMBRs should be applicable to highly concentrated particulate waste stream like municpal sludge where the membrane can decouple the solids. Greater assessment of vacuum driven immersed membrane is possible, combining external or immersed membranes with retained biomass.

Reactor design, control of membrane fouling, economic feasibility are the key research areas to be addressed.

References

Lio., B.Q., J.T. Kramer, and D.K. Balgley, "Anaerobic membrane bioreactors, and Research directions", published on line, 489–530, 2007

Martin et al., "Modeling the energy demands of aerobic and anaerobic membranes for waste water treatment" Environmental technology, 2011

Vanzyl, P.J., M.C. Wentzel, C.A. Ethamo, "Design and start up of a high rate anaerobic membrane bioreactor", Water Science….., 2008

13.8 Bioelectrochemical Reactors

In recent years, the removal of nitrate as a global soil and water contaminant has been increasingly considered. Various methods have been applied to remove this inorganic pollutant from water and waste water. Among them, an integrated bioelectro-chemical reactor system provides a novel method for the removal of water contaminants and waste water denitrification. The novelty of the system is improved biological denitrifiction by immobilizing auto hydrogenotrophic bacteria directly on the surface of a cathode, providing easy access to NO_3^- and H_2 as the electron acceptor and electron donor respectively. The system's effectiveness depends on the configuration of reactor and the operational and environmental parameters. Important design variables are: hydraulic retention time, electric current, pH and the carbon source.

13.9 Moss Bioreactors

A moss bioreactor is a photobioreactor used for the cultivation and propagation of mosses. It is normally used in molecular farming for the production of recombinant proteins using transgenic moss. In environmental science, moss bioreactors are used to multiply peat mosses. The moss consortium are employed to monitor air pollution.

Moss is a fungal photoautotrophic organism that has been kept in-vitro for research purposes since the beginning of 20th century.

The first moss bioreactor for the model organism "Physconitrella patens" was developed in 1990s to comply with the safety standards regarding the handling of genetically modified organisms and to gain sufficient biomass for experimental purposes.

The moss bioreactor is used to cultivate moss in a suspension culture in agitated and aerated liquid medium. The culture is kept with constant temperature, pH. The culture medium contains all nutrients and minerals needed for the growth of the moss.

For maximum growth rate, moss is kept at protonema stage by continuous mechanical disruption e.g. by using rotating discs. The cultivation chamber may consists of a column, a tube or exchangeable plastic bags.

Various biopharmaceuticals have already been produced in moss bioreactors. The recombinant protein can be directly purified from the culture medium. One example is the production method of factor H. This molecule is a part of the human complement system. Defects in the corresponding gene are associated with human diseases such as severe kidney disorders.

Biologically active "recombinant factor H" was produced in a moss bioreactor for the first time in 2011. The enzyme, α-galactosidase is now produced in moss bioreactor by the German Federal Institute of drugs and medical devices.

It will be tested as enzymes replacement therapy in the treatment of Fabry's disease.

References

Decker, E.L. and Reski, R "The moss Bioreactor", Current opinion Plant Biotechnology 7 (2), 166–170, April, 2004

Hohe, A. and Reski, R. "Control of growth and differentiation of bioreactor culture of physoconitrella by environmental parameters", Plant cell, Tissue and organ culture, 81(3) 307–311, 2005

Reutter, K and Reski, R. "Production of heterologous protein in bioreactor culture of fully differentiated moss plants", 81 (3), 142–147, 1976.

13.10 Continuous Oscillatory Baffled Reactor (COBR)

A continuous oscillatory baffled reactor (COBR) is a specially designed reactor to achieve plug flow under laminar flow conditions. Achieving plug flow has previously been limited to either a large number of CSTRs in series or conditions with high turbulent flow.

The technology incorporates annular baffles to a tubular reactor framework to create eddies when liquid is pumped up through the tube. Likewise when the liquid is in a down stroke through the tube, eddies are created on the other side of the baffles. Eddy generation on both sides of the baffles create very effective mixing while still maintaining plug flow. By using COBR potentially high yields of product can be made with greater control and reduced waste.

Design:

A standard COBR consists of 10 – 150 mm Id tube with equally spaced baffles throughout. There are typically two pumps in a COBR: one pump is reciprocating and generate continuous oscillatory flow and a second pump creates net flow through the tube.

The design offers a control over mixing intensity that a conventional tubular reactor cannot achieve. Each baffled cell acts as a CSTR, because a secondary pump is creating a net laminar flow; much larger residence times can be achieved relative to turbulent flow system. It has been found that spacing of baffles 1.5 times tube diameter apart has the most effective mixing condition.

Biological Applications:

The low shear rate and enhanced mass transfer provided by the COBR makes it an ideal reactor for various biological processes viz. cultivation of animal cells and plant cells.

For the case of mass transfer, COBR fluid mechanics allows for an increase in oxygen gas residence time. Moreover, vortices created in the COBR causes a gas bubble break up and increase in surface area for high oxygen gas transfer. It is easy to scale up the process also.

Limitation:

Additional complexity in COBR design is to be noted compared to other reactors. For bioprocessing, the fouling of the baffles and internal surface is a problem to be taken into consideration.

13.11 Single-use Bioreactor

A single-use bioreactor or disposable bioreactor consists of a disposable bag instead of a culture vessel.

This refers to a bioreactor in which the lining in contact with cell culture will be plastic and this lining is encased with a more permanent structure using either a rocker or a cuboid or cylindrical steel support.

Single-use bioreactors are widely used in the field of mammalian cell culture and are now replacing conventional bioreactors.

Types of single-use bioreactors:

There are two types of single-use bioreactors. One type uses stirrers like conventional bioreactors. But in this case stirrers are integrated into plastic bag. The closed bag and the stirrer are pre-sterilized. The bag is mounted in the bioreactor and the stirrer is connected to a driver mechanically or magnetically.

Other type single-use bioreactors are agitated by a rocking motion.

Both the stirred and rocking single use bioreactors are used upto a scale of 1000 litres volume.

A single-use bioreactor is equipped with a disposable bag which is usually made of three layer plastic foil. One layer is made from polyethylene tetrephthalate or LDPE to provide mechanical stability. A second layer is made of PVA or PVC, which acts as a gas barrier. Finally a contact layer is made from PVA or PP.

Single-use Process:

Single-use process steps available are media and buffer preparation, cell harvesting, filtration, purification and virus inactivation.

Advantages and Disadvantages:

Compared to conventional bioreactor systems, single-use bioreactor has some advantage. Single-use technologies reduce cleaning and sterilization demands. Cost savings are more than 60% compared to fixed asset stainless steel bioreactors.

Application:

The single-use bioreactors reduce the risk of cross contamination and enhance biologically process safety. They are specially suitable for any kind of pharmaceutical product. The limiting factor for the use of such reactors is the available oxygen transfer, k_L (the liquid phase mass transfer coefficient) for the specific phase area, resulting in the value of k_La (the volumetric mass transfer coefficient). Theoretically it can be enhanced by higher stirrer speed or higher rocking frequency.

13.12 Algae Fuel

Algae fuel or algal biofuel is an alternative to liquid fossil fuels that uses algae as its source of energy rich oils.

Algae fuels are an alternative to common known biofuel sources such as corn and sugarcane. The energy crisis and the world food crisis have greater interest in algae culture (farming algae) for making biodiesel and, other biofuels using land unsuitable for agriculture. "Solazyme" and "Saphire Energy" already began commercial sales of algae biofuels during 2012 to 2013. Algenol (a company) produced biofuels commercially in 2014.

Algae fuels include Biodiesel, Biobutanol, Biogasoline, methane, bioethanol, biogasoline jet fuels by hydrotreating. Algal cultivation has been carried out in closed loop system, photobioreactors, open pond, algae turf scrubber.

The lipid or oily part of the algae biomass can be extracted and converted into biodiesel through a process similar to that used for any other vegetable oil or converted in a refinery into a new fuel in replacement for petroleum based fuels.

Alternatively, following lipid extraction, the carbohydrate content of algae can be fermented for bioethanol or butanol fuel.

Algae can be used to produce "green diesel" (also known as renewable diesel, by hydrotreating vegetable oil, or hydrogen derived renewable diesel), through a refinery process that breaks molecules down into shorter hydrocarbon chains used in diesel engines without modification of the normal engine.

Rising jet fuel prices are putting severe pressure on airline companies creating an incentive for algae jet fuel research. US military has already planned to start large scale jet fuel production from algae ponds. The projects run by the companies like 'SAIC' and 'General Atomics' are expected to produce 1000 gallons of jet fuel per acre per year from algae ponds.

Regarding the species of algae, research into algae for the mass production of oil focuses mainly on 'micro algae' (organism capable of photo synthesis that are less than 0.4 mm diameter including the diatoms and cyanobacteria) as opposed to micro algae such as 'seaweed'. Micro algae has been prefered due largely to their less complex structure, fast growing rates, high oil content. Research is going on for utilizing other varieties of algae.

Algal cultivation is an important factor for production of algae fuels. Algae grows much faster than food crops and can produce hundred times more oil per unit area than conventional crops such as rapeseed, palms, soyabeans, or jatropha. Algae have a harvesting cycle of 1–10 days, their cultivation permits serveral harvests in a very short time frame. Most research on algae cultivation has focussed on growing algae in clean but expensive photobioreactors or in open ponds which are cheap to maintain but prone to contamination. The closed-loop system (not exposed to air) avoid the problem of contamination by other organisms blown in by air. Several experiments have found that CO_2 from a smoke track works well for growing algae. Other systems are open pond, algal turf scrubber, which have been commercially used for algae production.

13.13 Algae PARC

Algae Parc is a multidisciplinary research program which develops cost effective and sustainable micro algae methods outdoor.

Research program involves integration of biological and engineering aspects of cultivation and biorefinery. Algae Parc houses a range of different research facilities as mentioned below:

- Fully equipped laboratories including a variety of analytical equipments and different seized controlled photobioreactors.
- Pilot scale cultivation systems consisting of

1. Open pond
2. Flat panel photo-bioreactor
3. Horizontal tubular photobioreactor
4. Vertically stacked 3D tubular photobioreactor

Algae Parc Refinery program has been initiated by Wageningen University and research centre in cooperation with eleven industrial partners and University of Twente (Netherlands). At Algae park there are three large European projects and expected participation in several others. The different projects cover different products like bioplastics, biofuels and feed additives.

Research programs in Algae Parc of Netherlands includes:

(a) Biorefinery and product development
(b) Chain analysis and design
(c) Cellular processes
(d) Strain improvement
(e) Cultivation optimization
(f) Scale up

13.14 Research Projects on Algae Fermentation

(i) It includes solar energy fixation and conversion with algal bacterial systems. Research has been undertaken on methane fermentation of algae as part of an integrated biological process of solar energy fixation and conversion to natural gas along with purification of waste water.
Ref. Ph.D. thesis, California University, Barkley, Title: Solar energy fixation and conversion with algal bacterial system.

(ii) Ethanol production from marine algal hydrolysates by using Escherichia Coli. Ethanol was produced by fermentation of algal hydrolysates. 10% of S. Cerevisiae or E.Coli culture was added to 50 g/L pretreated algal biomass supplemented with concentrated LB (concentration of 10 g/L peptone, 5 g/L yeast extract, 10 g/L Nacl)
Reference: Bioresource Technology Vol. 102, issue 16 August, 2011 p. 7466–7469

(iii) Simultaneous Saccharification of Cassava starch and fermentation of algae for Biodiesel production.
A simple approach to produce biodiesel from cassava starch was established which successfully integrates the simultaneous saccharification and heterotrophic algal fermentation in an identical system (e.g. batch reactor). Batch experiments were investigated to verify the feasibility of raw starch substrates fermentation for microalgal oil.
Reference: Lu, Y. Ding, Y. I. Wu, Q. J. Appl. Phycology, Vol. 23, 115–121, 2011

(iv) Effects of hydrothermal depolymerisation and enzymatic hydrolysis of algae biomass on yield of methane fermentation process
Study was undertaken to determine the effects of preliminary hydrothermal depolymerization and enzymatic hydrolysis of macro algae biomass on the yield of methane fermentation process.
Reference: Anna Grala et al., Pol. J. Environmental study, Vol. 21, No. 2, 363–368, 2012

(v) Fermentative and photo chemical production of hydrogen in algae
Influence of hydrogen ion concentration – contrary to the behaviour of respiration and photosynthesis, the rate of fermentation in the algae is clearly dependent on the hydrogen ion concentration in the suspension of cells. After 2 hours of fermentation in nitrogen, the metabolism of those algae which were found to be capable of photoreduction with hydrogen changes in such a way that molecular hydrogen is released from the cell in addition to carbon dioxide.
Reference: Gaffron, H. and Rubin, J., J. of General Physiology, Vol. 26 No. 2, 219–240, 1942

(vi) Ethanol production from carbon dioxide by Fermentative microalgae
Ethanol productivity consists of micro algae cultivation and algal cell harvest, self fermentation of algae and ethanol extraction processes. The system seems simple and less energy consuming compared with conventional processes.
Reference: Hirayama, S. et al, Studies in surface Science Catalysis, Vol. 114, 657–660, 1998.

(vii) Acetone, butanol and ethanol production from waste water algae

The fermentation of carbohydrates present in algal biomass to C2, C3 and C4 compounds including acetone, butanol and ethanol are achieved in the saccharolytic clostridium. Batch fermentations were performed with 10% algae as feed stock. Clostridum saccharoper butylacetonicum can produce Acetone, Butanol and ethanol using waste water algae.
Reference: Bioresource Technology Vol. 111, 471–495, May 2012.

13.15 Mathematical Modelling of Algae Fermentation

Mathematical model was developed to describe microbial growth, lipid production and glycerine consumption under photo heterotrophic conditions based on logistic, LeudeKing – Piret like equations:

Logistic equation:

$$\frac{dX}{dt} = \mu \times \left(1 - \frac{X}{X_m}\right), \text{ for cell mass growth}$$

for product formation, Leudeking Piret equation may be used:

$$\frac{1}{X}\frac{dP}{dt} = \propto \mu + \beta$$

All experiments were conducted in a 2L batch reactor without considering CO_2 effect on algae's growth and lipid production.

Biomass and lipid production increased with glycerine as carbon source and were well described by the logistic and Leude King – Piret equations respectively.

Sensitivity analysis was applied to examine the effects of certain important variables on model parameters. Results showed that the initial concentration of glycerine (So) was the most significant factor for algae growth and lipid production. This model is applicable for prediction of the growth of other single cell algae species, but model testing is recommended before scaling up the fermentation process.

Reference: Bioresource Technology, Vol. 102, 3 February, 3077–3082, 2011

For the kinetic study on light-limited batch cultivation of photosynthetic cells, the relative importance of exponential and linear growth phases during light-limited batch cultivation of 'Chlorella pyrenoidosa', and 'spirulina platenosis' cells, was investigated. The relationships among the specific growth rates, the linear growth rates and the final cell concentrations during algal cultivation in various types and sizes of photobioreactors were studied. No good correlations were observed between exponential growth rate, linear growth rate and final cell concentration. The existence of the various growth phases during light limited batch cultivation of photosynthetic cells could be predicted by a simple mathematical model. The model predicts that the linear growth phase is relatively larger than the exponential growth phase. During light limited batch cultivation of photosynthetic cells, the linear growth rate is thus a better growth index than the specific growth rate.

APPENDICES

Appendix 1

Multiple Choice Questions (MCQ)

GATE 2016

(1/14) Which one of the following statements is not true?

(a) In competitive inhibition, substrate and inhibitor compete for the same active site of an enzyme.

(b) Addition of a large amount of substrate to an enzyme cannot overcome uncompetitive inhibition

(c) A transition state analogous in enzyme catalysed reaction increases the rate of product formation

(d) In non-competitive inhibition, K_M of an enzyme for a substrate remains constant as the concentration of the inhibitor increases

(2/14) Prandtl number is the ratio of

(a) Thermal diffusivity to momentum diffusivity

(b) Mass diffusivity to momentum diffusivity

(c) Momentum diffusivity to thermal diffusivity

(d) Thermal diffusivity to mass diffusivity

(3/17) Fed batch cultivation is suitable for which of the following

P. Processes with substrate inhibition

Q. Processes with product inhibition

R. High cell density cultivation

(a) P and Q only (b) P and R only (c) Q and R only (d) P, Q, R

(4/18) Which one is involved in the treatment of industrial effluent

(a) Primary (b) Secondary (c) Tertiary (d) Quaternary

(5/21) Which one of the following is the most suitable type of impeller for mixing high viscosity (viscosity > 10^5 cp) fluids?

(a) Propeller (b) Halical ribbon

(c) Paddle (d) Flat blade turbine

(6/28) Which of the following events occur during the stationary phase of bacterial growth

P. Rise in cell number stops

Q. Spore formation in some gram positive bacterial such as Bacillus subtilis (Itallics)

R. Cell size increases in some gram-negative bacteria such as Escherichia Coli

S. Growth rate of bacterial cells nearly equals their death rate
T. Decrease in peptidoglycan cross-linking
(a) P, Q and S only (b) P, S. and T only
(c) Q, R and S only (d) P, R and T only

(7/40) In animal cell culture, a CO_2- enriched atmosphere in the incubator chamber is used to maintain the culture pH between 6.9 and 7.4. Which one of the following statement is correct?
(a) Higher the bicarbonate concentration in the medium, higher should be the requirement of gaseous CO_2
(b) Lower the bicarbonate concentration in the medium, higher should be requirement of gaseous CO_2
(c) Higher the bicarbonate concentration in the medium, lower should be the requirement of gaseous CO_2
(d) CO_2 requirement is independent of bicarbonate concentration in the medium

(8/41) Choose the correct combination of True (T) and False (F) statements about micro carriers used in animal cell culture
P. Higher Cell densities can be achieved using micro carriers
Q. Micro carriers increase the surface area for cell growth
R. Microcarriers are used for both anchorage and non-anchorage dependent cells
S. Absence of surface charge on microcarriers enhances attachment of cells
(a) P-T, Q-F, R-T and S-F (b) P-T, Q-T, R-F and S-F
(c) P-F, Q, F, R-T and S-T (d) P-F, Q-T, R-F and S-T

(9/49) A bioreactor is scaled up based on equal impeller tip speed. Consider the following parameters for small and bio bioreactors:

Parameters	Small bioreactors	Large bioreactors
Impeller speed	N_1	N_2
Diameter of impeller	D_1	D_2
Power consumption	P_1	P_2

Assuming geometric similarity and the bioreactors are operated in the turbulent regime, what will be P2 / P1?
(a) $(D_1 / D_2)^2$ (b) $(D_2 / D_1)^2$
(c) $(D_1 / D_2)^5$ (d) $(D_2 / D_1)^5$

Answers of MCQ

GATE 2016

(1/14) (c) 2/16 (c) Since $Pr = No = \dfrac{\mu}{k/C_p} = \dfrac{\textit{momentum diffusity}}{\textit{Thermal diffusity}}$

(3/17) (b) (4/18) (b) (5/21) (b)

(6/28) (b) (7/40) (a) 5% CO_2 enriched air is used to maintain medium pH at 7.3. A carbonate buffer, HCO_3^{2-} / H_2CO_3 is used to control pH. But bicarbonate is consumed by cells. CO_2 enriched air is used to balance carbonate equilibrium.

(8/41) (b) (9/49) (b), Scale-up criterion is based on a equal impeller tip speed, So $N_2 D_{I_2} = N_1 D_{I_1}$

Or, $\dfrac{N_2}{N_1} = \dfrac{D_{I_1}}{D_{I_2}}$ where D_I is the impeller diameter

Now the power equation is

$P = N_P \rho N^3 D_I^5$, So $P \propto N^3 D_I^5$

where N_P is the power number, which is constant, ρ is also constant

$$P_2 \big/ P_1 = \frac{N_2^3 \, D_{I_2}^5}{N_1^3 \, D_{I_2}^5} = \left(\frac{D_{I_1}}{D_{I_2}}\right)^2 \left(\frac{D_{I_2}^5}{D_I^5}\right) = \left(\frac{D_{I_2}}{D_{I_1}}\right)^2$$

MCQ

GATE 2015

(1/11) Which one of the following relations holds true for the specific growth rate, (μ) of a microorganism with the death phase?

(a) $\mu = 0$ (b) $\mu < 0$

(c) $\mu = \mu max$ (d) $0 < \mu < \mu max$

(2/18) Which one of the following graphs represents uncompetitive inhibition?

Live inhibition

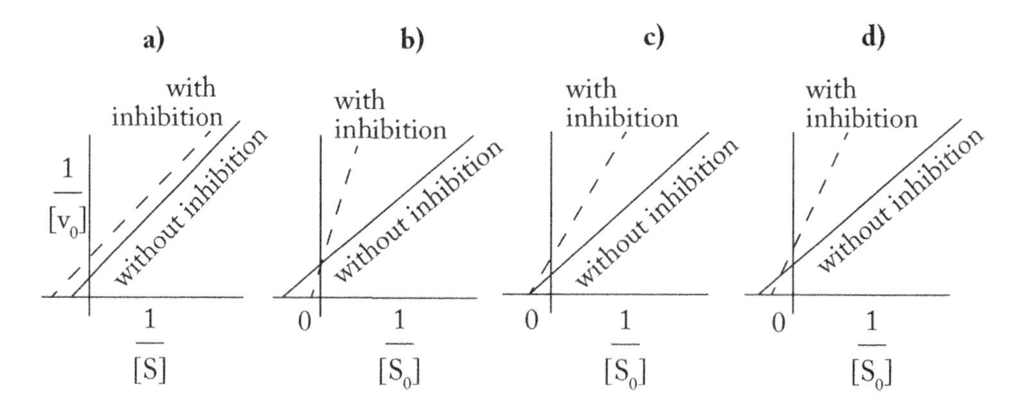

(3/21) Which one of the following is not a product of denitrification in *Pseudomonas*
(a) N_2 (b) N_2O (c) NO_2- (d) NH_4+

(4/23) Which one of the following is a second generation genetically engineered crop
(a) Bt brinjal (b) Roundip soyabean
(c) Golden rice (d) Bt rice

(5/27) In a fed-batch culture 200 gL^{-1} glucose solution is added at a flow rate of 60 Lh^{-1}. The initial culture volume (at quasi steady state) and initial cell concentration are 600L and 20gL^{-1} respectively. The yield coefficient ($Y_{X/S}$) is 0.5 g. cell / g cell substrate^{-1}. The cell concentration (gL^{-1}) at quasi steady state at t=8h is
(a) 40 (b) 52 (c) 60 (d) 68

(6/39) The diameters of a large and a small vessel are 1.62 m and 16.2 cm respectively. The vessels are geometrically similar and operated under similar volumetric agitated power input. The mixing time in the small vessel was found to be 15S. Determine the mixing time (in seconds) in the large vessel.
(a) 15 (b) 30 (c) 61 (d) 122

(7/46) Biomass is being produced in a continuous stirred tank bioreactor of 750L capacity. The sterile feed containing 8 gL^{-1} glucose as substrate was fed at a flow or 150Lh^{-1}. The microbial system follows Monod model will $\mu m = 0.4\ h^{-1}$, $Ks = 1.8\ gL^{-1}$ and $Y_{X/S} = 0.5$ g cell mass. gm substrate $^{-1}$. Determine the cell productivity ($gL^{-1}h^{-1}$) at steady state:
(a) 0.85 (b) 0.65 (c) 0.45 (d) 0.25

Answers

GATE 2015 (MCQ)

(1/11) (b) Since there is no growth of cells and the cells are consumed by the cells, during death phase, μ is less than zero

(2/18) (a) Uncompetitive inhbitions bind to ES Complex and have no affinity for the enzyme itself. The net effect is a reduction in both v_m and k_m.

(3/21) (d) Nitrogen-containing organic compounds are oxidized biologically to ammonium, which is further oxidized to nitrite and nitrate by nitrosomomas and nitrobacter respectively. The conversion of ammonium to nitrate by microbial action is called nitrification.

(4/23) (c) 2nd generation crops are the value added crops i. e. increased nutritional value, instead of insect resistance (BT) and herbilide tolerance (HT).

(5/27) (b) For quasi steady state

$$X_T = X_{T_0} + FY_{X/C}S_O t$$

$$V = V_O + Ft$$

$$= 600L + 50Lh^{-1} \times 8h = 1000L$$

$$X_{T_0} = 209L^{-1} \times 600L = 12000g$$

$$X_T = 12000 + 50Lh^{-1} \times 0.5 \times 200gL^{-1} \times 8h = 52000$$

$$X = \frac{52000}{1000} = 52gL^{-1}$$

(6/39) (c) $\frac{t_2}{t_1} = \left(\frac{D_2}{D_1}\right)^{1/6}\left(\frac{N_1}{N_2}\right)^{2/3}$

Where t_1 & t_2 are the mixing time for small reactor and large reactor respectively. 1 refers to the small reactor, 2 refers to the bigger one. D = diameter of vessel, N = rpm.

We know: $P_1/V_1 = P_2/V_2$ (constant power input)

Where P refers to agitation power and V refers to volume.

So $N_1^3D_1^2 = N_2^3D_2^2$

Or $N_1/N_2 = (D_2/D_1)^{2/3} = (10)^{2/3} = 4.6415$

Where $D_2/D_1 = 162$ cm / 16.2 cm = 10

$t_2 = 15\ (10)^{1/6}\ (4.6415)^{2/3} \approx 61$

(7/46) (b) : $D = F/V = \frac{150Lh^{-1}}{750L} = 0.2h^{-1}$

$$S = \frac{DK_S}{\mu m - D} = \frac{(0.2)1.5}{0.4 - 0.2} = 1.5gL^{-1}$$

$$X = Y_{X/S}(S_O - S) = 0.5(8 - 1.5) = 3.25gL^{-1}$$

$$\text{Cell productivity } (XD) = 3.25 \times .2 = 0.65gL^{-1}h^{-1}.$$

MCQ

GATE 2014

(1/6) The unit for specific substrate consumption rate in a growing culture is
(a) g/Lh (b) g/h (c) g/gh (d) gmoles/Lh

(2/8) In a batch culture of *Penicillium chrysogenum*, the maximum penicillin synthesis occurs during the

(a) lag phase (b) exponential phase
(c) stationary phase (d) death phase

(3/10) Which of the following is employed for the repeated use of enzymes in bioprocesses?

(a) polymerization (b) immobilization
(c) ligation (d) isomerization

(4/11) Since mammalian cells are sensitive to shear, scale-up of a mammalian cell process must consider, among other parameters, the following (given N = rotations/time, D = Diameter of impeller)

(a) πND (b) πN^2D (c) πND^2 (d) none of these

(5/25) The growth medium for mammalian cells contains serum. One of the major functions of serum is to stimulate cell growth and attachment. However, it must be filter sterilized to

(a) remove large proteins
(b) remove collagen only
(c) remove mycoplasma and microorganisms
(d) remove foaming agents

Answers

GATE 2014

(1/6) (c) Specific substrate consumption rate is defined as

$$\frac{1}{X}\frac{dS}{dt}\left(\frac{L}{g}\cdot\frac{g}{Lh}\right) \text{ or } \frac{g}{gh}$$

(2/8) (c) During stationary phase Penicillin Chrysegenum synthesizers penicillin at maximum concentration and penicillin is a non-growth associated product.

(3/10) (b) immobilization

Immobilized enzymes have stable and steady catalytic activity compared to free enzymes. Repeated use of free enzymes may have reduced activity due to structural change of the enzyme. This is minimized in the case of immobilized enzymes.

(4/11) (a) πND. One of the important scale-up criteria for mammalian cells is the constant impeller tip speed, πND.

(5/25) (c) remove mycoplasma and microorganisms.

Since serum in growth medium of animal cell culture is susceptible to heat, it can not be heat sterilized. So it is filter sterilized to remove unwanted mycoplasma and microorganisms.

MCQ

GATE 2013

(1/13) The catalytic efficiency for an enzyme is defined as

(a) k_{cat} (b) v_{max}/k_{cat} (c) k_{cat}/k_m (d) k_{cat}/v_{max}

(2/34) The activity of an enzyme was measured by varying the concentration of the substrate (S) in the presence of three different concentrations of inhibition (I), 0, 2, 4 mM. The double reciprocal plot given below suggests that the inhibitor (I) exhibits

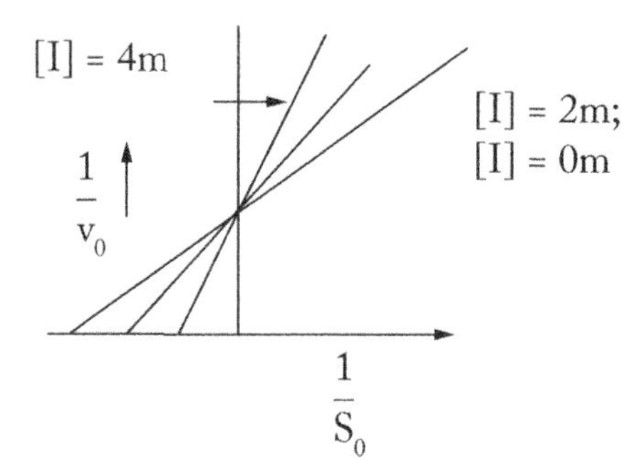

(a) substrate inhibition (b) uncompetitive inhibition
(c) mixed inhibition (d) competitive inhibition

Answers

GATE 2013

(1/13) (c) k_{cat}/k_m is often called specificity constant, used to compare the relative rates of an enzyme activity on competitive substrates. In the mechanism of enzymatic reaction,

$$E + S \underset{k_{-1}}{\overset{k_1}{\rightleftharpoons}} E^S \xrightarrow{k_2} E + P$$

k_2 is also known as $k_{cat}/k_m = k_2 = v_{max}/[E_0]$

(2/34) (d) Competitive inhibition. The net effect of competitive inhibition is to increase the value of km, reducing the rate of enzymatic reaction.

With increase in the concentration of the inhibitor (I), the rate of reaction decrease and the slope of the plot, (k_m/v_{max}) increases

$$K_{m\,app} = K_m\left(1 + \frac{[I]}{K_I}\right)$$

MCQ

GATE 2012

(1/17) The activity of an enzyme is expressed in International Units (IU). However, the S. I. unit for enzyme activity is katal. One katal is

(a) 1.66 × 104 IU (b) 60 IU

(c) 6 × 107 IU (d) 106 IU

(2/18) Identify the statement that is NOT applicable to an enzyme catalysed reaction:

(a) Enzyme catalysis involves propinquity effects

(b) The binding of a substrate to the active site causes a strain in the substrate

(c) Enzymes do not accelerate the rate of reverse reaction

(d) Enzyme catalysis involves acid-base chemistry

(3/46) β-galactosidase bound to DEAE-Cellulose is used to hydrolyze lactose to glucose and galactose in a plug flow bioreactor with a packed bed of volume 100 litres and a voidage (∈) of 0.55. An enzymatic reaction is described by the following rate expression

$$v = \frac{v_m S}{K_m + S + \frac{S^2}{K_S}}$$

which one of the following curves represents this experiment?

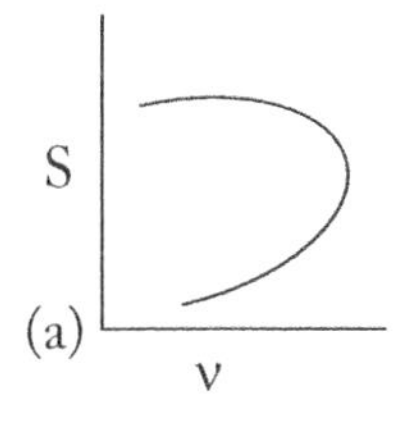

(a)

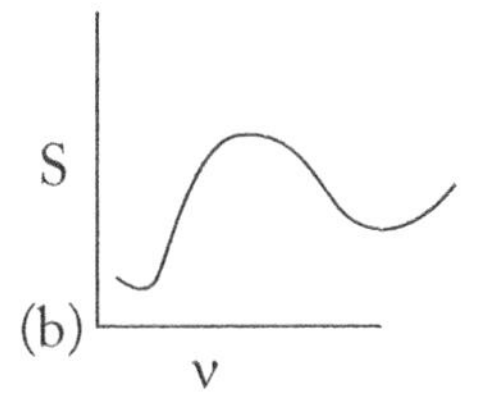

(b)

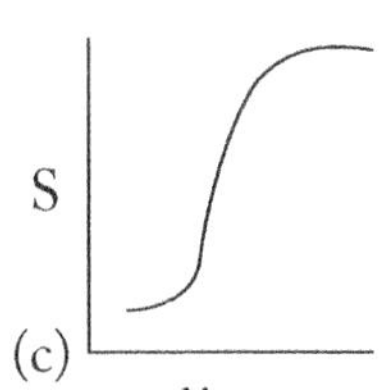

(c)

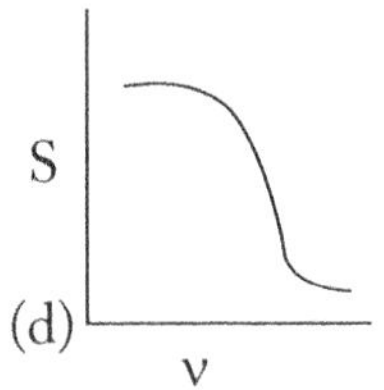

(d)

(4/54) The K'_M and v'_{max} for the immobilized enzymes are $0.72 gl^{-1}$ and $18 gl^{-1}h^{-1}$ respectively. The lactose concentration in the feed stream is $20 gl^{-1}$ and a fractional conversion of 0.90 is desired.

Diffusion limitation may be ignored.

The residence time required for the steady state reactor operation will be

(a) 0.1h (b) 0.4h (c) 1.0h (d) 1.1h

(5/55) The feed flow rate required for the above bioconversion

(a) 50lh^{-1} (b) 55 lh^{-1} (c) 137lh^{-1} (d) 550 lh^{-1}

Answers

GATE 2012

(1/17) (c) 6×10^7 IU

One Katal is 6×10^7 IU in international unit to express the activity of enzyme

(2/18) (c)

That Enzymes do not accelerate the rate of reverse reaction is not true. Enzymes accelerate both the rates of forward and backward reactions. The velocity of back reaction may be expressed as

$$v = k_{-2}[E][P] \text{ where the reactions are } ES \underset{k_{-2}}{\overset{k_2}{\rightleftharpoons}} E + P$$

(3/46) (a)

The equation given is for the rate with substrate inhibition,

$$v = \frac{v_m S}{K_m + S + S^2/K_S}$$

Upto certain values of S, the rate increases. Above that of S, the rate is reduced due to substrate inhibition

(4/54) (d) 1.1h

The packed bed design equation is

$$K_M ln\frac{S_O}{S} + S_O - S = v_{max}\tau$$

Data given:

$$K_M = 0.72 gl^{-1}$$

$$v_{max} = 18 gl^{-1}h^{-1}$$

$$S_O = 20 gl^{-1}$$

$$S = 0.1 \times 20 = 2 gl^{-1}$$

Substitute the values

$$0.72\, ln\frac{20}{2} + 20 - 2 = 18\tau$$

Or $\tau \approx 1.1 hr$

The feed flow rate for the above conversion should be

$$F = \frac{V}{\tau} = \frac{100}{1.1} = 91 Lh^{-1}$$

None of the four data match the calculated value. However, considering flow through the void space

$$F = \frac{V\epsilon}{\tau} = \frac{100 \times 0.55}{1.1} = 50 Lh^{-1}$$

So the correct answer is 9a) 50 Lh^{-1}

MCQ

GATE - 2011

(1/3) Apoptosis is characterized by
(a) necrosis
(b) programmed dell death
(c) membrane leaky syndrome
(d) cell-cycle arrest process

(2/5) The product commercially produced by animal cell culture is
(a) insulin
(b) tissue plasminogen activator
(c) interferon
(d) hepatitis B vaccine

(3/12) In balanced growth phase of a cell
P. all the compounds of a cell grow at the same rate
Q. specific growth determined by cell number or cell mass would be the same
R. the growth rate is independent of substrate
S. the growth rate decreases with decreasing substrate concentration
(a) P, Q and S only
(b) Q, R and S only
(c) P, Q and R only
(d) P only

(4/15) Substrate consumption in lag phase of microbial growth is primarily used for
P. turn- over of the cell material
Q. maintenance of the intra cellular pH
R. motility
S. increase in cell number
(a) P, Q and S only
(b) Q, R and S only
(c) P, Q and R only
(d) S only

(5/16) Washout (as defined by D = μ_{max}) of a continuous stirred tank fermenter is characterized by (X=biomass, S=substrate concentration in the bioreactor, So = substrate concentration in the feed, P = product concentration in the bioreactor)
(a) X = 0, S = 0, P = 0
(b) X = 0, S = So, P = 0
(c) X = 0, S < 0, P = 0
(d) X < 0, S < 0, P < 0

(6/27) Match the microbial growth characteristics in Group I with the corresponding features in Group II

Group I

P. Growth associated product formation

Q. Non-growth associated product formation

R. Product inhibition

S. Substrate inhibition

Group II

1. Specific growth rate decreases with increasing product formation
2. Specific product formation rate is constant
3. Specific product formation rate is proportional to specific growth rate
4. Specific growth rate decreases with increasing substrate concentration

	P	Q	R	S
(a)	1	2	4	3
(b)	3	2	1	4
(c)	2	1	3	4
(d)	2	3	4	1

(7/33) In a well aerated and agitated microbial culture, the supply of oxygen is equal to demand (uptake) of the growing culture. The k_{La} for such a system will be (k_{La} = volumetric mass transfer coefficient, C* = dissolved oxygen concentration in liquid in equilibrium with gaseous oxygen, C = instantaneous value of dissolved oxygen centration) 'r'= specific oxygen uptake rate per unit weight of cells, X = dry weight of the cells per unit volume)

(a) (rx) / (C * - C) (b) (r) / X (C* - C)

(c) (C* - C) / (rx) (d) (X) / r (C* - C)

(8/34) Structured William's model

P. can describe the changes in intracellular components of the cell during growth

Q. can not describe the death phase of the cells

R. can describe the variation of size of cells in different phases of growth

S. can not describe the lag period of growth

Which one of the following is correct?

(a) P, Q and S only (b) P, Q, and R only

(c) Q, R and S only (d) P, R and S only

(9/44) Maximum specific growth rate (μ_{max}) of a microorganism is calculated by taking the(ln = loge, x = biomass, t = time)

(a) slope of ln X vs t of the growth cycle

(b) slope of ln X vs t during the exponential growth phase

(c) slope of X vs t

(d) slope of X vs t during the exponential phase of growth

Common data for question:
A microorganism grows in a continuous chemostat culture of $60m^3$ working volume with sucrose as the growth limiting nutrient at dilution rate, $D = 0.55h^{-1}$. The steady state biomass concentration is 4.5 kg dry biomass m^{-3} and the residual sucrose concentration is 2.0 kgm^{-3}. The sucrose concentration in the incoming feed medium is 10.0 kgm^{-3}.

(10/50) What would be the yield, YX/S (Kg biomass/Kg substrate)
(a) 0.562 (b) 0.462 (c) 0.362 (d) 0.162

(11/51) What would be the sucrose concentration in the input feed for the output to be 45 kg biomass / hr
(a) 3.225 kgm^{-3} (b) 4.425 kgm^{-3} (c) 5.115 kg/m^{-3} (d) 6.525 kgm^{-3}

Answers

GATE 2011

(1/3) (b) programmed cell death
Apoptosis is a form of programmed cell death. Gross shear damage occurs when cell lyse, the process is called necrosis. More subtle effects of shear include changes in physiology and induce apoptosis or programmed cell death.

(2/5) (b) tissue plasminogen activator
Tissue plasminogen activator (t-PA), a naturally occurring protein, catalyzes the conversion of the inactive proenzyme plasminogen. This enzyme is found in blood and tissues from various organs. t-PA is used to treat people, having a stroke caused by blood clot.

(3/12) (a) P, Q and S only
In the balanced growth conditions, all component of a cell (protein, DNA, RNA, lipids etc) will grow at the same rate. This condition is true in CSTR system. (P) Specific growth rate (μ) based on cell number or cell mass would be the same. (Q) the growth rate decreases with decresing substrate concentration.
$dX/dt = \mu_{max} \times S/(K_S + S)$

(4/15) (c) P, Q and R
In lag phase, cells cannot divide. Synthesis of enzymes and other molecules formation occur in this phase. Substrate consumed by microbes in lag phase cause motility, maintenance of intra cellular pH, production of materials required for growth in the upcoming phases.

(5/16) (b) X = 0, S= So and P = 0
When dilution rate, D is equal to μmax. Washout of cells occur and the substrate concentration reaches the initial value and no product is formed in this washout condition

(6/27) (b) P-3, Q-2, R-1, S-4

For growth associated product formation, specific product formation rate is proportional to specific growth rate, 1/xdp/dt= qp =αμ

For nongrowth associated product formation, $\frac{1}{X}\frac{dP}{dt} = q_P = \beta$ where β is a constant.

Product induced inhibition can be seen at increase in product concentration with decrease in μ. This phenomena is vice versa with substrate inhibiton.

(7/33) (a) rx / (C*-C)

The supply of oxygen or O_2 mass transfer rate (OTR) – k_{La} (C_L*-C_L) which is equal to oxygen uptake rate per dry wits of cells = rx

k_{La} (C_L* - C_L) = rx

or k_{La} = rx / (C_L* - C_L) where C_L* is the solubility of O_2 in water

(8/34) (b) P, Q, and R only

Structured William model can describe the changes in intracellular components of cell during growth (P). The model cannot describe the cell behaviour during the death phase (Q) where cells are decomposed. But the model can explain the variation of size of cells in different phases of growth (R).

(9/44) (b) Slope of ln X vs t during the exponential growth phase.

X = X0 exp (μmax t)

The cell growth in the exponential phase is described by the equation. Taking logarithm ln X = ln Xo + μ_{max} t. The slope of ln X vs t will give μ_{max}.

(10/50) (a) 0.562

We know,

$$X = Y_{X/S}\left(S_O - S\right)$$

$$\text{Or } Y_{X/S} = \frac{X}{S_O - S} = \frac{4.5}{10-2} = 0.562$$

(11/51) (b) 4.425 ks/m^3

Given $V = 80m^3$, $D = 0.55h^{-1}$

$$D = \frac{F}{V} \text{ So } F = VD = 80m^3 \times .55h^{-1} = 33m^3 / hr$$

$$X' = 45\frac{Kg}{hr} biomass, X = \frac{X'}{F} = 45\frac{Kg}{hr} \times \frac{1}{33}\frac{hr}{m^3}$$

$$= 1.3636\frac{Kg}{m^3}$$

$$X = Y_{X/S}\left(S_O - S\right)$$

$X_{X/S} = 0.562$, $S = 2Kg/m^3$
1.3636 = 0.562 (S_o - 2)
Solving So = 4.426 Kg/m^3

MCQ

GATE 2010

(1/1) Hybridoma technology is used to produce
(a) monoclonal antibodies
(b) polyclonal antibodies
(c) both monoclonal and polyclonal antibodies
(δ) β Cells

(2/12) Interferon-β is produced by
(a) bacteria infected cells
(b) Virus infected cells
(c) both virus and bacteria infected cells
(d) fungi infected cells

(3/19) In a chemostat operating under steady state, a bacterial culture can be grown at dilution rate higher than maximum growth rate by
(a) partial cell recycling
(b) using sub-optimal temperature
(c) pH cycling
(d) Substrate feed rate recycling

(4/22) The degree of inhibition for enzyme catalysed reaction at a particular inhibitor concentration is independent of So.
The inhibitor follows
(a) competitive inhibition
(b) mixed inhibition
(c) uncompetitive inhibition
(d) non-competitive inhibition

(5/27) An immobilized enzyme being used in a continuous plug flow reactor exhibits an effectiveness factor (η) of 1.2. The value of η being greater than 1.0 could be apparently due to
(a) substrate inhibited with internal pore diffusion
(b) external pore diffusion limitation
(c) sigmoidal kinetics
(d) unstability of the enzyme

(6/28) A roller bottle culture vessel perfectly cylindrical in shape having inner radius (r) = 10 cm and length (l) = 20 cm was fitted with a spiral film of length (L) = 30 cm and width (w) = 20 cm. If the film can support 10^6 anchorage dependent

cells per cm^2, the increase in the surface area after fitting the spiral film and the additional number of cells that can be grown respectively are
(a) 1200 cm^2 and 12×10^7 cells (b) 600 cm^2 and 6×10^7 cells
(c) 600 cm^2 and 8300 cells (d) 1200 cm^2 and 8300 cells

(7/45) Thermal death of microorganisms in the liquid medium follows first order kinetics. If the initial cell concentration in the fermentation medium is 10^8 cells/ml and the final contamination level is 10^{-3} cells, for how long should $1m^3$ medium be treated at 120° (thermal deactivation rate constant = 0.23/min) to achieve acceptable load?
(a) 48 min (b) 11 min (c) 110 min (d) 20 min

Common data for questions (8/48 & 8/49)
A culture of Rhizobium is grown in a chemostat ($100m^3$ bioreactor). The feed contains 12g/L sucrose for the organism is (X_0 = 0.2 g/L) and $\mu_m = 0.3h^{-1}$

(8/48) The flow rate required to steady state concentration of sucrose as 1.5 g/L in the bioreactor will be
(a) $15m^3h^{-1}$ (b) $26m^3\ h^{-1}$ (c) $2.0\ m^3\ h^{-1}$ (d) $150\ m^3h^{-1}$

(9/49) If $Y_{X/S}$ = 0.04g/g for the above culture and steady state cell concentration in the bioreactor is $4gL^{-1}$ the resulting substrate concentration will be
(a) 2g/L (b) 8g/L (c) 4g/L (d) 9g/L

Answers

GATE 2010

(1/1) (a) monoclonal antibodies
Hybridoma technology is used to combine the properties of the specific antigen activity of β cells with the immortality property of Myeloma M (Cancerous) cells. Antibodies produced are all of single specificity and are called monoclonal in nature. Mab is used for determining nature of a primary tumour.

(2/12) (c) both virus and bacteria infected cells
Interferon - β is produced mainly by fibroblasts and some epithelial cell types. The synthesis can be induced by common inducers like interferons, viruses, double stranded RNA, bacteria and also by some cytokines.

(3/19) (a) partial cell recycling
For the chemostat operating at the steady state with sterile feed, D = μ.
When D = μ_{max}, cell washout occurs at the condition with cell recycle, μ can be given as
$\mu = (1+\alpha - \alpha C)D$.

where α is the recycle ratio based on volumetric flow rates, C is the ratio of cells in the recycle stream and that of exit stream. Since C > 1, α (1-C) < 0, then μ < D. So a chemostat can be operated at D greater than μ, when cell recycle is used.

(4/22) (d) Non-competitive inhibition
In non-competitive inhibition, inhibitors bind on sites other than the active site of enzyme forming an dead-end complex, regardless of whether a substrate molecule is bound or not. The net effect is the reduction of v_{max} and k_M remains constant. High substrate concentration would not overcome non-competitive inhibition.

(5/27) (c) Sigmoidal Kinetics
Sigmoidal kinetics are shown by allosteric enzymes and given by Hills equation,
(- ds/dt =vmax(S)^n/[KM" + (S)^n]
where n is called the Hills' constant. When n < 1, there is negative effectiveness factor. When n > 1, the effectiveness factor is greater than unity

(6/28) (b) 600 cm^2 and 6×10^7 cells
Inner radius (r) = 10 cm, Length (L) = 20 m
Length of spiral film (L) = 30 cm, Width (w) = 20 cm
The increase in surface area after filling the spiral film = 30 cm × 20 cm= 600 cm^2
Then the film will add = 10^5 cells/cm^2 × 600 cm^2 = 6×10^7 cells

(7/45) (c) 110 min
No = 10^8 cells/ml × 10^6 ml ($1m^3$) = 10^{14} cells
After sterilization, N = $10^{-3} \times 10^6 = 10^3$ cells

$$ln\frac{N}{No} = -kdt, kd = 0.23min^{-1}$$

$$\text{Or } \frac{ln\frac{No}{N}}{kd} = t$$

$$\text{Or } t = \frac{ln\frac{10^{14}}{10^3}}{0.23} = \frac{25.328}{0.23} = 110min$$

(8/48) (b) $26m^3h^{-1}$
We know, for sterile feed

$$D = \frac{F}{V} = \mu = \frac{\mu mS}{K_S + S}$$ = (0.3*) 1.5 /(0.2 + 1.5) =0.2648 hr-1

$K_S = 0.2g / L$, F= Vμ = (100)X 0.2648 =26.5m^3.

(9/49) (a) 2g/L
We know, X = $Y_{X/S}$(So –S) ....A
Given: X = 4g/L, $Y_{X/S}$ = 0.4g/g

So = 12 g/L
Putting the values of X, $Y_{X/S}$ & So in equation A
$4 = 0.4 (S_0 - S) = 0.4(12 - S)$
Solving for S, S = 2g/L

MCQ

GATE 2009

(1/6) Culture vessels in which physical, physiochemical and physiological conditions as well as cell concentration are kept constant is known as
(a) Cell Concentration (b) Biostat
(c) Batch bioreactor (d) Incubator

(2/15) Which of the following statements are true about bioreactors?
P. Continuous bioreactors provide less degree of control and uniform product quality than batch bioreactors
Q. Batch bioreactors are ideally suited for reaction with substrate inhibition
R. Choice of a bioreactor is dictated by kinetic consideration
S. Fed batch bioreactors are also called semibatch bioreactors
(a) P, Q (b) Q, S (c) R, S (d) P, R

(3/19) Match the products in group 1. Their producer organism in group 2

Group1	Group 2
P. Ethanol from glucose	1. Aspergillus niger
Q. Probiotics	2. Leuconostic mesenteroids
R. Citric acid	3. Saccharomyces Cerevisiae
S. Saur Krant	4. Bifido bacterium

(a) P-1, Q-3, R-2, S-4 (b) P-3, Q-4, R-1, S-2
(c) P-3, Q-4, R-2, S-1 (d) P-1, Q-4, R-3, S-2

Answers

GATE 2009

(1/6) (b) Biostat
Biostat maintains constant for all conditions mentioned in the question

(2/15) (c) R, S
Choice of Bioreactor depends on the kinetic parameters or the system (R). Fed-batch reactors are semibatch reactors because there is intermittent feeding and intermittent discharging

(3/19) (b) P-3, Q-4, R-1, S-2

Produced

Ethanol from glucose → Saccharomyces Cerevisiae (P-3)

From

Probiotics → Bifido bacterium (Q-4)
Citric Acid → Aspergilus niger (R-1)
Sauer krant → Leuconostoc mesenteroides

MCQ

GATE 2008

(1/1) Diauxic pattern of biomass growth is associated with
(P) multiple lag phases
(Q) Sequential utilization of multiple substrates
(R) Simultaneous utilization of multiple substrates
(S) Absence of lag phase
(a) P, R (b) P, Q (c) R, S (d) Q, S

(2/5) Which of the following techniques is best suited for immobilizing an affinity ligand
(a) Physical adsorption (b) Gel entrapment
(c) Cross linking with a polymer (d) Covalent linkage to a spacer arm

(3/13) Some living cells (e. g. plant cells) have the capacity to give rise to whole organism. The term used to describe the property is
(a) morphogenesis (b) androgenesis
(c) totipotency (d) organogenesis

(4/16) Match the bioreactor components in group 1 with the most appropriate function given in group 2

Group 1	Group 2
P. Marine type impeller	1. Recirculation of medium
Q. Draft tube	2. Aeration of medium
R. Diaphragm valve	3. Animal cell culture
S. Sparger	4. Sterile operation

	P	Q	R	S
(a)	4	2	1	3
(b)	3	1	4	2
(c)	3	4	2	1
(d)	2	1	4	3

(5/17) Evaluate the Michaelis constant for the following lipase catalysed trans-esterification reaction for the production of biodiesel

$$vegetable + lipase \underset{k_2}{\overset{k_1}{\rightleftharpoons}} oil - lippase\ complex \xrightarrow{k} Biodesel + glycerol$$

where $k_1 = 3 \times 10^8\ M^{-1}S^{-1}$
$k_2 = 2 \times 10^3 S^{-1}$, $k = 4 \times 10^3\ S^{-1}$
(a) 4.2×10^{-3} M (b) 14.0×10^{-4} M
(c) 6.4×10^{-6} M (d) 1.4×10^{-4} M

(6/18) In a chemostat, evaluate the dilution rate at the cell wash condition by applying Monod model with the given set of data:
$\mu max = 1h^{-1}$, $Y_{X/S} = 0.5$ g/g, $K_S = 0.2$ g/L
So = 10 g/L
(a) 1.00 hr^{-1} (b) $0.49h^{-1}$ (c) $0.98h^{-1}$ (d) $1.02h^{-1}$

(7/19) Match the product in group 1 with their producer organisms given in group 2

Group 1	Group 2
P. Ethanol	1. Streptomyces orientalis
Q. L-lysin	2. Sacchamyes cerevisiae
R. Biopesticide	3. Corynobacterium glutamicum
S. Vancomycin	4. Bacillus thuringiensis

	P	Q	R	S
(a)	2	3	4	1
(b)	3	4	1	2
(c)	4	1	2	3
(d)	2	1	4	3

(8/21) A bacterial culture with an approximate biomass composition of $CH_{1.5}O_{0.5}NO_2$ is grown aerobically on a defined medium containing glucose as a sole carbon source and ammonia being the nitrogen source. In this fermentation, biomass is formed with a yield coefficient of 0.35 gm dry cell weight per gm of glucose and acetate is a product with a yield coefficient of 0.1 gm acetate per gm of glucose. The respiratory coefficient for the above culture will be
(a) 0.90 (b) 0.95 (c) 1.00 (d) 1.05

(9/22) A bacterial culture having specific oxygen uptake rate of 5 m mol O_2 (g. DCW)$^{-1}$ hr^{-1} is being grown aerobically in a fed batch bioreactor. The maximum value of the oxygen transfer coefficient is $0.18S^{-1}$ for the starred tank bioreactor and the critical dissolved oxygen concentration is 20% of the saturation concentration (8 mg/L).

The maximum density to which the cells can be grown in the fed-batch process without the growth being limited by the oxygen transfer, is approximately

(a) 14 g/L (b) 28 g/L (c) 32 g/L (d) 65 g/L

Common data for questions: 10/23 and 11/24

An enzyme (24000Da) undergoes first order deactivating kinetics while catalyzing a reaction according to Michaelis-Mentern Kinetics (k_M = 10-4M). The enzyme has turn over number of 10^4 moles substrate / min-(molecule enzyme) and a deactivation constant (k_d) of 1 min^{-1} at the reaction condition. The reaction mixture initially contains 0.6 mg/L of active enzyme and 0.02M of the substrate.

(10/23) The time required to convert 10% of the substrate will be approximately.

(a) 16 min (b) 24 min (c) 32 min (d) 8 min

(11/24) The maximum possible conversion for the enzymatic reaction will be

(a) 100% (b) 50% (c) 25% (d) 12.5%

Statement for linked answer questions

A double reciprocal plot was created from the specific growth rate and the limiting substrate concentration data obtained from a chemostat experiment. A linear regression gave values of 1.25 hr and 100 mg-hr-l^{-1} for the intercept and slope, respectively.

(12/27) The respective values of the Monod Kinetic constants, μ_m/hr^{-1} and K_S(mg/L) are as follows:

(a) 0.08, 8 (b) 0.8, 0.8 (c) 0.8, 80 (d) 8, 8

(13/28) The same culture (with μ_m and K_S values as computed above) is cultivated in a 10-litre chemostat being operated with a 50 ml/min of substrate. Assuming an overall yield coefficient of 0.3 g dry wt of cells/g-substrate, the respective values of the outlet biomass and substrate are

(a) 15 g/L, 48 mg/L (b) 15 g/L, 0.48 g/L
(c) 48 g/L, 15 g/L (d) 4.8 g/L, 4.8 g/L

Answers

GATE 2008

(1/1) (b) P, Q

With two substrates in a batch reactor with microbial cells, both substrates are not consumed simultaneously. The preferred substrate is consumed first and then the less preferred substrate leading to two lag phase (P). The phenomenon is called Diauxic growth.

(2/5) (d) Covalent linkage to a spacer arm.
In affinity chromatography for the purification of a target molecular, a ligand having affinity for the target molecules is covalently attached to an insoluble support for capturing the target molecules from feed solution containing many undesired components. This is carried out in an affinity column. The attached molecules are then eluted from the column by buffer solutions.

(3/13) (c) totipotency
Totipotency is the capacity to regenerate whole plants from undifferentiated cells under aseptic environmental conditions. This capacity is essential to plants micropropagation and is often associated with secondary metabolite formation.

(4/16) (b) P-3, Q-1, R-4, S-2
In animal cell culture, main type impeller is used for less sheart in the reactor (P-3)
In air lift fermenter, draft tube is used for recirculation of medium (Q-1)
Diaphragm valve is used for sterile operation in a bio-reactor (R-4)
Sparger is used for aeration of medium for all aerobic fermentation (S-2)

(5/17) (d) 1.4×10^{-4}M

$$K_m = \frac{k_{-1} + k_2}{k_1}$$

$$= \frac{4\times 10^4 s^{f} + 2\times 10^3 s^{-1}}{3\times 10^8 M^{-1} S^{-1}} = \frac{42\times 10^3}{3\times 10^8} = 14\times 10^{-5} M$$

$$= 1.4\times 10^{-4} M$$

(6/18) (c) $0.98h^{-1}$
At washout, X = 0, = $Y_{X/S}$ (So-S) = 0
$Y_{X/S} \neq 0$, S = So

$$D = \mu = \frac{\mu_{max} S_O}{K_S + S_O} = \frac{1\times 10}{0.2 + 10} = 0.98 hr^{-1}$$

(7/19) (a) P-2, Q-3, R-4, S-1
Ethanol (P) – Saccharomyces Cerevisiae (2)
L-lysin (Q) – Loryne bacterium glutamicum (3)
Biopesticide (R) - Bacillus thuringiensis
Vancomycin (S) – Stepto mylesorientalis

(8/21) (c) 1.00

$$Y_{X/S} = 0.35 g / g, Y_{P/S} = 0.1 \frac{g.acasate}{g.gluem}$$

Respiratory Coefficient (Quotient)

$$q_{O_2} = \frac{Moles\ of\ CO_2\ formed}{Moles\ of\ O_2\ consumed} = \frac{5}{4.95} = 1.01\ (f/d)$$

$$aC_6H_{12}O_6 + bNH_3 + dO_2 \rightarrow CH_{1.8}O_{0.5}N_{0.2} + eH_{20} + fCO_2$$

Component balance:

C: 6a = 1 + f

H: 12a + 3b = 1.8 + 2e

O: 6a + 2d = 0.5 + e + 2f

N: b = 0.2

12a + 0.6 = 1.8 + 2e

Or 12a = 1.2 + 2e

6a = 1 + f

if a = 1, f = 5

e = 5.4, d = 4.95, b = 0.2, a = 1 $q_{o_2} = \frac{f}{d}$

(9/22) (b) 26 g/L

$$k_{La}\left(C_L^* - C_L\right) = q_{o_2}^0 X$$

$$\left(3600 \times .18\right) hr^{-1}\left(\frac{8}{32}\frac{m.moL}{L} - 0.2 \times \frac{8}{32}\frac{m.moL}{L}\right)$$

$$= 5\frac{m\ moL}{\left(g.DCW\right)\left(hr\right)} \times \left(X\right)\frac{g.DCW}{L}$$

$$\text{or } X = \frac{648 \times 0.8 \times 0.25}{5} = 25.92 \approx 26 g / L$$

(10/23) $v_{max} = 10^4 \times \frac{0.6 \times 10^{-3}}{24000} = 2.5^{-r}10^{-4}\frac{moles\ S}{(l)(m)}$

For deactivated enzyme

$$K_M ln\frac{S_O}{S} + S_O - S = -\frac{v_{max}}{k_d}\left(e^{-kd'_1} - 1\right)$$

$$\text{Or } 10^{-4}\, ln\frac{.02}{.018} + 0.02 - .018 = -\frac{2.5 \times 10^{-4}}{0.1}\left(e^{-0.1'-t}\right)$$

$$\text{Or } 0.8042 = -e^{-0.1'+t}$$

Solving for t = 1.63/0.1 = 16.3 min

(11.24) For high conversion, the rate is 1st order with respect to S

$$ln\frac{S_O}{S} = -\frac{v_{max}}{K_M k_d}\left(e^{-kd'} - 1\right)$$

Putting the value, $S = 4.5 \times 10^{-5}, \delta = \frac{0.02 - 4.5 \times 1.5}{.02} \approx 1$ t = 16mm

So the conversion. is 100%.

(12/27) (c) 0.8, 80

$$\mu = \frac{\mu_{max} S}{K_s + S}, \quad or \quad \frac{1}{\mu} = \frac{K_s S}{\mu_{max}} + \frac{1}{\mu_{max}}$$

$$\mu_{max} = \frac{1}{1.25} = 0.8 hr^{-1}$$

$$Slope = \frac{K_S}{\mu_{max}} = 100 \frac{mg - hr}{L}$$

$$K_S = 100 \times .80 = 80 \frac{mg}{L}$$

(13/28) (a) 15 g/L (x), 48 mg/L

$$V = 10L, F = \frac{50\,ml}{min} = 50 \times 60 \frac{ml}{hr} = 3L / hr$$

$$D = F / V = \frac{3}{10} = 0.3 hr^{-1}$$

From the previous problem

$\mu m = 0.8\,\text{hr}^{-1}$, $K_s = 80\,mg / L = .08\,g / L$

$$S = \frac{DK_S}{\mu m - D} = \frac{(0.3)(.08)}{0.8 - 3} = .048\,g / L = 48\,mg / L$$

X = Yx/s(S0- S)

$$= 0.3(50 - .048) = 14.98 \approx 15\,g / L.$$

MCQ

GATE 2007

(1/1) The specific growth rate (μ) of a microorganism in death phase is

(a) 0 (zero) (b) μmax (c) less than zero (d) greater than zero

(2/2) Which of the following reagents is used for harvesting anchorage dependent animal cells from culture vessels

(a) Trypsin/Collagenase (b) Trypsin / Collagen

(c) Collagen / Fibronectin (d) DMSO

(3/10) Match items in group 1 with correct examples from those in group 2

Group 1	Group 2
P. Catabolic product	1. Griseofulvin
Q. Bioconversion	2. Bakers yeast
R. Biosynthetic product	3. 6-Aminopenicillanic acid
S. Cell mass	4. Ethanol

(a) P-4, Q-3, R-2, S-1 (b) P-3, Q-4, R-1, S-2

(c) P-4, Q-3, R-1, S-2 (d) P-1, Q-4, R-3, S-2

(4/11) A biomedical solution to reduce oxides of nitrogen and carbon in flue gases is to integrate flue gas emission to

(a) micro-algal culture (b) fish culture
(c) mushroom culture (d) sericulture

(5/18) Match each parameter in group-1 with the appropriate measuring device in group-2

Group 1	Group 2
P. Pressure	1. Photometer
Q. Foam	2. Rotameter
R. Turbidity	3. Diaphragm gauge
S. Flow rate	4. Rubber Sheathed electrode

(a) P-3, Q-4, R-1, S-2 (b) P-1, Q-3, R-2, S-4
(c) P-4, Q-1, R-3, S-3 (d) P-1, Q-2, R-3, S-4

(6/19) Main functions of baffles in a bioreactor are:

P. to prevent vortex
Q. to increase aeration
R. to reduce interfacial area of oxygen transfer
S. to reduce aeration rate

(a) P, Q (b) Q, R (c) R, S (d) P, S

(7/20) How many kilograms of ethanol is produced from 1 kilogram of glucose in ethanol fermentation

(a) 2.00 (b) 0.20 (c) 0.51 (d) 0.05

Linked answer questions:
Q 8/25 to Q. 9/26
In a fed batch reactor glucose solution is added with a flow rate of $2m^3$/day. The initial volume the culture is $6m^3$.

(8/25) The volume of culture at this end of the 2nd day neglect loss due to vaporization

(a) $6m^3$ (b) $8m^3$ (c) $10m^3$ (d) $12m^3$

(9/26) What would be dilution rate of the system at the end of the second day

(a) 2.00 (b) 0.20 (c) 0.02 (d) 0.01

Answers

GATE 2007

(1/1) (c) less than zero

Specific growth rate is defined as the increase in cell per unit time per unit cell mass (μ= 1/x dx/dt).

In death phase no cellular growth is observed, but the cell population decreases as the cells consume themselves due to exhaustion of carbon source. So the rate becomes negative and so μ becomes less than zero

(2/2) (a) Trypsin / Collagenase
Trypsin, collagenase or pronase usually in combination with EDTA causes cells to detach from the growth surface, thus harvesting of anchorage dependent cells from the culture vessel is facilitated.

(3/10) (c) P-4, Q-3, R-1, S-2
Catabolic product is ethanol, 6-Amino penicillanic acid is produced by bioconversion biosynthetic product is Griseo fulvin. Baker's yeast is nothing but cell mass produced by yeast.

(4/11) (a) micro-algal culture
Flue gas containing various amounts of CO_2 can be fed directly to microalgal culture which utilize CO_2 to produce sugar in presence of sunlight, sugar is the carbon source for the growth of algae.
Other combustion products such NO_X or SO_X can be used as nutrients for microalgae. Microalgae culture may yield high value commercial products, which offset the capital and operation cost of the process.

(5/18) (a) P-3, Q-4, R-1, S-2
Pressure is measured by diaphragm gauze (P-3)
Foam is eliminated by rubber sheathed electrode (Q-4)
Turbidity is measured by photometer (R-1)
Flow rate is measured by rotameter (S-2)

(6/17) (a) P, Q
Baffles in a bioreactor are used to prevent vortex. Consequently they also improve mixing and increase aeration

(7/20) (c) $10m^3$
$V = Vo + Ft = 6m^3 + 2m^3/day \times 2\ days = 10m^3$

(9/26) (b) 0.20
Dilution roe at the end of the 2nd day
$D = F/V = 2m^3/day\ /\ 10m^3 = 0.2\ day^{-1}$

MCQ

GATE 2006

(1/3) To promote attachment and spreading of anchorage dependent animal cells, the surface of the cultural vessel needs to be coated with
(a) trypsin (b) collagen (c) pronase (d) polyglycol

(2/5) Which of the following statements is CORRECT about immobilized plant cultures?

(a) It is possible to use high cell densities

(b) Cells remain active for long periods

(c) Cells products or inhibitors can be removed easily

(d) It provides low shear resistance to cells

(3/7) Identify the natural plant growth regulators from the following list

(P) Zeatin

(Q) Benzylammino purine (BAP)

(R) Indole-3 acetic acid (IAA)

(S) 2, 4 dichlorophenoxy acetic acid

(4/10) During cultivation of microorganisms in fermentation, various parameters are controlled by appropriate sensor (probe). Match each probe in group 1 with appropriate response mechanism in group 2?

Group 1 (Proble)	Group 2 (responses)
P. Thermistor	1. Activation of acid alkali pump
Q. Oxygen electrode	2. Activation of vegetable oil pump
R. Metal rod	3. Activation of hot cold water pump
S. pH electrode	4. Increase / decrease in stirrer motor speed

(a) P-2, Q-3, R-1, S-4

(b) P-1, Q-2, R-4, S-3

(c) P-3, Q-2, R-4, S-1

(d) P-3, Q-4, R-2, S-1

(5/14) Immobilization of enzymes using entrapment method requires

(P) Photo sensitive polyethylene glycol dimethyl acrylate

(Q) CNBr activation of sepharose

(R) Polyfunctional reagent like hexamethylene diisocyanate

(S) radiation of polyrange alcohol

(a) P, Q (b) R, S (c) P, S (d) Q, S

(6/15) Which one of the following mono layer culture systems have the highest surface area

(a) Roux bottle

(b) spiralled roller bottle

(c) Hollow fiber

(d) Plastic bag / film

(7/18) Match items in group 1 with correct options from group 2

Group 1	Group 2
P. Amperometric biosensor	1. Light beam
Q. Evanescent biosensor	2. Flux of redox electrons
R. Calorimetric biosensor	3. Field effect transisters
S. Potentiometer biosensor	4. Exothermic reaction

(a) P-3, Q-4, R-2, S-1

(b) P-2, Q-1, R-4, S-3

(c) P-3, Q-2, R-4, S-1

(d) P-2, Q-4, R-3, S-1

Statement for linked answer questions 8/27 & 8/28

A bioreactor of working volume 50m3 produces a metabolite (X), in batch culture under given operating conditions from a substrate (S). The final concentration of metabolite (X) at the end of each run was 1.1 kgm^{-3}. The bioreactor was operated to complete 70 runs in each year.

(8/27) What will be annual output of the system (production of metabolite (X)) in kg per years?
(a) 55 (b) 3850 (c) 45.5 (d) 77

(9/28) What will be the overall productivity of the system in $kgyear^{-1}m^{-3}$
(a) 19250 (b) 38.50 (c) 3850 (d) 77

Answers

GATE 2006

(1/3) (b) Collagen
Collagen is widely used as a thin layer on tissue culture surfaces to enhance the attachment and proliferation of a variety of cells viz epithetial cells, fibroblasts, hepatocyles etc.
Moreover, collagen I can self assemble into 3D supermolecular gel in vitro, making it an ideal biological scaffold.

(2/5) (d) It provides low shear resistance to cells.
Immobilized plant cells can have high cell density if porous supports are used. The activity of the cells are longer under immobilized conditions.
The statement that it provides low shear resistance to cells is not correct. Due to immobilization, cells can withstand high shear.

(3/7) (c) P, Q (Zeatin, Indole-3-acetic acid)
Zeatin is a plant hormone obtained from purine adenine. It promotes growth of lateral buds and stimulates cell division to produce bushier plants
Indole-3-acetic acid (IAA) is the most common, naturally occurring plant hormone of the auxin class.

(4/10) (d) P-3, Q-4, R-2, S-1
Thermisters monitor temperature and activate hot/cold water pump. The oxygen electrode measures the concentration of dissolved oxygen in the liquid medium and is used to regulate oxygen level which increases with stirrer speed for more-aeration.

(5/14) (c) P, S (Photosensitive polyethylene glyclol dimethacrylate, radiation of polyvinyl alcohol)
Photo sensitive polyethylene glycol dimethyl acrylate is used for free radical copolymer cross-linking reactions which fix enzyme with solid matrix.

Polyvinyl alcohol is used for formation of spherical structural beads where cells are encapsulated. After this process, polyvinyl alcohol is degraded by gamma radiation.

(6/15) (c) hollow fiber
Hollow fibers have large surface area in a small volume (high surface-volume ratio) and provide large cell mass concentration. Cells grow around the fibers which are arranged like shell and tube heat exchangers

(7/18) (b) P-2, Q-1, R-4, S-3
Amperometric biosensors are based on current generated by flux of redox electrons when a potential is applied between two electrodes (P-2)
Evanescent wave biosensors are built with the optical fibers which detect the refractive index of the light beam.
Calorimetric biosensors can analyse concentration by measuring the rate of reactions on the basis of heat released by exothermic reactions.
Potentiometric biosensors use ionselective electrodes which transduce the biological reaction into an electrical signal.
Ion selective electrodes produce ion-selective field effect transistors (ISFETs). Biosensors based on ISFETs use enzyme-linked system.

(8/27) (b) 3850 kg/year
1.1 kg/(m^3)(run) × 50m^3 × 70 runs/yr
= 3850
(9/28) (d) 77 kg/year, m3
1.1 kg/(m^3)(run) × 70 runs/yr
=77 kg/m^3 yr→productivity of the system

Appendix 2 : Short Questions

GATE 2016

(1/20) The power required for agitation of non-aerated medium in a fermentation is ....

Operating conditions are as follows:

Fermentation diameter = 3m, Number of impellers = 1, Mixing speed = 300 rpm, Diameter of Rushton turbine = 1m, viscosity of broth = 0.001 ρa s, density of the broth = 1000 kg/m^3, Power number = 5
Solution:
Assuming Impeller based Reynolds number, R_{eI} is in the turbulent range, Power number is given as

$$N_P = \frac{P}{\rho N^3 D_I^5}$$

Or $P = N_P \rho N^3 D_I^5$

Or P $= (5)(1000)\left(\frac{300}{60}\right)^3 (1)^5 = 625000W = 625\,kW$

(2/35) The equilibrium potential of a biological membrane for Na^+ is 55mv at 37°c. Concentration of Na^+ inside the cells is 20 mv. Assuming the membrane is permeable to Na+ only, the Na^+ concentration outside the membrane will be ....mv

Faraday constant – 23,062 cal $v^{-1}mol^{-1}$

Gas constant, R=1.98 cal $mol^{-1}k^{-1}$

Solution:

$$E_{eq}(Na^+) = \frac{RT}{ZF} ln \frac{(Na^+)_O}{(Na^+)^{in}}$$

Give $E_q Na^+ = 55mv = 55 \times 10^{-3} V$

R $= 1.98$ cal $mol^{-1} K^{-1}, T = 37 + 273 = 300K$

F=23062 cal V^{-1} mol^{-1}, Z=1

$[Na^+]_i = 20mM$

$$\text{So } 55 \times 10^{-3} = \frac{(1.98)(300)}{(1)(23062)} ln \frac{(N_A^+)_O}{20}$$

$$\text{Or } 2.135 = ln \frac{(N_A^+)_O}{20}$$

$[Na^+]_O = 169mM$

(3/42) In an assay of the type II dehydrogenase of molecular mass 18kDa, it is found that the Vmax of the enzyme is 0.0134μmol min^{-1} when 1.8 g enzyme is added to the assay mixture. If the Km for the substrate is 25μM, the kcat / Km ratio will be .... 10^4 $M^{-1}S^{-1}$

Solution:

$V_{max} = 0.0134 \mu_{mol} min^{-1} = 2.233 \times \ 10^{-10} mol\, S^{-1}$

$$E_t = \frac{1.8}{18} = 0.1\mu_{mol} = 0.1 \times 10^{-6} mol$$

$$K_{cat} = \frac{V_{max}}{E_t} = \frac{2.233 \times 10^{-10}}{0.1 \times 10^{-6}} = 2.233 \times 10^{-3} S^{-1}$$

$$\frac{K_{cat}}{K_m} = \frac{2.233 \times 10^{-3} S}{25 \times 10^{-6} M} = 0.00893 \times 10^3 M^{-1} S^{-1}$$

(4/47) Saccharomyces cerevisae is cultured in a chemostat (continuous fermentation) at a dilution rate of $0.5h^{-1}$. The feed substrate concentration is $10gL^{-1}$. The biomass concentration in the chemostat at steady state will be .... gL^{-1}.
Assumptions:

Feed is sterile, maintenance is negligible and maximum biomass yield with respect to substrate is 0.4(biomass per gm, gluclose). Microbial growth kinetics is given by
$\mu = \mu m\ S/(K_S + S)$
where μ is the specific growth rate (h^{-1}), $\mu_m = 0.7h^{-1}$, $K_S = 0.3\ gL^{-1}$ and S is the substrate concentration (gL^{-1})
Solution:
Given: $D = 0.5\ h^{-1}$, $S_o = 10\ gL^{-1}$,

$$\mu m = 0.7h^{-1}, K_S = 0.3gL^{-1}, Y_{X/S} = 0.4$$

$$S = \frac{DK_s}{\mu m - D} = \frac{(0.5)(0.3)}{0.7 - .5} = 0.75gL^{-1}$$
$$X = Y_{X/S}(S_O - S) = 0.4(10 - 0.75) = 3.7gL^{-1}$$

(5/48) Decimal reduction time of bacterial spores is 23 min at 121°C and the death kinetics follow first order. One litre of medium containing 10^5 spores per ml, was sterilized for 10 mins at 121°C in a batch sterilizer. The number of spores in the medium after sterilization (assuming destruction of spores in heating and cooling period is negligible) will be .... $\times 10^7$.
Solution:

$$D_R = \frac{2.303}{k_d} \text{ Or } k_d = \frac{2.303}{23(DR)} = 0.1min^{-1} \text{ at } 131^\circ C$$

Where $D_R = 23$ min
$N_o = 10^5 \times 1000 = 10^8$ spores

$$ln\frac{N_O}{N} = k_d t = 0.1 \times 10 = 1$$

$$\frac{N_O}{N} = e' = 2.718$$

Or $N = \dfrac{N_O}{2.718} = 3.678 \times 10^7$ spores. where $N_O = 10^8$ spores

(6/50) An enzyme converts substrate A to produce B. At a given liquid feed rate, 25 $Lmin^{-1}$ and the feed substrate concentration of 2 mol L^{-1}, the volume of continuous stirred tank reactor needed for 95% conversion will be .... L.

Given the rate equation,

$$-r_A = \frac{0.1C_A}{1+0.5C_A}$$

where –rA is the rate of reaction in mol L-1min-1 and CA is the substrate concentration in mol L-1.

Solution:

Design equation of CSTR

$$F(C_{AO} - C_A) = (-r_A)V_R$$

$C_{AO} = 2moLL^{-1}$, at 95% conversion, $C_A = 0.05 \times 2 = 0.10$ mol L^{-1}

F = 25 L min^{-1}

$$r_A = \frac{0.1C_A}{1+0.5C_A}$$

$$V_R = \frac{F(C_{AO} - C_A)(1+0.5C_A)}{0.1C_A} = \frac{25(2-0.1)(1+(0.5)(.1))}{(0.1)(0.1)} = 4987.5L$$

(7/52) An enzyme is immobilized on the surface of a non-porous spherical particle of 2 mm diameter. The immobilized enzyme is suspended in a solution having a bulk substrate concentration of 10mM. the enzyme follows first order kinetics with rate constant $10S^{-1}$ and the external mass transfer is 1 cmS^{-1}.

Assume steady state condition where in rate of enzyme reation ($mmolK^{-1}S^{-1}$). at the surface is equal to mass transfer rate ($mmolL^{-1}S^{-1}$). The substrate concentration at the surface of the immobilized particle will be .... mM.

Solution:

For a spherical particle

$$a\left(\frac{Surface\ area}{Volume}\right) = \frac{4\pi r^2}{\frac{4}{3}\pi r^3} = \frac{3}{r} = \frac{3}{0.1cm} = 30cm^{-1}$$

$$k_L = 1Sec^{-1}, S_O = 10mM$$

$$\frac{v_{max}}{K_M}(first\ order\ rate\ constor) = 10S^{-1}$$

$$k_L a(S_O - S_S) = \frac{v_{max}}{K_M} S_S$$

$$(1)(30)(10 - S_S) = 10S_S$$

$$3(10 - S_S) = S_S$$

Or $4S_S = 30$ Or $S_S = 7.5mM$ (millimolar)

Where Ss is the surface concentration of the substrate.

Short Questions

GATE 2015

(1/31) The K_i of novel competitive inhibitor designed against an enzyme is 2.5μM. The enzyme was assayed in the absence or presence of the inhibitor (5μM) under identical conditions.

The Km in the presence of the inhibitor was found to be 30μM. The Km in the absence of inhibitor is .... μM.

Solution.

$1/v$= Km/vmax * 1/S + 1/vmax (1)

$$\frac{1}{v}=\frac{K_{mI}}{v_{max}}\left[1+\frac{I}{K_I}\right]\frac{1}{3}+\frac{1}{\mu_{max}} \quad (2)$$

Subtracting eqn. (2) from eqn. (1)

$$\frac{Km}{S}\cdot\frac{1}{v_{max}}=\frac{K_{m_i}}{v_{max}}\left[1+\frac{I}{K_I}\right]\frac{1}{S}$$

$$\text{Or } K_m=K_{mI}\left[1+\frac{I}{K_I}\right]=30\left[1+\frac{5}{2.5}\right]=90\mu M$$

(2/35) A synchronous culture containing 1.8×10^5 monkey kidney cells was seeded into three identical flasks. The doubling time of these cells is 24hrs. After 24 hrs, the cells from all these flasks were pooled and dispensed equally into each well of three 6-well plates. The number of cells in each well will be ....$\times 10^4$

Solution:

Total cells after 24 hrs = $1.8 \times 10^5 \times 2 = 3.6 \times 10^5$

Total number of wells = $6 \times 3 = 18$, so, number of cells in each well

= $3.6 \times 10^5/18 = 2 \times 10^4$ Ans.

(3/43) Oxygen transfer was measured in a stirred tank bioreactor using dynamic method. The dissolved tension was found to be 80% air saturation under steady state conditions. The measured oxygen tensions at 7s and 17s were 55% and 68% air saturation respectively. The volumetric mass transfer coefficient k_{La} is .... S^{-1}

Solution:

$$k_{La}=\frac{ln\left(\frac{C_L^*-C_{L_1}}{C_L^*-C_{L_2}}\right)}{t_2-t_1}$$

$$\text{Or } k_{La} = \frac{ln\left(\frac{80-55}{80-68}\right)}{17-7} = 0.0784S^{-1}$$

(4/44) Samples of bacterial culture taken at 5 pm and then the next day at 5am were found to have 10^4 and 10^7 cell mL^{-1} respectively.
Assuming that both samples were taken during log phase of cell growth, the generation time of bacterium will be .... h.
Solution:

$$G(generation\ time) = \frac{t}{ln\frac{N_2}{N_1}} = \frac{12hr}{ln\frac{10^7}{10^4}} = 1.737hr$$

Short Questions

GATE 2014

(1/30) A T-flask is seeded with 10^5 anchorage-dependent cells. The available area of T-flask is $25cm^2$ and the volume of medium is 25ml. assume that the cells are rectangles of size $5\mu m \times 2\ \mu m$
If the cells growth to a monolayer confluence after 50h, the growth rate is number of cells / $(cm)^2$(hr) is .... 10^5
Solution:
Seeding of the flask with no of anchorage dependent cells = 10^5
Area of the Tflask available = 25 cm^2
Volume = 25ml
Size of cells = $5\mu m \times 2\mu m = 10\mu m^2 = 10 \times 10^{-8} cm^2$/area
Covered by a single cells
Total cells in a monoloyer = $25/10 \times 10^{-8} = 2.5 \times 10^8$ cells
So the growth rate of cells
= 2.5×10^8 cells / ($25cm^2$) (50h) = 2×10^5 cells / (cm^2) (hr)

(2/3) Consider a continuous culture provided with sterile feed containing 10mM glucose. The steady state cell density and substrate concentration at three different dilution rates are given in the table below

Dilution rate, h^{-1}	Cell density gL^{-1}	Substrate concentration mM
0.05	0.245	0.067
0.5	0.205	1.667
5	0	10

The maximum specific growth rate μm (in h^{-1}) will be ....

Solution:
For sterile feed D = μ

$$D = \mu = \frac{\mu_m S}{K_S + S}$$

Taking the reciprocal

$$\frac{1}{D} = \frac{K_S}{\mu m}\frac{1}{S} + \frac{1}{\mu m} \text{ or } y = mx + C$$

$$C = 1/\mu_m$$

Since the last one is the first washout condition, we take the two, sets,
20 = m 17.54 + C/........ (1)
-2 = m (0.6) + C........... (2)

---

18 = 16.94 m
m = 1.06

1/D	1/S
20	17.54
2	0.6

Putting the value of m in equation (1)
20 = (1.06) (17.54) + C
Solving for C = 1.4076

Or $\mu = \frac{1}{C} = 0.71h^{-1}$

Short Questions

GATE 2013

(1/21) A callus of 5g dry weight was inoculated on semisolid medium for growth. The dry weight of the callus was found to increase by 1.5 fold after 10 days of incubation. The growth index of the culture is ....
Solution:
Initial dry weight = 5g
After inoculation and after 10 days incubation, dry weight of cells = 5 × 1.5 = 7.5 gms
Growth index of callus =(7.5 – 5)/ 5 = 0.5 Ans.

(2/23) A batch bioreactor is to be scaled up from 10 to 10, 000 litres. The diameter of the large bioreactor is 10 times that of small bioreactor. The agitator speed of the bioreactor is 450 rpm.
Determine the agitator speed as that of the small bioreactor ....

Solution:
Scale up factor = $(10,000/10)^{1/3} = 10$
D_1 = diameter of the small reactor
$10D_1$ = diameter of the larger reactor
Impeller diameter of the small bioreactor, $D_I = 03D_I$
$D_{I2} = 0.3\ (10D_1) = 3D_1$
Scale up criterion is equal impeller tip speed
$N_1 D_{I_1} = N_2 D_{I_2}$
$(450)\ 0.3D_1 = N_2\ 3D_1$
$N_2 = 450 \times 0.3/3 = 45$ rpm

(3/46) The maximum cell concentration (gL^{-1}) expected in a bioreactor with initial cell concentration of $1.7gL^{-1}$ and initial glucose concentration of 125 gL^{-1} is ($Y_{X/S} = 0.6$ gcells/gsubstrate) is ....
Solution:
$X_{max} = Xo + Y_{X/S}\ (So - 0) = 1.7 + 0.6(125\text{-}0) = 76.7\ gL^{-1}$

Short Questions

GATE 2012

Short Questions nil

GATE 2011

Short Questions-----nil

GATE 2010---------nil

GATE 2002

(1/29) (a) The volume of a chemostat system is 1000L. the feed flow rate to the reactor is 200L/h and the glucose concentration in the feed is 5 g/L. Determine cell and glucose concentration in the effluent of the reactor under steady state conditions. Use the following constants for the cells
$\mu_{max} = 0.3h^{-1}$, $K_S = 0.1g/L$
$Y_{X/S} = 0.4$ g dry wt cell / g. glucose
(a) Find out the dilution rate which gives maximum biomass productivity
Solution: (a)
$D = F/V = 200\ L/h/\ 1000L = 0.2h^{-1}$

$$S = \frac{DK_S}{\mu_m - D} = \frac{0.2 \times .1}{0.3 - 0.2} = 0.2g/L$$

$$X = Y_{X/S}(S_O - S) = 0.4(5 - 0.2) = 1.92 g / L$$

(b) $$D_{max} = \mu_{max}\left[1 - \sqrt{\frac{k_S}{K_S + S_O}}\right] = 0.3\left[1 - \sqrt{\frac{0.1}{0.1 + 5}}\right] = 0.258 h^{-1}$$

Appendix – 3

Table 3.1
Parameters, Dimensions, SI units

Symbol	Definition	Dimensions	SI units
C_A	Concentration of A	$L^{-3}M$	Kgm^{-3}
C_P	Specific heat capacity	$L^2T^{-2}\theta^{-1}$	$Jkg^{-1}K^{-1}$
E	Activation energy	$L^2MT^{-2}N^{-1}$	$Jmol^{-1}$
E_K	Kinetic energy	L^2MT^{-2}	J
F	Shear force	LMT^{-2}	N
h	Specific enthalpy	L^2T^{-2}	Jkg^{-1}
h	Heat transfer coefficient	$MT^{-3}\theta^{-1}$	$Wm^{-2}k^{-1}$
Δhc	Molar heat of combustion	$L^2MT^{-2}N^{-1}$	$Jmol^{-1}$
H	Enthalpy	L^2MT^{-2}	J
ΔHr	Heat of reaction	L^2MT^{-2}	$Jmol^{-1}$
k	Thermal conductivity	$LMT^{-3}\theta^{-1}$	$Wm^{-1}k^{-1}$
k_L	Liquid phase mass transfer coefficient	LT^{-1}	ms^{-1}
M	Torque	L^2MT^{-2}	Nm
P	Pressure	$L^{-1}MT^{-2}$	Pa
μ	Viscosity	$L^{-1}MT^{-1}$	PaS
τ	Shear stress	$L^{-1}MT^{-2}$	Pa

Appendix – 4

Table 4.1
CGS and SI Prefixes

Factor	Prefix	Abbreviation	Factor	Prefix	Abbreviation
10^{12}	tera	T	10^{-1}	deci	d
10^9	Giga	G	10^{-2}	centi	cm
10^6	Mega	M	10^{-3}	milli	mm
10^3	Kilo	K	10^{-6}	micro	μ

10^2	hector	H	10^{-9}	nano	n
10^1	deka	da	10-12	Pico	P
			10^{-15}	femto	f
			10^{-18}	atto	a

Table 4.2
Values of Gas Constants:

R	Units
8314	J / (kgmole) /ok
1.98	Cal/ (g. mole) /ok
82.056×10^{-3}	m^3 – atm/ (kgmole) /ok
1.9858	Btu/ (lbmode) / oR
1545.3	ft-lbf / (lbmode)/ (oR)
7.7045×10^{-4}	hp-hr/ (lbmode) (R)
5.8198×10^{-4}	kWh/(lbmole) (oR)/

Appendix – 5

Michaelis's constants for some enzymes & substrate systems

Table 5.1
(Ref: Attkinson B. and F. Mavitiena, Biochemical Engg and Biotechnology Handbook, Macmillan, Basingstaks, 1991)

Enzyme	Source	Substrate	Km (mM)
1. Alcohol dehydrogenase	Saccharomyces Cerevisiae	Ethanol	13.0
2. α-Amylase	Bacillus Stearothermophilus	Starch	1.0
3. β-Amylase	Sweet potato	Amylose	0.07
4. Aspartase	Bacilus Cadaveris	L-Aspartate	30.0
5. β- galactosidase	Escheria Coli	Lactose	3.85
6. Glucose oxidase	Aspergillus niger Penicillium notatum	D-glucose D-glucose	33.0 9.6
7. Histidase	Pseydomonas fluorescens	L-Histadine	8.9
8. Invertase	S Cerevisiae Neurospora Crasa	Sucrose Sucrose	9.1 6.1

9. Lactate dehydrogenase	Bacillus subtilis	Lactate	30.0
10. Penicillinase	Bacillus Licheniformis	Benzyl Penicillin	0.049
11. Urease	Jack bean	Urea	10.5

Appendix – 6

Effective Diffusivities of solutes in Biological gels in aqueous solution (Ref. : Doran, P. M., Bioprocess Engg., Principles, Academic Press, 2006)

Table 6.1

Solute	Gel	Concentration %	Temp °C	Diffusivity, m^2/s
1. Gluclose	Ca-alginate	2	25	6.1×10^{-10}
2. Ethanol	Ca-alginate	2	25	1.0×10^{-9}
3. Sucrose	Gelatin	3.8	5	2.09×10^{-10}
4. Sucrose	Gelatin	7.6	5	1.35×10^{-10}
5. Lactose	Gelatin	25	5	0.37×10^{-10}
6. L-Tryptiphan	Ca-alginate	2	30	6.67×10^{-10}
7. Lactate	Agar (1%) Containing 1% Ehrlich Ascites fumon ells	--	37	1.4×10^{-9}
8. Oxygen	Microbial aggregates (Trickling filter)	--	25	0.82×10^{-9}
9. Nitrate	Compressed film of nitrifying organisms	--	30	1.4×10^{-9}
10. Ammonia	Compressed film of nitrifying organisms	--	30	1.3×10^{-9}

Appendix – 7

Respiration rates of Microorganisms, Plant and animal cells

Table 7.1

Organism	qo_2 mmol o_2 g dr cell / hr	Relation between qo2, X, k_{La}
1. Bacteria: E. Coli Azotobacter sp. Streptomycles sp.	10-12 30-90 2-4	$k_{La} (C_L^* - C_L)$ $= q_{o2} X$
2. Yeast Saccharomyces Cerevisiae	8	where k_{La} = volumetric mass transfer, coefficient

3. Mold: Penicillin sp. Aspergillus niger	3-4 3.0	C_L^* = solubility of oxygen (mg/L) C_L = critical (mg/L)
4. Plant Cells: Acer pseudo-platamus (sycanmore) Saccharun (Sugarcane) Catharanthus rosens	 0.2 1-3 0.2	Oxygen concentration, X = cells mass (g/L)
5. Animal Cells Hela	0.4 mmolO2/lh/ 10^6 cells / ml	
Diploid embryo, WI – 38	0.15 mmol/lh / 10^6 cell/ml	
General (Animal Cells)	$0.06 - 0.2 \times 10^{-12}$ mol O_2/(hr) (cell)	

Appendix – 8

Dimensionless Groups

Symbol	Name	Definition
C_D	Drag Coefficient	$2 F_D g_c / \rho u_\emptyset^2 A_P$
f	Fanning Friction factor	$\Delta pgcD/(2L\rho \overline{V}^2)$
J_H	Heat transfer factor	$h/cpG(cp\mu/k)2/3\ (\mu w/\mu)^{0.14}$
J_M	Mass transfer factor	$K\overline{M}/G\ (\mu/Dv\rho)^{2/3}$
NP_O	Founier number	$\alpha\ t/r^2$
NF_r	Frounde number	v^2/gL
NG_r	Grashof number	$gD^3\rho_L(\rho_L - \rho_g)/\ \mu_L^2$
$N'G_Z$	G_Z for mass transfer	$m/\rho DvL$
NM_a	Mach number	u/a, where a is the sound velocity
NN_a	Nusselt number	hD/k
N_P	Power number	$P/\rho N^3Di^5$
NP_e	Peclet number	uL/D_L, where D_L is the dispersion coefficient
NP_r	Prandtl number	$cp\mu/k$
NR_e	Reynolds number	$Du\rho/\mu$
N_s	Separation number	Ut u0/gDp
N_{Sc}	Schmidt No.	$\mu/\rho Dv$
N_{sh}	Sherwood No	KcD/Dv
N_{we}	Weber No.	$D_p\rho/\ \sigma g_c$
N_{Ar}	Arrhenius No.	E/RTs
ND_a	Damköhler No.	$Vmax/(k_L Sb)$

Appendix - 9

Henry's Constant(H) for gases dissolved in water at 298K (25°C)

Table 9.1

Gas	H/bar	Gas	H/bar
Acetylene	1,350	Helium	126,800
Air	72,950	Hydrogen	71,600
Carbon dioxide	1,670	Hydrogen sulphide	55,200
Carbon monoxide	54,600	Methane	41,850
Ethane	30,600	Nitrogen	87,650
Ethylene	11,550	Oxygen	44,380

Table 9.2 **Solubility of some gases, C_L* in water at 25°C**

Gas	p_A, partial press, bar	H, bar	x_A* mol fraction	C_L*, mol/L
O_2	0.202	44, 380	4.55×10^{-6}	0.25×10^{-3}
5% CO_2 in air	0.05	1, 670	3.0×10^{-5}	1.6×10^{-3}
CH_4	1.013	41, 850	2.42×10^{-3}	1.3×10^{-3}

Based on Henry Law: $p_A = x_A{}^*H$
where x_A* is the mole fraction of component A in water (mol O_2 / mol H_2O)

Appendix - 10

Table 10.1
Some primary metabolites of industrial importance

Metabolites	Organism	Uses
Riboflavin (Vitamin B_2) Ashiya	Eremothecium ashbyii, Ashbya, Gossypii	Food and animal feed supplement
β-carotene	Blakes lea spp., Choanephora spp.	Precursor of vitamin A, colouring agent in food industry food supplement
Xanthan gum, dextran	Xanthomonas Campestris Leuconstoc mesenteroides	Food, pharmaceutical, textile industries, useful for thickening stiffening and setting properties
Acetic acid	Saccharomyces ellip- soideus or saccharomyces cerevisiae plus acetobactor spp. Clostridium spp.	Vinegar – preservative in food industry, chemical feed stocks & polymer industry
Acetone, Butanol	Clostridium acetobutylicum	Solvents, thinners, synthetic polymers

Ethanol	Saccharomyces cerevisiae, zymomonas mobilis, clostridium thermocellium and other clostridium supp.	Alcoholic bevereages, solvents in chemical industry, fuel extender
Glutamate	Carynebacterium glutamicum, Brevibacterium spp.	Food industry as flavour – acetentuaring agent
Lysine, threonine	Corynebacterium glutamicum, Brevibacterium flavum, Escherichia coli	Essential amino acids added to supplement low grade protein
Citric acid	Aspergillus niger, Candida spp.	Acidulant in food industry, pharmaceuticals esters used as plasticizers
Guanylic, inosinic and xanthylic acids	Bacillus subtillis, corynebacterium glutamicium	Food industry as flavour accentuaring agent
Itaconic acid	Aspergillus terreus, Aspergillus itaconicus	Synethetic fiber, resin manufacture, copolymers (e. g. styrene butadiene)
Fumaric acid	Rhizopus arrhizus, Rhizopus nigricans	Plastics and food industries
Cyanocobalamin (Vitamin B_{12})	Propionic bacterium shermanii, Pseudomonas denitrificans	Food and animal feed supplement

APPENDIX - 11

Some Mathematical formulae

A. Differentiations"

1) $\frac{d}{dx}\left(x^{n}\right)=nx^{x-1}$

2) $\frac{d}{dx}\left(e^{Ax}\right)=Ae^{x}$

3) $\frac{d}{dx}\left(\frac{u}{C}\right)=\frac{1}{C}\frac{du}{dx}$

Where u is a function of x

4) $\frac{d}{dx}\left(\frac{c}{u}\right)=c\frac{d}{dx}\left(\frac{1}{u}\right)=-\frac{C}{u^{2}}\frac{du}{dx}$

5) $\frac{d}{dx}\left(\frac{u}{v}\right)=\frac{v\frac{du}{dv}-u\frac{dv}{dx}}{v^{2}}$

6) $\frac{d}{dx}\left(lnx\right)=\frac{1}{x}$

(7) $\frac{d}{dx}\left[(Av)\cdot lnx\right]$

$= Ax\frac{1}{x} + \ln xA$

$= A(1 + lnx)$

B. Integration:

1) $\int \frac{dx}{x} = lnx + K$

2) $\int x^{n} = \frac{x^{n+1}}{n} + K$

3) $\int x^{-n} = -\frac{x^{-n+1}}{n} + K$

4) $\int \frac{dx}{A + Bx} = \frac{1}{B} ln(A + Bx) + K$

5) $\int \frac{dx}{(A-x)(B-x)} = \frac{1}{(B-A)} ln\frac{(B-x)}{(A-x)} + K$

C. Taylor Series Expansion

Taylor's Theorem is the basis of many numerical methods. It defines the function x = f(t) at a point t = a+h in terms of the function and its derivatives at the point a

$$f(a+h) = f(a) + hf'(a) + \frac{h^2 f^{II}(a)}{2!} + \frac{h^3 f^{III}(a)}{3!} + \frac{h^4 f^4(a)}{4!} \quad \text{.................. (C-1)}$$

Similar equation are defined for function of more than one variable such as

$$f(x,y) - f(x_S, y_S) = \frac{\partial f(x_S, y_S)}{dx}\delta x + \frac{\partial f(x_S, y_S)}{\partial y}\delta y \quad \text{.................. (C-2)}$$

neglecting the higher derivatives.

The rate for first irreversible reaction can be given as

$$r(X,T) = k_o exp\left(-\frac{E}{RT}\right)C_{A_O}(1-X) \quad \text{.................... (C-3)}$$

The deviation of r from the steady state, rs may be expanded to yield

$$r(X,T) - r(X_S,T_S) = \frac{\partial r}{\partial X}(X - X_S) + \frac{\partial r}{\partial T}(T - T_S) \quad \text{.................... (C-4)}$$

and the derivatives are:

$$\frac{\partial r}{\partial X} = -C_{A_0} k(T_s) \quad \text{.......................... (C-5)}$$

$$\frac{\partial r}{\partial T} = -\frac{E}{RT_S^2} k(T_S) C_{A_0} (1-X) \quad \text{...................... (C-6)}$$

D. Linear Regression (Least square method)

To evaluate the constants of a linear equation by least square method is given as

y = mx + c .... (1)

or c + mx − y = 0 .... (2)

for n number of x, y data to be fitted to the above equation. We get the following equations from equations (2)

nc + mΣx - Σy = 0 .... (3)

$c\Sigma x + m\Sigma x^2 - \Sigma xy = 0$ .... (4)

From these two equations, m and c can be evaluated.

Example:

x	0.5	1.0	1.5	2.0	2.5
y	0.30	0.80	1.30	1.90	2.5

Fit the above data to the linear equation:

Solution:

Σx = 7.5, Σy = 6.8

$\Sigma x^2 = 13.75$, Σxy = 12.95, n = 5

Substituting these values in equations (3) & (4), we get

5c + 7.5m = 6.8 .... (5)

7.5c + 13.75m = 12.95 .... (6)

The constants c and m can be determined from equations (5) & (6) by the use of determinants

$$C = \frac{\begin{bmatrix} 6.8 & 7.5 \\ 12.95 & 13.7 \end{bmatrix}}{\begin{bmatrix} 5 & 7.5 \\ 7.5 & 13.75 \end{bmatrix}} = \frac{93.16 - 97.12}{(68.75 - 56.25)} = 0.317$$

$$m = \frac{\begin{bmatrix} 5 & 6.8 \\ 7.5 & 12.95 \end{bmatrix}}{(68.75 - 56.25)} = \frac{64.75 - 51}{12.5} = 1.1$$

So the equation is

y = 1.1x + 0.317

E. Differentiation by three point formula:

The slopes, Δy/Δx at each of these three pints equally spaced at interval h are given in terms of the ordinates.

Thus

$$y_0' = \frac{-3y_0 + 4y_1 - y_2}{2h}$$

$$y_1' = \left(-y_2 + y_3\right)/2h$$

$$y_2^{II} = \left(y_0 - 4y_1 + 3y_2\right)/2h$$

This may be repeated for as many groups of three equally spaced points.

Example:

x	0	0.5	1.0	1.5	2.0
y	1.0	1.5	1.6	1.8	2.3
dy/dx from 3 point	0.050	0.10	0.25	0.62	1.234

F. Numerical solutions of ordinary Differential Equations / Numerical Integration

G. Numerical Integration by Simpson's Rule

1) For three pairs of interval or three point

$$\int_{x_0}^{x_e+2h} y dx = \frac{h}{3}\left(y_o + 4y_1 + y_2\right)$$

2) For even n

$$\int_{x_0}^{x_e+2h} y dx = \frac{h}{3}\left(y_o + 4y_1 + y_2 + 4y_3 + 2y_4 + 4y_5 + y_n^{+}....\right)$$

Or $\int_{x_0}^{x_e+2h} y dx = \frac{h}{3}\left[y_o + 4\left(y_1 + y_3 + \ldots y_{n-1}\right) + 2\left(y_2 + y_4 \ldots + \ldots y_{n-2}\right) + y_n\right]$

Example: Integrate $\int_0^{0.6} \left(\frac{1+x_A}{1-x_A}\right)^{1/2} dx$

Solution: using 3-point formula

$X_A\left[\left(1+x_A\right)/\left(1-x_A\right)\right]^{1/2}$

$01\left(y_0\right)$

$0.31.362\left(y_1\right)$

$0.62.0\left(y_2\right)$

$$I = \frac{h}{3}\left(y_0 + 4y_1 + y_2\right)$$

$$= \frac{0.3}{3}\left[1 + 4\left(1.362\right) + 2\right] = 0.8448$$

II. Euler's Method

This is the simplest and crudest method. A differential equation of first order may be written as

$dy/dx = f(x, y)$ .... (A)

The integral of the above equation may be given symbolically as

$y = F(x)$

We can evaluate Δy as

$\Delta y = (dy/dx)_o \, \Delta x_1$

So that $y1 \approx yo + (dy/dx) \, \Delta x$

The value of y corresponding to

$x_2 = (x_1 + h)$ and, $x_3 = (x_2 + h)$

$y_2 \approx y_1 + (dy/dx)_1 \, h$

$y_3 \approx y_2 + (dy/dx)_2 \, h$ etc.

By taking h small enough and proceeding in this manner, we can tabulate the integral of A as a set of corresponding values of x and y.

III. Modified Euler method

Starting with the initial value y_o, an approximate value of y_1 is obtained from the relation

$y_1^{(1)} \approx y_o + (dy/dx)_o \, h$

To get an approximate value at the end of the first interval, we have

$(dy/dx)_1^{(2)} = f(x_2, y_1^{(1)})$

Now the average of dy/dx at the end of the interval xo to x

or

$$\Delta y \approx \frac{\left(\frac{dy}{dx}\right)_0 + \left(\frac{dy}{dx}\right)_1^{(1)}}{2} . h$$

This values of Δy is more accurate than the previous one, $(dy/dx)_o \, h$

Now, the second approximation for y_1, is

$$y_1^{(2)} \approx y_0 + \frac{\left(\frac{dy}{dx}\right)_0 + \left(\frac{dy}{dx}\right)_1^{(1)}}{2} \cdot h$$

This improved value of $y_1^{(2)}$ is substituted in the equation (A) to get a second approximation for (dy/dx) or

$$\left(\frac{dy}{dx}\right)_1^{(2)} = f\left(x_1, y_1^{(2)}\right)$$

The third approximation for y, is then

$$y_2^{(3)} = y_0 + \frac{\left(\frac{dy}{dx}\right)_0 + \left(\frac{dy}{dx}\right)_1^{2}}{2} \cdot h$$

This process is repeated until no change is produced in the value of y to the number of digits retained.

IV. Runga – Kutta method

This is the most powerful method for the solution of ordinary differential equations. The method can be used for several types of differential equations such as i) first order equation, 2) second order equation & 3) simultaneous equations of the first order. Here only the formula for first order equation has been presented.

- First-order Equation

Let $dy/dx = f(x, y)$ and let h denotes the interval between equidistant values of x.

If the initial values are x_o, y_o, the final increment in y is computed from the following formulas

$k_1 = f(x_o, y_o)h$

$k_2 = f(x_o + h/2, y_o + k_1/2)h$

$k_3 = f(x_o + h/2, y_o + k_2/2)h$

$k_4 = f(x_o + h, y_o + k_3)\ h$

$\Delta y = 1/6\ (k_1 + 2k_2 + 3k_3 + k_4)$

Then $x_1 = x_o + h$, $y_1 = y_o + \Delta y$

The increment in y for the second interval is computed by similar manner by the formulas

$k_1 = f(x_1, y_1)h$

$k_2 = f(x_1 + h/2, y_1 + k_1/2)h$

$k_3 = f(x_1 + h/3, y_1 + k_2/2)h$

$k_4 = f(x_1 + h, y_1 + k_3)\ h$

$\Delta y = 1/6\ (k_1 + 2k_2 + 3k_3 + k_4)$

and $x_2 = x_1 + h$, $y_2 = y_1 + \Delta y$

To find out the nth interval, we have to substitute x_{n-1}, y_{n-1} in the expressions of k_1, k_2...... etc.

H. Some Statistical formulae

The standard deviation of a set of N numbers X_1, X_2............ X_N is denoted by S and defined as

$$S = \sqrt{\frac{\sum_{2=1}^{N}\left(X_i - \bar{X}\right)^2}{N}} = \sqrt{\frac{\sum\left(X - \bar{X}\right)^2}{N}}$$

Where $\bar{X}$ is the mean (arithmetic)

$$\bar{X} = \frac{(X_1 + X_2 + \ldots X_N)}{N} = \frac{\sum X_N}{N}$$

Mean deviation or M. D

$$\sum_{j=1}^{N} \frac{|X_i - \bar{X}|}{N} = \frac{\sum |X - \bar{X}|}{N}$$

Variance:

The variance of a set of data is defined as the square of the standard deviation and is given by S^2

$$S^2 = \frac{\sum (X - \bar{X})^2}{N}$$

$$\text{or } S^2 = \frac{\sum f_i (X - \bar{X})^2}{N} = \sqrt{\frac{f(X - \bar{X})^2}{N}}$$

where X_1, X_2, ...X_N occur with frequencies f1, f2, ... fn

$$\text{or } S^2 = \frac{\sum X^2}{N} - \left(\frac{\sum X}{N}\right)^2$$

- **Normal distribution or Gaussian Distribution:**

$$Y = \frac{1}{\sqrt{2\pi}} e^{-\frac{1}{2}z^2}$$

when the variable X is expressed in terms of standard units,

$$Z = (X - \mu) / \sigma$$

where μ = mean, σ = standard deviation

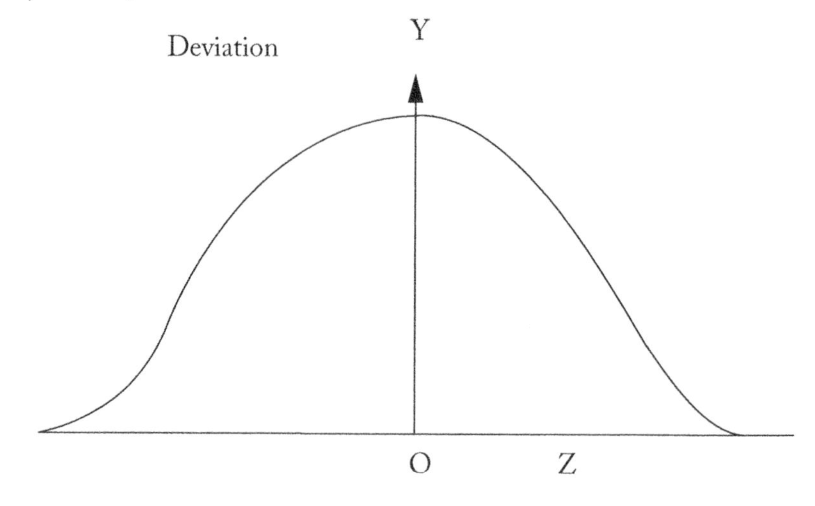

I. Error Integral

The Error Integral or function, erf is defined as

$$erf(y) = \frac{2}{\sqrt{\pi}} \int_0^y e^{-x^2}\, dx$$

$$erf(\pm \infty) = \pm 1$$

$$erf(0) = 0$$

$$erf(-y) = -erf(y)$$

Values of erf(y) are given in standard mathematical table.

J. Exponential Integrals

$$E_i(x) = \int_{-\infty}^{x} \frac{e^u}{u} = 0.5772 + lnx + x + \frac{x^2}{2.2!} + \frac{x^3}{3.3!} + \ldots$$

$$e_i(x) = \iint_x^{\infty} \frac{e^{-u}}{u} du = -0.577 - lnx + x - \frac{x^2}{2.2!} + \frac{x^3}{3.3!}$$

$$\text{For } x \geq 10 \begin{cases} E_i(x) = e^x \left[\dfrac{1}{x} + \dfrac{1}{x^2} + \dfrac{2!}{x^3} + \dfrac{3!}{x^4} \right] \\ e_i(x) = e^{-x} \left[\dfrac{1}{x} - \dfrac{1}{x^2} + \dfrac{2!}{x^3} - \dfrac{3!}{x^4} + \ldots \right] \end{cases}$$

List of Symbols

Symbol	Definition	SI Units
a	Area per unit volume	m^{-1}
A	Area per unit volume	m^{-1}
A	Arrhenius constant	S^{-1}
A_i	Inner surface area	m^2
A_o	Outer surface area	m^2
B_i	Biot number	-
C	Concentration	$Kg\ m^{-3}$
C_A, C_B, C_C	Concentrations of components A, B, C	$Kg\ m^{-3}$
C_{AO}	Initial concentration of component A	$Kg\ m^{-3}$
C_e	Exit concentration	$Kg\ m^{-3}$
C_b	Bulk concentration of a component	$Kg\ m^{-3}$
C_L	Concentration of gaseous component in liquid phase	$Kg\ m^{-3}$
C_L^*	Saturated concentration of a gaseous component in water	
C_s	Surface concentration	$Kg\ m^{-3}$
C_p	Specific heat	$JKg^{-1}\ K^{-1}$
D	Dilution rate	S^{-1}
D	Diameter	m
D_a	Damkohler number	-
D_b	Bubble diameter	m
D_{AB}	Molecular diffusely of A in B	$m^2\ s^{-1}$

Symbol	Definition	SI Units
D_e	Effective diffusivity	$m^2\ s^{-1}$
D_I	Impeller diameter	m
D_k	Knudsen diffusivity	$m^2\ s^{-1}$
D_L	Dispersion coefficient	$m^2\ s^{-1}$
D_p	Bubble diameter	m
d_p	Particle diameter	m
dt	Tube diameter	m
E	Activation energy	$Jmol^{-1}$
Ed	Activation energy for deactivation	$Jmol^{-1}$
f	Function	-
f^+	Fraction of plasmid carrying cells	-
f^-	Fraction of plasmid free cells	-
F	Volumetric flow rate of liquid	$m^3\ s^{-1}$
F_g	Volumetric flow rate of gas	$m^3\ s^{-1}$
g	Acceleration due to gravity	$m\ s^{-2}$
Gr	Grashof number	-
ΔG°	Standard free energy change	$Jmol^{-1}$
h	Specific enthalpy	JKg^{-1}
h	Heat transfer coefficient	$W\ m^{-2}\ k^{-1}$
h_c	Heat transfer coefficient for cold fluid	$W\ m^{-2}\ k^{-1}$
h_d	Fouling factor	$W\ m^{-2}\ k^{-1}$
H	Henry's constant	bar
H	Enthalpy	J
H	Height	m
ΔHR°	Standard heat of reaction	$Jmol^{-1}$
J_A	Molal flux of component A	$mol\ m^{-2}\ s^{-1}$
J_D	Mass transfer correlation factor	-

Symbol	Definition	SI Units
J_H	Heat transfer correlation factor	-
k	Thermal conductivity	$Wm^{-1} k^{-1}$
k	Reaction rate constant	-
k	Mass transfer coefficient	$m s^{-1}$
k_1	First order rate constant	S^{-1}
k_2	Second order rate constant	$m^3 mol^{-1} S^{-1}$
k_o	Zero order rate constant for cell mass	$mol m^{-3} s^{-1}$
kg	Gas phase mass transfer coefficient	ms^{-1}
K_o	Overall mass transfer coefficient	s^{-1}
k_L	Liquid phase mass transfer coefficient	ms^{-1}
k_d	First order decay or death constant	s^{-1}
K	Constant of integration	-
K	Consistency under for power law of non-Newtonian fluids	$Pa S^n$
K	Reaction equilibrium constant	-
Km	Michaelis Menten constant	$Kg m^{-3}$
K_s	Saturation constant	$Kg m^{-3}$
L	Length or film thickness	m
m_s	Maintenance coefficient	s^{-1}
$\bar{M}$	Mass flow rate	$Kg m^{-3}$
$\bar{M}_c$	Mass flow rate of cold fluid	$Kg m^{-3}$
$\bar{M}_h$	Mass flow rate of hot fluid	$Kg m^{-3}$
n	Number of generation	-
N	Stirred spead	s^{-1}
N_o	Initial number of viable cells	-
N	Number of cells at any instant	-
N	Number of discs in RBC	-
N_p	Power number	-

Symbol	Definition	SI Units
N_u	Nusselt number	-
p	Probability of plasmid loss or probability of survival of spores after sterilization	-
P	Product concentration	$Kg\ m^{-3}$
P_o	Power without sparging	W
Pg	Power with sparging	W
Pr	Prandtle number	-
$\bar{q}$	Heat flow	Wm^{-2}
Q	Volumetric flow rate	$m^3\ S^{-1}$
Q	Heat	J
$\bar{Q}$	Rate of heat flow	W
r	Radius	m
r_A	Rate of reaction of component A	moles $m^{-3}\ S^{-1}$
R	Universal gas constant	$Jmo^{-1}\ k^{-1}$
R	Radius	m
r_{max}	Maximum substrate utilization rate	$Kg\ m^{-3}\ s^{-1}$
RQ	Respiratory quotient	-
R_i	Inner radius	m
R_o	Outer radius	m
r	Recycle ratio	-
S	Substrate concentration	$Kg\ m^{-3}$
Sc	Schmidt number	-
S_o	Initial substrate	$Kg\ m^{-3}$
S_b	Bulk surface concentration	$Kg\ m^{-3}$
S_x	External surface	m^2
S_s	Substrate concentration at the solid surface	$Kg\ m^{-2}$
Sh	Sherwood number	-

Symbol	Definition	SI Units
t	Time	S
t ½	Half-life time	S
t_b	Batch reaction time	S
t_1	Time at the end of the heating period	S
t_2	Time at the end or holding period	S
t_c	Circulation time	S
t_d	Doubling time	S
t_m	Mixing time	S
T	Temperature	°C
T_c	Cold fluid temperature	°C
T_h	Hot fluid temperature	°C
T_s	Steam temperature	°C
ΔTL	Logarithmic mean temperature difference	°C
u_L	Linear liquid velocity	m S^{-1}
v_g	Gas superficial velocity	m S^{-1}
ut	Terminal velocity of a particle	m S^{-1}
U	Overall heat transfer coefficient	Wm^{-2} k^{-1}
v	Volumetric rate of enzymatic reaction	Kg m^{-3} s^{-1}
v_{max}	Maximum rate of enzymatic reaction	Kg m^{-3} s^{-1}
V	Volume	m^3
V_G	Volume of the gas	m^3
V_L	Volume of the liquid	m^3
V_S	Volume of solid	m^3
V_T	Total volume	m^3
W	Weight of catalyst	Kg
W	Width of impeller	m
x	Distance	m

Symbol	Definition	SI Units
X	Cell mass concentration or deviation variable	Kg m^{-3}
X_o	Initial cell mass concentration	Kg m^{-3}
X_T	Total cell mass	Kg
X_{max}	Maximum cell mass concentration	Kg m^{-3}
Y	Deviation variable	m
Y_A	Mole fraction of component A in gas mixture	-
$Y_{P/S}$	Yield coefficient for product formed/substrate consumed	–kg/kg
$Y_{P/X}$	Yield coefficient for product formed/cell mass produced	–kg/kg
$Y_{X/S}$	Yield coefficient for cell mass produced/substrate consumed	–kg/kg
$Y_{X/O}$	Yield coefficient for cell mass produced/mass of oxygen consumed	kg/kg
Z	Distance	

	Greek Symbols	
α	Recycle ratio	
β	Dimensionless parameter, S_O/K_M	-
γ	Ratio of excess sludge flow to feed flow rate	
δ	Fractional conversion of substrate, $(S_O - S)/S_O$	
Δ	Difference	
$\in$	Porosity	
$\in_G$	Fractional gas hold up	-
$\in$	Rate of turbulent energy dissipation	Wkg^{-1}
η_s	Effectiveness factor for a spherical particle	-
η_C	Effectiveness factor for a cylindrical particle	-
η_L	Effectiveness factor for a flat slab	-
θ	Dimensionless time, $t/\bar{t}$	-
λ	Kolmogorov scale or length	m

Symbol	Definition	SI Units
μ	Viscosity	PaS
μ_b	Viscosity of the bulk fluid	Pa.s
μ_w	Viscosity of the fluid near the wall	PaS
μ_{max}	Maximum specific growth rate	h^{-1}
ρ	Density	Kg m^{-3}
ρ_L	Density of liquid	Kg m^{-3}
ρ_f	Density of fluid	Kg m^{-3}
ρ_p	Particle density	Kg m^{-3}
σ	Surface tension	Nm^{-1}
σ	Standard deviation	-
Σ	Summation	-
τ	Average residence time	S
τ	Shear stress	P_a
ϕ	Thiele modulus	-
τ_o	Yield stress	P_a
Φ	Observable thiele modulus	-
Ψ	Volume fraction of solids	-
$\in$	Porosity	-
$\in_\mu$	Microporosity	-
$\in_M$	Macroporosity	-

Index

P

R

S